International Association of Fire Chiefs

Fire Officer

Principles and Practice
FOURTH EDITION

Michael J. Ward
MGA, MIFireE, FACPE

JONES & BARTLETT
LEARNING

Jones & Bartlett Learning
World Headquarters
5 Wall Street
Burlington, MA 01803
978-443-5000
info@jblearning.com
www.psglearning.com

International Association of Fire Chiefs
4025 Fair Ridge Drive
Fairfax, VA 22033
www.IAFC.org

National Fire Protection Association
1 Batterymarch Park
Quincy, MA 02169-7471
www.NFPA.org

Jones & Bartlett Learning books and products are available through most bookstores and online booksellers. To contact Jones & Bartlett Learning Public Safety Group directly, call 800-832-0034, fax 978-443-8000, or visit our website, www.psglearning.com.

Substantial discounts on bulk quantities of Jones & Bartlett Learning publications are available to corporations, professional associations, and other qualified organizations. For details and specific discount information, contact the special sales department at Jones & Bartlett Learning via the above contact information or send an email to specialsales@jblearning.com.

Copyright © 2021 by Jones & Bartlett Learning, LLC, an Ascend Learning Company, and the National Fire Protection Association.

All rights reserved. No part of the material protected by this copyright may be reproduced or utilized in any form, electronic or mechanical, including photocopying, recording, or by any information storage and retrieval system, without written permission from the copyright owner.

The content, statements, views, and opinions herein are the sole expression of the respective authors and not that of Jones & Bartlett Learning, LLC. Reference herein to any specific commercial product, process, or service by trade name, trademark, manufacturer, or otherwise does not constitute or imply its endorsement or recommendation by Jones & Bartlett Learning, LLC and such reference shall not be used for advertising or product endorsement purposes. All trademarks displayed are the trademarks of the parties noted herein. *Fire Officer: Principles and Practice, Fourth Edition* is an independent publication and has not been authorized, sponsored, or otherwise approved by the owners of the trademarks or service marks referenced in this product.

There may be images in this book that feature models; these models do not necessarily endorse, represent, or participate in the activities represented in the images. Any screenshots in this product are for educational and instructive purposes only. Any individuals and scenarios featured in the case studies throughout this product may be real or fictitious, but are used for instructional purposes only.

The publisher has made every effort to ensure that contributors to *Fire Officer: Principles and Practice, Fourth Edition* materials are knowledgeable authorities in their fields. Readers are nevertheless advised that the statements and opinions are provided as guidelines and should not be construed as official policy. The recommendations in this publication or the accompanying resource manual do not indicate an exclusive course of action. Variations taking into account the individual circumstances and local protocols may be appropriate. The publisher disclaims any liability or responsibility for the consequences of any action taken in reliance on these statements or opinions.

19013-7

Production Credits
General Manager and Executive Publisher: Kimberly Brophy
Vice President, Product Operations: Christine Emerton
Executive Editor: Bill Larkin
Senior Managing Editor: Donna Gridley
Senior Editor: Amanda Mitchell
VP, Sales, Public Safety Group: Phil Charland
Project Specialist: John Fuller
Director of Marketing Operations: Brian Rooney
Production Services Manager: Colleen Lamy

VP, Manufacturing and Inventory Control: Therese Connell
Composition: S4Carlisle Publishing Services
Cover Design: Kristin E. Parker
Text Design: Scott Moden
Rights Specialist: John Rusk
Senior Media Development Editor: Troy Liston
Cover Image (Title Page, Part Opener, Chapter Opener): © Glen E. Ellman.
Printing and Binding: LSC Communications
Cover Printing: LSC Communications

Library of Congress Cataloging-in-Publication Data unavailable at time of printing.

6048

Printed in the United States of America
23 22 21 20 19   10 9 8 7 6 5 4 3 2 1

Brief Contents

SECTION 1
Fire Officer I

CHAPTER **1** The Fire Officer I as a Company Supervisor — 3

CHAPTER **2** Understanding Leadership and Management Theories — 39

CHAPTER **3** Leading a Team — 73

CHAPTER **4** Community Relations and Risk Reduction — 115

CHAPTER **5** Fire Department Administration — 127

CHAPTER **6** Preincident Planning and Code Enforcement — 157

CHAPTER **7** Command of Initial Emergency Operations — 195

CHAPTER **8** Safety and Risk Management — 223

SECTION 2
Fire Officer II

CHAPTER **9** The Fire Officer II as a Manager — 251

CHAPTER **10** Applications of Leadership — 271

CHAPTER **11** Managing Community Risk Reduction Programs — 299

CHAPTER **12** Administrative Communications — 315

CHAPTER **13** Fire Cause Determination — 333

CHAPTER **14** Managing Major Incidents — 353

APPENDIX **A:** Preparing for Promotion — 375

APPENDIX **B:** NFPA 1021, *Standard for Fire Officer Professional Qualifications, 2020 Edition* Correlation Guide — 387

GLOSSARY — 399

INDEX — 408

Contents

Acknowledgements xiv
New for the *Fourth Edition* xix

SECTION 1
Fire Officer I

CHAPTER 1
The Fire Officer I as a Company Supervisor — 3

Introduction 4
Fire Officer Qualifications 4
Roles and Responsibilities of the Fire Officer I and Fire Officer II 5
The Fire Service in the United States 7
   History of the Fire Service 8
   Communications 8
   Paying for the Fire Service 8
Fire Department Organization 10
   Source of Authority 10
   Chain of Command 11
   National Incident Management System 12
   Management Principles 12
     Unity of Command 12
     Span of Control 12
     Division of Labor 12
     Discipline 13
   Other Views of Organization 13
     Function 13
     Geography 13
     Staffing 13
The Functions of Management 13

Rules and Regulations, Policies, and Standard Operating Procedures 14
Establishing a Strong Supervisor/Employee Relationship 16
Positive Labor-Management Relations 17
   The International Association of Fire Fighters 17
   National Volunteer Fire Council 18
The Fire Officer's Role as a Supervisor 18
The Fire Officer's Tasks 18
   The Beginning of Shift Report 19
   Notifications 19
   Decision Making and Problem Solving 20
   Example of a Typical Fire Station Workday 21
   Example of a Typical Volunteer Duty Night 21
The Transition from Fire Fighter to Fire Officer 22
The Fire Officer as Supervisor-Commander-Trainer 23
   Supervisor 23
     Unpopular Orders and Directives 23
   Commander 23
   Trainer 24
     Developing a Personal Training Library 24
     Know the Neighborhood 25
     Use Problem-Solving Scenarios 25
The Fire Officer's Supervisor 26
Integrity and Ethical Behavior 26
   Integrity 27
   Ethical Behavior 27
Workplace Diversity 28
   The Fire Officer's Role in Workplace Diversity 29
     Actionable Items 29

Hostile Workplace and Sexual Harassment	29
Handling a Harassment or Hostile Workplace Complaint	30
Credentialing and Fire Officer Development	31
The IAFC Company Officer Leadership Program	32

CHAPTER 2
Understanding Leadership and Management Theories — 39

Introduction	40
Introduction to Leadership and Management	40
Leadership and Management Theories	41
Peter Northouse's Classification Approach	41
The Trait Approach	41
The Skills Approach	41
The Behavioral Approach	42
Blake and Mouton's Managerial Grid	42
Impoverished Management	42
Authority-Compliance Management	43
Country Club Management	43
Middle-of-the-Road Management	44
Team Management	44
Situational Leadership	45
Transformational Leadership	46
Authentic Leadership	46
Servant Leadership	46
Adaptive Leadership	47
Followership	47
Power as a Leadership Resource	47
Leadership Challenges	48
Fire Station as Municipal Work Location versus Fire Fighter Home	48
Leadership in the Volunteer Fire Service	48
U.S. Marine Corps Leadership Principles	49
Fourteen U.S. Marine Corps Leadership Traits	49
Human Resources Management	52
Utilizing Human Resources	53
Mission Statement	53
Getting Assignments Completed	53
Effective Company Officer Delegation	56
Origins of Crew Resource Management	56
Researching and Validating CRM Concepts	57
Human Error	57
Active Failures and Latent Conditions	57
Error Management Model	58
The CRM Model	58
Communication Skills	59
Assertive Statement Process	59
Teamwork	59
Leadership	60
Mentoring	60
Handling Conflict	60
Responsibility	60
Self-Assessment of Each Team Member	61
Task Allocation	61
Critical Decision Making	62
Situational Awareness	63
Maintaining Emergency Scene Situational Awareness	63
Postincident Analysis	64
Conducting a Postincident Analysis	65
Documentation and Follow-Up	66

CHAPTER 3
Leading a Team — 73

Introduction	74
The Communication Cycle	74
The Message	75
Ensure Accuracy	75
The Sender	75
The Medium	75
Overcoming Environmental Noise	76
The Receiver	77
Feedback	77
Active Listening	77
Stay Focused	78
Emergency Incident Communications	78
Using the Communications Order Model	78
Initial Situation Report to Dispatch	78
Other Responding Units	79
Radio Reports	79
Project Mayday Radio Phrases	80
Keep Your Supervisor Informed	80

How Quickly Do You Get Bad News?	80
The Grapevine	81
Supervisory Tasks	**81**
Grievance Procedure	81
Sample Step 1	82
Sample Step 2	82
Sample Step 3	82
Sample Step 4	82
Making Decisions	82
Define the Problem	83
Generate Alternative Solutions	84
Select a Solution	84
Implement the Solution	84
Evaluate the Results	86
Assigning Tasks to Unit Members	**86**
Supervising a Single Company	86
Closeness of Supervision	87
Standardized Actions	87
Command Staff Assignments	87
Assigning Tasks During Emergency Incidents	88
Emergency Scene Leadership	88
Assigning Tasks During Nonemergency Incidents	89
Safety Considerations	90
Fire Officer Training Responsibilities	**90**
The Four-Step Method of Skill Training	90
Step 1: Preparation	92
Step 2: Presentation	92
Step 3: Application	92
Step 4: Evaluation	92
Four Levels of Fire Fighter Skill Competence	93
Unconscious Incompetence	94
Conscious Incompetence	94
Conscious Competence	94
Unconscious Competence	95
Mentoring	95
Provide New or Revised Skill Sets	95
Ensure Competence and Confidence	95
When New Member Training Is On-the-Job	96
Skills That Must Be Learned Immediately	96
Skills Necessary for Staying Alive	97
Developing a Specific Training Program	98
Assess Needs	98
Establish Objectives	98
Develop the Training Program	98
Deliver the Training	98
Evaluate the Impact	99
Addressing Member-Related Problems	**99**
Complaints, Conflicts, and Mistakes	99
Managing Conflict	100
Personnel Conflicts and Grievances	100
Conflict Resolution Model	101
Investigate	102
Take Action	102
Follow-Up	102
Understanding Emotional Confrontations	102
Behavioral and Physical Health Issues	104
Awareness	104
Substance Abuse	104
Family and Marital Problems	105
Financial Problems	105
Resiliency to Stress	106
Moral Injury	106
Acute Stress Disorder	106
Posttraumatic Stress Disorder	106
Behavioral Health Resources for Fire Officers	107
Employee Assistance Programs	107

CHAPTER 4
Community Relations and Risk Reduction — 115

Introduction	116
Understanding the Community	116
Community Risk Reduction	117
Step One: Identify Risks	118
Step Two: Prioritize Risks	118
Step Three: Develop Strategies and Tactics to Mitigate Risks	118
Step Four: Prepare the CRR Plan	119
Step Five: Implement the CRR Plan	119
Step Six: Monitor, Evaluate, and Modify the Plan	119
Public Education	119
Community Emergency Response Team	120
CERT Course Schedule	120

Responding to Public Inquiries	121
Citizen Concerns	122
Customer Service versus Customer Satisfaction	122
Keep the Complainant Informed	123
Follow Up	124

CHAPTER 5
Fire Department Administration — 127

Introduction	128
Policy Recommendations and Implementation	128
Recommending Policies and Policy Changes	129
Implementing Policies	129
The Budget Cycle	130
Base and Supplemental Budgets	132
Making a Budget Request	132
Project Description and Justification	134
Purchase Information	134
Total Cost of Project	134
Revenue Sources	134
Local Government Sources	135
Volunteer Fire Departments	135
Bingo and Other Activities	135
Real Estate and Portfolio Management	136
Lower Revenue Options	136
Defer Scheduled Expenditures	136
Privatize or Contract Out Elements of the Service Provided	137
Regionalize or Consolidate Services	137
Reduce the Workforce	137
Reduce the Size of the Fire Department	137
Human Resources	138
Evaluation	139
Starting the Evaluation Process with a New Fire Fighter	139
Providing Feedback after an Incident or Activity	140
Positive and Corrective Disciplinary Actions	141
Positive Discipline	141
Corrective Discipline	142
Employee Documentation and Record Keeping	145
Record-Keeping Systems	145
Inventory	145
Inspection and Maintenance	145
Training and Credential Records	146
Individual Exposure History and Medical Records	146
Reporting	147
Types of Reports	147
Written Reports	147
Using Information Technology	150

CHAPTER 6
Preincident Planning and Code Enforcement — 157

Introduction	158
The Fire Officer's Role in Community Fire Safety	159
Built-in Fire Protection	159
Common Built-in Fire Protection Code Violations	159
Preincident Planning	160
A Systematic Approach	161
1: Identify Physical Elements and Site Considerations	161
2: Identify Occupancy Considerations	161
3: Identify Water Supply and Fire Protection Systems	162
4: Special Considerations—Special Hazards	162
5: Identify Emergency Operation Considerations	165
6: Document Findings	166
Understanding Fire Codes	166
Building Code versus Fire Code	167
State Fire Codes	167
Local Fire Codes	167
Model Codes	167
Retroactive Code Requirements	168
Understanding Built-in Fire Protection Systems	168
Water-Based Fire Protection Systems	168
Fire Pumps	170
Special Extinguishing Systems	170
Carbon Dioxide	170
Dry or Wet Chemical	171
Halon	171
Foam Systems	171
Fire Alarm and Detection Systems	171

Understanding Fire Code Compliance Inspections	171
Fire Company Inspections	171
Classifying by Building or Occupancy	172
Construction Type	172
Occupancy and Use Group	174
Assembly	174
Business	174
Educational	175
Industrial	175
Institutional	175
Mercantile	175
Residential	175
Storage	176
Mixed Occupancies	176
Special Properties	176
NFPA 704 Marking System	176
Preparing for an Inspection	177
Reviewing the Fire Code	177
Review Prior Inspection Reports, Fire History, and Preincident Plans	177
Coordinate Activity with the Fire Prevention Division	177
Arrange a Visit	179
Assemble Tools and References	179
Conducting the Inspection	179
Meet with the Representative	180
Inspecting from the Outside In, Bottom to Top	180
Exit Interview	180
Writing the Inspection/Correction Report	181
General Inspection Requirements	181
Access and Egress	181
Exit Signs and Emergency Lighting	181
Portable Fire Extinguishers	182
Built-in Fire Protection Systems	182
Special Hazards	182
Selected Use Group–Specific Concerns	182
Public Assembly	182
Business	183
Educational	183
Factory Industrial	183
Institutional	183
Mercantile	184

Residential	184
Special Properties	184
Manufacturing	184
Storage	184
Mixed	185
Legal Considerations During Investigations	185
Searches	185
Securing the Scene	185
Evidence	186
Protecting Evidence	186
Requesting an Investigator	187

CHAPTER 7
Command of Initial Emergency Operations — 195

Introduction	196
The History of the Incident Command System	196
FIRESCOPE and Fire Ground Command	197
Developing One National System	197
Incident Command and the National Incident Management System Model	197
ICS Training	197
Levels of Command	198
The Fire Officer's Role in Incident Management	199
Initial Incident Command	199
Responsibilities of Command	200
Establishing Command	200
Initial Operational Modes	200
Investigation Mode	201
Fast-Attack Mode	201
Command Mode	201
Functions of Command	202
Command Safety	202
Fire Fighter Accountability	202
Personnel Accountability Report	203
Fire Research and Risk/Benefit Analysis	203
Fire Behavior Graph	203
Flow Path	205
Modern versus Legacy Single Family Dwellings	205
Risk/Benefit Analysis	206
IAFC Rules of Engagement	207

Sizing Up the Incident	208
Prearrival Information	208
On-Scene Observations	208
Lloyd Layman's Five-Step Size-Up Process	209
Facts	209
Probabilities	209
Situation	210
Decision	210
Plan of Operation	210
National Fire Academy Size-Up Process	210
Phase One: Preincident Information	210
Phase Two: Initial Size-Up	211
Phase Three: Ongoing Size-Up	211
Developing an Incident Action Plan	211
Incident Priorities	212
Tactical Priorities	212
RECEO-VS	212
S.L.I.C.E.-R.S.	212
Location Designators	214
Tactical Safety Considerations	214
Scene Safety	215
Rapid Intervention Crews	216
Transfer of Command	217

CHAPTER 8
Safety and Risk Management — 223

Introduction	224
Fire Officer's Role in Risk Management	224
Community Fire Risk	224
Five Principal Risk Management Steps	225
Step 1: Identify Risk Exposure	225
Step 2: Evaluate Risk Exposure	225
Step 3: Rank and Prioritize Risks	225
Step 4: Determine and Implement Risk Management Control Actions	225
Step 5: Evaluate and Revise Risk Control Actions	225
Fire Fighter Injury and Death Trends	226
Everyone Goes Home	226
National Firefighter Near Miss Reporting System	227
Reducing Deaths from Sudden Cardiac Arrest	227
Reducing Deaths from Suicide	228

Reducing Deaths from Cancer	228
Reducing Deaths from Motor Vehicle Collisions	229
Reducing Deaths from Fire Suppression Operations	230
Maintaining Crew Integrity	230
Air Management	231
Teams and Tools	231
Directing Members during Training Evolutions	231
Reducing Deaths during Training	232
Incident Safety Officer	233
Incident Safety Officer and Incident Management	233
Qualifications to Operate as an Incident Safety Officer	234
Typical Incident Safety Officer Tasks	234
Assistant Incident Safety Officers at Large or Complex Incidents	235
Incident Scene Rehabilitation	235
Creating and Maintaining a Safe Work Environment	236
Safety Policies and Procedures	236
Emergency Incident Injury Prevention	236
Physical Fitness	236
Personal Protective Equipment	236
Fire Station Safety	237
Clothing	237
Housekeeping	238
Lifting Techniques	238
Infection Control	238
Infectious Disease Exposure	239
Accident Investigation	239
Accident Investigation and Documentation	239
Postincident Analysis	241

SECTION 2
Fire Officer II

CHAPTER 9
The Fire Officer II as a Manager — 251

Introduction	252
Requirements for Fire Officer II	252

Additional Roles and Responsibilities for Fire Officer II ... 252
Authorization to Function as a Fire Department ... 253
  State Police Powers ... 253
  Administrative Law ... 253
Supervising Multiple Companies ... 254
  Determining Task Assignments ... 254
  Assigning Resources ... 255
Challenges for the Captain ... 256
  Fire Fighter Emotional Health ... 256
  Supervision and Motivation ... 256
  Increase in Non-fire Incidents ... 256
  Deterioration of the Built Environment ... 258
  Catastrophic Weather Events ... 258
  Diversity ... 258
Working with Other Organizations ... 258
Legislative Framework for Collective Bargaining ... 259
  Norris-LaGuardia Act of 1932 ... 259
  Wagner-Connery Act of 1935 ... 260
  Taft-Hartley Labor Act of 1947 ... 260
  Landrum-Griffin Act of 1959 ... 260
  Collective Bargaining for Federal Employees ... 260
  State Labor Laws ... 261
    Right to Work ... 261
Organizing Fire Fighters into Labor Unions ... 262
  Federal Labor–Management Conflicts ... 262
  Postal Workers' Strike: 1970 ... 262
  PATCO Air Traffic Controllers' Strike: 1981 ... 262
Labor Actions in the Fire Service ... 262
  Striking for Better Working Conditions: 1918–1921 ... 263
  Striking to Preserve Wages and Staffing: 1931–1933 ... 263
  Staffing, Wages, and Contracts: 1973–1980 ... 263
  Negative Impacts of Strikes ... 264
  The Wisconsin Challenge ... 264
The Growth of IAFF as a Political Influence ... 264

CHAPTER 10
Applications of Leadership ... 271

Introduction ... 272
Fire Officer II as Transformational Leader ... 272
  Transactional Relationship Versus Transformational Leadership ... 272
Evaluating Job Performance of Assigned Members ... 273
  Establishing Annual Fire Fighter Goals ... 273
  Keeping Track of Every Fire Fighter's Activity ... 273
  Informal Work Performance Reviews ... 274
Correcting Unacceptable Behavior ... 274
  Using Just Culture to Identify Source of Behavior ... 274
    Human Error ... 274
    At-Risk Behavior ... 275
    Reckless Behavior ... 275
  Advance Notice of a Substandard Employee Evaluation ... 275
  Work Improvement Plan ... 275
Negative Discipline: Correcting Unacceptable Behavior ... 277
  The Progression of Negative Discipline ... 277
  Formal Written Reprimand ... 278
  Predisciplinary Conference ... 278
  Alternative Disciplinary Actions ... 279
  Suspension ... 279
  Termination ... 280
Documenting Fire Fighter Performance ... 280
  Mid-Year Review ... 280
  Six Weeks Before the End of the Annual Evaluation Period ... 280
  Evaluation Errors ... 282
    Leniency or Severity ... 282
    Personal Bias ... 282
    Recency ... 282
    Central Tendency ... 282
    Frame of Reference ... 282
    Halo and Horn Effect ... 282
    Contrast Effect ... 282
  Conducting the Annual Evaluation ... 282

The Four Borders of Human Resources	283	Building Stock	301
Federal and State Laws	283	Public Safety Response Agencies	301
Labor Contract	283	Community Service Organizations	302
Jurisdiction Regulations	283	Hazards	302
Fire Department Policies	284	Economic Factors	302
Using Just Culture Concepts to Improve Health and Safety	284	Past Loss/Event History	302
		Critical Infrastructure Systems	302
Analyzing Near-Miss Reports	284	Determining Risk Likelihood and Impact Consequences	302
HFACS Level 1: Unsafe Acts	284	Developing a Community Risk Reduction Plan	303
HFACS Level 2: Preconditions to Unsafe Acts	285	The Five E's of Prevention and Mitigation	303
HFACS Level 3: Unsafe Supervision	285	Writing the Community Risk Reduction Plan	303
HFACS Level 4: Organizational Influences	285	Cooperating with Allied Organizations	304
Data Analysis	285	Implementing and Monitoring the CRR	304
Designing Better Systems	286	Media Relations	305
Controlling Contributing Factors	286	Fire Department Public Information Officer	305
Adding Barriers	286	Step 1: Build a Strong Foundation	306
Adding Recovery	286	Step 2: Proactive Outreach	307
Adding Redundancy	286	Step 3: Measured Responsiveness	307
Professional Development	287	Press Releases	307
Training Versus Education	287	The Fire Officer as Spokesperson	307
Academic Accreditation	287	Be Prepared	308
Accreditation of Skill Certification	288	Stay in Control	308
National Board on Fire Service Professional Qualifications	289	Look and Act the Part	308
		It Is Not Over Until It Is Over	309
International Fire Service Accreditation Congress	289	Social Media Outreach	309
		Social Media Challenges	309
Building a Professional Development Plan	290		
CPSE Fire Officer Credential	290		
NFA Managing Officer Program	290		
IAFC Company Officer Leadership Symposium	291		

CHAPTER 11
Managing Community Risk Reduction Programs — 299

CHAPTER 12
Administrative Communications — 315

Introduction	300
Managing Community Risk Reduction Implementation	300
Gathering Community Risk Assessment Data	300
Demographics	300
Geography	300

Introduction	316
Written Communications	316
Informal Communications	316
Formal Communications	317
Standard Operating Procedures	317
Standard Operating Guidelines	317
General Orders	317
Announcements	317

Legal Correspondence	317	Suspicious Fire Considerations	334
Writing a Report	318	Fire Company Fire Cause Evaluation	335
Recommendation Report	319	Finding the Area of Origin	335
Presenting a Report	319	Fire Patterns	336
Navigating the Budgetary Process: A Case Study	319	Evaluating the Cause of the Fire	336
Developing a Budget Proposal	320	Source and Form of Heat Ignition	337
Overview of the Alternative Response Unit Program	320	Material First Ignited	338
ARU Annual Personnel and Operating Expenditures	320	Ignition Factor and Cause	338
		Fire Analysis	338
Personnel Expenses	320	Conducting Interviews	338
Operating Expenses	320	Vehicle Fire Cause Determination	339
ARU Capital Budget	321	Wildland Fire Cause Determination	340
ARU Vehicle	321	Fire Cause Classifications	341
ARU Administrative Office	321	Accidental Fire Causes	341
ARU Uniforms and Protective Clothing	321	Natural Fire Causes	342
ARU Cost Recovery and Reduction	321	Incendiary Fire Causes	342
The Purchasing Process	323	Undetermined Fire Causes	343
Petty Cash	323	Indicators of Incendiary Fire Causes	343
Purchase Orders	323	Accelerants and Trailers	343
Requisitions	323	Multiple Points of Origin	344
The Bidding Process	323	Arson	344
Revenue Sources	324	Arson Motives	344
Grants	325	Profit	344
Nontraditional Revenue Sources	325	Crime Concealment	344
Reimbursement for Response	326	Excitement	346
Expenditures	326	Spite/Revenge	346
Personnel Expenditures	326	Extremism	346
Operating Expenditures	326	Vandalism	346
Capital Expenditures	327	Documentation and Reports	346
Bond Referendums and Capital Projects	328	Preliminary Investigation Documentation	346
Recommending Change	328	Legal Proceedings	347
Identify the Problem	329	After the Fire Officials Are Gone	347
Explore Alternatives and Select Solution	329		
Implement and Evaluate	329		

CHAPTER 13
Fire Cause Determination — 333

Introduction	334
The Fire Officer's Role in Fire Cause Determination	334

CHAPTER 14
Managing Major Incidents — 353

Introduction	354
National Response Framework	354
The Stafford Act	354
Incident Action Plan	357
Staffing the Incident Management System	357
Command Staff	357

Safety Officer	358
Liaison Officer	358
Public Information Officer	358
General Staff Functions	358
Operations	358
Planning	359
Logistics	359
Finance/Administration	360
Tactical-Level Incident Management	360
Divisions, Groups, and Units	360
Divisions	360
Groups	361
Units	361
Division/Group/Unit Supervisor Responsibilities	361
Branches	361
Fire Officer Greater Alarm Responsibilities	362
Staging	362
Task-Level Incident Management	362
Task Forces and Strike Teams	363
Greater Alarm Infrastructure	363
Selected Structure Major Incidents Considerations	363
Single-Family Dwellings	363
Mid-Rise Multifamily Dwellings	364
High-Rise Considerations	364
Postincident Analysis	365
Preparing Information for an Incident Review	365
Conducting a Critique	366
Documentation and Follow-Up	368

APPENDIX A: Preparing for Promotion — 375

Introduction	375
The Origin of Promotional Examinations	375
Sizing Up The Promotional Process	375
Postexamination Promotional Considerations	376
When Fire Officers Are Voted In	376
Preparing a Promotional Examination	376
Charting the Required Knowledge, Skills, and Abilities	377
Promotional Examination Components	377
Multiple-Choice Written Examination	377
Assessment Centers	378
In-Basket Exercises	378
Emergency Incident Simulations	379
Interpersonal Interaction	380
Writing or Speaking Exercise	381
Technical Skills Demonstration	381
Mastering the Content and the Process of a Promotional Exam	381
Building a Personal Study Journal	381
Preparing for Role Playing	384

APPENDIX B: NFPA 1021, *Standard for Fire Officer Professional Qualifications, 2020 Edition* Correlation Guide — 387

Glossary	399
Index	408

Acknowledgements

© Glen E. Ellman.

The Jones & Bartlett Learning Public Safety Group, the National Fire Protection Association, and the International Association of Fire Chiefs would like to thank the author, contributor, and reviewers of *Fire Officer: Principles and Practice, Fourth Edition*.

About the Author

Michael J. Ward, MGA, MIFireE, FACPE, worked in an urban county fire and rescue department in suburban Washington DC, with assignments in fire suppression, paramedicine, academy, hazardous materials, code enforcement, and EMS administration. During that time, Ward was simultaneously employed as both a community college fire science faculty member and as a state instructor for Fire and EMS Programs. Ward went on to become a Member of the Institution of Fire Engineers (MIFireE) and earned his master's degree in State and Local Government.

More recently, Ward was hired as an Assistant Professor of Emergency Medicine as well as the Emergency Health Services program director at a university medical center in Washington DC. As a member of a consulting firm, Ward then served as the interim director of two hospital-based paramedic and transport services. In 2018, Ward achieved Fellow (executive) credential with the American College of Paramedic Executives (FACPE).

Ward retired as a Captain II.

Contributor

Kelli J. Scarlett
University of Maryland University College
Adelphia, MD

Reviewers

Robert Alley
Chief
Gerton Fire and Rescue
Gerton, NC

Shane L. Anderson, NREMT, BS, MA
Crowder College
Neosho, MO
Monett Fire Department
Monett, MO

Chris Angermuller, MPA, CFO, CTO, MIFireE
Deputy Chief
Grand Junction Fire Department
Grand Junction, CO

Michael Athey
Shepherdstown West Virginia Fire Department
Shepherdstown, WV

Michael Azevedo
Assistant Fire Chief
Carmel Fire and Rescue
Carmel, ME

Dean Russell Bailey
Firefighter/EMT Training Coordinator
City of Galena Park Fire Department
Galena Park, TX

Bobby Barton
Captain
Murphy Volunteer Fire Department
Murphy, NC

Dan Bell
Training Officer
Whitchurch-Stouffville
Ontario, Canada

Tyler H. Bergemann
Lieutenant, Gaylordsville Volunteer Fire Department
Gaylordsville, CT

Deputy Fire Marshall, City of Danbury Fire Department
Danbury, CT

Tom Berry
Captain
Huguenot Volunteer Fire Department
Powhatan, VA

Brian K. Blomstrom, MPA, CFO, CTO, PEM
Greenville Department of Public Safety
Greenville, MI

Laurence P. Brandoli, MEd (Ret.)
Springfield Technical Community College
Springfield, MA

David L. Bullins, EFO, CFO, MS
Director of Public Safety
Mitchell Community College
Statesville, NC

Melvin Byrne (Ret.)
Ashburn Volunteer Fire and Rescue Department
Ashburn, VA

Richard Carroll
Cleveland Community College
Shelby, NC

Daniel Casner
Haz-Mat Instructor
State of California
Upland, CA

Jordan J. Collins
Training Officer / Fire Equipment Operator
Lafourche Parish Fire District #3 / City of Houma Fire Department
Cut Off, LA

James M. Craft, Jr.
Captain (Ret.)
Dallas Fire-Rescue Department
Plano, TX

Josh Crisp
Gaston College Regional Emergency Services Training Center
Dallas, NC

Frank DeFrancesco, CFO, MA
Captain
Hernando County Fire Rescue
Brooksville, FL

Blake J. Deiber, BS
Western Technical College
Sparta, WI

Keith A. Deubell
Lieutenant
Horry County Fire Rescue
Myrtle Beach, SC

Michael Todd Eddy, BS, NREMT-P
Alabama Fire College
Tuscaloosa, AL

Robert Edmiston
Chief of Health and Safety
US Army Garrison Daegu
Daegu Metropolitan City, South Korea

Chief Doug Eggiman, BS, AS
Midway Fire Rescue
Pawleys Island, SC

David English
Captain
Richardson Fire Department
Richardson, TX

Danny Evans
Captain
Saginaw Township Fire Department
Saginaw, MI

Billie E. Floyd, Jr.
City of North Myrtle Beach Dept. of Public Safety
Myrtle Beach, SC

Robert Fash
Specialist, Emergency Services
Public Fire Protection
National Fire Protection Association
Quincy, MA

Brian Focht
Willow Grove Fire Company
Willow Grove, PA
PECO Fire Academy
West Conshohocken, PA

Ken Fowler, MSL
Louisiana State University Fire and Emergency Training Institute
Baton Rouge, LA

Michael Free
Battalion Chief
Port Arthur Fire Department
Port Arthur, TX

Jan F. Ganzel
Blackman-Leoni Dept. of Public Safety
Jackson, MI

Michael Gladieux
Lieutenant
City of Royal Oak Fire Department
Royal Oak, MI

Gary Graf
Assistant Chief/Training Officer
Pacific Fire District
Pacific, MO

Brian Grant, FO, NRP
Northern Arizona Consolidated Fire District #1
Kingman, AZ

Robert Andrew Halpin
Captain (Ret.)
High Point Fire Department
High Point, NC

Michael Harper
Captain (Ret.)
Lincoln Park Fire Department
Lincoln Park, MI

Robert L. Havens
Deputy Fire Chief/Deputy EMC
Port Havens Fire Department
Arthur, TX

J. Brad Haywood
Fairfax County Fire and Rescue Department
Fairfax, VA

Robert C. Hecker, NRP
Training & Safety Officer (Ret.)
St. Tammany Parish Fire Training
Mandeville, LA

James R. Holland
Battalion Chief
City of Watertown Fire Department
State Fire Instructor
New York State Office of Fire Prevention and Control
Watertown, NY

Randy Hunter
Captain
Bluffton Township Fire District
Bluffton, SC

Shawn S. Kelley
International Association of Fire Chiefs
Chantilly, VA

Theodore G. Kolb II
Captain
City of Watertown Fire Department
Watertown, NY

Carl Magann
Mendocino Community College
Ukiah, CA

Robert Maibach
Chief
Goose Creek Rural Fire Department
Goose Creek, SC

Matt Mead
Assistant Fire Chief
Shepherd Tri-Township Fire Department
Shepard, MI

Ralph McNemar
Program Coordinator
West Virginia University-Fire Service Extension
Beckley, WV

Andrew Mihans
Captain
Arlington Fire District
Poughkeepsie, NY

James P. Moore
Fire Chief (Ret.)
Crystal Lake, IL
Assistant Director
University of Illinois Fire Service Institute
Champaign, IL

Brandon Nargessi
Captain
Clarkdale Fire Department
Clarkdale, AZ

Hugh O'Callaghan, EFO, MA
Assistant Fire Chief
West Hartford Fire Department
Hartford, CT

Greg Palmer, MA
North Carolina Office of State Fire Marshal
Raleigh, NC

Jonathan Parrott
Assistant Chief
Englewood Fire Department
Englewood, TN

Joseph W. Pfaff
Stoddard-Bergen Fire Department
Stoddard, WI

Brian Piercy
Training Officer
Ashford North Cove Fire Department
Marion, NC

Michael Pruitt
Glade Hill Fire Department
Glade Hill, VA

Joe Pulvermacher, IAFC EFO Section-Board of Directors, EFO, CFO
Fire Chief
City of Fitchburg
Fitchburg, WI

Joe Ramsey
Battalion Chief
Winston-Salem Fire Department
Winston-Salem, NC

David Rogers
Assistant Fire Chief
Tallassee Fire Department
Tallassee, AL

Joseph Scaglione
Lieutenant
City of Meriden Fire Department
Meriden, CT

Peter Silva, Jr.
Chief of Training (Ret.)
Madison College
Madison, WI

Trevor Sommerfeld
Captain
Calgary Fire Department
Calgary, Alberta, Canada

Todd Spruill
Captain
Richmond Fire and Emergency Services
Richmond, VA

Greg Summitt
Training Chief
Kannapolis Fire Department
Kannapolis, NC

Howard Sykes
Chief
Lebanon Volunteer Fire Department
Durham, NC

Kevin Thompson
Captain
Comstock Fire Department
Kalamazoo, MI

Brent M. Turner, AS, BA
SCFA Instructor, FO I, II, and III Instructor
South Carolina State Fire/Fire Academy
Columbia, SC

Gerald "Red" Van Ert, Jr., MS, MPA
Altoona Fire Department
Altoona, WI

Brian A. Wade, BS, CFO
Training Specialist
North Carolina Office of State Fire Marshal
Kinston, NC

Beverley E. Walker, BS, MPA, EFO
East Georgia State College
Swainsboro, GA

Terry L. Watts, BAS
Shelby County Fire Department
Memphis, TN

John West
Molalla Fire
Molalla, OR

Allan Wickline
Adjunct Faculty
The Community College of Allegheny County, Boyce Campus
Monroeville, PA

Orlando P. Willis
Fire Chief
VictoryLand Fire/Rescue and EMS
Shorter, AL

Stephen Wise
Lieutenant
Spotsylvania County Fire Rescue
Spotsylvania Courthouse, VA

Donald Wollenbecker
Mt. Pleasant Fire Rescue
Summerville, SC

Donnie L. Woods III, BS
NSA Naples, Italy Fire & Emergency Services
Naples, Italy

James Woolf
Twinsburg Fire Department
Streetsboro, OH

Shannon Young
Battalion Chief
Appleton Fire Department
Appleton, WI

Fourth Edition Focus Group

Chris Bachman
Deputy Chief
Pike Township Fire Department
Indianapolis, IN

Tim Baker
Director
Lansing Community College
Regional Firefighter Training Center
Lansing, MI

Charlie Butterfield, MEd, NRP, CFO
Deputy Chief of Operations
Meridian Fire Department
Meridian, ID

Tania Daffron
Battalion Chief
City of Bloomington Fire Department
Bloomington, IN

Joseph Edwards
Tennessee Fire and Codes Academy
Bell Buckle, TN

Chief Doug Eggiman, BS, AS
Midway Fire Rescue
Pawleys Island, SC

Kevin J. Fedrizzi
Division Chief of Training
Meridian Fire Department
Meridian, ID

Val Gale
Assistant Chief—Operations
Chandler Fire Department
Chandler, AZ

Bill Gardner, CFO, CFE, EMT-P
Fire Chief/Emergency Management–Homeland Security Coordinator
Leander Fire Department
Leander, TX

Robert L. Havens, EFO, CFE, FSCEO, CFPS
Deputy Fire Chief/Deputy EMC
Port Arthur Fire Department
Port Arthur, TX

Donnie Hurd II
Captain/Licensed Paramedic
Trophy Club Fire Department
Trophy Club, TX
Adjunct Instructor
Fire Service Training Center
Tarrant County College
Fort Worth, TX

Jack Johnson
Captain
Grand Rapids Fire Department
Grand Rapids, MI

Jennifer Johnson
Battalion Chief (Ret.)
Kansas City Kansas Fire Department
Kansas City, KS

James P. Moore
Fire Chief (Ret.)
Crystal Lake, IL
Assistant Director
University of Illinois Fire Service Institute
Champaign, IL

Jon Perry
Training Chief
Idaho Falls Fire
Idaho Falls, ID

David Piper
Battalion Chief–Operations
Orland Fire Protection District
Orland Park, IL

Zackary Riddle, MPA
Deputy Chief
Oak Lawn EMS Fire Rescue
Oak Lawn, IL

Darrell Rutledge
Coordinator–Fire Service Training Center
Tarrant County College–Northwest
Fort Worth, TX

Michael Simmons
Director, Kilgore College Fire Academy
Chief of Operations (Ret.)
Kilgore Fire Department
Kilgore, TX

Michael Skaza
Assistant Chief
West Rutland Fire Department
West Rutland, VT

John F. Sullivan, MPA, EFO, CFO, MFireE
Vice-chair
International Association of Fire Chiefs Safety, Health & Survival Section
Chief of Department
Brookline Fire Department
Emergency Management Director
Town of Brookline
Brookline, MA

Kilipaki Vaughan
Deputy Fire Chief–Ka Pounui Kinaiahi
Kauai Fire Department
Lihue, HI

Mark A. Vroman, MBA, EFO
Battalion Chief
Meridian Township Fire Department
Okemos, MI

James M. Zeeb
Assistant Chief
Chief of the Training Division
Kansas City Kansas Fire Department
Kansas City, KS

New for the Fourth Edition

The fourth edition of *Fire Officer: Principles and Practice* was significantly updated and reorganized to better serve the Fire Officer I and Fire Officer II. The content meets and exceeds the job performance requirements for Fire Officer I and II in the 2020 Edition of NFPA 1021, *Standard for Fire Officer Professional Qualifications*. The new edition places emphasis on the understanding and application of leadership and management theories and traits, challenges, and how we use leadership to reach objectives. The basic principles and methods of community risk reduction are also discussed.

In addition, the program now has two distinct sections: Section One includes eight chapters, which set the foundation for Fire Officer I knowledge and understanding. Section Two comprises six chapters, which encompass the higher-level competencies required for Fire Officer II. This new organization will allow you the flexibility to teach your Fire Officer I and II course(s) exactly the way you wish.

SECTION 1

Fire Officer I

CHAPTER **1** **The Fire Officer I as a Company Supervisor**

CHAPTER **2** **Understanding Leadership and Management Theories**

CHAPTER **3** **Leading a Team**

CHAPTER **4** **Community Relations and Risk Reduction**

CHAPTER **5** **Fire Department Administration**

CHAPTER **6** **Preincident Planning and Code Enforcement**

CHAPTER **7** **Command of Initial Emergency Operations**

CHAPTER **8** **Safety and Risk Management**

CHAPTER 1

Fire Officer I

The Fire Officer I as a Company Supervisor

KNOWLEDGE OBJECTIVES

After studying this chapter, you will be able to:

- Identify the qualifications of a Fire Officer I. (**NFPA 1021: 4.1.1**) (pp. 4–5)
- Describe the roles and responsibilities of the Fire Officer I. (**NFPA 1021: 4.1.1**) (pp. 5–6)
- Describe the fire service in the United States. (pp. 7–10)
- Explain fire department organizational structure. (**NFPA 1021: 4.1.1**) (pp. 10–13)
- Identify the standard incident management system used by the fire service. (**NFPA 1021: 4.1.2**) (p. 12)
- Identify and describe the four management principles. (**NFPA 1021: 4.1.1**) (pp. 12–13)
- Explain the function of each management position in a department. (**NFPA 1021: 4.4.4**) (pp. 13–14)
- Explain the roles of rules and regulations, policies, and departmental operating procedures. (**NFPA 1021: 4.1.1**) (pp. 14–16)
- Identify practices that contribute to strong employee–supervisor relationships. (**NFPA 1021: 4.1.1**) (pp. 16–17)
- Recognize the value of positive labor–management relations. (**NFPA 1021: 4.1.1**) (p. 17)
- Describe the impact of the International Association of Fire Fighters on fire fighters and emergency medical services personnel. (pp. 17–18)
- Describe the fire officer's role as a supervisor. (**NFPA 1021: 4.1.1**) (p. 18)
- Describe the fire officer's basic tasks, including routine administrative functions. (**NFPA 1021: 4.1.1, 4.4.2**) (pp. 18–22)
- Identify how relationships to fellow fire fighters and to the organization change when transitioning from fire fighter to fire officer. (pp. 22–23)
- Compare and contrast the fire officer's three distinct roles of supervisor, commander, and trainer. (pp. 23–26)
- Describe the activities a fire officer performs to maintain an effective working relationship with his or her supervisor. (p. 26)
- Explain the importance of integrity and ethical behavior. (**NFPA 1021: 4.1.1**) (pp. 26–28)
- Describe how officers can act to maintain workplace diversity. (**NFPA 1021: 4.1.1**) (pp. 29–31)
- Compare and contrast the two organizations that offer fire credentialing, the Center for Public Safety Excellence and the National Fire Academy. (p. 31)

SKILLS OBJECTIVES

There are no skills objectives for this chapter.

ADDITIONAL NFPA STANDARDS

- **NFPA 1001**, *Standard for Fire Fighter Professional Qualifications*
- **NFPA 1041**, *Standard for Fire and Emergency Services Instructor Professional Qualifications*

You Are the Fire Officer

It is 45 minutes before "A" shift will start its day at City Fire and Rescue Department. The newly promoted A shift lieutenant is meeting with the newly promoted captain in the fire station kitchen. They are planning their first day. The captain has been with the department for 15 years, spending 8 years as a truck company lieutenant and a training officer. The lieutenant has been with the department for 7 years; most of that time was on engine companies.

The battalion chief that runs the district is coming by at 10:00 AM to meet with the new officers.

1. What are the responsibilities of a supervising fire officer?
2. How does a new supervising officer work within the department's organizational structure?
3. How can you convert the theories and concepts memorized for a promotional exam into effective unit officer activities?

Access Navigate for more practice activities.

Introduction

Organizations such as the **National Fire Protection Association (NFPA)** and the International Code Council (ICC) issue voluntary consensus-based codes and standards that set the expected organization, construction, performance, and operation of many aspects of fire service operations. This text provides information to meet the criteria outlined in National Fire Protection Association (NFPA) 1021, *Standard for Fire Officer Professional Qualifications*, at the Fire Officer I and Fire Officer II levels. NFPA 1021 defines four levels of fire officer.

The Fire Officer I level is the first step in a progressive sequence and is generally associated with an officer supervising a single fire company or apparatus. A Fire Officer I, or lieutenant, could also be assigned to supervise a small administrative or technical group. The next step, Fire Officer II, generally refers to the senior non–chief officer level in a larger fire department. An officer at this level could be the overall supervisor of a multiunit fire station. A Fire Officer II, or captain, could also be in charge of a larger group performing a specialized service or a significant administrative section within the fire department.

Fire Officer III and IV generally refer to chief officer positions. An individual who is qualified at the Fire Officer III level might work as a battalion or district chief in a large department and possibly as a deputy or assistant chief in a smaller organization. Personnel at the Fire Officer IV level tend to be fire chiefs or hold senior positions in charge of a major component of the fire department.

The Fire Officer I student will most benefit from this textbook if there is an existing understanding of the regulatory environment that guides emergency services work. This includes an understanding of negligence, duty to act, standard of care, tort immunity, types of laws (statutes, regulations, etc.), role of the Occupational Safety and Health Administration (OSHA), impact of NFPA standards on OSHA and standard of care, and sexual harassment.

An officer is responsible for being a leader and supervisor to a crew of fire fighters, managing a budget for the station, understanding the response district, knowing departmental operational procedures, and being able to manage an incident. The officer must also understand fire prevention methods, fire and building codes and applicable ordinances (laws enacted by a local government or jurisdiction), and the department's records management system. At each higher level, there are increasing requirements for knowledge, leadership, and management skills.

Section 1 of this textbook provides information to meet the training and performance qualifications specified in NFPA 1021 at the Fire Officer I level. Section 2 of this textbook provides information to meet the training and performance qualifications at the Fire Officer II level. Fire Officer III and IV are covered in another publication (*Chief Officer: Principles and Practice* by David Purchase).

Fire Officer Qualifications

The **Fire Officer I** classification is bestowed upon an individual who supervises a single fire suppression unit or a small administrative group within a fire department. At this level, emphasis is placed on accomplishing the department's goals and objectives by working through subordinates to achieve desired results. The Fire Officer I must be able to prioritize multiple demands on the time of the company or work group members and to delegate tasks to subordinates. These demands may be related to emergency operations, nonemergency tasks, supervisory duties, or administrative functions.

The Fire Officer I performs administrative duties and supervisory functions that are related to a small group of fire department members. Typical administrative duties include record keeping, managing projects, preparing budget requests, initiating and completing station and apparatus maintenance requisitions, and conducting preliminary accident investigations. Supervisory functions include making work assignments and ensuring that health and safety procedures are followed. Nonemergency duties could include developing preincident plans, providing company-level training, participating in risk reduction activities and programs, and responding to community inquiries.

Emergency duties include supervising a group of fire fighters who are performing company-level tasks, functioning as the initial arriving officer at an emergency scene, performing a size-up, establishing the **incident command system (ICS)**, developing and implementing an incident action plan, deploying resources, and maintaining personnel accountability. Once the emergency incident has been mitigated, the Fire Officer I is expected to conduct a preliminary investigation to determine the origin and cause, secure the scene to preserve evidence, and conduct a postincident analysis. Fire Officer I candidates are also required to meet all of the requirements of Fire Fighter II as defined in NFPA 1001, *Standard for Fire Fighter Professional Qualifications*, and of Fire Instructor I as defined in NFPA 1041, *Standard for Fire and Emergency Services Instructor Professional Qualifications*.

The International Association of Fire Chiefs (IAFC) uses another term to distinguish the different company officers. The IAFC calls the Fire Officer I level a **supervising fire officer** within its *Officer Development Handbook*. In this textbook, the Fire Officer I will be referred to as a **lieutenant**.

The **Fire Officer II** classification begins with meeting all of the requirements for Fire Officer I as defined in NFPA 1021. As is true for the Fire Officer I, the duties of the Fire Officer II can be divided into administrative, nonemergency, and emergency activities.

Administrative duties include evaluating a subordinate's job performance, correcting unacceptable performance, and completing formal performance appraisals on each member. Other duties include developing a project or divisional budget, including the related activities of purchasing, soliciting and awarding bids, and preparing news releases and other reports to supervisors.

Nonemergency duties include conducting inspections to identify hazards and address fire code violations; reviewing accident, injury, and exposure reports to identify unsafe work environments or behaviors; and taking approved action to prevent reoccurrence of an accident, injury, or exposure. Other duties could include developing a preincident plan for a large complex or property; developing policies and procedures appropriate for this level of supervision; analyzing reports and data to identify problems, trends, or conditions that require corrective action; and then developing and implementing the required actions.

Emergency duties include supervising a multiunit emergency operation using the ICS and developing an operational plan to deploy resources to mitigate the incident safely. The ICS defines the roles and responsibilities to be assumed by personnel and the operating procedures to be used in the management and direction of emergency operations. The Fire Officer II is also expected to determine the area of origin and preliminary cause of a fire and to develop and perform a postincident analysis of a multicompany operation.

The IAFC calls the Fire Officer II level a **managing fire officer** within its *Officer Development Handbook*. In this textbook, the Fire Officer II will be referred to as a **captain**.

FIRE OFFICER TIP

Company Officer, Defined

A company officer is the individual responsible for command of a company, a designation not specific to any particular fire department rank (can be a fire fighter, lieutenant, captain, or chief officer, if responsible for command of a single company) (Reproduced from NFPA's Glossary of Terms, 1026, © 2013 NFPA, Copyright © 2013 National Fire Protection Association. All rights reserved.)

Roles and Responsibilities of the Fire Officer I and Fire Officer II

The roles and responsibilities of a fire officer differ from those of a fire fighter. Understanding these roles is essential for the new fire officer to succeed.

A Fire Officer I has the following roles and responsibilities:

- Supervises and directs the activities of a single unit
- Instructs members of the company regarding departmental operating procedures, including

- duty assignments and giving special instructions when fighting fires
- Responds to alarms for fires, vehicle extrications, hazardous materials incidents, emergency medical incidents, and other emergencies as required
- Assumes command of emergency scenes, per the ICS; analyzes situations; and determines proper procedures until being relieved by a higher-ranking officer
- Assures the safety and well-being of the fire fighters assigned to the company, in quarters, in travel, and when operating on incidents
- Administers emergency medical first aid and cardiopulmonary resuscitation (CPR), and attends to victims until primary medical personnel arrive
- Oversees routine and preventive maintenance and makes periodic inspections of the assigned apparatus
- Receives direction and instruction from the fire captain and battalion chief regarding station operations, apparatus, grounds and building maintenance, and overall fire scene action
- Provides training to crew members regarding the apparatus operations, including leading practical training exercises; participates in departmental in-service training and drills
- Evaluates employee performance and conducts performance reviews
- Reads, studies, interprets, and applies departmental procedures, technical manuals, building plans, and so on
- Completes and maintains manual or computer records and prepares necessary reports on incidents, accidents, and personnel training
- Performs preincident planning activities, including touring and studying businesses for physical layout, possible hazards, location of water sources, exposure problems, potential life loss, and other factors
- Conducts occupancy inspections
- Determines a preliminary origin and cause of a fire
- Participates, prepares, and delivers various community risk reduction programs regarding fire prevention and safety and conducts tours of the fire station as required
- Assists in fire safety inspections of public and private buildings or property
- Participates in and oversees the periodic inspection and testing of equipment, such as hoses, ladders, and engines
- Works directly in firefighting activities; utilizes tools, equipment, portable extinguishers, hoses, ladders, and other items as necessary
- Takes appropriate action on the maintenance needs of equipment, buildings, and grounds
- Supervises and performs maintenance and cleaning work on fire equipment, buildings, and grounds

A Fire Officer II has the following roles and responsibilities:

- Accomplishes the goals and objectives of the fire department through the work of supervisors and subordinates
- Supervises and directs the activities of multiple units, some through direct supervision and others through the efforts of supervisors
- Assures crew members' compliance of departmental operating procedures, including duty assignments and giving special instructions when fighting fires
- Responds as the senior fire official to alarms for fires, vehicle extrications, hazardous materials incidents, emergency medical incidents, and other emergencies as required
- Assumes command of emergency scenes, per the ICS; analyzes situations; and determines proper procedures until being relieved by a higher-ranking officer
- Administers emergency medical first aid and CPR, and attends to victims until primary medical personnel arrive
- Assures the safety and well-being of the fire fighters under the officer's command, in quarters, in travel, and when operating on incidents
- Working through supervisors, assures that routine and preventive maintenance is accomplished, including periodic inspections of the assigned apparatus
- Receives direction and instruction from the battalion chief and administrative officers regarding station operations, grounds and building maintenance, and overall fire scene action
- Manages crew training regarding the apparatus operations, including practical training exercises; participates in departmental in-service training and drills

- Evaluates supervisor and employee performance and conducts performance reviews
- Reads, studies, interprets, and applies departmental procedures, technical manuals, building plans, and so on
- Completes and maintains manual or computer records and prepares necessary reports on incidents, accidents, and personnel training
- Performs preincident planning activities, including touring and studying businesses for physical layout, possible hazards, location of water sources, exposure problems, potential life loss, and other factors
- Conducts occupancy inspections
- Determines a preliminary origin and cause of a fire
- Identifies, develops, and manages various community risk reduction programs regarding fire prevention and safety and conducts tours of the fire station as required
- Assists in fire safety inspections of public and private buildings or property
- Manages the process of periodic inspection and testing of equipment, such as hoses, ladders, and engines
- Works directly in firefighting activities; utilizes tools, equipment, portable extinguishers, hoses, ladders, and other items as necessary
- Takes appropriate action on the maintenance needs of equipment, buildings, and grounds
- Supervises and performs maintenance and cleaning work on fire equipment, buildings, and grounds

The transition to fire officer is a big step in a fire fighter's career. It involves not only increased responsibility, but also a different role from that of a fire fighter. The officer is a part of management and is responsible for the conduct of others. The officer has to apply policies, procedures, and rules to subordinates and to different situations. Doing so successfully means being consistent and fair and not playing favorites. These changes often involve difficult adjustments for the new officer. As an officer, you will be required to take actions that might not make you happy or popular, but they are your responsibility. It is like the role of a parent—often difficult, but ultimately rewarding.

The Fire Service in the United States

The U.S. fire service originated as communities of citizens who responded when a fire broke out. Fighting fires was considered a civic duty in early America, and no compensation was provided for such activities. Citizens volunteered their time to answer the call of public service, and each fire fighter had a regular occupation that provided a living. Over time, however, the fire service has evolved into many different methods of providing personnel when the alarm sounds.

Many fully volunteer departments are composed of members who are notified when an alarm occurs. These men and women drop whatever they are doing to respond to the emergency. Some volunteer departments have a sufficient number of personnel and volume of calls that members are scheduled to be on standby or present at the fire station for specific shifts, according to a duty roster. In both forms of fully volunteer departments, the personnel are not paid for their services.

Some departments have moved away from the purely volunteer method of staffing due to an increasing number of alarms and a decreasing availability of volunteers. Some departments provide an incentive for fire fighters by paying them for each response to an alarm. These departments are termed *paid on call* or use part-time paid personnel.

Recruitment and retention are a challenge in volunteer fire departments that respond to several calls each day, particularly in areas where few individuals are available to serve as volunteers. The demands often exceed the amount of time volunteers are able to commit to the fire department, even if compensation is provided. A combination department uses full-time career personnel along with volunteer or paid-on-call personnel. This system usually provides faster response times because some personnel are on duty at the stations, ready to respond immediately. Frequently, the full-time staff consists of a minimum number of fire fighters, allowing for apparatus to respond and handle routine emergencies, such as requests for medical assistance, motor vehicle collisions, and incipient fires.

FIRE OFFICER TIP

Fire Fighters Determine Fire Officer Success
Throughout this text, practical advice and information are provided on teamwork and communication. Even the best and brightest officers will fail if the company does not support the supervisor. New fire officers must dedicate considerable effort to develop the members of the unit into a well-prepared team. Fire fighters want to do the best job they can; the fire officer provides the opportunities to meet that need.

The volunteer or paid-on-call staff are dispatched as a backup force when an incident exceeds the capabilities of the full-time personnel, such as a working structure fire.

A career department is staffed by full-time, paid personnel whose regular job is working for the fire department. These departments are typically found where the level of risk and call volumes require personnel to be on duty at the station at all times.

Although various forms of staffing fire department organizations are commonly used, most discussions divide fire fighters into two categories: career and volunteer. NFPA provides a statistical snapshot with its U.S. Fire Department Profile report. There are 1.06 million fire fighters in the United States. Of this total, approximately 30 percent are full-time career fire fighters and 70 percent are volunteers, a group that includes both part-time and paid-on-call fire fighters. Almost 42 percent of the volunteer fire fighters have more than 10 years of experience.

There are 29,819 fire departments in the United States. Combination departments include varying proportions of career and volunteer members and can be mostly career or mostly volunteer. Emergency medical services (EMS) at the first responder or emergency medical technician (EMT) level is provided by 46 percent of the fire departments. Paramedic-level emergency medical services are provided by 16 percent of the fire departments.

Most career fire fighters (72 percent) work in communities with populations of 25,000 or more. Most volunteer fire fighters (95 percent) work in fire departments that protect small or rural communities with populations less than 25,000 (Evarts and Stein, 2019). For calendar year 2016, 64.4 percent of fire department responses were for medical calls and 3.8 percent were for fire calls (NFPA, 2018).

Private industry and nongovernmental organizations may operate their own fire brigades or facility emergency response teams to protect factories, processing plants, large private facilities, and isolated communities. Although many of these groups are established to handle incipient fire situations, some are organized along the lines of a fire department, including fire officers.

History of the Fire Service

The first paid fire department in the United States was established in 1679 in Boston, Massachusetts, which also had the first fire stations and fire engines. The first organized volunteer fire company was established in Philadelphia, Pennsylvania. The Union Fire Company was formed in 1735, under the leadership of Benjamin Franklin. Franklin recognized the many dangers of fire and continually sought ways to prevent it. For example, he developed the lightning rod to help draw lightning strikes—a common cause of fires—away from homes. Another early volunteer fire fighter, George Washington, imported one of the first hand-powered fire engines from England, which he donated to the Alexandria (Virginia) Fire Department in 1765 (Bugbee, 1971).

The fires that have resulted in many deaths have helped to improve the requirements for better building standards and fire protection systems (**TABLE 1-1**).

Communications

Communications are vital for a fire officer to coordinate the firefighting efforts effectively. During the firefight, officers must be able to communicate with fire fighters or summon additional resources. Improvements in communication systems are tied to the history of the fire service. Today's two-way radios enable fire units and individual fire fighters to remain in contact with one another at all times. Before electronic amplification and two-way radios became available, the chief officer would shout commands through his trumpet (**FIGURE 1-1**). The chief's trumpet, or bugle, eventually became a symbol of authority. Although chief officers no longer use trumpets for communicating, the use of trumpets to symbolize the rank of chief signifies the chief's need to communicate clearly.

Paying for the Fire Service

Many early volunteer fire departments were funded by donations or subscriptions, and many volunteer departments still rely on this source of revenue to purchase equipment and pay operating expenses. The first fire wardens were employed by communities and paid from community funds.

Fire insurance companies were established in England soon after the Great Fire of London in 1666, to help residents and business owners cope with the financial loss from fires. The companies would collect fees (premiums) from homeowners and businesses and pledged to repay the owner for any losses due to fire.

Because the insurance companies could save money if the fire was put out before much damage was done, they agreed to pay fire companies for trying to extinguish fires. Early insurance companies marked the homes of their policyholders with a plaque, or fire mark, displaying the name or logo of the insurance company (**FIGURE 1-2**). Most fire

TABLE 1-1 Historical Large Life-Loss Fire in the United States

Date	Event	Number of Deaths
October 8–10, 1871	Great Chicago Fire	300 (est.)
October 8, 1871	Peshtigo, WI	1200 (est.)
December 5, 1876	Conway's Theatre Brooklyn, NY	285
December 30, 1903	Iroquois Theatre Chicago, IL	602
January 12, 1908	Rhodes Opera House Boyertown, PA	170
March 4, 1908	Lakeview Grammar School Collinwood, OH	175
March 25, 1911	Triangle Shirtwaist Factory New York, NY	146
May 15, 1929	Cleveland Clinic Hospital Cleveland, OH	125
April 21, 1930	Ohio State Penitentiary Columbus, OH	320
April 23, 1940	Rhythm Club Natchez, MS	207
November 28, 1942	Cocoanut Grove Night Club Boston, MA	492
July 6, 1944	Ringling Brothers Barnum & Bailey Circus Hartford, CT	168
December 7, 1946	Winecoff Hotel Atlanta, GA	119
December 1, 1958	Our Lady of the Angels School Chicago, IL	95
November 23, 1963	Golden Age Nursing Home Fitchville, OH	63
May 28, 1977	Beverly Hills Supper Club Southgate, KY	165
November 21, 1980	MGM Grand Hotel Las Vegas, NV	85
March 25, 1990	Happy Land Social Club New York, NY	87
February 20, 2003	The Station Nightclub West Warwick, RI	100

Reproduced from NFPA's report "Deadliest single building or complex fires and explosions in the U.S." Copyright © 2008 National Fire Protection Association.

FIGURE 1-1 The chief's trumpet, used for amplification before electronic devices, eventually became a symbol of authority in the fire service.
© Jones & Bartlett Learning.

FIGURE 1-2 Fire marks were originally symbols affixed to the front of a building designating the insurance company responsible for covering that structure.
© Jones & Bartlett Learning. Photographed by Glen E. Ellman.

companies were loosely governed, and more than one company might show up to fight the same fire. If two fire companies arrived at a fire, however, a dispute might arise over which company would collect the money. This type of conflict hastened many jurisdictions to begin assuming the responsibility for providing fire protection. Today, local tax revenues pay for most career fire departments and support many volunteer organizations.

Real estate taxes are the primary source of revenue for fire departments. The amount of the tax and the allocation of resources are determined by the elected local officials or a board of directors.

Fire Department Organization

As soon as the scope of firefighting exceeded one person and one bucket, there was a need for an organization to focus individual efforts and provide structure for the endeavor. The model adopted reflects the unique characteristics of the community and the conditions that resulted in the organization of a fire department.

Organizations will evolve as the department grows, leaders change, and the community determines which services and activities are needed by the fire department. This section examines the formal conditions and practices found in most departments.

Source of Authority

Governments—whether municipal, county, state, provincial, or national—are charged with protecting the welfare of the public against common threats. Fire is one such peril; an uncontrolled fire threatens everyone in the community. Citizens accept certain restrictions on their behavior and pay taxes to protect themselves and support the common good. People charged with protecting the public are given certain authority to enable them to perform effectively. For example, fire departments can legally enter a locked home without permission to extinguish a fire and protect the public. Extinguishing a fire is considered to be an important measure to protect the community.

In most areas, the fire service draws its authority from the governing entity responsible for protecting the public from fire—whether it is a town, a city, a county, a township, or a special fire district. The head of the fire department, usually the **fire chief**, is accountable to the leader of the governing body, such as the city council, the county commission, the mayor, or the city manager. Because of the relationship between a fire department and a local government, fire fighters should consider themselves civil servants, working for the tax-paying citizens who fund the fire department.

Federal and state governments also grant authority to fire departments and operate their own fire departments and fire protection agencies to protect federal or state properties, particularly for military installations and wildland areas. Some private corporations have government contracts to provide fire protection services or offer subscription services to private property owners.

Most urban and suburban fire departments are organized by a jurisdiction or county government. Typically, these agencies fall under the organizational umbrella as a department, just like the police department, the public works department, and the human resources department.

Another form of organization that is usually similar to that of a jurisdiction or county department is a fire protection district. A fire protection district is a special political subdivision that can be established by a state or a county, with the single purpose of providing fire protection within a defined geographic area. Its operation is overseen by a fire district board that is usually elected by the voters in the district. A fire district operates very much like a school district and has the ability to set a tax rate, collect taxes, and issue bonds (Smeby, 2013).

In some states, a volunteer fire department can be established by a charter and is independent of any local government body. The fire department may be a private association rather than a governmental entity. This type of department is not funded directly through taxes, although it may receive a grant or contract with the local government to provide services. Funding can also be obtained from fund-raising events, donations, subscriptions, and fees for service.

Chain of Command

The organizational structure of a fire department consists of a **chain of command**. The ranks may vary in different departments, but the basic concept is generally the same. The chain of command creates a structure for managing the department as well as for directing fire-ground operations. Fire fighters usually report to a supervising officer, or lieutenant, who is responsible for a single fire company (e.g., an engine company) on a single shift. Some fire departments have only one officer rank, whereas others have two or more officer ranks (sergeants, lieutenants, and captains).

When there are two levels of officers in a fire department, the senior officer has more authority, functioning as a managing fire officer. A managing fire officer, or captain, could be directly responsible for supervising a fire company on one shift and also responsible for coordinating all of the company's activities with other shifts. A managing fire officer could also be in charge of all of the companies on one shift in a multiunit fire station.

Supervising and managing fire officers report directly to an administrative fire officer. In a large organization, there are often several levels of administrative fire officers, usually called chiefs. **Battalion chiefs**, or district chiefs, are responsible for managing the activities of several fire companies within a defined geographic area, usually in more than one fire station. A battalion chief is usually the officer in charge of a single-alarm working fire.

Above battalion chiefs in the fire department hierarchy are **assistant or division chiefs**, and/or deputy chiefs. Officers at these levels are usually in charge of major functional areas within the department, such as training, emergency operations, support services, and fire prevention. They can also have responsibility for relatively large geographic areas, including several battalions or districts. These officers report directly to the chief of the department.

The fire chief (or chief of the department) is the executive fire officer who has overall responsibility for the administration and operations of the department. The fire chief can delegate responsibilities to other members of the department but is still responsible for ensuring that these activities are properly carried out.

The chain of command is used to implement department rules, policies, and procedures. This organizational structure enables a fire department to determine the most efficient and effective way to fulfill its mission and to communicate this information to all members of the department (**FIGURE 1-3**). Using the chain of command ensures that a given task is carried out in a uniform manner (Kimball, 1966).

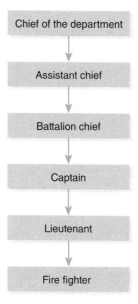

FIGURE 1-3 The chain of command ensures that the department's mission is carried out effectively and efficiently.
© Jones & Bartlett Learning.

National Incident Management System

The National Incident Management System (NIMS) was established by the Federal Emergency Management Agency (FEMA) and includes the ICS. NIMS applies to all incidents, regardless of cause, size, location, or complexity. It is the standard for emergency management used by all public agencies, nongovernmental organizations, and commercial organizations in the United States for both planned and emergency events.

NIMS is designed to expand or contract based on the size and the complexity of the event. It uses a standardized set of concepts and principles for all threats, hazards, and events. It is applied to all mission arenas: prevention, protection, mitigation, response, and recovery (FEMA, 2017).

Management Principles

The fire department uses a paramilitary style of leadership. Most fire departments are structured on the basis of four management principles:

1. Unity of command
2. Span of control
3. Division of labor
4. Discipline

Unity of Command

Unity of command is the management concept that each fire fighter answers to only one supervisor, and each supervisor answers to only one boss, and so on (**FIGURE 1-4**). In this way, the chain of command ensures that everyone is answerable to the fire chief and establishes a direct route of responsibility from the chief to the fire fighter.

At the fire ground, all functions are assigned according to incident priorities. A fire fighter with more than one supervising officer during an emergency could be overwhelmed with various conflicting assignments. The incident priorities may not get accomplished in a timely and efficient manner. Unity of command is designed to prevent such problems.

Span of Control

Span of control refers to the maximum number of personnel or activities that can be effectively controlled by one individual (usually three to seven). Many experts believe that span of control should extend to no more than five people, but this number can change, depending on the assignment or the task to be completed. A fire officer must recognize his or her own span of control to be effective.

Division of Labor

Division of labor is a way of organizing an incident by breaking down the overall strategy into smaller tasks. Some fire departments are divided into units based on function. For example, the functions of engine companies are to establish water supplies and flow water; truck companies perform forcible entry, rescue, and ventilation functions. Each of these functions can be divided into multiple assignments, which can then be assigned to individual fire fighters. With division of labor, the specific assignment of a task to an individual makes that person responsible for completing the task and prevents duplication of job assignments.

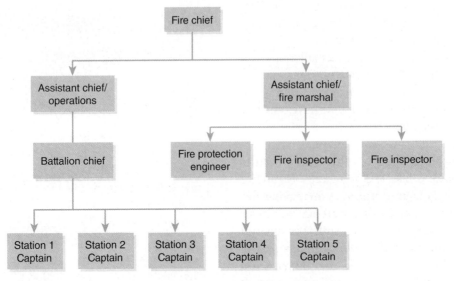

FIGURE 1-4 The organization of a typical fire department.
© Jones & Bartlett Learning.

Discipline

Discipline means guiding and directing fire fighters to do what their fire department expects of them. Discipline encompasses behavioral requirements, such as always following orders from superior officers and performing up to expectations. Standard operating procedures, suggested operating guidelines, policies, and procedures are all forms of positive discipline because they outline how things are to be done, and usually how far a person can go without requesting further guidance. Counseling sessions, formal reprimands, or suspension of duty are corrective disciplines taken to discourage inappropriate behavior or poor performance. Firefighting demands strong discipline if all personnel are to operate safely and effectively.

Other Views of Organization

There are several different ways to look at the organization of a fire department, including in terms of function, geography, and staffing.

Function

Fire departments can be organized along functional lines. For example, the training division is responsible for leading and coordinating department-wide training activities. Likewise, engine companies and truck companies have certain defined functional responsibilities at a fire. Hazardous materials squads have different functional responsibilities that support the overall mission of the fire department.

Geography

Each fire department is responsible for a specific geographic area. Fire stations are located throughout a community to ensure a rapid response time to every area, and each station is responsible for its own specific geographic area. This arrangement also enables the fire department to distribute and use specialized equipment efficiently throughout the community.

Staffing

A fire department must have sufficient trained personnel available to respond to a fire at any hour of the day, every day of the year. Staffing issues affect all fire departments—career departments, combination departments, and volunteer departments. In volunteer departments, it is important to ensure that enough responders are available at all times, but particularly during the day. In the past, when many people worked in or near the communities where they lived, volunteer response time was not an issue. People often now have longer commutes to their places of employment and work longer hours, so the number of people available to respond during the weekday can be severely limited (Gratz, 1972). Members of volunteer departments are also challenged with ever-increasing time demands to obtain additional emergency service certifications, licenses, and continuing education. Although some of this training can be accomplished while on duty at the fire station, other activities will require additional time at an academy, a specialized training location, or a regional testing facility.

In combination departments, the challenge is to make sure the appropriate mix of qualifications, certifications, and capabilities is present to deliver prompt and safe response. Within the mix of volunteer and career members there needs to be a credentialed fire company commander, a qualified apparatus operator, and structural or wildland fire fighters. In many communities, the career employee will have multiple certifications, such as apparatus operator and paramedic, and is expected to fill any position that is not covered by a volunteer member.

In career departments, the challenge is to ensure that all of the assigned positions are covered by appropriately qualified personnel. Although each career position has a home fire station and fire company, it is common for as much as one-third of the on-duty workforce to be assigned to a different fire company or a different fire station. The company officer needs to make work assignments based on individual qualifications, experience, and capability.

In all three types of departments, the company officer functions as the staffing coordinator and gatekeeper, implementing the department's rules, regulations, and procedures to ensure proper fire company staffing.

The Functions of Management

Fire officers are managers, and like all other managers, they have specific functions that they must perform, regardless of the size or type of organization. The four functions of management were originally identified by Henri Fayol and published in the *Bulletin de la Société de l'Industrie Minérale* in 1916; Constance Storrs provided the English translation in the 1949 textbook *General and Industrial Management*. These four functions are as follows:

1. Planning
2. Organizing
3. Leading
4. Controlling

Planning means developing a scheme, program, or method that is worked out beforehand to accomplish an objective. The fire officer is responsible for developing a work plan for a fire company or an administrative work group. The fire officer develops plans to achieve departmental, work unit, and individual objectives. Short-range planning covers developing a plan that extends up to a year. Medium-range planning covers planning that is 1 to 3 years in advance. Long-range planning covers events longer than 3 years in advance.

Planning includes establishing goals and objectives and then developing a way to meet and evaluate those goals and objectives. This task could be as simple as planning the daily activities for the fire company or as complex as developing an annual budget. Planning can also include emergency activities, such as developing strategies and tactics and an incident action plan.

Organizing means putting resources together into an orderly, functional, structured whole. The fire officer takes the available people, equipment, structure, and time and develops them into an orderly, functional, and structural unit in order to implement the plan and deliver the expected services. Fire officers decide which companies will perform certain duties when they arrive on an emergency scene. They also decide which station duties each member of the crew will perform.

Leading means guiding or directing in a course of action. The act of **leadership** is a complex process of influencing others to accomplish a task. Leading is the human side of managing. It includes motivating, training, guiding, and directing employees.

Controlling means restraining, regulating, governing, counteracting, or overpowering. Fire officers implement the controlling function when they consider the impact on the budget before making purchases, when they conduct employee performance appraisals, and when they ensure compliance with departmental policies (Drucker, 1974).

Fire officers use the functions of management to get work accomplished by and through others. The four functions constitute a continuous cycle—they are never truly "finished." Although fire officers at all levels use all four functions, each level may use them to different degrees.

Rules and Regulations, Policies, and Standard Operating Procedures

Fire officers must thoroughly know the department's regulations, policies, and standard operating procedures. This knowledge is essential to ensure a safe and harmonious working environment. Although a fire fighter is required to follow all regulations, policies, and procedures, the fire officer must not only follow these directives, but also ensure compliance with them by subordinates. The transition to fire officer is often a difficult one because of the new role the officer is now required to fill. For officers to enforce the organizational rules, they must understand the differences among rules and regulations, policies, and standard operating procedures.

Rules and regulations are developed by various government or government-authorized organizations to implement a law that has been passed by a government body. Rules may also be established by the local jurisdiction that sets the conditions of employment or internally within a fire department. For example, an organization may have a rule stating that employees with more than 15 years of service will receive 12 days of vacation. A fire department may have a rule that requires all members to wear their seat belts when riding in vehicles. Rules and regulations do not leave any room for latitude or discretion.

Policies are developed to provide definite guidelines for present and future actions. Fire department policies outline what is expected in a described situation. Policies often require personnel to make judgments and to determine the best course of action within the stated policy. Policies governing parts of a fire department's operations may be enacted by other government agencies, as in the case of personnel policies that cover all employees of a city or county. An example of a policy is one stating that the fire officer shall ensure that station sidewalks are maintained to provide safety from slips and falls during snow and ice accumulation. Because it gives the officer latitude in determining how to ensure the safety of pedestrians, this directive is a policy.

Standard operating procedures (SOPs) are written organizational directives that establish or prescribe specific operational or administrative methods to be followed routinely for the performance of designated operations or actions. SOPs are developed within the fire department, are approved by the chief of the department, and ensure that all members of the department approach a situation or perform a given task in the same manner. SOPs provide a uniform way to deal with emergency situations, enabling different stations or companies to work together smoothly, even if they have never worked together before. These procedures are vital because they enable everyone in the department to function properly and know what is expected for each task. Fire officers must learn and frequently review departmental SOPs. An example of an SOP is a statement of the step-by-step process

Voice of Experience

In 1994, early in my career, I, like many young fire fighters, had aspirations of someday pursuing a promotion as a company officer. I had always felt that I was good at my job and was an easy employee to manage. I was a hard worker who reported to work early and easily found tasks to keep myself occupied in the downtime around the station. Basically, I felt that I did everything I could to be the kind of fire fighter I myself would want to manage as a company officer.

I believed that I was on the path toward advancement, so I continued what I was doing and waited for that "someday" to come. Years went by, and I watched with discouragement as my peers (some with less seniority) were easily promoted to various positions within the department, all while I continued to wait for my time to come. I soon came to realize that my "seniority"-based path to promotion was not working.

I sought advice from my company officer. After receiving some brutally honest constructive criticism, I learned that while I was just waiting for my turn for promotion, others were taking educational classes through both formal and informal opportunities to further their knowledge, increase their skills, and fine-tune their abilities. My co-workers recognized opportunities and began practicing for the positions they wanted to be promoted to, and I did not.

I was initially upset about the criticism, but I soon realized that I was the only one to blame. I had the same opportunities that my co-workers did; however, I had failed to recognize and act on them. This wake-up call changed my perception and gave me a clearer understanding of what I needed to do to better prepare for promotion. I took this lesson to heart and began recognizing and taking advantage of multiple opportunities. I started practicing to become a company officer and prepared myself for this promotion by furthering my education, increasing my skills, and fine-tuning my abilities. This desire and motivation paid off, and I was ultimately promoted to the rank of captain.

As an aspiring company officer, it is imperative that you continually seek opportunities to improve your knowledge, skills, and abilities. Self-improvement also consists of looking to others for advice and being willing to accept constructive criticism. This is a small but important part of the preparation process. I often remind myself of the famous quote from Henry Hartman: "Success always comes when preparation meets opportunity." Preparation for promotion takes a lot of effort, but when your preparation meets opportunity, the possibilities are endless.

Congratulations for seizing this opportunity, and I wish you luck and good fortune in your journey.

Aaron R. Byington, MA, NRP
Captain
Layton City Fire Department
Layton City, Utah

(procedure) to be used whenever vertical ventilation is required.

Some fire departments prefer the term *suggested* or *standard operating guidelines (SOGs)* instead of SOPs because conditions often require the fire officer to use personal judgment in determining the most appropriate action for a given situation. The term SOG suggests that a specific step-by-step procedure should be used, but it allows the officer to deviate from this procedure if the conditions warrant doing so. The distinction between an SOP and an SOG is very subjective (Griffiths, 2017).

FIRE OFFICER TIP

Guide, Procedure, Order, or Directive?

The collection of SOPs, SOGs, general orders, directives, or informational bulletins can become overwhelming. The collection can be considered the administrative history of the fire department, with some items linked to a bad outcome or high-profile event. The administrative concept is that general orders, directives, or informational bulletins are short-term items, lasting less than a year. The standard operating procedures or standard operating guides, on the other hand, are permanent, similar to the administrative law regulations used by the town, city, region, or state.

Establishing a Strong Supervisor/Employee Relationship

The basis for a strong, positive, and effective supervisor/employee relationship is open, honest, and constant communications between the fire officer and the fire fighter. Once those communication paths are opened and trusted, they will prove beneficial through good times and tough times alike. Some form of agreement, compromise, or answer can be found when the fire officer is open, honest, and constant.

Key recommendations that form the foundation of any strong supervisor/employee relationship between a fire officer and a fire fighter include the following:

1. Schedule regular one-on-one meetings with each member of your company. Such routine contact establishes a personal connection and trust between the fire officer and each fire fighter. During the meeting, discuss job performance and expectations on the part of both people involved. Give guidance and coaching where necessary.
2. Schedule regular meetings with the company as a whole. Use this time to discuss new policies and procedures, any concerns about station procedures (e.g., checking out apparatus, housekeeping, kitchen duty), and upcoming personnel or policy changes. This is also a good opportunity to obtain feedback and input from the company members. Keeping an open line of communications is critical to an effective supervisor/employee relationship.
3. If a disagreement arises, work together to articulate the concern and to develop possible solutions. When both parties decide together what they want, they are usually successful in attaining a mutually acceptable goal. When only one side decides what they want, the success rate drops off sharply.
4. If the personal and professional relationship between you and a fire fighter is rocky from the beginning and both of you have decided to work to improve it, start by listing the areas in which you can succeed together. Set goals and deadlines. Start with goals that are easily attainable ("low-hanging fruit"), and build on those successes. Ask yourself if the goal you are defining will have a positive impact on the company and the other station personnel.

Maintaining a good relationship does not mean that you will always agree on everything; rather, it means that you "agree to disagree" when you have discussed the issue and cannot find middle ground. In some cases, a third party will need to be brought in to help mediate the discussions to arrive at an acceptable solution.

To trust each other's intentions, the company officer and the fire fighter must be honest and up-front. Sometimes the supervisor/employee relationship can take a detour when either side holds back information, exaggerates, or deceives. Unfortunately, suspicion usually leads to acrimony, and acrimony tends to lead to even more suspicion. In any long-term relationship, bluffing or threats can only prevail for so long and are generally counterproductive. Whenever a company officer feels that communications are not going well, the officer should focus on bringing them back into alignment by discussing feelings and concerns in an open exchange.

The progression of the firefighting profession is just that: professionals working together for the common good and common goals. A cooperative, collaborative supervisor/employee relationship is the profession at its best. Chapter 3, *Leading a Team*, provides specific tools and techniques to support a collaborative fire officer/fire fighter relationship. With hard work and open, honest communication, fire officers and fire fighters can continue to make the relationship better and raise the level of all fire service personnel's professionalism, trust, and stature with those whom they serve.

Positive Labor-Management Relations

A healthy labor-management relationship is essential to produce positive outcomes and avoid the strife and consequences of a confrontational climate. Successful relationships are built on trust, respect, and open lines of communication. Each side must be willing and able to focus on the mutual benefits of a positive relationship or face the consequences of a negative outcome.

The root cause of almost every labor disturbance is a failure to manage the relationship between labor and management properly. The traditional way of thinking was based on the premise that either labor or management must score a victory over the other side to settle every point. A philosophical shift in labor-management relationships is moving away from confrontational strategies and toward cooperative relationships, often through mediation. **Mediation** is the intervention of a neutral third party in an industrial dispute. As with many traditions in the fire service, the ability and the willingness to change were produced by necessity.

Tremendous amounts of time, energy, and money can be wasted in the process of two sides trying to overwhelm each other instead of working together. A poor labor-management relationship usually produces casualties on both sides. Both fire chiefs and union presidents can lose their positions of power and influence in the aftermath of such conflicts. Positive relationships based on mutual respect and understanding are much more likely to produce positive results.

Some fire departments exist merely to meet an internal determination of what makes up the basic requirements for public safety. Others are committed to excellence and continual progress. Most observers agree that the most successful and progressive fire departments put significant effort into managing their labor-management relationships instead of engaging in continual confrontations and power struggles.

In some instances, management uses the moral ethos of public service as leverage against labor, and vice versa. Public support is usually viewed as vital by both sides because elected officials, who represent the public, have ultimate control over the economic and policy issues.

The International Association of Fire Fighters

The largest fire service labor organization in the United States, the International Association of Fire Fighters (IAFF), represents 316,000 fire fighters and paramedics in the United States and Canada (IAFF, 2019) (**FIGURE 1-5**). The IAFF has provided a century of support and advocacy for its members. This powerful organization has impacted many aspects of a fire fighter's job.

The IAFF was established on February 28, 1918. The three principal objectives of the IAFF at that time were to obtain pay raises; to establish the two-platoon, or 12-hour workday, schedule; and to ensure that appointments and promotions were based on individual merit, not political affiliation.

The IAFF is unique among labor organizations in its dominance in representing a profession. In other public service professions, such as law enforcement and education, two or more national labor organizations compete to represent the employees. Although the International Brotherhood of Teamsters and the American Federation of State, County, and Municipal Employees also organize fire fighters, they represent relatively few fire service personnel and have little influence over the firefighting profession as a whole.

Labor advocacy has improved the quality of protective clothing, the safety of firefighting equipment, the content of training programs, response to fire fighter addiction and posttraumatic stress disorder

FIGURE 1-5 The International Association of Fire Fighters (IAFF) is the largest fire service labor organization.
Courtesy of IAFF.

FIGURE 1-6 Description of the NVFC Health and Safety accomplishments, edited from the 2018 Annual Report Infographic. (NVFC, 2019).
National Volunteer Fire Council (2019) Annual Report Infographic: 2018 by the Numbers. Accessed May 20, 2019 https://www.nvfc.org/wp-content/uploads/2019/01/2018-Annual-Report-Infographic.pdf

(PTSD), and advanced techniques of emergency incident operations. The generalized process for handling a grievance, has also emerged from the labor-management relationship. This topic is covered further in Chapter 3, *Leading a Team*.

National Volunteer Fire Council

The National Volunteer Fire Council (NVFC) is the leading nonprofit membership association representing the interests of the volunteer fire, EMS, and rescue services. The NVFC serves as the voice of the volunteer in the national arena and provides resources, programs, education, and advocacy for first responders across the nation (**FIGURE 1-6**). The mission of the NVFC is to provide a unified voice for volunteer fire/EMS organizations. NVFC members include 24,551 individuals, departments, state associations, and corporations. NVFC programs include the following:

- Fire Corps
- Heart-Healthy Firefighter Program
- National Junior Firefighter Program
- Make Me a Firefighter volunteer recruitment program
- Share the Load
- Wildland Fire Assessment Program

The Fire Officer's Role as a Supervisor

One duty of a fire officer is to supervise the activities of subordinate fire fighters. The basic authority of a supervisor and the duties of subordinates are defined by the personnel rules of the city or governmental organization, as well as the specific rules, regulations, and procedures of the fire department. In addition, when a collective bargaining agreement is in effect, other details of the relationship are spelled out in the labor contract or memorandum of understanding (MOU). Supervisors are expected to follow all of the established rules and procedures in assigning duties and all other aspects of the relationship with their subordinates. In many cases, it is a significant challenge for a newly promoted fire officer to learn which rules and regulations apply to which situation and how they are interpreted and applied.

In most organizational structures, there is a clear distinction between labor (the workers) and management (the managers and supervisors). The managers and supervisors represent the organization, and the union represents the workers. If any doubt or disagreement about the application or interpretation of the contract arises, a process should be in place for labor and management representatives to meet and resolve the problem.

This line between labor and management is complicated in fire departments because the first-level supervisors are often members of the same collective bargaining unit as the fire fighters they supervise. As a consequence, the fire officer's relationship to the organization is often covered by the same contract that the officer has to follow and enforce. The formal line between labor and management is often established at a higher level, such as administrative fire officer. In a few jurisdictions, the supervisory and managing fire officers are members of a labor bargaining unit that is separate from that of the fire fighters.

As a supervisor, a fire officer is generally the first point of contact between the fire fighters and the fire department organization (**FIGURE 1-7**). If a disagreement occurs about the interpretation or application of a work rule that is covered by the contract, the first-level supervis is the individual who should have the first awareness of the problem and the first opportunity to resolve the problem. An officer who is a member of the same bargaining unit as an individual who is dissatisfied must clearly understand the established problem-solving processes.

The Fire Officer's Tasks

For this edition of the textbook, we received survey responses from 204 battalion-level commanders who responded to the following questions:

- What is the most important leadership concept you want your new first-line supervisor to know?

FIGURE 1-7 A fire officer is generally the first point of contact between the workers and the organization.
© Jones & Bartlett Learning. Photographed by Glen E. Ellman.

FIGURE 1-8 Providing the beginning of shift report is a vital company officer task.
© Jones & Bartlett Learning. Photographed by Glen E. Ellman.

- What is the second most important leadership concept you want your new first line supervisor to know?
- What leadership activity do you want all of your first-line supervisors to do more of?
- What leadership activity do you want your first-line supervisors to stop doing?

The results fell into four task areas: beginning of shift report, notifications, decision making, and problem solving.

The Beginning of Shift Report

Fire officers should provide a prompt and accurate report at the start of the workday (**FIGURE 1-8**). This report is provided from each work location to the battalion or district chief within the first quarter hour of the reporting time. The format of the report may be electronic, paper, or verbal. Some departments use sophisticated online staffing systems, such as Tele-Staff, that provide real-time scheduling that conforms to department operational requirements. The chiefs rely on this information to make staffing adjustments at the beginning of the shift. An accurate report is needed to ensure that adequate staffing and equipment are in place and ready for the balance of the shift (**FIGURE 1-9**).

The report provides the on-duty staffing information and sick leave list and identifies any positions that need to be filled for that day. The positions are a priority because someone who worked the previous shift must remain on duty until a relief person shows up to fill any position that remains vacant. The battalion chief moves available staff to cover the vacancies noted on the beginning of shift reports, and any vacancies that remain have to be covered by fire fighters working overtime. It takes time to reassign on-duty personnel, call the overtime personnel, and get everyone to their work locations so the holdover personnel can be released. It is maddening when a new fire officer sheepishly calls the administrative fire officer two hours into a work shift to report that a fire fighter from the previous shift is still on involuntary holdover because someone failed to report the position vacancy on the beginning of shift report. By the time the position is covered, the fire fighter on involuntary holdover may have remained on duty for several more hours.

In addition, the beginning of shift report notes the location and condition of all of the apparatus or rolling stock, such as a reserve pumper that has been loaned to another station or an ambulance at the shop. Finally, the report provides the chief with any "must know" information that will require immediate attention.

Some reports include anticipated staffing for the next day. This information allows the chief to make assignments in anticipation of the known vacancies in systems that lack an online staffing program.

Notifications

The new supervising fire officer must make prompt notifications. This was the second most important issue noted by the administrative fire officers in the previously mentioned survey. Some types of information must be passed up the chain of command quickly. For example, all injury and infectious disease exposure reports must be processed without delay. If a fire fighter is exposed to a possible bloodborne pathogen early

Today's Date is: May 29

	Total	Paramedics	Fire Officer	EMS Officer	Prearranged Callback	Annual Leave	Vacancies	Detail Out-of-Operations	Injury Lv or Light Duty	LWOP	Fire OIC	EMS OIC
Minimums	10	3	2	1							Fire OIC	EMS OIC
Today's staffing	9	3	1	1	1	1	0	1	0	0	Smyth	Willow
Next day staffing	8	2	1	0							Smyth	????

Today's shortage	Why?	PM Surplus	Next Day's Shortage	Why?
Engine Officer	O/R	none	Engine Officer	Off Rep
			Medic Officer	Leave

Sick Leave	Detailed Out of Ops	Next Day APPROVED Leave
None	Capt. Johnson	FF Tolliver
		Lt. Willow

Injury Leave/Light Duty
None

		Vehicle Status
	Engine 7746	Eng 46
Messages for the Chief:	Rescue 7099	Res 46
Vehicle 7234 overdue for preventative maintenance	Medic 6322	Med 46
Rescue 46 thermal imager broken	Reserve Engine 7234	Eng 35
Furnace malfunctioning	Reserve Medic 4276	Med 11
Fire Chief at 46 for dinner @ 1830	Battalion 9 5040	shop
	Reserve Suburban 5107	BC 09

FIGURE 1-9 Example of a jurisdiction beginning of shift report.
© Jones & Bartlett Learning.

FIRE OFFICER TIP

When in Doubt, Notify

When you start working with a battalion chief or supervisor, it is better to over-notify rather than under-notify. Learn how your boss wants to get time-sensitive items, like a fire fighter on-duty injury, as well as routine items, like the status of a delegated task.

Saturday morning, the exposure report cannot sit on the fire officer's desk until Monday morning before it gets to the battalion chief. The designated infection control officer must be informed immediately so that he or she can get patient information while it can still be easily obtained. The same priority applies to any information the chief needs to know about when it is current, particularly before someone at a higher level calls to ask about the issue. Many chiefs call this the "no surprises" rule.

Decision Making and Problem Solving

The third and fourth issues cited by the administrative fire officers—decision making and problem solving—are of equal importance. Some chiefs complain that new supervising fire officers hesitate to make decisions. They seem to want the chief to make the hard decisions and to enforce unpopular rules. Chiefs typically want new officers to run their companies and make the decisions that are within their scope of

FIGURE 1-10 Discuss possible solutions to a problem with the chief.
© Jones & Bartlett Learning. Photographed by Glen E. Ellman.

responsibility. Chiefs are available for consultation, but they expect their officers to run the fire stations.

Fire officers should not complain about problems without proposing any solutions; they need to think through problems and contribute to solutions. The most valuable proposals from officers will consider the larger picture of how a possible solution would affect the rest of the department (**FIGURE 1-10**).

Example of a Typical Fire Station Workday

A supervising fire officer is responsible for accomplishing the fire department mission through the efforts of the fire fighters under his or her command. At the company level, facilitating this outcome requires a balance of management and leadership skills. The officer must organize the work and provide the leadership to ensure that work gets done safely and effectively.

The fire department has an agency-wide mission that is translated into annual goals. These goals are used to develop annual, quarterly, and monthly objectives for each fire company. The administrative fire officer and the supervising or managing fire officer meet regularly to set these objectives and to review progress toward their achievement. The monthly goals show up as the planned activities on the fire company daily planner.

The following is an example of a 24-hour shift in a fire station. The starting point is a schedule that ensures that all of the required tasks are completed in a logical order. The fire officer has to anticipate that emergency incidents will alter the workday and require adjustments in the schedule.

0700 Line-up and equipment check. Send beginning of shift report to chief.

0800 Dust and vacuum all carpeted areas. Empty trash, run dishwasher, sweep all tile floors.

0830 Physical training and skill drill.

1100 Heavy cleaning (while still in physical training clothes):
- Monday: Air out bunkroom and rotate mattresses; clean all windows.
- Tuesday: Clean utility rooms and shop area.
- Wednesday: Clean and inventory EMS equipment, self-contained breathing apparatus (SCBA), and decontamination areas.
- Thursday: Move recyclables outside for pickup, clean weight room and lockers.
- Friday: Scrub kitchen and clean out refrigerators.

1130 Scrub bathrooms after fire fighters clean up from physical training.

Noon Lunch

1330 Scheduled productivity activity (e.g., fire safety inspections, school visits, inside or outside training).

1800 Dinner, followed by kitchen clean-up. Run dishwasher.

1930 Individual study time, occasional fire safety inspections (nightclubs) or drills.

2130 Remove all trash, tidy up day room, and make final pass through the kitchen.

Special station activities
- First and third Thursdays of the month: Scrub apparatus bay floor and clean windows.
- Second and fourth Fridays of the month: Wax kitchen floor and hallway.
- January: Community CPR and AED training.
- February: Safety officer inspection of facility, apparatus, and personal protective equipment (PPE).
- March and September: Steam-clean carpeted areas.
- April: Fire chief's annual inspection of the fire station.
- May: EMS Week open house.
- October: Fire Prevention open house.
- December: Holiday community party.

Example of a Typical Volunteer Duty Night

The following example of a volunteer duty night is less detailed than the previous schedule, but incorporates the same essential tasks as a 24-hour shift in a municipal fire department, including equipment maintenance, training, and station maintenance. Many departments use a day book or monthly calendar to plan the duty crew activities.

1800 Evening duty crew starts. Equipment check.

1900 Dinner, followed by kitchen clean-up. Run dishwasher.

1930 Classroom session, skill drill, or community outreach activity.

2230 Remove all trash, tidy up, and make final pass through the kitchen. Take clean dishes and cups out of dishwasher and put away.

General station, company evolutions, and apparatus cleaning tasks are conducted on weekends. One section of the fire station gets a major cleaning two or three times each year (e.g., wax floors, scrub apparatus floors). Specialized heavy cleaning, such as steam-cleaning carpets and waxing apparatus, is scheduled throughout the year.

An important role of the volunteer supervisory officer is the accurate entry and maintenance of participation logs. Volunteer fire departments track each member's contribution in emergency response, training, administrative activities, and public outreach. This is a vital measurement of each member's activity that is used in activity reports and calculating length of service award program benefits.

The Transition from Fire Fighter to Fire Officer

There are four times in a fire fighter's career when a major change occurs in how the individual relates to the formal fire department organization. The first change occurs when the fire fighter completes the probationary training period. Completing the initial training and probationary period is a major milestone—one that is marked in many departments by a change in helmet color or shield. The second change takes place when the fire fighter successfully completes a promotional process and starts working as a fire officer or supervisor. The third event is when the fire officer advances through the promotional process and starts working as a chief officer. The fourth event is when a fire fighter retires.

In all four situations, a significant change occurs in the individual's relationship to the organization and to the other members of the fire department. Part of this change is the individual's sphere of responsibility within the formal organization. As a fire fighter, the individual shares the sense of mutual responsibility that prevails among the crew members on the rig. They work together to accomplish fire-ground tasks and look out for one another.

A promotion from fire fighter or driver/operator to company-level officer is a large step. The company-level officer is directly responsible for the supervision, performance, and safety of a crew of fire fighters. He or she has a sacred duty to ensure that all of the fire company team members remain safe when operating in a hostile or hazardous work environment. The company-level officer functions as a working supervisor, sharing the hazards and work conditions with the fire company team (**FIGURE 1-11**).

This change in the sphere of responsibility often requires the new officer to change some on-duty behaviors or practices. The formal fire department organization considers a fire officer to be the fire chief's representative at the work location. This role creates an expectation that the fire officer will behave in a way that is appropriate for a first-line supervisor. Notably, behavior that was acceptable for a fire fighter may be unacceptable for a fire officer.

For example, consider a fire fighter who is known for developing elaborate practical jokes within the fire station environment. As a new fire officer, this individual would have to consider the impact that being seen as a practical joker would have on his or her ability to function as an effective supervising fire officer. Conversely, the new fire officer needs to consider how to respond to pranks and verbal jabs from fire fighters. What may have been an appropriate response from another fire fighter may no longer be the best response from an officer. Wearing the fire officer badge enhances the effect and consequences of any action or response.

FIGURE 1-11 The company-level officer is responsible for the supervision, performance, and safety of a crew of fire fighters.

Courtesy of Captain David Jackson, Saginaw Township Fire Department

Promotion to a chief officer rank changes the individual's relationship to the organization and the members to an even greater degree. A chief officer has less of a hands-on role than a fire officer of a company. The chief officer is typically directly responsible for several fire companies and must depend on fire officers to provide direct supervision over crews performing fire-ground tasks, often in hazardous conditions. In many cases, the chief officer works outside the hazardous area, but is still responsible for whatever happens inside that area.

The Fire Officer as Supervisor-Commander-Trainer

In his book *Effective Company Command*, Los Angeles County Fire Captain James O. Page divided the fire officer's duties into three distinct roles: supervisor, commander, and trainer (Page, 1973).

Supervisor

In the supervisor role, the fire officer functions as the official representative of the fire chief. That means the fire chief expects that every fire officer will issue orders and directives and conduct business in a way that meets the chief's objectives. As part of that role, the fire officer is expected to supervise the fire company in a manner consistent with the rules and regulations of the fire department. For example, if all fire companies are expected to spend Tuesday afternoon conducting fire safety inspections in commercial properties, the fire officers are responsible for ensuring that their individual fire stations complete this task.

Unpopular Orders and Directives

On occasion, a fire officer may be required to issue and enforce unpopular orders (**FIGURE 1-12**). Even if the officer disagrees with a particular directive, the formal organization requires and expects the officer to carry out that directive to the best of his or her ability. A fire officer can improve his or her effectiveness in handling an unpopular order by determining the story or history behind the directive, which would enable the fire officer to put the order in perspective.

When faced with an unpopular order, the fire officer should express any concerns and objections with his or her supervisor in private. This is the time for the fire officer to discuss suggestions for modifying or reversing the order. Occasionally, special circumstances may make the order difficult to implement, and the

FIGURE 1-12 An officer sometimes has to issue and enforce unpopular orders.
© Jones & Bartlett Learning. Photographed by Glen E. Ellman.

FIGURE 1-13 At the scene of an emergency incident, the fire officer may function as the initial incident commander.
© Jones & Bartlett Learning. Photographed by Glen E. Ellman.

supervisor might be able to authorize adjustments. Once the meeting with the officer's supervisor is over, the formal organization expects the officer to enforce the order as issued or amended.

Telling the fire fighters that their officer does not agree with an order undermines the officer's authority and supervisory ability. The fire chief expects an officer to perform the required supervisory tasks, and the fire fighters must understand that the fire officer does not make all the rules or have a choice about which ones to enforce. Enforcing unpopular orders is part of the job.

Commander

When operating at the scene of an emergency incident, the fire officer is expected to function as a commander and to exercise strong direct supervision over the company members. In some cases, the fire officer could be responsible for directing the actions of additional resources, or he or she might function as the initial incident commander (**FIGURE 1-13**).

Functioning as the initial incident commander on a major emergency is one of the higher-profile roles of a fire officer. The ability to bring order out of the chaos of an emergency incident is an art that requires a well-developed skill set. The fire officer needs to be clear, calm, and concise in the initial radio transmissions. The communication of incident size-up information must be consistent with the organization's requirements and ICS.

Developing a command presence is a key part of mastering the art of incident command. Command presence is the ability of an officer to project an image of being in control of the situation. To be a successful leader, the officer must convince others to follow by demonstrating the ability to take charge and make the right things happen. A fire officer who is going to establish command upon arriving at an emergency incident should have a detailed knowledge of the responding companies; a mastery of the local procedures; and the ability to issue clear, direct orders. Fire fighters are aggressive, action-oriented people who want to make a difference. It is important that a new fire officer develop a command presence to focus the efforts of this action-oriented team in a constructive direction (**FIGURE 1-14**).

Trainer

The fire officer has the responsibility of making sure the fire fighters under his or her command are confident and competent in their skills. The company-level officer is responsible for the performance level of the fire company and must establish a set of expectations that the company will perform at the highest level possible (**FIGURE 1-15**). In a large department, one fire company may need to have a higher level of specialized skill in one area than in another. For example, a fire company may have a specialty assignment or a co-located specialized unit that the company staffs when it is requested. Technical rescue rigs, decontamination units, foam pumpers, mass-casualty units, and mobile command posts are examples of co-located specialized units.

In addition, the fire company's response district may require a higher level of fire fighter skill or knowledge. Consider hose handling skills. A fire company in a high-rise district would require different hose handling skills and competencies than those required of a company in a mountainous wildland interface area. Although both companies need to demonstrate familiarity with the basics of operating a hose line from a standpipe, the high-rise fire company should have a much higher level of expertise and competence in this skill. Conversely, the mountain fire company usually has a higher level of expertise and competence in rope rescue evolutions.

The company-level officer plays a key role in developing these competencies within the company. James O. Page made three specific recommendations to assist fire officers in this task: develop a personal training library, know the neighborhood, and use problem-solving scenarios (Page, 1973).

Developing a Personal Training Library

Page's personal library starts with a three-ring notebook with subject matter tabs. The subject tabs could

FIGURE 1-14 It is important for the fire officer to develop a command presence.
Courtesy of William Moreland.

FIGURE 1-15 The fire officer must ensure that fire fighters are confident and competent in their skills.
© Glen E. Ellman.

match the topic headings in NFPA 1001, reflect the recruit school curricula, or represent a personal list of important topics. Every time the fire officer attends a training event, the notes from that session are placed into the three-ring binder. All related handouts, product information sheets, and other related items are also placed into the three-ring binder. When the officer is preparing to present a class that covers a specific topic, his or her personal training library is the first stop.

Leo D. Stapleton, a retired Boston fire commissioner and author of *Thirty Years on the Line,* advocated that officers maintain a personal journal where they record information about the incidents they run and the issues they handle. For working incidents, that would include what was encountered, what was learned, and how to handle it the next time. Stapleton encouraged officers to make these entries as soon as possible after the incident, while the information is vivid in the officer's memory (Stapleton, 1983).

Today, a fire officer can create an electronic version of Page's three-ring binder and Stapleton's journal with the use of a tablet or smartphone. As part of creating this personal electronic library, the fire officer can collect PowerPoint presentations and video clips, download files from the Internet, and convert various paper media into Adobe Portable Document Format (.pdf) files. A digital projector allows all these materials to be used in classroom sessions.

In addition to storing information in a three-ring notebook or on a digital device, Page and Stapleton recommended that the fire officer obtain personal copies of the textbooks and references used in fire fighter training and promotional examinations. You can highlight, tab, and write in this copy of the book until it becomes your personal reference. You can write notes in the margins and even note disagreements with the author's statement or point of view to personalize the learning experience. These kinds of annotations are especially valuable when you are reading a book in preparation for a promotional exam (**FIGURE 1-16**).

Know the Neighborhood

Fire fighters should have a detailed knowledge of the environment they protect. That requires going out on inspections to walk through each nonresidential structure in the jurisdiction, from the roof through the sub-basement; these walkthroughs reinforce the written preincident plan or diagrams. During the walkthrough, with the permission of the facility representative, the fire officer should take pictures to capture significant details. A detailed overhead view of many communities can be obtained from aerial photographs, digital maps, or Google Earth. This aerial

FIGURE 1-16 Highlight, tab, and write in your personal copies of textbooks used for training.
© Jones & Bartlett Learning. Photographed by Glen E. Ellman.

view is especially valuable for apartment complexes, business parks, and retail areas. The maps and photos can be used to locate access routes, plan the placement of first-alarm apparatus, and identify exposure problems.

Knowing the neighborhood may include working with the building owners and occupants to practice incident action plans and procedures. Fire officers should strive to establish good working relationships with every building manager or emergency program manager in their response district.

Use Problem-Solving Scenarios

The fire officer can help the company members become more skilled and knowledgeable by providing opportunities to use their problem-solving skills. Instead of reading the code regulations to provide training for his company members, Page would present fact-based situations and require them to use the code to solve the problem. This technique forced the fire fighters to identify the occupancy use group, identify the issues, look up the applicable regulations, and make decisions. This is an excellent way for adults to learn concepts, regulations, and decision-making skills. The same technique can be used for fire-ground hydraulics, technical rescue scenarios, and incident management procedures.

In addition, this problem-solving approach can be used in reviews of preincident action plans. The fire officer can construct various emergency incident scenarios for training sessions, based on actual buildings and potential situations. Some computer-based fire incident simulators can accept digital pictures taken during walkthrough visits and use them to create customized emergency incident scenarios.

The Fire Officer's Supervisor

Every fire officer has a supervisor. A fire officer's supervisor is usually a command-level officer (a battalion chief, a district chief, or a battalion commander) who supervises numerous fire companies within a geographic area. The battalion chief's supervisor could be a deputy chief or an assistant chief, who reports directly to the fire chief. In the same manner, the fire chief would report to the mayor, the city manager, the commissioner of public safety, or an individual at an equivalent level. Under the chain of command, the fire chief's orders and directives are passed down through the deputy chief or assistant chief to the battalion or district chief, who then ensures that the fire officers enact those orders or directives.

Regardless of the organizational structure, every fire officer has an obligation to work effectively with a supervisor. Three activities are necessary to ensure a good working relationship:

- Keep your supervisor informed.
- Make appropriate decisions at your level of responsibility.
- Consult with your supervisor before taking major disciplinary actions or making policy changes.

No supervisor likes surprises. Part of the supervisor/subordinate relationship is to make sure the supervisor is not surprised or blindsided (**FIGURE 1-17**). For example, if a fire company vehicle collides with a civilian vehicle at 9:30 PM, the chief needs to immediately be notified. It would be poor form for the chief to find out about the accident from a story on the 11 o'clock news.

Supervising fire officers should not hesitate to make decisions appropriate for their level of responsibility. This means that problems should be addressed and situations resolved where and when they occur. If a fire officer has the authority to solve a problem, he or she should not wait for the supervisor to arrive to solve it.

This level of authority usually covers fire station-level activities such as maintenance, training, and community outreach. At fire stations with rotating career work shifts, the expectation is that most issues between the shifts will be resolved at the supervising

FIGURE 1-17 Always keep your supervisor informed.
© Jones & Bartlett Learning. Photographed by Glen E. Ellman.

fire officer level. A fire officer who is working in a staff position should make the decisions appropriate for that position. Volunteer officers should make decisions appropriate for their level within the organization.

Some issues require the fire officer to consult with a supervisor before making a decision or taking action. Policy changes, for example, cannot be enacted within an administrative vacuum. If a decision will have an impact that goes beyond the fire officer's scope of authority, it is time to move the discussion up to the next level in the chain of command.

This policy also applies before major disciplinary actions are taken. Consultation with a supervisor is required by the municipal personnel regulations to ensure that all major discipline is delivered in a consistent and impartial manner. This is also a recommended practice in most volunteer fire departments.

Integrity and Ethical Behavior

The formal organization provides the new supervising fire officer with the symbols of power and authority that are associated with the badge, insignia, and distinctive markings on the officer's helmet. The individual, however, needs to provide the core values of integrity and ethical behavior that, combined with the formal symbols, create an effective fire officer. An unethical fire officer is ineffective and damages the department's reputation. If their behavior goes uncorrected, an unethical and corrupt fire officer will corrode the department's ability to deliver services and maintain public trust.

> ### FIRE OFFICER TIP
>
> **One Captain's Expectations and Requirements to a New Fire Fighter**
>
> This document was provided to new fire fighters assigned to a senior captain's team:
>
> We expect you to be the best fire fighter on the best shift. We want other fire fighters at other stations to be envious of your position. Give 100 percent effort all of the time.
>
> Everyone will make mistakes. Own up to your blunder. Don't blame others. Always try to do the right thing. I will back you up if you can explain and justify why you did something wrong.
>
> When you go on a call, treat everyone like you would treat your Momma. I don't care if you have gotten up three times after midnight, even if it's a snivel call, treat the public with respect.
>
> This is a dangerous job. Use your common sense and experience to make decisions. If you don't know the answer, ask.
>
> We risk a lot to save a lot and risk little to save little. You will see many disturbing things on the street that will make your stomach turn. Learn how to deal with the stress of the job.
>
> Inside a burning building, if you think you are in trouble, call a mayday. Don't worry about peer pressure. If you think you are in trouble, you are. Get help; don't let your pride kill you.
>
> When given a difficult task, be able to grasp the detail of the order and put it into a safe practice to complete the task. This will apply under normal day-to-day routine situations and emergency conditions.
>
> Communicate properly and cordially, especially when dealing with the public. They are the ones who support us. Lead by example 24/7, not just at work.

Integrity

Integrity refers to the complex system of inherent attributes that determine a person's moral and ethical actions and reactions, including the quality of being honest.

The fire officer should "walk the talk" and demonstrate the behaviors that he or she says are important (**FIGURE 1-18**). If the company officer says that physical fitness is important, then the fire fighters should see their officer performing physical fitness training during the workday.

Integrity can be demonstrated by a steadfast adherence to a moral code. Such a code combines a fire officer's internal value system and the fire department's official organizational value system. Formal

FIGURE 1-18 The fire officer should demonstrate the behaviors that he or she says are important.
© Jones & Bartlett Learning. Photographed by Glen E. Ellman.

organizations publish their expectations as a code of ethics, a code of conduct, or a list of value statements.

Ethical Behavior

The fire officer position provides a wide range of opportunities to demonstrate ethical behavior. The fire officer demonstrating **ethical behavior** makes decisions and models behavior consistent with the department's core values, mission statement, and value statements.

Inappropriate behavior by individuals in power or holding the public trust is a frequent target of attention for the media, and the fire service is not exempt from such scrutiny. Most of the time fire officers make ethical decisions, choosing to do the right thing; however, sometimes officers make unethical choices (which may then influence fire fighters to make unethical choices). When they engage in such behavior, officers' poor choices often appear in the newspaper and have very negative consequences for both the individual and the fire service organization.

Ethical choices are based on a value system. The officer must consider each situation, often subconsciously, and make a decision based on his or her values. If the organization has clear values that are part of a strong organizational culture, the officer uses the organization's value system. If the values are not clear, the individual substitutes his or her own value system.

The key to improving ethical choices is to have clear organizational values. This can be accomplished by:

- Having a code of ethics that is well known throughout the organization
- Selecting employees who share the values of the organization

- Ensuring that top management exhibits ethical behavior
- Having clear job goals
- Having performance appraisals that reward ethical behavior
- Implementing an ethics training program

Even at the company level, these values can be implemented to help prevent undesirable ethical choices. One way to help judge a decision is to ask yourself three questions:

1. What would my parents and friends say if they knew?
2. Would I mind if the newspaper ran it as a headline story?
3. How does it make me feel about myself?

Asking these questions can help prevent an event that could devastate the department's and the fire officer's reputation for years to come.

Fire department activity tends to be high profile, regardless of the task. Even a trip to the grocery store to pick up dinner draws public attention. The fire officer should act as if someone is always documenting his or her actions when out of the fire station.

Workplace Diversity

Diversity is a characteristic of a fire workforce that reflects differences in terms of age, cultural background, race, religion, sex, and sexual orientation. Issues of diversity are guided by the civil rights of Americans. Those rights are established by federal laws, which are enforced in the workplace by the Equal Employment Opportunity Commission (EEOC). Title VII of the Civil Rights Act of 1964 covers state and local governments, schools, colleges, and unions. This law states that it is illegal for an employer to:

(1) fail or refuse to hire or discharge any individual, or otherwise discriminate against any individual with respect to compensation, terms, conditions, or privileges of employment because of such individual's race, color, religion, sex, or national origin, or
(2) limit, segregate, or classify employees or applicants for employment in any way that would deprive any individual of employment opportunities or otherwise adversely affect status as an employee because of such individual's race, color, religion, sex, or national origin.

The Equal Employment Opportunity Act of 1972 amended the Civil Rights Act of 1964 and expanded its coverage to include almost all public and private employers with 15 or more employees. In general, the 1972 act covers volunteer fire departments and other nonprofit emergency service organizations.

The Civil Rights Act of 1991 provides additional compensatory and punitive damages in cases of intentional discrimination under Title VII and the Americans with Disabilities Act of 1990. The changes introduced in the 1991 act increased the number of discrimination lawsuits filed against organizations.

Many fire departments have made changes to their recruitment, hiring, and promotion practices to comply with the various civil rights laws. Some departments took action to diversify their workforces without the prompting of the courts. Where this did not occur, the legal remedies have ranged from consent decrees, in which the fire department agrees to accomplish specific diversity goals within a specific time, to court orders outlining specific hiring practices.

Diversity, as applied to fire departments, means the workforce should reflect the community it serves. In all cases, the overarching goal is for the fire department to reflect the diversity of the community. Consider the example of a fire department that is 90 percent Caucasian in a community where the overall population is 40 percent African American, 20 percent Latino, and 20 percent Asian American. This fire department population is not reflecting the broader community.

If the department has agreed to an EEOC consent decree, it has promised the court that it will work to hire qualified individuals who reflect the community. A consent decree can require a variety of activities, including community outreach, job fairs, pre-employment preparation, and peer group coaching. The court formally meets with the fire department representative periodically to see how well the department is progressing.

Some departments operate under a specific court-mandated hiring process. Using our earlier example, the court could require the department to hire two African Americans, one Latino, and one Asian American before it can hire one additional Caucasian. This requirement would remain in effect until the hiring diversity goal is accomplished.

Expiration of a court order does not relieve the fire department of its charge to maintain diversity. One large fire department had a qualified applicant pool that included 23 percent candidates representing a protected class when a consent decree was terminated in 1999. By 2008, that representation shrunk to seven percent of the qualified applicant pool, resulting in another inquiry by the court.

The Fire Officer's Role in Workplace Diversity

Jack W. Gravely is a lawyer, subject-matter expert, and trainer on workplace diversity. He has been a frequent speaker at public safety agencies on issues of racial and cultural diversity.

When Gravely started teaching diversity training classes to local government leaders, the emphasis was on the language of the regulations and their potential impact. In a 2015 presentation, Gravely pointed out that today's supervisor has the benefit of four decades of equal employment opportunity (EEO)/affirmative action (AA) court decisions to guide decision making. Gravely noted that thousands of court decisions have created a large body of case law that presents a generally consistent policy on how a supervisor should behave in the workplace. Gravely recommended that a fire officer focus on actionable items and the definition of a hostile work environment. According to the United States Department of Labor, a **hostile work environment** can result from the unwelcome conduct of supervisors, co-workers, customers, contractors, or anyone else with whom the victim interacts on the job, and the unwelcome conduct renders the workplace atmosphere intimidating, hostile, or offensive. In 2017, the EEOC filed 184 lawsuits and 84,254 workplace discrimination charges.

Actionable Items

An **actionable item** is an employee behavior that requires an immediate corrective action by the supervisor. Dozens of lawsuits have shown that failing to act when these situations occur is likely to create a liability and a loss for the department. The best example is the use of certain words in the workplace. An employee's use of derogatory or racist terms about people from other ethnicities, religions, or genders requires immediate corrective action by the supervisor. Regardless of the conditions, context, or situation, such words are inappropriate and represent a potential million-dollar liability to the organization.

The fire officer must act immediately in these situations. That means speaking with the offending fire fighter in private, and counseling the fire fighter that the use of such words is unacceptable in the fire station or in any situation where the individual is representing the department (either on duty or off duty but in uniform **FIGURE 1-19**). The fire officer should provide the fire fighter with the fire department's or municipality's EEO/AA policy statement and, if applicable, the code of conduct. The fire officer should also maintain a formal or informal record of the counseling session.

FIGURE 1-19 Discuss inappropriate behaviors privately.
© Jones & Bartlett Learning. Photographed by Glen E. Ellman.

If any doubt arises about whether the message was fully understood, the fire officer should ensure that a higher-level supervisor is informed of the actions that have been taken.

The same policies should apply to fire fighters who are regularly assigned to the fire company, fire fighters detailed in or visiting the fire station, and other uniformed or civilian members of the fire department. Case law has shown that use of unacceptable language requires an immediate response. An officer's failure to act has been interpreted as official condoning or encouragement of such behavior.

Although enough case law is available to provide general guidelines and some specific examples, the subject of what constitutes harassment remains a dynamic aspect of the work environment. The fire officer needs to stay informed about the organization's EEO/AA and diversity policies. Most large organizations have a diversity or EEO/AA office that can provide up-to-date information and answer questions.

Hostile Workplace and Sexual Harassment

In 1999, the EEOC amended the guidelines on sexual harassment. The amended regulations broadened the types of harassment that are considered illegal and included a requirement that employers have a duty to maintain a harassment-free work environment. The standard for evaluating sexual harassment is what a "reasonable person" in the same or similar circumstances would find intimidating, hostile, or abusive. The 1999 guidelines clearly state that employers are liable for the acts of those who work for them if the organization knew or should have known about the conduct and took no immediate, appropriate corrective action. That is why a fire officer must immediately respond to any utterances of offensive or derogatory language in the work environment.

Sexual harassment is unwelcome sexual advances, requests for sexual favors, and other verbal or physical harassment of a sexual nature. The abuse of power is an essential component of sexual harassment. The EEOC guidelines state that verbal and physical conduct of a sexual nature is harassment when the following conditions are present:

- The employee is made to feel that he or she has to endure such treatment to remain employed.
- Whether the employee submits to or rejects such treatment is used when making employment decisions.
- The employee's work performance is affected.
- An intimidating, hostile, or offensive work environment is present (EEOC, 1990).

The term *hostile work environment* can be used to describe a broad range of situations in which an employee is subject to discrimination in the workplace. The generalized hostile workplace definition can apply to a variety of circumstances that do not necessarily include a specific abuse of supervisory power.

Jack Gravely believes that hostile workplace complaints will shape workplace diversity. There are few quid pro quo sexual harassment cases—cases in which a supervisor specifically promises a work-related benefit in return for a sexual favor. Instead, the trend is toward more complaints about a hostile work environment, which can involve sexual issues. The 2017 EEOC report notes an increased percentage of retaliation complaints.

Many examples can be cited of hostile workplace issues that resulted in large court-directed settlements for the complainant. In one municipality, a series of lawsuits for sexual harassment and discrimination from 2006 to 2018 cost the city $30 million in settlements and legal fees (McGuire et al., 2018).

Handling a Harassment or Hostile Workplace Complaint

Fire fighters who want to initiate a harassment complaint have a choice of three methods for doing so. They can start with the federal government, they can start with the local government, or they can start within the fire department—it is their choice. If the process starts within the fire department, a fire officer may be the first formal point of contact. In this case, the fire officer would function as the first step of a multistep procedure.

The fire officer should know the department's procedure for handling a harassment or hostile workplace complaint. The fire officer's designated role in conducting an investigation of an EEO complaint depends on the procedures adopted by the jurisdiction or the fire department. In some cases, an officer's role is limited to starting the process by making appropriate notifications. Many local governments have specially trained EEO staff who conduct an investigation. In other cases, the fire officer might be required to perform the initial investigation and submit a report. Following are general guidelines:

- Keep an open mind. Many fire officers have a difficult time believing that discrimination or harassment could be happening right under their noses. Failure to investigate a complaint, however, is the most common reason a local government is found liable. Every complaint must be investigated and documented. Do not come to any conclusions until your investigation is complete. Follow your organization's procedures and inform your supervisor.
- Treat the person who files the complaint with respect and compassion. Employees often find it extremely difficult to complain about discrimination or harassment. When an employee comes to you with concerns about discrimination or harassment, be professional, and be understanding.
- Do not blame the person filing the complaint. Case law and current practice dictate that it is the complainant who determines whether the situation is hostile or is harassment. Blaming the person bringing the complaint to you is the second most common way that local governments lose harassment or hostile workplace complaints.
- Do not retaliate against the person filing the complaint. It is against the law to punish someone for complaining about discrimination or harassment. The most obvious forms of retaliation are termination, discipline, demotion, or threats to do any of these things. More subtle forms of retaliation could include changing the shift hours or work location of the accuser, even if the intent is to remove the alleged victim from the problem.
- Follow established procedures. Local government personnel regulations generally provide a detailed procedure on how to handle harassment or hostile workplace complaints. Follow the procedure used in your jurisdiction.
- Interview the people involved. An initial investigation usually involves conducting

interviews with the people who are involved in the situation, typically starting with the person who made the complaint. The interviewer needs to find out exactly what the employee is concerned about. Get details: what was said or done, when and where, and who was present. Then talk to any employees who are being accused of discrimination or harassment. Get details from them as well. Be sure to interview any witnesses who may have seen or heard any problematic conduct. Take notes about your interviews and gather any relevant documents.

- Look for corroboration or contradiction. Discrimination and harassment complaints often involve "he said/she said" situations. The accuser and the accused may offer different versions of an incident, leaving you with no way of knowing who is telling the truth. In such a case, you may have to turn to other sources for clues. Witnesses may have seen part of an incident. In some cases, documents, such as e-mails and posted notes, may prove one side to be right.
- Keep it confidential. A discrimination complaint can polarize a workplace. The unique team nature of 24-hour fire service work creates a close environment where it is difficult to maintain confidentiality. The fire officer must insist on and enforce confidentiality during the investigation.
- Write it all down. Take notes during all interviews. Before the interview is over, go back through your notes with the interviewee to ensure accuracy. Keep a journal of the investigation. Write down the steps you have taken to get at the truth, including dates and places of interviews. Keep a list of all documents that are reviewed. Document any action taken against the accused or the reasons for deciding not to take action. Anticipate that all of this written record will be used in any subsequent civil service or court actions.
- Cooperate with government agencies. If the fire fighter files a complaint with another government agency (either the federal EEOC or an equivalent state agency), that agency may investigate. Notify your supervisor as soon as you receive a call or visit from an EEOC investigator. You will probably be asked to provide certain documents, to give your side of the story, and to explain any efforts you made to deal with the complaint yourself. Be cautious, but cooperative.

Regardless of where the complaint is filed, the fire chief is required to take corrective action if the investigation confirms that the complaint has merit. If the department concludes that some form of discrimination or harassment occurred, formal corrective action could include mandatory training, work location transfer, or demotion. Termination may be proposed for more egregious kinds of discrimination and harassment, such as threats, stalking, or repeated unwanted physical contact.

Credentialing and Fire Officer Development

The foundation of company officer practice started with the big-city fire company foreman of the 1880s and used the company commander experience from the 20th century wars to evolve practices and procedures. These elements were codified when the NFPA established the professional qualification standards, adopting NFPA 1021 for fire officers in 1976. The IAFC expanded company officer development, updated in 2010, providing an *Officer Development Handbook* to encourage company officers to acquire the appropriate levels of training, experience, self-development, and education throughout their professional journey to prepare for the Chief Fire Officer (CFO) designation, the pinnacle of professional development. There are two organizations that offer fire credentialing, the Center for Public Safety Excellence (CPSE) and the National Fire Academy (NFA).

The CPSE is a not-for-profit 501(c)(3) corporation that is an international technical organization that works with the most fire and emergency service agencies and most active fire professionals. The CPSE mission is to lead the fire and emergency service to excellence through the continuous quality improvement process of accreditation, credentialing, and education. CPSE provides accreditation to fire agencies and credentialing to individuals. The Commission on Professional Credentialing (CPC) offers five distinct designations covering the various levels and specialties of fire officers. CPC provides (1) an application process that officers use to develop their portfolio, (2) training and support while developing their portfolio, and (3) access to experienced peer reviewers. Fire Officer is one of the designations, open to all junior officers, company officers, or those who have served on an intermittent acting status for 12 months. The CPC model looks at the whole officer:

1. Education
2. Experience

3. Professional development (training and certifications)
4. Professional contributions and recognitions
5. Professional memberships and affiliations
6. Technical competence (depending on the credential, 7 to 20 different competencies)
7. Community involvement

The process includes a self-assessment, professional portfolio, peer review, and interview process. There are over 450 CPC-credentialed Fire Officers (CPSE, 2019).

The NFA is a component of the United States Fire Academy (USFA) that works to enhance the ability of fire and emergency services and allied professionals to deal more effectively with fire and related emergencies. Free training courses and programs are delivered at a campus in Emmitsburg (Maryland), online, and through designated state fire agencies.

The NFA's Managing Officer Program is a multi-year curriculum that introduces emerging emergency service leaders to personal and professional skills in change management, risk reduction, and adaptive leadership. Acceptance into the program is the first step in your professional development as a career, volunteer, part-time, or combination fire/EMS manager, and includes all four elements of professional development: education, training, experience, and continuing education. There are three elements of the program:

1. Five prerequisite courses (online and classroom deliveries in your state)
   a. Q0890, Introduction to Emergency Response to Terrorism
   b. ICS-100, Introduction to ICS for Operational First Responders
   c. ICS-200, Basic NIMS ICS for Operational First Responders
   d. IS-700a, National Incident Management System (NIMS), An Introduction
   e. IS-800c, National Response Framework, An Introduction
2. Four courses at the NFA in Emmitsburg, Maryland
   a. Application of Community Risk Reduction
   b. Applications of Leadership in the Culture of Safety
   c. Analytical Tools for Decision-Making
   d. Training and Professional Development Challenges for Fire and Emergency Service Leaders
3. A community-based capstone project
   a. Lessons learned from one of the core courses required in the Managing Officer Program.
   b. Experiences of the Managing Officer as identified in the *IAFC Officer Development Handbook*, Second Edition.
   c. An issue or problem identified by your agency or jurisdiction.
   d. Lessons learned from a recent administrative issue.
   e. Identification and analysis of an emerging issue of importance to the department.

A certificate of completion for the Managing Officer Program is awarded after the successful completion of all courses and the capstone project (FEMA, 2019).

The IAFC Company Officer Leadership Program

Designed for crew leaders, senior station leadership, sergeants, lieutenants, and captains, the IAFC Company Officer Leadership (COL) program is a three (3)-level program that provides what company officers need and what chief officers expect. The program is based on NFPA 1021 *Standard for Fire Officer Professional Qualifications,* (Fire Officer I and II) and the *IAFC Officer Development Handbook*. There are three levels for the company officer program; each level consists of 21 contact hours:

- Administration and Human Relations — 3 hours
- Leadership — 3 hours
- Community Risk Reduction — 3 hours
- Operations — 3 hours
- Safety, Health, and Wellness — 3 hours
- Elective — 3 hours
- Social Learning and Networking — 3 hours

You Are the Fire Officer Conclusion

The chief arrives and has the newly promoted captain and lieutenant assemble in the shift captain's office. Closing the door, the chief welcomes them and shares his expectations, outlining the responsibilities of a supervising fire officer. "Make sure the crews are properly trained in all the tools and equipment. I expect them, and you, to stay physically and mentally prepared to work at an emergency incident. Keep the apparatus, tools, and station in great shape. Train every day. Get intimately familiar with the target hazards."

"Do not surprise me. If there is an injury, property damage accident, or incident, I need to hear about it directly from you and not from headquarters, or the News at Noon. We will work these issues out together." Pointing to the "Everyone Goes Home" poster from the National Fallen Firefighters Foundation, the chief concludes the orientation with two directives:

1. Everyone has their seat belt on before the rig starts to respond.
2. The rigs come to a complete stop at every stop sign and red-light intersection.

The supervising fire officer is responsible for the safety and performance of the fire fighters that work under the supervisor, either as a member of a fire company or a specialized or administrative team. The organizational structure of the fire department provides the supervising fire officer with a chain of command that describes the level of authority and responsibility at each rank. The theories and concepts studied for a promotional exam provide the new fire officer with a toolbox of supervisory, leadership, and management tools that can be used to meet the department's goals, objectives, and tasks.

After-Action REVIEW

IN SUMMARY

- At the Fire Officer I level, emphasis is placed on accomplishing the department's goals and objectives by working through subordinates to achieve desired results.
- The Fire Officer I performs administrative duties and supervisory functions that are related to a small group of fire department members.
- The roles and responsibilities of a fire officer differ from those of a fire fighter. Understanding the new role is essential.
- Most fire department organizations divide fire fighters into two categories: career and volunteer.
- Source of authority, chain of command, and the NIMS models help departments focus individual efforts and provide structure.
- Most fire departments are structured on the basis of four management principles: unity of command, span of control, division of labor, and discipline.
- The four functions of management are planning, organizing, leading, and controlling. Fire officers use the functions of management to get work accomplished by and through others.
- The fire fighter is required to follow all regulations, policies, and procedures. The fire officer must not only follow these directives, but also ensure compliance with them by subordinates.
- The basis for a strong, positive, and effective supervisor/employee relationship is open, honest, and constant communications between the fire officer and the fire fighter.
- The root cause of almost every labor disturbance is a failure to manage the relationship between labor and management properly.
- One duty of a fire officer is to supervise the activities of subordinate fire fighters. In most organizational structures, there is a clear distinction between labor (the workers) and management (the managers and supervisors). The managers and supervisors represent the organization, and the union represents the workers.

- Important leadership concepts and activities include the beginning of shift report, notifications, decision making, and problem solving.
- A significant change occurs when the fire fighter transitions to a fire officer. The company-level officer is directly responsible for the supervision, performance, and safety of a crew of fire fighters.
- In the supervisor role, the fire officer functions as the official representative of the fire chief. When operating at the scene of an emergency incident, the fire officer is expected to function as a commander and to exercise strong direct supervision over the company members.
- The company-level officer is responsible for the performance level of the fire company.
- A fire officer's supervisor is usually a command-level officer (a battalion chief, a district chief, or a battalion commander) who supervises numerous fire companies within a geographic area.
- The fire officer should demonstrate the behaviors that he or she says are important.
- Diversity, as applied to fire departments, means the workforce should reflect the community it serves.
- There are two organizations that offer fire credentialing, the Center for Public Safety Excellence and the National Fire Academy.

KEY TERMS

Actionable item Employee behavior that requires an immediate corrective action by the supervisor; dozens of lawsuits have shown that failing to act in the face of such behavior will create a liability and a loss for the department.

Assistant or division chiefs A midlevel chief who often has a functional area of responsibility, such as training, and who answers directly to the fire chief.

Battalion chiefs Usually the first level of fire chief. These chiefs are often in charge of running calls and supervising multiple stations or districts within a city. A battalion chief is usually the officer in charge of a single-alarm working fire.

Captain The second rank of promotion in the fire service, between the lieutenant and the battalion chief. Captains are responsible for managing a fire company and for coordinating the activities of that company among the other shifts.

Chain of command A rank structure, spanning the fire fighter through the fire chief, for managing a fire department and fire-ground operations.

Controlling Restraining, regulating, governing, counteracting, or overpowering.

Discipline The guidelines that a department sets for fire fighters to work within.

Diversity A characteristic of a fire workforce that reflects differences in terms of age, cultural background, race, religion, sex, and sexual orientation.

Division of labor Breaking down an incident or task into a series of smaller, more manageable tasks and assigning personnel to complete those tasks.

Ethical behavior Decisions and behavior demonstrated by a fire officer that are consistent with the department's core values, mission statement, and value statements.

Fire chief The top position in the fire department. The fire chief has ultimate responsibility for the fire department and usually answers directly to the mayor or other designated public official.

Fire Officer I The fire officer, at the supervisory level, who has met the job performance requirements specified in NFPA 1021 for Level I. (Reproduced from NFPA 1021, Standard for Fire Officer Professional Qualifications, Copyright © 2003, National Fire Protection Association, All Rights Reserved.)

Fire Officer II The fire officer, at the supervisory/managerial level, who has met the job performance requirements specified in NFPA 1021 for Level II. (NFPA 1021)

Hostile work environment An environment that can result from the unwelcome conduct of supervisors, co-workers, customers, contractors, or anyone else with whom the victim interacts on the job, and the unwelcome conduct renders the workplace atmosphere intimidating, hostile, or offensive.

Incident command system (ICS) The combination of facilities, equipment, personnel, procedures, and communications operating within a common organizational structure that has responsibility for the management of assigned resources to effectively accomplish stated objectives pertaining to an incident or training exercise. (Reproduced from Technical Committee on Rescue Technician, Professional Qualifications, NFPA 1006, Second Draft Meeting, December 1-2, 2015, National Fire Protection Association.)

Leadership A complex process by which a person influences others to accomplish a mission, task, or objective and directs the organization in a way that makes it more cohesive and coherent.

Leading Guiding or directing in a course of action.

Lieutenant A company officer who is usually responsible for a single fire company on a single shift; the first in line among company officers.

Managing fire officer The description from the IAFC *Officer Development Handbook* for the tasks and expectations for a Fire Officer II. In this role, the company officer is encouraged to acquire the appropriate levels of training, experience, self-development, and education to prepare for the Chief Fire Officer designation.

Mediation The intervention of a neutral third party in an industrial dispute.

National Fire Protection Association (NFPA) The association that delivers information and knowledge through more than 300 consensus codes and standards, research, training, education, outreach, and advocacy; and by partnering with others who share an interest in furthering its mission in helping save lives and reduce loss with information, knowledge, and passion.

Organizing Putting resources together into an orderly, functional, structured whole.

Planning Developing a scheme, program, or method that is worked out beforehand to accomplish an objective.

Policies Formal statements that provide guidelines for present and future actions. Policies often require personnel to make judgments.

Rules and regulations Directives developed by various government or government-authorized organizations to implement a law that has been passed by a government body.

Span of control The maximum number of personnel or activities that can be effectively controlled by one individual (usually three to seven). (Reproduced from Technical Committee on Rescue Technician, Professional Qualifications, NFPA 1006, Second Draft Meeting, December 1-2, 2015, National Fire Protection Association.)

Standard operating procedures (SOPs) A written organizational directive that establishes or prescribes specific operational or administrative methods to be followed routinely for the performance of designated operations or actions. (Reproduced from Technical Committee on Rescue Technician, Professional Qualifications, NFPA 1006, Second Draft Meeting, December 1-2, 2015, National Fire Protection Association.)

Supervising fire officer The description from the IAFC *Officer Development Handbook* for the tasks and expectations for a Fire Officer I. In this role, the company officer is encouraged to acquire the appropriate levels of training, experience, self-development, and education to prepare for the Chief Fire Officer designation.

Unity of command The concept by which each person within an organization reports to one, and only one, designated person. (Reproduced from NFPA's Glossary of Terms, 1026, ©2013 NFPA, Copyright ©2013 National Fire Protection Association. All rights reserved.)

REFERENCES

Bugbee, Percy. 1971. *Man Against Fire: The Story of the National Fire Protection Association 1896–1971*. Boston, MA: National Fire Protection Association.

Center for Public Safety Excellence (CPSE). 2019. "How to get credentialed." Accessed June 21, 2019. https://cpse.org/credentialing/how-to-get-credentialed/.

Drucker, Peter F. 1974. *Management: Tasks, Responsibilities, Practices*. New York: Harper & Row.

Equal Employment Opportunity Commission (EEOC). 1990. "Policy Guidance on Current Issues of Sexual Harassment." Accessed June 21, 2019. https://www.eeoc.gov/policy/docs/currentissues.html.

Evarts, Ben, and Gary P. Stein. 2019. "U.S. Fire Department Profile." Quincy, MA: National Fire Protection Association. Accessed June 21, 2019. https://www.nfpa.org/News-and-Research/Data-research-and-tools/Emergency-Responders/US-fire-department-profile

Fayol, Henri. 1949. *General and Industrial Management*. Translated by Constance Storrs. London, Pitman.

Federal Emergency Management Agency (FEMA). 2017, October. *National Incident Management System, Third edition*. Washington, DC: Department of Homeland Security.

Federal Emergency Management Agency (FEMA). 2019, March. *Managing Officer Program Handbook*. Washington, DC: Department of Homeland Security.

Gratz, David B. 1972. *Fire Department Management: Scope and Method*. Encino, CA: Glencoe Press.

Griffiths, James S. (2017). *Fire Department of New York - an Operational Reference, Eleventh edition*. New York: James S. Griffiths.

International Association of Fire Fighters (IAFF). n.d. "About Us." Accessed May 20, 2019. http://client.prod.iaff.org/#page=AboutUs

Kimball, Warren Young. 1966. *Fire Attack 1: Command Decisions and Company Operations*. Boston, MA: National Fire Protection Association.

McGuire, Stephen J. J., Robbins, Patricia, Babaian, Oshin, Iniguez, Fernando, Koshkaryan, Vardui, Pelayo, Maribel, and

Sogomonnyan, Taguhi. 2018. "Los Angeles Fire Department: Diversity Ignites Discrimination." *Journal of Case Research and Inquiry* 4 (December): 129–157.

National Fire Protection Association (NFPA). 2008. "Deadliest Single Building or Complex Fires and Explosions in the U.S." Accessed June 10, 2019. https://www.nfpa.org/News-and-Research/Data-research-and-tools/US-Fire-Problem/Catastrophic-multiple-death-fires/Deadliest-single-building-or-complex-fires-and-explosions-in-the-US.

National Fire Protection Association (NFPA). 2018. "Fire Department Calls Table." Accessed March 18, 2019. https://www.nfpa.org/News-and-Research/Data-research-and-tools/Emergency-Responders/Fire-department-calls.

National Volunteer Fire Council (NVFD). 2019. "Annual Report Infographic: 2018 by the Numbers." Accessed May 20, 2019. https://www.nvfc.org/wp-content/uploads/2019/01/2018-Annual-Report-Infographic.pdf.

Page, James O. 1973. *Effective Company Command*. Alhambra, CA: Borden Publishing.

Smeby, L. Charles. 2013. *Fire and Emergency Services Administration: Management and Leadership Practices, Second edition*. Burlington, MA: Jones and Bartlett Learning.

Stapleton, Leo D. (1983). *Thirty Years on the Line*. Dover, NH: DMC Associates.

U. S. Department of Labor. 2012. "What Do I Need to Know about... Workplace Harassment?" Accessed March 21, 2019. https://www.dol.gov/agencies/oasam/civil-rights-center/internal/policies/workplace-harassment/2012.

U.S. Equal Employment Opportunity Commission. 2016. "EEOC Enforcement Guidance on Retaliation and Related Issues." Accessed March 19, 2019. https://www.eeoc.gov/laws/guidance/retaliation-guidance.cfm.

U.S. Equal Employment Opportunity Commission. 2018. "EEOC Releases Fiscal Year 2017 Enforcement and Litigation Data. https://www.eeoc.gov/eeoc/newsroom/release/1-25-18.cfm.

U.S. Equal Employment Opportunity Commission. n.d. "Sexual Harassment." Accessed March 19, 2019. https://www.eeoc.gov/laws/types/sexual_harassment.cfm.

Fire Officer in Action

When the captain or lieutenant is not on duty, a senior fire fighter is assigned to be the acting fire officer. The department expects acting fire officers to function in any of the roles assumed by a lieutenant or captain.

Back at the station, the acting officer functions as a substitute teacher.

1. The National Incident Management System is used when:
   A. managing any incident.
   B. there are more than four mutual aid fire companies at the incident.
   C. the state requests a federal Stafford Act response.
   D. an incident receives a presidential disaster declaration.

2. Fire departments are authorized to provide services by:
   A. evaluation from the International Association of Fire Chiefs.
   B. registration through the Center for Public Safety Services.
   C. a local or regional governmental entity.
   D. recognition by the Department of Homeland Security.

3. When determining if a decision is ethical, you should answer which of these questions?
   A. Will the impact of this decision generate a desired change?
   B. Does the decision help the fire department meet its objective?
   C. Does the decision meet a personal need or objective?
   D. Would I mind if the newspaper ran it as a headline story?

4. When each fire fighter answers to only one supervisor and each supervisor answers to only one boss, this describes:
   A. span of control.
   B. unity of command.
   C. professional discipline.
   D. division of labor.

Fire Lieutenant Activity

NFPA Fire Officer I Job Performance Requirement 4.4.4

Explain the purpose of each management component of the organization, given an organization chart, so that the explanation is current and accurate and clearly identifies the purpose and mission of the organization.

Application of 4.4.4

Identify who should be initially contacted for the following situations:

1. A fire fighter has deployment orders from his military reservist unit.
2. A fire fighter's CPR card expires next month.
3. The tip-to-base intercom system is not working.
4. You encounter an unpermitted business using over 100 gallons of flammable liquid in a residential townhouse.
5. A fire fighter fails to report for a shift exchange.

Access Navigate for flashcards to test your key term knowledge.

CHAPTER 2

Fire Officer I

Understanding Leadership and Management Theories

KNOWLEDGE OBJECTIVES

After studying this chapter, you will be able to:

- Compare and contrast leadership and management. (**NFPA 1021: 4.1.1**) (pp. 40–41)
- Identify leadership theories used in the fire service. (**NFPA 1021: 4.2.1, 4.2.2, 4.2.6**) (pp. 41–47)
- Explain the relationship between leadership and followership. (**NFPA 1021: 4.1.1**) (p. 47)
- Identify types of power used in leadership. (p. 47)
- Describe the leadership challenges related to the fire station work environment. (pp. 47-49)
- Describe the concept of the fire station as a business work location. (pp. 47–48)
- Describe the leadership challenges related to the volunteer fire service. (pp. 48–49)
- Define *human resources management*. (pp. 49, 51)
- Identify fire department human resources management functions. (**NFPA 1021: 4.2.6**) (pp. 49, 51–52)
- Explain the purpose of a mission statement. (**NFPA 1021: 4.4.4**) (p. 52)
- Define *delegation*. (p. 55)
- Explain the steps in effective delegation. (**NFPA 1021: 4.2.6**) (pp. 55–56)
- Explain the origins of crew resource management (CRM). (p. 56)
- Identify human factors that contribute to errors. (p. 57)
- Explain the Swiss cheese model of how defenses, barriers, and safeguards may be penetrated by an accident trajectory. (p. 57)
- Differentiate between active failures and latent conditions. (pp. 57–58)
- Define *error management*. (p. 58)
- Describe the six-point CRM model that can be used in the fire service. (pp. 58–66)
- Describe the benefits of a CRM approach. (p. 61)
- Differentiate between insubordination and assertive statements. (**NFPA 1021 4.2.6**) (p. 59)
- Compare and contrast the naturalistic decision-making and recognition-primed decision-making (RPD) models. (p. 62)
- Identify six steps to maintain situational awareness. (pp. 62–64)
- Explain the purpose of a postincident analysis. (**NFPA 1021: 4.6.3**) (p. 64)
- Identify the critical information to be gathered during a postincident analysis. (**NFPA 1021: 4.6.3**) (p. 65)
- Describe the role of the moderator in conducting a postincident analysis. (**NFPA 1021: 4.6.3**) (pp. 64–66)
- Describe documentation considerations for a postincident analysis. (**NFPA 1021: 4.6.3**) (p. 66)

SKILLS OBJECTIVES

After studying this chapter, you will be able to:

- Conduct a postincident analysis of a single company incident. (**NFPA 1021: 4.6.3**) (pp. 64–66)

You Are the Fire Officer

The battalion chief is conducting a 6-month probationary assessment of a lieutenant. The assessment includes a review of fire company performance as well as in-station activities. The chief notes that few target hazard walk-throughs were completed within in the district, and there was a reduced performance by the fire company crew on standardized evolutions at a recent fire academy evaluation.

The evaluation revealed that fire fighters are taking too much time and missing critical elements of fire hose deployment. The lieutenant had delegated the daily in-station training to the senior fire fighter who functions as the acting officer when the lieutenant is not there. The lieutenant is using a "circle of consensus" process that leads to a holistic development of internal satisfaction. The battalion chief is irritated by this response and suggests that the lieutenant focus on the internal satisfaction of the battalion chief by complying with departmental goals: Perform a standardized skill drill every day and a target hazard walkthrough every week.

1. Why are there so many management concepts that provide contradictory information?
2. How do these theories assist you in executing supervisory tasks?
3. Is there a way for you to evaluate a new management concept or procedure before using it at the fire station?

Access Navigate for more practice activities.

Introduction

Leadership is a complex process having multiple dimensions. Peter Northouse observes that scholars and practitioners have attempted to define leadership for more than a century without universal consensus. Four components are central to the leadership concept: (a) Leadership is a process, (b) leadership involves influence, (c) leadership occurs in groups, and (d) leadership involves common goals (Northhouse, 2013). Gary Yukl describes leadership as "the process by which a person influences others to understand and agree about what needs to be done and how to do it, and the process of facilitating individual and collective efforts to accomplish shared objectives" (Yukl, 2013).

This chapter discusses leadership and management theories, characteristics, and challenges. Crew resource management, a vital leadership responsibility, is also discussed.

Introduction to Leadership and Management

Today's fire officer is faced with challenges of reduced municipal funding, declining volunteer participation, unprecedented numbers of responses to medical emergencies, violent acts that create mass-casualty incidents, record-breaking wildland fires, new elements of our built environment that are producing large-loss fires, and hazards that impact fire fighter health and well-being. The fire officer needs to expand his or her personal leadership toolbox.

John Kotter first looked at leadership by studying big-city mayors. He described the **leadership** process as producing movement through three subprocesses:

1. Establishing direction: developing a vision of the future, often the distant future, along with strategies for producing the changes needed to achieve that vision
2. Aligning people: communicating the direction to those whose cooperation may be needed so as to create coalitions that understand the vision and that are committed to its achievement
3. Motivating and inspiring: keeping people moving in the right direction despite major political, bureaucratic, and resource barriers to change by appealing to very basic, but often untapped, human needs, values, and emotions

Kotter defines management as the process that ". . . create[s] orderly results which keep something working efficiently" (Kotter, 1990). Both leadership and management are needed to make a successful organization. **Management** looks at short time frames—a few months to the next fiscal year. Leaders are looking at longer time frames. These roles and responsibilities differ. (See **TABLE 2-1**.)

TABLE 2-1 The Difference between Management and Leadership

Management: Order and Consistency	Leadership: Change and Movement
Planning and budgeting • Establish agendas • Set timetables • Allocate resources	Establish direction • Create a vision • Clarify the big picture • Set strategies
Organizing and staffing • Provide structure • Make job placements • Establish rules and procedures	Aligning people • Communicate goals • Seek commitment • Build teams and coalitions
Controlling and problem solving • Develop initiatives • Generate creative solutions • Take corrective action	Motivating and inspiring • Inspire and energize • Empower others • Satisfy unmet needs

Modified from Kotter, J. P. (1990). *A Force For Change: How Leadership Differs from Management.* New York: The Free Press. Pages 3-8.

FIRE OFFICER TIP

Digging through the Management Toolbox

A new fire officer should consider senior fire officers to be valuable resources when developing his or her own toolbox of theories, concepts, and practices. Consider which theory, concept, or practice an admired and experienced fire officer would use in the same situation. If possible, ask senior fire officers for suggestions or feedback when considering which management tool to use to resolve an issue.

Managers accomplish activities and master routines, whereas leaders influence others and create visions for change. The difference is summarized by this statement: "Managers are people who do things right and leaders are people who do the right thing" (Bennis and Nanus, 1985).

Leadership and Management Theories

Theories of leadership continue to evolve and proliferate as the human condition changes. It is hard to keep track of the myriad theories and practices. For this textbook we are using the classification approach developed in Peter Northouse's textbook, *Leadership: Theory and Practice* (Northouse, 2013). The one component common to nearly all classifications is that leadership is an influence process that assists groups of individuals toward goal attainment. Leadership is defined as a process whereby one person influences a group of individuals to achieve a common goal.

Peter Northouse's Classification Approach

The Trait Approach

What type of person becomes a leader? Can we identify traits that lead to effective leadership? Northouse summarized 60 years of research into five major leadership traits:

- Intelligence
- Self-confidence
- Determination
- Integrity
- Sociability

Although the **trait approach** has benefited from decades of research and provided benchmarks for the development and identification of leaders, it exists in a vacuum. The trait approach does not look at the situational effects of leadership on the group. An example of **situational leadership** is the response by Mayor Rudy Giuliani to the 2001 terrorist attack in New York City (Forbes, 2011). In the chaos of the first hours after the attack, the mayor was visible, composed, vocal, and resilient. The fire officer needs to adopt the appropriate leadership style for the specific situation. "In times of consuming trauma, psychologists and historians say, a leader must speak with a trusted voice and sketch honestly the painful steps to safety" (Powell, 2007).

The Skills Approach

The **skills approach** also exclusively focuses on the leader, but it identifies a set of skills that can be developed. Robert Katz identified a Three-Skill Approach for personal development (Katz, 1955):

1. Technical
2. Human
3. Conceptual

FIGURE 2-1 shows the relationship of the three skills to the level of leadership.

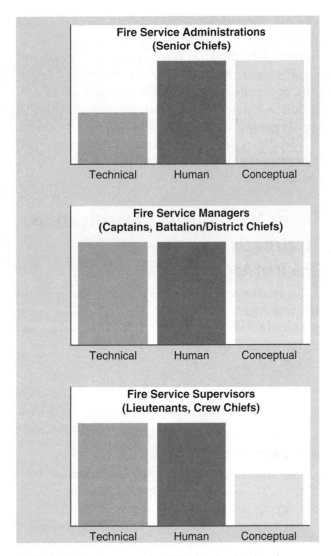

FIGURE 2-1 Amount of skill by leadership level.

The Behavioral Approach

The behavioral approach is different from the skills approach. The behavioral approach emphasizes the personality characteristics of the leader, and the skills approach emphasizes the capabilities of the leader. The **behavioral approach** focuses on what leaders do and how they act.

The behavioral approach falls into two general kinds of behaviors: task behaviors and relationship behaviors. *Task behaviors* facilitate goal accomplishment: They help group members to achieve their objectives. *Relationship behaviors* help followers feel comfortable with themselves, with each other, and with the situation in which they find themselves. The central purpose of the behavioral approach is to explain how leaders combine these two kinds of behaviors to influence followers in their efforts to reach a goal (Northouse, 2013).

Blake and Mouton's Managerial Grid

In the early 1960s, the Exxon Corporation hired behavioral scientists Robert Blake and Jane Mouton to perform a series of experiments designed to increase leadership effectiveness (Blake, Mouton, and Bidwell, 1962). In the 1990s, the grid theory that these researchers developed was applied to crew management of aircraft and aerospace teams (Blake and Mouton, 1982). This concept was later adopted by the fire service as "crew resource management," a method used to improve incident safety.

The grid theory assumes that every decision made and every action taken in the workplace are driven by people's values, attitudes, and beliefs. At the individual level, these values are based on two fundamental concerns that influence behavior—a concern for people and a concern for results. Blake and Mouton developed a survey document with 35 questions that measures a person's level of concern in each area. The results are plotted on an X-Y chart. Blake and Mouton described five behavioral models, or management styles, based on a person's position on this grid (**FIGURE 2-2**).

Impoverished Management

The impoverished, or indifferent, management style represents the lowest level of concern for both results and people. The key word for this style is *neutral*. An individual who demonstrates this style is the least visible person in a team; he or she is a follower who maintains distance from active involvement whenever possible. An indifferent manager carefully goes through the motions of work, doing enough to get by but rarely making a deliberate effort to do more.

The stereotypical image of this personality is the minion in a bureaucratic government agency where everyone is treated like a number. This sort of workplace allows the person to blend in without attracting attention. In fact, he or she often seeks work that can be done in isolation so as to carry on without being disturbed or noticed.

The indifferent manager relies heavily on instructions and process, depending on others to outline what needs to be done. Reliance on instructions avoids the need to take personal responsibility for results. If problems arise, the indifferent manager is often content to ignore or overlook them, unless the instructions specify how to react to that particular problem. The indifferent manager might point out the problem to someone else, but would be unlikely to offer a solution. With no instructions, the indifferent manager simply carries on with the attitude that "It is not my problem."

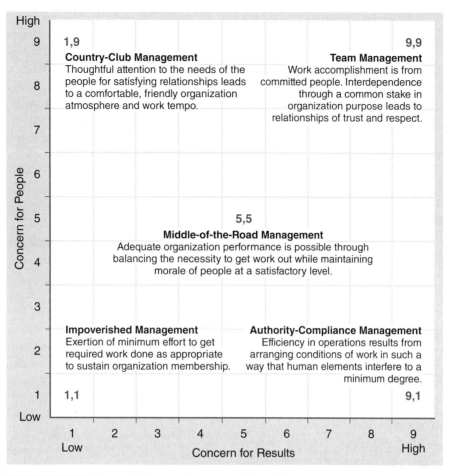

FIGURE 2-2 Blake and Mouton's managerial grid.
Data from Blake, Robert R. and Anne Adams. 1991. *Leadership Dilemmas—Grid Solutions*. Houston, TX: Gulf Professional Publications.

Authority-Compliance Management

An authority-compliance (or controlling) manager demonstrates a high concern for results, but a low concern for others. The high concern for results brings determination, focus, and drive for success. Consequently, such a person is usually highly trained, organized, experienced, and qualified to lead a team to success. Unfortunately, the low concern for others prevents the controlling manager from being aware of others involved in an activity, beyond what is expected of them in relation to results. The controlling manager expects everyone else to keep up with his or her efforts, and so moves ahead, intensely focused on results, often leaving others lost in the wake of his or her forceful initiative.

This type of leader is demonstrating the **autocratic** leadership style. In the context of the fire service, the autocratic style of leadership is required in two situations. The first situation occurs when the fire company is involved in a high-risk, emergency scene activity such as conducting a primary search during a structure fire. There is no time for discussion, and this is not the situation to experiment with alternative approaches.

The second situation occurs when the fire officer needs to take immediate corrective supervisory activity, such as during a "control, neutralize, command" response to a confrontation. In this scenario, the officer must be firmly in control of the situation.

Country Club Management

The country club (or accommodating) manager demonstrates a low concern for results and a high concern for other people. Such an individual maintains a heightened awareness of the personal feelings, goals, and ambitions of others, and always considers how proposed actions will affect them. The accommodating manager is approachable, fun, friendly, and always ready to listen with sympathy and encouragement.

Some cornerstone phrases of the accommodating attitude are "Let's talk about it," "What can I do to help?", and "Let me know what you think." The main weakness in this behavior lies in the focus of the ensuing discussions. Discussions with an accommodating manager tend to include an overwhelming emphasis on personal feelings and preferences, while avoiding concrete issues. The discussion itself becomes the goal, so conversations can meander in any direction instead of concentrating on solving the problem.

The controlling (authority-compliance manager) and accommodating (country club manager) styles are diametrically opposed in their perspectives. Each of these orientations leads in a narrow and singularly focused manner by ignoring the other primary concerns in the workplace.

Middle-of-the-Road Management

The middle-of-the-road, or status quo, manager believes there is an inherent contradiction between the concerns for results and for people but does not value one concern over the other. Instead, the status quo manager sees a high level of concern for either people or results as too extreme and tries to moderate both in the workplace.

The objective of the status quo manager is to play it safe and work toward acceptable solutions that follow proven methods. Such a politically motivated approach seeks to avoid risk by maintaining the tried-and-true course, following popular opinion and norms without pushing too hard in any direction.

Another key aspect of the status quo approach is the emphasis on maintaining popular status within the team and organization. This type of manager must be intelligent and informed enough to persuade people and companies to settle for a compromise—often less than they want and less than they could achieve. This requires being well liked, staying well informed, and effectively convincing people that the consequences are not worth the risk. On the surface, this might make the status quo manager appear unbiased and impartial, but the status quo approach actually represents a narrow view that underestimates people, results, and the power of change.

The status quo manager is demonstrating a **laissez-faire** leadership style, that moves the decision making from the fire officer to the individual fire fighters. The fire officer depends on the fire fighters' good judgment and sense of responsibility to get things done within basic guidelines. The laissez-faire approach is an effective leadership style when working with experienced fire fighters and when handling routine duties that pose little personal hazard.

Team Management

The team management (or sound) manager sees no contradiction in demonstrating a high concern for both people and results at the same time. He or she feels no need to restrain, control, or diminish the concerns for both people and results in a relationship. The consequence is a freedom to test the limits of success with enthusiasm and confidence. The sound attitude leads to more effective work relationships based on "what's right" rather than "who's right."

The team management model is preferred for a candidate who seeks to become a successful fire officer. The full integration of concerns for both people and results stands in contrast to the outcomes associated with the other styles. Each of the other models represents a weakness in one or both critical areas. The controlling (authority-compliance) manager feels that a high concern for results is more important than a high concern for people. The accommodating (country club) manager feels the reverse—namely, that a high concern for people is more important than results. The person employing the status quo (middle-of-the-road) approach feels that a high concern for either people or results is too risky, and prefers to maintain the current situation, holding the safe middle ground. The indifferent (impoverished) manager sees any major concern for people or results as unrealistic and too demanding.

The team management model demonstrates the **democratic** leadership style. A consultative approach takes advantage of all of the ingenuity and resourcefulness of the group in determining how to meet an objective or complete a task (**FIGURE 2-3**). The officer should use the democratic style of leadership when planning a project or developing the daily work plan of the company. This approach can also be used in some low-risk emergency scene operations.

Specialized and highly technical fire companies often use the democratic leadership approach when

FIGURE 2-3 The democratic leadership style uses the resourcefulness of the group in determining how to meet an objective.
© Jones & Bartlett Learning. Photographed by Glen E. Ellman.

faced with a complex or unusual emergency situation. They depend on the skills and experience of the individual team members to analyze the situation, consider the alternatives, and develop the incident action plan. Execution of the plan, however, often involves an autocratic command style.

> **FIRE OFFICER TIP**
>
> **Transitioning from Democratic to Autocratic Leadership**
>
> The increased deployment of specialized and highly technical emergency services creates a unique challenge for the fire officer. The best way to size up a complicated situation and develop an incident action plan is to use the democratic style of leadership, using the knowledge and experience of all responders to develop the best approach. Once the operation begins, however, the fire officer needs to assume the autocratic role to ensure the safe execution of the plan.

Situational Leadership

Building upon William Reddin's 3-D leadership concept (Reddin, 1967), Paul Hersey and Kenneth H. Blanchard introduced the concept of situational leadership in the context of managing organizational behavior (Hersey and Blanchard, 1969). The situational leadership approach stresses that leadership is composed of both a directive and a supportive dimension. To determine what is needed in a particular situation, a leader must evaluate her or his followers and assess how competent and committed they are to perform a given goal (Northouse, 2013). Based on the assumption that followers' skills and motivation vary over time, situational leadership suggests that leaders should change the degree to which they are directive or supportive to meet the changing needs of followers. Effective leaders are those who can recognize what followers need and then adapt their own style to meet those needs (Northouse, 2013).

Due to its direct and practical approach, situational leadership is popular in leadership training and organizational consulting venues. There are four distinct categories of directive and supportive behavior:

- S1, or high directive-low supportive, or "directing" style: Leader focuses on goal achievement by providing direct instruction on what is to be done, under close supervision, using very little supportive behavior.
- S2, or high directive-high supportive, or "coaching" style: Leader focuses on both the goal and meeting the follower's socioemotional needs. The leader gives encouragement and solicits follower input. The leader retains the S1 requirement to make the final decision on the what and how of goal achievement.
- S3, or high supportive-low directive, or "supporting" style: The leader uses supportive behaviors to bring out the follower's skills around the goal to be accomplished.
- S4, or low supportive-low directive, or "delegating" style: The group agrees on the goal, the leader allows the followers the responsibility to accomplish the goal. (Data from Northouse, 2013.)

While popular, situational leadership has few peer-reviewed research studies. The ambiguous conceptualization of the development models makes it difficult to perform direct analysis on results.

The fire officer needs to adopt the appropriate leadership style for the specific situation. During nonemergency situations, a participative leadership approach can develop group cohesiveness and productivity. In such a case, there is time for discussion and feedback before a decision is made. The consequences seldom involve imminent risk of injury or death.

In contrast, time is often of the essence at an emergency incident. Decisions are needed quickly and actions must be implemented in a timely manner. These situations do not allow a long discourse on what needs to be done or development of a group consensus. The fire officer must frequently make decisions with little or no input from subordinates, and every member of the company must be prepared to perform a specific set of tasks with a minimum of instruction.

When the fire officer gives an order to lay a supply line, the fire fighter who is assigned to catch the hydrant must know exactly what to do. While the fire fighter is dragging the hose around the hydrant, the fire officer should be gathering information to develop the next step in the incident action plan. The other fire fighter, who will be advancing a fire attack line, can anticipate which task comes next without needing a lengthy explanation from the fire officer. Success using an authoritative style comes when the officer has developed the trust and confidence of his or her subordinates before the incident. Officers who demonstrate sound knowledge of the technical aspects of firefighting and demonstrate trustworthiness are well prepared to lead fire fighters into a high-hazard task. Fire fighters do not trust the judgment of a fire officer who does not fully understand the job and has not made fire fighter safety a priority.

Transformational Leadership

Transformational leadership gives more attention to the charismatic and affective elements of leadership. It is a process that changes and transforms people. It is concerned with emotions, values, ethics, standards, and long-term goals. It includes assessing followers' motives, satisfying their needs, and treating them as full human beings (Northouse, 2019). Leaders demonstrate charismatic and visionary leadership in five ways:

1. Providing strong role models for the beliefs and values they want their followers to adopt
2. Being charismatic leaders who appear competent to their followers
3. Articulating ideological goals that have moral overtones
4. Communicating high expectations for followers and exhibit confidence in the follower's ability to meet these expectations
5. Arousing task-relevant motive in followers that may include affiliation, power, or esteem (Data from Northouse, 2013)

Jim Kouzes and Barry Posner interviewed 1300 senior- and middle-level managers in public and private sector organizations. They were able to identify five fundamental practices that enable transformational leaders to get extraordinary things accomplished (Kouzes and Posner, 2017).

1. Model the way
2. Inspire a shared vision
3. Challenge the process
4. Enable others to act
5. Encourage the heart

Transformational leadership utilizes a 360-degree feedback process and focuses on developing a compelling vision for the leader.

Authentic Leadership

A recent area of leadership research interest, authentic leadership does not have a clear definition. One viewpoint looks at the intrapersonal perspective, looking at the leader's self-knowledge, self-regulation, and self-concept. Additionally, there is a developmental perspective that looks at authentic leadership as being nurtured by the leader. Bill George has identified five dimensions of authentic leaders (George, 2003):

- Purpose
- Values
- Relationships
- Self-discipline
- Heart

Servant Leadership

Robert K. Greenleaf initiated the concept of servant leader in 1970. **Servant leadership** is an approach that focuses on leadership from the point of view of the leader and his or her behaviors. Servant leadership emphasizes that leaders be attentive to the concerns of their followers, empathize with them, and nurture them. Servant leaders put followers first, empower them, and help them develop their full personal capacities (Northouse, 2013).

Servant leadership has been a focused interest of the International Association of Fire Chiefs (IAFC) for fire officer development. Within the fire service, the dividends of using the servant leadership approach are realized when the leader gets buy-in among the group and furnishes them with a sense of belonging and ownership. A crew that feels their needs are being addressed will perform at a higher level than one that feels their leader is only thinking of himself or herself. The most valuable outcome is increased trust between the crew and the officer and between those who lead and those who are led. That trust is displayed on every shift by increased ownership: crew members fixing things they did not break, picking up trash that is not theirs, and finishing jobs left by the previous shift because they got dispatched to a late call for service (Jester, 2018).

Robert Greenleaf said listening is the premier skill of the servant leader:

> I have a bias which suggests that only a true natural servant automatically responds to any problem by listening first. When he is a leader, this disposition causes him to be seen as servant first. This suggests that a nonservant who wants to be a servant might become a natural servant through a long arduous discipline of learning to listen, a discipline sufficiently sustained that the automatic response to any problem is to listen first. (Greenleaf, 1970/2002)

Don Frick provides this servant leader listening checklist:

- Do my body and face show that I am involved in the conversation and interested in what the person is saying?
- Am I interrupting or hurrying the person along?
- Am I asking appropriate, open questions to draw the person out?
- Am I using my own words to clarify the person's message and reflect his or her feelings?

- Am I not judging, criticizing, analyzing, or trying to fix the person?
- Am I responding to feedback in a non-defensive manner? (Sipe and Frick, 2015)

Adaptive Leadership

Adaptive leadership is how leaders encourage people to adapt when confronted with problems, challenges, and changes. Adaptive leadership focuses on the activities of the leader in relation to the work of the followers when confronted with a technical or adaptive challenge.

Based on the work of Ronald Heifetz and colleagues, there are six leadership behaviors of adaptive leadership:

1. Get on the balcony.
2. Identify the adaptive challenge.
3. Regulate distress.
4. Maintain disciplined attention.
5. Give the work back to the people.
6. Protect leadership voices from below. (Heifetz, Linsky, and Grashow, 2009)

Adaptive leadership is a complex process that includes situational challenges, leader behaviors, and adaptive work.

Followership

Followership is a process whereby an individual or individuals accept the influence of others to accomplish a common goal. Followership involves a power differential between the follower and the leader (Yukl, 2013). Typically, followers comply with the directions and wishes of leaders—they defer to leaders' power.

The fire officer has to be both a leader and a follower. On the one hand, the officer leads the fire company to achieve the goals and objectives that have been established by the department or the jurisdiction. On the other hand, the officer has to follow leadership that comes from a higher level, even if it is not always pointing in a direction where the officer would prefer to go. In many cases, the officer is the messenger who has to deliver unpopular news to the company members to ensure compliance with instructions that came from a higher level. Chapter 1, *The Fire Officer I as a Company Supervisor,* discusses how the fire officer handles an unpopular order.

Followership is particularly important for a fire officer because subordinates are always aware of what the fire officer does. If the fire officer demonstrates selective following of orders from the fire chief, deciding which ones to follow and which ones to ignore, the fire officer sends a clear message to the company members: It is acceptable for the fire fighters to be selective in following orders from their fire officer. Followership is an important character trait that will serve the officer well in the future when dealing with others who have not followed the rules because they do not seem fair.

Power as a Leadership Resource

Power is the capacity of one party to influence another party. Social power is described as the result of the "target person['s]" response from the "agent" making a request (French and Raven, 1959).

- Legitimate power: The target person believes that the agent has the right to make the request and the target person has the obligation to comply. For example, under the Incident Management System, the incident commander has the legitimate power to reassign the ventilation sector.
- Reward power: The target person complies to obtain rewards believed to be controlled by the agent.
- Expert power: The target person complies due to a belief that the agent has special knowledge.
- Referent power: The target person complies due to admiration of or identification with the agent and seeks approval.
- Coercive power: The target person complies to avoid punishment believed to be controlled by the agent.

Yukl updated the French and Raven taxonomy to define two types of power: personal and positional. **Personal power**, which includes expert and referent power, reflects the effectiveness of the individual. **Positional power**, in contrast, is defined by the role an individual has within the organization. Legitimate, reward, and coercive power are the three examples of positional power. Yukl provides two additional position-based power descriptions (Yukl, 2013):

- Information power: Control over information. Unlike expert power, information power is based on the target person's assessment of the agent's ability to discover or obtain relevant information rapidly and efficiently, usually through a cultivated network of sources.
- Ecological power: Control over the physical environment, technology, or organization of work. The target person's behavior is based on perceptions of opportunities and constraints.

Leadership Challenges

The fire officer works in a dynamic environment with changing conditions and evolving organizational needs. The basic leadership concepts prevail when considering two unique challenges: the fire station as a work location and leading a volunteer fire company.

Fire Station as Municipal Work Location versus Fire Fighter Home

Fire fighters work rotating shifts together and prepare and share meals. There are days of dull routine, punctuated by episodes of intense excitement. These factors create a powerful and special community among fire fighters that is difficult to compare with any other workplace environment. The fire station becomes a "home away from home," and fire fighters become part of an extended family.

Although this workplace environment produces a special type of bonding among fire fighters, it can easily lead to a variety of productivity problems, as well as behavioral and personality traits that are often associated with a dysfunctional family. The resulting situations can be quite different from the problems that occur in a "normal" work environment, and the behavioral issues are often more severe. Case law and administrative actions reinforce the notion that the fire station is a local government facility, subject to the same rules and expectations as any other workplace. The formal organization expects fire fighters who are in the station at 3:00 AM to behave the same way as administrative services staff behaves at 3:00 PM.

A fire officer must balance the expectations of the employer with the realities of a fire station work environment and the desire to create an effective team. A certain amount of spirited behavior can be healthy, as long as basic rules are observed. The fire officer has to maintain order and ensure that whatever occurs can be explained and would be acceptable to a rational observer. The fire officer accomplishes this by taking the following steps:

- Educate the employees on the workplace rules and regulations that define expected behavior: Start with the local government's "code of conduct" or other documents that outline the chief administrative officer's expectations for all municipal employees. This information can usually be found in the municipality's mission statement, core values, or personnel regulations.
- Promote the use of "on-duty speech": The goal is not to change the thoughts or feelings of individual fire fighters, but rather to establish a workplace environment where certain behaviors and words are not used. Fire fighters can think what they want. However, while they are on duty, in the fire station, or in uniform, they cannot use certain words or phrases or act out certain behaviors.
- Be the designated adult: This action requires the fire officer to model appropriate behavior as well as encourage and enforce the same behavior by the fire fighters. The fire officer must identify and correct unacceptable workplace behavior whenever it is observed. Ignoring a problem is, in reality, permitting it to continue. The fire officer who is a candidate for promotion is expected to identify, explain, and enforce the limits of unprofessional behavior.

A company-level officer should make it a practice to walk around the fire station at various times during the workday to observe what is going on. This walk-around is more important when the officer is in a big station with multiple companies and in combination career–volunteer departments, where there is a constant flow of people coming into and out of the fire station. This practice is not designed to catch someone doing something wrong, but rather is intended to make sure that everything is functioning properly. The officer should routinely determine what the crew members are doing and check on the safety of the facility and equipment. At the same time, the officer should look out for unexpected situations and surprises. Having the reputation of knowing what is going on in the station and reacting to inappropriate situations goes a long way toward encouraging appropriate workplace behaviors.

FIRE OFFICER TIP

Demonstrating Values and Building Teams
Responding to emergencies and preserving the public trust are two values that every fire fighter needs to demonstrate. Within those values is wide latitude for team-building extracurricular activities.

Leadership in the Volunteer Fire Service

Fire officers leading volunteer fire companies have to rely on their leadership skills even more than their municipal counterparts do. Pride, group identity, and personal commitment are key factors that keep volunteers active and loyal to the organization—there is no paycheck that compels a volunteer to endure an unpleasant situation. Consequently, the volunteer fire officer must pay attention to the satisfaction level of every member, and be alert for issues that create conflict or frustration.

The relationship between the volunteer fire officer and fire fighters is more like that of a "mom-and-pop" family business than that of a municipal agency. Effective leadership is often the strongest force that influences members' performance and commitment to the organization. If the negative aspects outweigh the positive aspects, an individual can step aside or drop out.

Among the unique issues that a volunteer fire officer must consider are the following:

- Changes in employment, family situations, or child/elder care can profoundly affect the time a volunteer is able to devote to the fire department.
- Extensive training requirements can have an impact on volunteer availability. It is easy to assign a municipal employee to attend a 40-hour training class that occurs during the regular workweek, but it can prove difficult to accomplish the same training when the opportunities to do so are restricted to evenings and weekends.
- Interpersonal conflicts can develop between members. Conflicts can drive away members and erode fire company preparedness. The volunteer officer must act quickly when such problems are identified. The rights and reasonable expectations of individual members must be considered, without compromising the mission or the good of the organization.

A fire commissioner serving in a large county observed four phases of volunteer participation:

1. **Large loss of applicants during initial fire fighter training:** The candidates could not make the time commitment, were physically unprepared, or could not pass the cognitive certification exam.
2. **Small loss during the probationary period:** During the first year after joining the department, members are typically consistent participants in emergency and administrative activities.
3. **Moderate to high loss of fire fighters between the third and sixth years of membership:** The twenty-somethings are finishing college, getting married, having children, and building their careers. The ones who do not leave cannot devote as much time to the volunteer service as they once did.
4. **Recommitment between the 15th and 18th years of membership:** Those volunteer fire fighters who never left significantly increase their time with the department, and many who left start returning. The children are grown; the career is established. The commissioner noted that this group forms the core of the volunteer fire department.

According to this fire commissioner, his county reported working with a total of 1000 volunteers annually. It processed 250 applications per year and lost 250 members per year. Six hundred members had certification to operate as fire fighters, apparatus operators, company commanders, and chief officers. That group averaged 36 hours of service per volunteer per month. Within that group were 125 individuals who were the pillars of the volunteer organization, providing 80 to 200 hours of service per month.

A special concern in volunteer organizations is the political balance that results from electing officers. A conscientious volunteer officer has to use strong leadership and the courage of conviction to implement an unpopular policy.

U.S. Marine Corps Leadership Principles

The fire service is a decentralized organization, with fire stations spread through the community and a hierarchical command system. Fire companies respond from all corners of the community and assemble to handle an emergency incident. Work groups or task forces are created to resolve an organizational problem or handle a temporary need. These activities create unique leadership needs that are similar to military small group teams. To serve as an example, the U.S. Marine Corps provides a list of leadership principles and leadership traits that the fire officer may find worthwhile. **TABLE 2-2** lists the leadership principles; **TABLE 2-3** lists the leadership traits.

Fourteen U.S. Marine Corps Leadership Traits

The military expectation of leaders includes accomplishing the mission and training the next generation of leaders. The Marines identify 14 leadership traits as qualities of thought and action which, if demonstrated in daily activities, help Marines earn the respect, confidence, and loyal cooperation of other Marines.

TABLE 2-2 Eleven U.S. Marine Corps Leadership Principles	
Leadership Principle	Description
Be technically and tactically proficient	A technically and tactically proficient Marine knows his or her job thoroughly and possesses a wide field of knowledge. Before you can lead, you must be able to do the job. Tactical and technical competence can be learned from books and from on-the-job training.
Know yourself and seek self-improvement	This principle of leadership should be developed by the use of leadership traits. Evaluate yourself by using the leadership traits and determine your strengths and weaknesses. You can improve yourself in many ways.
Know your Marines and look out for their welfare	This is one of the most important of the leadership principles. A leader must make a conscientious effort to observe Marines and how they react to different situations. A Marine who is nervous and lacks self-confidence should never be put in a situation that requires an important decision. This knowledge will enable you as the leader to determine when close supervision is required.
Keep your Marines informed	Marines by nature are inquisitive. To promote efficiency and morale, a leader should inform the Marines in his or her unit of all happenings and give reasons why things are to be done. This is accomplished only if time and security permits. Informing your Marines of the situation makes them feel that they are a part of the team and not just a cog in a wheel. Informed Marines perform better. The key to giving out information is to be sure that the Marines have enough information to do their job intelligently and to inspire their initiative, enthusiasm, loyalty, and convictions.
Set the example	A leader who shows professional competence, courage, and integrity sets high personal standards for himself or herself before he or she can rightfully demand it from others. Your appearance, attitude, physical fitness, and personal example are all on display daily for the Marines and Sailors in your unit. Remember, your Marines and Sailors reflect your image!
Ensure assigned tasks are understood, supervised, and accomplished	Leaders must give clear, concise orders that cannot be misunderstood, and then closely supervise to ensure that these orders are properly executed. Before you can expect your Marines to perform, they must know what is expected of them. The most important part of this principle is the accomplishment of the mission.
Train your Marines as a team	Teamwork is the key to successful operations and is essential from the smallest unit to the entire Marine Corps. As a leader, you must insist on teamwork from your Marines. Train, play, and operate as a team. Be sure that each Marine knows his or her position and responsibilities within the team framework.
Make sound and timely decisions	The leader must be able to rapidly estimate a situation and make a sound decision based on that estimation. Hesitation or a reluctance to make a decision leads subordinates to lose confidence in your abilities as a leader. Loss of confidence in turn creates confusion and hesitation within the unit.
Develop a sense of responsibility among your subordinates	Another way to show your Marines you are interested in their welfare is to give them the opportunity for professional development. Assigning tasks and delegating authority promotes mutual confidence and respect between leader and subordinates. It also encourages subordinates to exercise initiative and to give wholehearted cooperation in accomplishment of unit tasks. When you properly delegate authority, you demonstrate faith in your Marines and increase authority, and you increase their desire for greater responsibilities.

Leadership Principle	Description
Employ your command in accordance with its capabilities	A leader must have a thorough knowledge of the tactical and technical capabilities of the command. Successful completion of a task depends upon how well you know your unit's capabilities. If the task assigned is one that your unit has not been trained to do, failure is very likely to occur. Failures lower your unit's morale and self-esteem. Seek out challenging tasks for your unit, but be sure that your unit is prepared for and has the ability to successfully complete the mission.
Seek responsibility and take responsibility for your actions	For professional development, you must actively seek out challenging assignments. You must use initiative and sound judgment when trying to accomplish jobs that are required by your grade. Seeking responsibilities also means that you take responsibility for your actions. Regardless of the actions of your subordinates, the responsibility for decisions and their application falls on you.

Sustaining the Transformation, U.S. Marine Corps, Department of Navy.

TABLE 2-3 Fourteen U.S. Marine Corps Leadership Traits

Leadership Trait	Description
Justice	Giving reward and punishment according to merits of the case in question. It is also the ability to administer a system of rewards and punishments impartially and consistently.
Judgment	The ability to weigh facts and possible solutions on which to base sound decisions.
Dependability	The certainty of proper performance of duty.
Initiative	Taking action in the absence of orders.
Decisiveness	Ability to make decisions promptly and to announce them in a clear, forceful manner.
Tact	The ability to deal with others without creating offense.
Integrity	Uprightness of character and soundness of moral principles. Integrity includes the qualities of truthfulness and honesty.
Enthusiasm	The display of sincere interest and exuberance in the performance of duty.
Bearing	Creating a favorable impression in carriage, appearance, and personal conduct at all times.
Unselfishness	Avoidance of providing for one's own comfort and personal advancement at the expense of others.
Courage	The mental quality that recognizes fear of danger or criticism, but enables a Marine to proceed in the face of it with calmness and firmness.
Knowledge	The range of one's information, including professional knowledge and an understanding of your Marines.
Loyalty	The quality of faithfulness to a Marine's country, corps, unit, seniors, subordinates, and peers.
Endurance	The mental and physical stamina measured by the ability to withstand pain, fatigue, stress, and hardship.

Sustaining the Transformation, U.S. Marine Corps, Department of Navy.

> **FIRE OFFICER TIP**
>
> **Managing Different Generations**
> Today's fire officer is tasked with managing four different generations, all with different needs.
> - **1997 to present: Post-millennials, or Generation Z**
>   - Most independent and success oriented
>   - Grew up with the Internet, will research issues or problems
> - **1981 to 1996: Millennials, or Generation Y**
>   - Want to know the "why"
>   - Not afraid to tell anyone what they think or feel
> - **1965 to 1980: Thirteeners, or Generation X**
>   - Early technology adaptors
>   - Independent and productive
> - **1946 to 1964: Baby Boomers**
>   - Goal-oriented
>   - Conflict avoiders

Human Resources Management

Most fire officers will find that their greatest challenge relates to managing people. It is the workers who get the job done. The managing or supervising fire officer is responsible for performing a set of functions that direct and coordinate these workers' efforts, provide them with the necessary tools and resources, and ensure that the outcome meets the standards. Fire department operations are labor-intensive ventures, with tasks being completed by skilled workers. To be effective as a manager, a fire officer must develop skills that are directly related to managing human resources.

The concept of management emerged from the Industrial Revolution. The introduction of steam power in the late 1700s led to the creation of large factories in Europe that were staffed with former agricultural and small-town residents who needed direction and supervision.

Human resources management focuses on the task of managing people using physical, financial, and time assets. Although some human resources functions affect career departments more than volunteer departments, all functions are required. Typical human resources management functions include the following:

- Human resources planning
- Employee (labor) relations
- Staffing
- Human resources development
- Performance management
- Compensation and benefits
- Employee health, safety, and security

Human resources planning is the process of having the right number of people in the right place at the right time who can accomplish a task efficiently and effectively. Most often, this includes forecasting future staffing needs and determining how those needs can be met. When a fire department projects the number of retirements for the next year and determines that a new recruit class should begin prior to those vacancies, the department is fulfilling the human resources planning function. Typically, this is not an activity that is conducted at the company level, other than determining needs while developing an incident action plan.

Employee relations include all activities designed to maintain a rapport with the membership. Most often, these actions are associated with working with labor organizations. Fire officers must know and understand all agreements between labor and the fire department. The fire officer should also be aware of the vast number of laws that regulate the relationship between employees and their employer. Ignorance of the law is no excuse, and fire officers who violate the law may quickly find themselves and their department faced with a federal complaint. (Chapter 9, *The Fire Officer II as a Manager*, focuses on this human resource function.)

Staffing is the process of attracting, selecting, and maintaining an adequate supply of labor. The fire service traditionally has been fortunate in its ability to attract applicants. However, with the increase in the services provided and the greater educational requirements, recruitment is becoming more difficult. Although some fire departments have developed programs to actively seek out qualified women and minorities, most fire departments have not.

Staffing also includes labor force reductions. There are many methods of shrinking the fire department's size; most fire departments have started with furloughs and attrition rather than terminations and focused on increasing work hours without increasing pay. The staffing function is typically accomplished at the organizational level.

Human resources development includes all activities to train and educate the employees. This function is heavily dependent on the fire officer at the company level. The development of employees begins when they first arrive at the department. The fire department orients the recruits to the fire department's methods of operation through a recruit class. The process continues once they arrive at their fire station assignment. The fire company officer will usually sit down with each recruit and orient him or her to the job and the ways things are done at the fire station.

Performance management, compensation and benefits, and employee health, safety, and security are

FIRE OFFICER TIP

Assigning a Mentor for Each Rookie

One effective method of orienting a rookie is to select a fire fighter who exemplifies good performance to act as a mentor to the new member. The fire officer asks the experienced fire fighter to become the "big brother" or "big sister" to the recruit. If the fire fighter accepts the responsibility, the recruit is then brought into the discussion so that the recruit knows that the mentor fire fighter is there to help with any questions the recruit might have. This gives the recruit an experienced fire fighter to use as a resource for questions.

---

covered in Chapter 10, *Applying Leadership and Management Theories*.

Utilizing Human Resources

Mission Statement

One basic principle is for the fire officer to know and understand the fire department's mission. Frequently, the fire department's mission is expressed through a written statement. The mission statement is a formal document that outlines the basic reason for the organization's existence and states how it sees itself. It is designed to guide the actions of all employees.

Getting Assignments Completed

One of the greatest demands on the fire officer is to make effective use of time. Fire officers will find that a great number of demands are made on the company's time. These demands include conducting public education, doing inspections, and undertaking other fire prevention efforts. They also include training and education of the crew members, and routine duties such as cleaning the station, doing paperwork, and maintaining the apparatus. And, of course, the company must respond to calls.

Some of these activities are known months in advance; others may require the immediate response of the crew. The daily schedule of some crews is in a constant state of flux due to the call volume. Other crews may only be interrupted occasionally. No matter what the situation, the fire officer can ensure maximum efficiency by applying good time-management skills.

The first duty is to determine which activities are to be completed, when they must be completed, and how long it will take to complete them. The fire officer can then identify what needs to be done during the shift, the week, the month, and the year. Items that must be completed during the shift are a higher priority than those that must be completed next week. For example, completing the daily log is more important than completing a weekly inspection.

Occasionally, there is insufficient time to complete all of the required tasks during the shift. When this occurs, the fire officer must determine the fire department's priorities. This step allows the fire officer to determine which activities must be completed and which will have to wait. For example, meeting the deadline for the employee's timesheets may be a higher priority than getting the fire engine to the service center for an oil change.

Because many activities, such as an emergency call, are not known until they occur, the fire officer must plan ahead and build in the expected interruptions. The fire officer must not wait until the last minute to have inspections completed because he or she may have calls that prohibit this from happening. The sooner the scheduled activity is completed, the more flexibility the fire officer has.

Once the activities have been prioritized and it is determined when they must be accomplished, the fire officer must develop a plan that lays out how the activities will be accomplished. Some activities might require the entire company to be involved in the activity, whereas others might require only a part of the crew to complete them. An example of splitting up the crew to accomplish multiple goals simultaneously would be sending one crew member to the store, while a second member cleans the fire station kitchen, and a third member completes the weekly inventory of supplies.

Having many tasks to coordinate can easily become overwhelming. One method to assist in making sure that activities are accomplished is to place all scheduled events on a monthly calendar. Inspections, public education, and special training should be noted, along with employee leave time. The calendar provides a visual method of tracking upcoming events.

Another method is to create a "daily" file. Within this file a page describes each activity and specifies when it is to be completed. The file is organized from the earliest-due activity to complete to the latest. This method allows the fire officer to determine what needs to be done quickly without having to look through a pile of papers.

One of the best tools to improve time efficiency is **delegation**. Delegation allows subordinates to complete tasks they are capable of performing. These duties should be ones that allow the subordinate to grow. An example is the fire officer delegating the responsibility for ordering supplies to a fire fighter. Delegation allows the fire officer to focus on duties that cannot be delegated, such as performance appraisals.

Once a task is assigned, the fire officer must provide the fire fighter with regular follow-up and feedback.

Voice of Experience

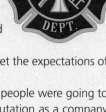

One of the most challenging times in one's career is promotion. I had worked hard to achieve my goal and then it was like the proverbial "dog chasing and catching a car." What was I going to do now?

I had mixed feelings now that I was a company officer: Would I be able to meet the expectations of my fire fighters? Of my fellow officers? Of my chief officers? Of myself?

Now that I was "the guy in charge" I had to come to grips with the fact that people were going to be coming to me for answers, and those answers could make or break my reputation as a company officer and a leader. I was also going to have to give orders, and I was expected to meld several individuals into a unit that could effectively manage details as minor as kitchen duties or as complex as multialarm fires.

To work effectively with fire fighters, company officers should be consistent, fair, and resist micromanaging. If these goals are accomplished, management of the company should be easier.

One of the most important things a new officer can do is to be consistent. Do the same things the same way. Always strive to do the right thing—this will show your fire fighters that your intentions are to be a good company officer.

It is imperative to be fair. Resist the temptation to be everyone's buddy, and resist the temptation to be a taskmaster. Remember, your fire fighters are looking to you to set the standards of the company.

It is easy to micromanage. Resist this at all costs. Let your fire fighters take on new responsibilities and help guide them. Your job as the company officer is to develop new company officers. To get the best out of your company you have to help the members of your company realize their strengths and help correct weaknesses. Never pass on an opportunity to teach and mentor. Don't forget to praise, and don't forget to discipline. You will make mistakes, and how you handle those mistakes will go a long way in how you are perceived as a leader.

After a structure fire or any other incident, I would always bring the company together and start by outlining the mistakes I felt I had made. This helped the other members to open up, and productive discussions would follow. Show your fire fighters that you support and care for them. Take the time to get to know them. This helps when trying to motivate them.

Before my promotion I had composed a list of "What I would do if I were an officer." I made it a habit to refer to this list at least once a month to help keep myself in check. Here is my list:

- Stay real.
- Be fair.
- Do right.
- Listen.
- Don't ask anyone to do anything you wouldn't do.
- Be fast to praise; be slow to discipline.
- Teach.
- Never be satisfied.
- Lead by example.
- Get feedback.

Voice of Experience

My promotion (and subsequent promotions) were always challenging and rewarding personally and professionally, but the biggest reward is having a former fire fighter tell you that it was a pleasure serving in your company.

Stephen McClure
Operations Chief (ret.), Charleston (WV) Fire Department
Director, Jackson County EMS
Ripley, West Virginia

This can take many forms. At the station level, a verbal progress report is most often given. On more formal projects, the progress report may be in writing to provide long-term documentation.

FIRE OFFICER TIP

Consistency Counts

Given a choice, fire fighters prefer a company officer who is consistent. The brilliant officer who is inconsistent creates concern and doubt in the fire fighter.

Effective Company Officer Delegation

There are seven steps in effective delegation:

1. Define your desired results.
2. Select the appropriate fire fighter.
3. Determine the level of delegation.
4. Clarify expectations, and set parameters.
5. Give authority to match the level of responsibility.
6. Provide background information.
7. Arrange feedback during the process.

FIRE OFFICER TIP

Delegating Routine Activities

A goal of an effective fire officer is to push decision making to the lowest possible level. Because a fire officer gains experience as a leader and builds confidence and trust with a team, many routine activities can be delegated. For example, one fire fighter could be assigned to manage in-station training activities, another could oversee routine station maintenance, and a third could keep preincident plan files updated. The fire officer continues with the morning line-up and evening check-up, but the individual fire fighters perform their assignments autonomously. This sort of delegation allows the officer to focus on activities that require his or her personal attention.

Determining the level of delegation in Step 3 relates to the amount of decision-making authority provided to the fire fighter. The fire fighter has five options:

1. Take action independently—no need to report back to the officer.
2. Take action and report back to the officer when done (as with a task assignment within the incident command system).
3. Recommend action that the company officer must approve.
4. Provide two or more recommended actions from which the company officer will choose.
5. Provide information about the pros and cons of different recommendations.

Delegation is a challenging task to a new fire officer. The first times that delegated tasks are assigned, you should plan to closely monitor the results. Although the person with the delegated task has all of the authority, the company officer making the assignment still retains the responsibility for the task to be correctly completed.

Origins of Crew Resource Management

One of the most unique and vital leadership responsibilities is the fire officer's role in crew resource management. **Crew resource management (CRM)** is a behavioral approach to reducing human error in high-risk or high-consequence activities. CRM requires a focused attention on your crew and an openness to receive concerns and time-sensitive information. Let us see how this responsibility evolved by looking at the sentinel event that got the commercial airline industry's attention.

On December 28, 1978, United Airlines Flight 173 was making a routine flight from Denver, Colorado, to Portland, Oregon, with 189 passengers and crew onboard. During the final approach, an unfamiliar "thump" was felt as the landing gear deployed, and the "gear down and locked" light on the cockpit instrument panel did not illuminate. The captain decided to circle the airport while he, the first officer, and the flight engineer attempted to figure out the problem. The flight engineer told the captain that 15 minutes of circling would run the plane really low on fuel.

One hour passed as Flight 173 circled the Portland area. Aviation tradition held that the captain was the infallible head of the ship and was never questioned. While the crew continued to fuss with the lights, the plane's engines coughed, sputtered, and finally went quiet. In the ensuing crash, 10 people were killed, including the flight engineer, and 23 were critically injured. The McDonnell Douglas DC-8 aircraft used by Flight 173 was a fully functional, mechanically sound airframe that crashed because the humans flying the machine became over-engrossed in a burned-out light bulb. In response to this incident, a behavioral modification training system known as crew resource management was developed in a 1979 National Aeronautics and Space Administration (NASA) workshop examining the role of human error in aviation accidents.

Resistance to mandatory CRM training continued until United Airlines Flight 232 experienced a catastrophic failure of its center engine in 1989. All three hydraulic lines necessary for controlling flaps, rudders, and other flight controls were severed. This damage robbed the crew of both the primary and redundant safety features that are built into every airframe. When these conditions were presented within a flight simulation exercise, it always resulted in an unrecoverable spin with all lives lost.

The flight crew and a check ride pilot, using engine controls alone, managed to bring the crippled plane into the Sioux City, Iowa, airport. Of the 295 people on board, 184 survived the fiery crash onto the runway. The crew attributed their success to CRM training as they initiated behaviors to overcome the five factors that contribute to human error. United Airlines Flight 232 was the landmark event that validated CRM's worth (IAFC, 2002).

Researching and Validating CRM Concepts

Part of the 80 percent reduction in the aviation industry's accident rate is attributed to the development, refinement, and system-wide adoption of CRM. "CRM is designed to train team members how to achieve maximum mission effectiveness in a time-constrained environment under stress. That is a concept with nearly universal utility and timeless applicability" (Kern, 2012). Dr. Kern received the Distinguished Leadership Award from *Aviation Week and Space Technology* in 2003 after making controversial decisions to ground nine air tankers during the 2002 U.S. wildland fire season after fatal accidents involving a Lockheed C-130A and a Consolidated PB4Y2, to bar future aerial firefighting contracts for these two airplane types, and to restrict aircraft operations in cooperation with other firefighting agencies. At the time, Kern was the national aviation director of the U.S. Forest Service.

Prof. Robert Helmreich and his staff at the University of Texas Human Factors Research Project (HFRP) developed CRM. They showed the dramatic value of applying CRM to aerospace, aviation, the military, maritime operations, and the medical profession. Their project started in the early 1980s as the Aerospace Crew Research Project, exploring the relationship among personality, group culture, and performance (Helmreich, Merritt, and Wilhelm, 1999).

Human Error

Gordon Dupont spent seven years reviewing aviation accidents as a technical investigator for the Canadian Aviation Safety Board. As the special programs coordinator for Transport Canada, he was responsible for developing programs that would reduce maintenance error. During his study, Dupont considered the similarities between errors that occur in the cockpit and those that occur in the maintenance hangar. Dupont's "dirty dozen" are a comprehensive list of reasons and ways that humans make mistakes (Dupont, 1999):

1. Lack of communication
2. Complacency
3. Lack of knowledge
4. Distraction
5. Lack of teamwork
6. Fatigue
7. Lack of resources
8. Pressure
9. Lack of assertiveness
10. Stress
11. Lack of awareness
12. Norms

Whereas Dupont considered the human factor in the error-making process, Dr. James Reason looked at the systems approach to human error management:

> High technology systems have many defensive layers: some are engineered (alarms, physical barriers, automatic shutdowns, etc.), others rely on people (surgeons, anaesthetists, pilots, control room operators, etc.), and yet others depend on procedures and administrative controls. Their function is to protect potential victims and assets from local hazards. Mostly they do this very effectively, but there are always weaknesses. (Reason, 2000)

Reason points out that each layer of defense is more like a slice of Swiss cheese than a solid barrier. The presence of a hole in one defensive layer does not create a bad outcome event, but when the holes in all of the levels of defense align, a bad or catastrophic outcome becomes more likely (**FIGURE 2-4**).

Active Failures and Latent Conditions

James Reason cites two reasons that explain why holes appear in the layers of defense: active failures and latent conditions.

Active failures are the unsafe acts committed by people who are in direct contact with the situation or system. They have direct and short-lived effects on the integrity of the defenses. Not wearing a seat belt while in a moving vehicle is an example of an active failure.

Latent conditions are the inevitable "resident pathogens" within the system. These conditions have two kinds of adverse effects. First, they can translate into error-provoking conditions within the local workplace. Examples include time pressure, understaffing,

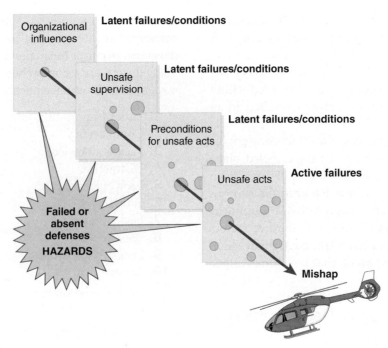

FIGURE 2-4 The Swiss cheese model of how defenses, barriers, and safeguards may be penetrated by an accident trajectory.
Modified from Reason J. "Human Error." Cambridge University Press, Cambridge, UK. 1990.

FIRE OFFICER TIP

Creating a Safety Culture

Every fire department should take the following five steps to create a safety culture:
1. Provide honest sharing of safety information without the fear of reprisal from superiors.
2. Adopt a nonpunitive policy toward errors.
3. Take action to reduce errors in the system. Walk the talk.
4. Train fire fighters in error avoidance and detection.
5. Train fire officers in evaluating situations, reinforcing error avoidance, and managing the safety process.

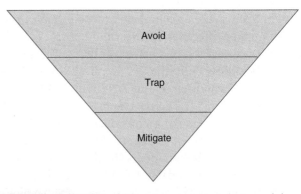

FIGURE 2-5 Helmreich's error management model.
Courtesy of Dr. Robert Helmreich.

inadequate equipment, fatigue, and inexperience. Second, they can create long-lasting holes or weaknesses in the defenses. Examples include untrustworthy alarms, unworkable procedures, and design and construction deficiencies.

Latent conditions may lie dormant within the system for many years before they combine with active failures and local triggers to create an accident opportunity. The ancient planes used for wildland firefighting that were grounded by Dr. Kern in 2002 provide an example of latent conditions.

Error Management Model

CRM is an error management model that incorporates three activities: avoidance, entrapment, and mitigating consequences. Error avoidance provides the greatest opportunity for trapping and preventing errors from becoming a catastrophe. Errors that are not avoided are trapped at the second level. Errors that slip through the first two levels require mitigation. Mitigation is the action taken by emergency responders to minimize the effect of an emergency on a community (**FIGURE 2-5**).

The CRM Model

The fire service CRM model covers six areas: communication skills, teamwork, task allocation, critical decision making, situational awareness, and postincident analysis. Everyone within the department needs to recognize the following facts:
- *No one* is infallible.
- Humans create technology; therefore, technology is fallible.

- Catastrophes are the result of a chain of events.
- Everyone has an obligation to speak up when they see something wrong.
- People who work together effectively are less likely to have accidents.
- For the team to become more effective, every member of the team must participate.

Communication Skills

Communication is the successful transfer and understanding of a thought from one person to another. Review of airline disaster cockpit recordings shows that flight crews typically engaged in several forms of miscommunication that contributed to catastrophic events. These problems have included misinterpretations of take-off, altitude, and landing instructions; airlines that fostered and condoned "fighter pilot" mentalities in their captains; a lack of assertiveness on the part of crew members who recognized problems before the captain; and distractions in the cockpit that inhibited focused communication, such as idle chit-chat during preflight checks.

Developing a standard language and teaching appropriate assertive behavior are the keys to reducing errors resulting from miscommunication. In the aviation application, CRM advocates maintaining a "sterile cockpit"—that is, a cockpit environment where all communication and focus are devoted to flight operations.

When a fire apparatus is responding to an alarm, the crew members are sitting in an enclosed box; they are surrounded by switches, gauges, wheels, pedals, and noise; and they are moving down the road. For these machines to move efficiently and safely, the people inside them need to focus on the mission and communicate effectively. The entire crew should be exchanging only information that is pertinent to responding to and arriving safely at the scene of the alarm.

The CRM-enriched environment generates a climate where the freedom to question is encouraged. Members are encouraged to speak up respectfully when they see something that causes concern, by using clear, concise questions and observations. A discrepancy between what is going on and what should be occurring is often the first indication of an error.

For many organizations, the freedom for all members to question something that appears unusual or is not understood is a delicate subject. Members need to speak directly in a manner that does not challenge the authority of a superior.

Inquiry and advocacy are discrete, learnable skills that promote synergy between the mechanical element and the human players in a scenario. *Inquiry* is the process of questioning a situation that causes concern. *Advocacy* is the statement of opinion that recommends what the person believes is the proper course of action under a specific set of circumstances. Using the two skills effectively requires practice and patience on the part of all crew members. The key factor to keep in mind is that communication should not focus on *who* is right, but rather on *what* is right.

One effective tactic is to use specific buzzwords, such as "red light" and "red flag," to signal discomfort with a situation. These terms are cues that open the door to inquiry and advocacy. They should be reserved for situations that involve an immediate risk of injury.

Assertive Statement Process

Todd Bishop, from the Error Prevention Institute, teaches a five-step assertive statement process that encompasses the communications steps of inquiry and advocacy (IAFC, 2002). The assertive statement runs like this:

1. *Use an opening/attention getter.* Address the other individual: "Hey, Chief," or "Bill," or whatever appropriate moniker is used to get the individual's attention.
2. *State your concern.* Use an owned emotion: "That smoke is really pushing from those windows. I have a bad feeling about it."
3. *State the problem as you see it.* "It looks like it is going to flash. I think any crews entering that place are going to take a beating."
4. *State a solution.* "Why don't we vent the roof and give the place a minute or two to vent before we send in the attack team?"
5. *Obtain agreement or buy-in.* "Does that sound good to you?"

The inquiry and advocacy process and the assertive statement are essential components of the communication segment of CRM. They are the toughest lessons to impart and adopt because they often require a wholesale change in interpersonal dynamics. Once mastered, however, inquiry and advocacy can enhance performance, prevent mishaps, and save lives.

Effective listening is an important CRM communication skill. One effective listening technique is to purposely refrain from making any response or counterargument until the other individual has drained his or her emotional bubble. (Chapter 3, *Leading as a Team*, contains a detailed explanation of this process.)

Teamwork

CRM promotes members working together for the common good. Achieving this level of cohesiveness

FIGURE 2-6 The triangle of leadership.
© Jones & Bartlett Learning.

requires developing effective teams through buy-in by all members, leaders, and followers, in the effort to be efficient and safe. Leaders will always be in charge, but they must also be open to suggestion and constructive criticism. The incident command system provides the formal structure for CRM at the task, tactical, and strategic levels.

Leadership

Fire officers formally exercise leadership of the team through a combination of rank and authority. For officers to become truly effective leaders, they must earn the trust and respect of their subordinates and demonstrate the skills of effective leadership. These three components comprise the triangle of leadership (**FIGURE 2-6**).

The informal authority to lead is derived through respect. True respect is based on three competencies: personal, technical, and social. *Personal competence* refers to an individual's own internal strengths, capabilities, and character. *Technical competence* refers to an individual's ability to perform tasks that require specific knowledge or skills. *Social competence* refers to the person's ability to interact effectively with other people. All three competencies are essential for leaders to function effectively in a CRM environment.

The fact that a leader must demonstrate effective social skills does not mean that a leader must always be deferential. Social competence requires knowing how to speak to people respectfully, whether praising or chastising them.

Mentoring

Mentors pay special attention to helping others develop their skills. CRM provides an effective vehicle for leaders to impart knowledge and skills to their subordinates. Leading by example is a highly effective technique. Crews notice a supervisor's personal habits, pick up on his or her thought processes, and constantly evaluate the leader's performance. Traits that are admired and respected by fire fighters have a major influence on their future behavior. The crew usually sees more than the supervisor may think. Leading by example requires constant effort on the part of the leader, with no opportunities to turn the positive role model on or off at different times.

Human error is a normal, inevitable occurrence. A leader must be willing to admit to making a mistake. Superior leaders readily admit when they have made mistakes, accept responsibility, and focus their attention on moving forward. This trait creates an environment that fosters open communication, promotes safety, and encourages subordinates' belief in the team.

Mentoring also requires sharing knowledge. Knowledge is often closely associated with power. Leaders who are insecure about their positions tend to withhold knowledge and fail to share lessons as they are learned. Mistakes are often repeated in such an environment, and errors that could have been avoided are permitted to occur. A fire officer must maintain technical competence, stay on top of technological advances, and ensure that others have the information that will keep them from being killed or injured.

Technical competency requirements vary from position to position. The street-level fire fighter does not expect the fire chief to pull hose and throw ladders, but the chief must be able to command a fire, mass-casualty incident, or weapons of mass destruction event with the same degree of skill that the fire fighter is expected to use to pull hose and throw ladders.

Handling Conflict

The focal point of CRM in conflict resolution is to focus on *what* is right, not *who* is right. This approach allows the leader to focus on the best, safest outcome. The fire officer needs to establish an open climate for error prevention. The leader who perceives a subordinate's comment or question as a threat to his or her own authority is part of the problem instead of an essential component of the solution. Leaders are required to keep their own egos in check. A leader who rules by intimidation can often win an argument, but the forces of nature always prevail over words. That is, an officer can order a crew to advance into a dangerous position, but the building will not follow an order to remain standing. An officer who fails to listen or refuses to listen is simply dangerous.

Responsibility

Skeptics suggest that CRM advocates "management by committee," in which leadership gives way to consensus. In reality, that is not an accurate statement.

Leaders need to keep their eyes and ears open, especially for input from subordinates. For decision making to be efficient, someone must be in charge—that is, someone must have ultimate responsibility for decisions and outcome.

In the cockpit, the captain retains ultimate authority for decision making and ultimate responsibility for getting the plane from airport to airport. Subordinates are encouraged to provide input, but the final decision rests with the recognized authority. Lines of authority are maintained and that ultimate authority rests with the legitimate ranking officer.

Mission analysis requires fire service leaders to look at all situations with a risk-versus-gain mentality. Chief Alan Brunacini's mantra—risk a lot to save a savable life, take a calculated risk to save savable property, and risk nothing to save what is already lost—establishes a concrete foundation on which leaders can manage emergency operations (Brunacini, 2004). CRM exhibits core values of trust, respect, safety, and mission (LeSage, Dyar, and Evans, 2011) (**FIGURE 2-7**).

Self-Assessment of Each Team Member

All followers should perform a self-assessment of their ability to function as part of a team. The self-assessment should consider four critical areas:

1. Physical condition: People in good physical condition are more aware, alert, and oriented to their surroundings. Maintaining good physical health is critical to the success of any endeavor; however, it is absolutely critical in the fire service.
2. Mental condition: People are constantly pulled in a variety of directions; however, the challenging and dangerous environments where fire fighters must perform demand their full attention. Fire fighters must ask themselves, "Am I free of distractions that could divert my attention from the task at hand?"
3. Attitude: To be an effective team member, a fire fighter must be willing to follow orders and be part of a cohesive team.
4. Understanding human behavior: The effectiveness of CRM is based on understanding human behavior and interpersonal dynamics in a team environment. To be effective using CRM, the team members must understand how their individual behavior relates to one another's actions and to the team as a whole.

To be effective team players and maximize CRM benefits, each individual must have the following characteristics:

- A healthy appreciation for personal safety
- A healthy concern for the safety of the crew
- A respect for authority
- A willingness to accept orders
- A knowledge of the limits of authority
- A desire to help their leader be successful
- Good communication skills
- The ability to provide constructive, pertinent feedback
- The ability to admit errors
- The ability to keep one's ego in check
- The ability to balance assertiveness and authority
- A learning attitude
- The ability to perform demanding tasks
- Adaptability

These qualities of a good follower are remarkably similar to those required of effective leaders. Effective followership enhances leadership, which in turn promotes effective teams. Teams that practice CRM make fewer mistakes and are able to recognize and correct errors before they cause tragic outcomes.

Task Allocation

Task allocation refers to dividing responsibilities among individuals and teams in a manner that allows them to be accomplished effectively. Task overload occurs when the fire officer exceeds his or her capacity to manage all of the simultaneous functions and responsibilities. Safety is compromised with task overload.

Think about the individual who is driving down the interstate at rush hour, texting with a colleague.

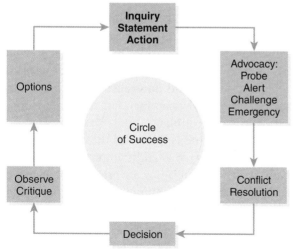

FIGURE 2-7 CRM circle of success.
© Jones & Bartlett Learning.

He looks up and realizes that traffic has stopped. His smartphone falls to the floor as he slams on the brakes, too late to avoid a collision. Task overload has occurred—with disastrous results.

Knowing one's own limits and the capacity of the team is the first step in the CRM task allocation phase. Everyone has a point at which outside stimuli override the ability to process information and perform effectively. If resources are not available to carry out all tasks at once, then tasks must be prioritized to identify those that can be done safely and effectively.

Fire officers fall into three categories when it comes to multitasking ability. Some officers are reluctant to admit that they are always overwhelmed and believe that they become more effective as the situation becomes more hectic. A second group of officers become overwhelmed before the full complexity of the event is even recognized. The third group of leaders effectively assesses the incident, calls for additional resources early, and manages to stay ahead of the incident and balance the span of control.

NASA noted that pilots believed they were capably handling multiple tasks, even when mistakes began to appear and performance deteriorated (Loukopoulos, Dismukes, and Barshi, 2009). Pilots were so busy that they failed to recognize when the situation was getting out of control.

Each fire officer has to evaluate his or her personal capacity to manage complicated situations and identify weak spots. Training and practice can improve performance, but everyone still has limitations. When these are known, fire officers can take action early to compensate for them.

The fire officer must also know the crew's limits. In some fire departments, the same crew works together as a team for long periods, allowing the fire officer to evaluate strengths and weaknesses and focus on team development. In contrast, this understanding can be difficult to develop when the make-up of a crew changes from day to day or the crew is assembled at the time of alarm.

The strengths of individuals and crews should be evaluated and enhanced in multiple nonemergency settings. Performance is enhanced through training classes, live training exercises, tabletop modeling, didactic presentations by experts, mentoring, and exchanging (**FIGURE 2-8**).

Critical Decision Making

Although CRM promotes the concept of team involvement in all aspects of an operation, emergency scenes often demand rapid decision making by the crew leader. Input is welcome from all team members; however, the final responsibility for decision making

FIGURE 2-8 Training exercises help the officer determine the strengths of individuals.
© Jones & Bartlett Learning. Photographed by Glen E. Ellman.

resides with the crew leader, and experience and training often play pivotal roles in successful outcomes.

Gary Klein, a cognitive psychologist who studies decision making under pressure, found that fire officers and military combat officers rely on past experiences when developing plans of action. Urban fire commanders told Klein that they made fire-ground decisions based on previous experiences, rather than referring to traditional decision-making models. The commanders reported that they often did not even consider the options available; instead, they simply looked at the incident, recalled previous experiences of a similar nature, and applied the processes that had been used to mitigate the previous incidents successfully. Many of the fire officers interviewed by Klein had tremendous urban firefighting experience during the 1970s and 1980s.

Klein observed that the fire-ground and combat commanders must often make decisions in environments that are time compressed and dynamic, involve missing or ambiguous data and multiple players, and lack real-time feedback (Zsambok and Klein, 2014). His research identified two decision-making models: **recognition-primed decision making (RPD)**, which describes how commanders can recognize a plausible plan of action, and **naturalistic decision making**, which describes how commanders make decisions in their natural environment.

Decision making is improved through gaining experience, training constantly, improving communication skills, and engaging in preincident planning. Practicing enhanced communication skills and taking advantage of improved crew interaction at all levels provide the following benefits:

- Problem identification is enhanced.
- Supervisors at all command levels maintain better incident control.

- Situational awareness is improved.
- Hazards are more rapidly identified.
- Resource capability is rapidly assessed.
- Potential solutions are more rapidly developed.
- Decision making is improved.
- Surprises and unanticipated problems are reduced.

Situational Awareness

Situational awareness is the ongoing activity of assessing what is going on around you during the complex and dynamic environment of a fire incident. NASA has expanded this definition to "the awareness of acknowledging and assessing; [situational awareness] is the basis for choosing courses of action of the situation both now and in the future" (Reinhart, 2007).

Situational awareness affects performance and decision making. When such awareness is maintained, operations are completed without a hitch. When it is not maintained, errors occur, performance suffers, and catastrophic events often result. Situational awareness is a human factors element that is somewhat difficult to explain but is all too easy to recognize after an error occurs. It also carries over into nonincident operations. Failing to note at shift change that the self-contained breathing apparatus (SCBA) cylinders are down to 1500 psi, for example, is a situational awareness failure that can have unfortunate consequences.

One of the common human behavior factors that leads to a loss of situational awareness is the tendency to ignore or disregard information that is out of context. If an incident commander is busy thinking about an attack strategy, a brief observation or comment about unusual smoke conditions can be easily brushed aside and forgotten. A critical radio message might be acknowledged without really being understood or placed in context in the situation at hand. Chief Rich Gasaway, PhD points out that this occurs at most fire incidents (SAMattersTV, 2018).

The fire officer needs to constantly check and cross-check the situation and operational performance. This constant review helps ensure that the individual and the entire team are focused on concluding the incident in an efficient and injury-free fashion.

Maintaining Emergency Scene Situational Awareness

Six steps should be followed to maintain emergency scene situational awareness:

1. **Fight the fire:** All members of the crew must focus their attention on the details of the incident, while keeping the larger picture in mind. Crash investigations show that air crews have sometimes become distracted by events that took their attention away from flying the plane. In some cases, these distractions resulted in CFIT (controlled flight into terrain) events.
2. **Assess problems in the time available:** Emergency scenes do not always permit leisurely, time-unconstrained periods for decision making. Nevertheless, there has to be a balance between rushing headlong into a burning building and waiting until every possible hazard is evaluated before attacking an incipient fire. Seasoned incident commanders know that taking an extra 30 to 60 seconds to absorb and process as much information as possible often results in better and more confident decisions.
3. **Gather information from all sources:** Information gathering at the emergency scene, by necessity, has to be a rapid process. Of course, one person cannot see everything, know everything, hear everything, or smell everything. Using the rest of the crew as information resources, as well as additional arriving command officers and people familiar with the building or terrain, should enhance

---

FIRE OFFICER TIP

CRM Contributes to the "Miracle on the Hudson"
Hitting a flock of geese on takeoff from New York's LaGuardia Airport completely shut down both engines on U.S. Airways Flight 1549's Airbus A320 in 2009. Jerry Mulenberg, writing in NASA's *ASK Magazine*, describes the sequence of events:

With more than 19,000 hours of flight time, including flying gliders, Captain Chesley (Sully) Sullenberger's experience and training helped prepare him for the once-in-a-lifetime decision he faced. In less than 3 minutes, Captain Sullenberger:

- Requested permission to return to LaGuardia—approved (normal procedure)
- Performed engine restart procedure—engines did not restart (multiple attempts would have resulted in the same outcome and time was short)
- Asked for alternate airport-landing location in New Jersey—granted (wanted to avoid densely populated area near LaGuardia)
- Made final decision to land in the Hudson River—successful in saving all onboard (co-pilot provided airspeed and altitude readings for pilot to glide aircraft in)

During a talk about his experiences, Captain Sullenberger mentioned that he owed a lot to the CRM training he went through (Mulenberg, 2011).

decision making and help keep situational awareness current.
4. **Choose the best option:** Once all of the factors have been weighed, choose the option that maximizes results and minimizes risk.
5. **Monitor results and alter the plan as necessary:** Maintaining situational awareness requires continual evaluation of the effectiveness of decisions. Having a "plan B" or "plan C" ready for implementation may be necessary. An extra set of eyes at a strategic location on the incident scene can often provide valuable information.
6. **Beware of situational awareness loss factors:** The following loss factors must be taken into account:
   - Ambiguity: An event, order, or message that has more than one potential meaning.
   - Distraction: Anything that takes attention away from the larger mission. Distraction can come in the form of outside influences (the slamming of a door down the hall while you are watching television) or inside influences (cultural and personal biases).
   - Fixation: Tunnel vision.
   - Overload: More things are happening than one person can process. For the fire officer, the first few minutes after arrival at an exceptionally chaotic scene can lead to overload.
   - Complacency: When humans perform the same process repeatedly, they tend to become bored due to the hypnotic effect of repetitive action. Inattentiveness breeds carelessness. The perception or description of any structure fire as routine is the pinnacle of complacency.
   - Improper procedure: Advancing a handline into a structure fire before ventilating and positioning apparatus downwind at a hazardous materials release are examples of improper procedures. Immediate action is required to correct these mistakes.
   - Unresolved discrepancy: Suppose you are heading westbound on an interstate and you see a vehicle approaching eastbound in the same lane you are traveling in. There is a brief period where your eyes send the signal to the brain, but you mutter, "I don't believe this." Indications and observations that are in conflict with expectations must not be ignored. Fire fighters must be trained to recognize unresolved discrepancies as needing attention immediately.
   - Nobody fighting the fire: An incident commander may find his or her attention drawn to the actions of one specific company, causing the leader to ignore the rest of the incident. Emergency scene commanders cannot allow themselves to be drawn away from the big picture.

An aviation practice that can help to maintain situational awareness is to say the checklist out loud when preparing for a low-frequency/high-risk task. Memorizing a checklist and going through it by yourself is not as safe or as effective as two people looking at a checklist and speaking the items out loud. This practice has also been used to reduce paramedic medication errors.

FIRE OFFICER TIP

Mental Joggers

A solid strategy for improving situational awareness requires the use of mental joggers for the individual and crew. The crew mental joggers ask the following:
- What do we have here?
- What's going on here?
- How are we doing?
- Does this look right?

The personal mental joggers ask the following:
- What do I know that they need to know?
- What do they know that I need to know?
- What do we all need to know?

Postincident Analysis

Some form of after-action review or postincident analysis should be conducted at the company level after every call in which the company performs emergency operations. In most cases, this can be an informal discussion conducted by the company officer to review the incident, discuss the situation, and evaluate team performance. Every situation should be viewed as a potential learning experience, and the company officer should provide feedback to the crew members to reinforce positive performance and identify areas where there is room for improvement. When a deficiency is noted, a plan for addressing the problem should be developed at the same time.

Situations that involve multicompany operations should be reviewed from the perspective of how well the overall team performed, in addition to individual fire officers conducting company-level reviews. The format of the multicompany review necessarily depends on the nature and the magnitude of the incident, whether it was a one-room house fire that was addressed by three or four companies or a five-alarm blaze that involved dozens of companies and activated every component of the fire department. It is relatively easy to gather the companies that were involved in a one-alarm fire to conduct a basic review of the incident soon after the event. The critique of a large-scale incident, by comparison, can be a major event involving extensive planning, scheduling, and preparations.

Conducting a Postincident Analysis

The actual postincident analysis should be scheduled as soon as convenient after the event, while allowing adequate time to gather and prepare the necessary information. It is best to capture the information while it is still fresh in the minds of the participants. With a small-scale incident, it may be possible to have all of the involved companies in attendance at the analysis so that all of the crew members can be involved in the discussion and learning experience. The attendance at a critique of a larger-scale incident could be limited to company officers and chief officers; however, it is a good idea to invite the entire crew of any company that played a significant role or was involved in some unusual occurrence.

Start the analysis with an overview presentation of the background and basic information about the incident, including the timeline and the units that were dispatched. The first-arriving officer should then be asked to describe the situation as it was presented on arrival and the actions that were taken. This description should include the first officer's role as the initial incident commander as well as the tactical operations performed by that company.

Each successive company should then take a turn explaining what they saw and what they did. An effective visual method is to draw a plot layout of the area on a dry-erase board and then have the various company-level officers locate where they positioned their apparatus and where they operated, and identify which actions they took. This input should gradually produce a complete diagram of the incident. Photographs, videos, and radio recordings can be inserted at pertinent points in the discussion.

During this process, the moderator should keep the analysis directed at key factors, including what the initial strategy was and how it changed (if such changes occurred), how the command structure was developed, how resources were allocated, and which special or unusual problems were encountered. The discussion should also focus on standard operating procedures, including whether they were followed and how well they worked in relation to the actual situation. The officer acting as moderator should help the participants differentiate between problems that occurred because procedures were not followed and

TABLE 2-4 Postincident Analysis Questions

- Did the preincident plan provide accurate and useful information?
- Are there factors that could have or should have been addressed by fire prevention before the incident?
- Were the appropriate units dispatched based on procedures and the information that was received?
- Were the units dispatched in a timely manner?
- Was the appropriate information obtained and transmitted to the responding units?
- What was the situation on arrival?
- What was the initial strategy as determined by the initial incident commander?
- How did the strategy change during the incident?
- How was the incident command structure developed?
- Were the resources provided adequate for the situation?
- How were the resources allocated and assigned?
- Were standard operating procedures followed?
- Do any standard operating procedures need to be changed?
- Which unusual circumstances were encountered, and how were they addressed?
- Is additional training needed?
- Did all support systems function effectively?

areas where procedures should be changed or updated (**TABLE 2-4**). The best way to evaluate the effectiveness of procedures is to determine whether following them actually produced the anticipated results. If a deviation from standard procedures occurred, the reasons and the impact should be discussed.

Once every company-level officer has had a chance to make a presentation, the officer directing the analysis should provide his or her perspective on and assessment of the operation, including both positive and negative factors. This officer should make a point of

FIRE OFFICER TIP

Addressing Poor Performance

Postincident analysis should always focus on lessons and positive experiences; however, in some situations, errors must be addressed. The serious discussion between the fire officer and his or her immediate supervisor should occur in private, not in front of a room full of observers. At the company level, poor performance by one particular crew member should be discussed one-on-one in the office, not at the kitchen table. Areas where the whole crew needs to make improvements should be discussed with the entire crew.

FIRE OFFICER TIP

Building Trust in the New Fire Officer

It is vital for a new supervisor to make an effort to build trust with the workers in his or her company. The failure to establish trust from the beginning of the relationship can quickly undermine a supervisor's career—especially when that manager is a new fire officer. Fire fighters make or break a new officer, and they will make every effort to protect themselves from an officer who appears to be unsafe, unstable, or unprepared for the job.

How can the new fire officer build trust? Here are five suggestions:

1. **Know the fire officer job:** Both administrative and tactical aspects.
2. **Be consistent:** Strive to provide a measured response to any problem, emergency, or challenge.
3. **Walk your talk:** Actions speak much louder than words.
4. **Support your fire fighters:** Make sure you help meet their physiological, safety, security, and order needs.
5. **Make fire fighters feel strong:** Help them become competent and confident in their emergency service skills. Show how they can control their destiny.

evaluating the outcome in terms of where and when the fire was stopped and how much damage occurred. If the outcome was positive and everything met expectations, praise should be widely distributed. Where there is room for improvement, that fact should be noted in terms of valuable lessons learned.

Have the attendees identify the positive and negative points and list the actions that should be taken to improve future performance. These points should be listed on the board for everyone to see as the final product of the analysis.

Documentation and Follow-Up

The last step in conducting a postincident analysis is to write a summary of the incident for departmental records. The results of this review should be documented in a standard format; many fire departments use a form to document the essential findings. Use of a form allows the information to be stored in a consistent manner and ensures that all relevant information is collected. The form should also list recommendations for changes in procedures, if necessary. Care should be taken in writing the postincident analysis, as it may be considered a public record and available to the public.

The completed package should then be forwarded to the appropriate section or sections within the department for review and follow-up. The training division should receive a copy of every incident review to determine training deficiencies and needs. Recommendations for policy changes should also be forwarded to the appropriate personnel for consideration. The completed report should include all of the basic incident report information included in the National Fire Incident Reporting System (NFIRS) as well as the operational analysis information.

You Are the Fire Officer Conclusion

The chief concludes the 6-month assessment meeting. "A challenge when considering new management concepts is making sure you are still meeting the departmental needs. The Circle of Consensus may be a useful tool as long as it results in meeting departmental goals. I was disappointed that your delegation of the in-station training did not appear to include monitoring the activity. The senior fire fighter had the authority to conduct the in-station training but you retained the responsibility. The fire fighter is sharp, but he is inexperienced as a trainer. I need you to help this fire fighter become a better in-station trainer."

The chief pauses and sits back in the chair. "Our challenge is to get the best from our people as effectively as possible. You know how you feel when you first arrive at a working structure fire, with smoke showing from many windows and a report of people trapped? You want to do everything at once and use every tool carried on the rig. There are not enough people or time to use all of the tools to complete critical fire-ground tasks. The smart company officer sizes up the situation and selects the best tool to accomplish the task."

The chief ends on a final note: "Management concepts are like the tools on the truck. You have collected a large set of management ideas, concepts, and practices in your management toolbox. You need to pick the best idea, concept, or practice that accomplishes the objective without pummeling fire fighters or eroding your authority. You will see new management tools during your tenure as a company officer. Like new firefighting tools, they will have high promise. Until you evaluate how each new management tool works by understanding its history and development and trying it out in a training session, it remains an unknown resource."

After-Action REVIEW

IN SUMMARY

- John Cotter describes the leadership process as producing movement through three subprocesses: establishing direction, aligning people, and providing motivation and inspiration.
- Management tends to look at short time frames; leaders look at longer time frames.
- Peter Northouse's classification approach includes the trait approach, the skills approach, and the behavioral approach.
- The grid theory of management assumes that every decision made and every action taken in the workplace is driven by people's values, attitudes, and beliefs. Blake and Mouton described five behavioral models based on a person's position on this grid:
  - Impoverished management
  - Authority-compliance management
    - Demonstrates an autocratic leadership style
  - Country club management
  - Middle-of-the-road management
    - Demonstrates a laissez-faire leadership style
  - Team management
    - Demonstrates a democratic leadership style.
- Situational leadership is based on the assumption that followers' skills and motivation vary over time, and leaders should change the degree to which they are directive or supportive to meet the changing needs of followers.
- Transformational leadership is concerned with emotions, values, ethics, standards, and long-term goals.
- Servant leadership emphasizes that leaders be attentive to the concerns of their followers, empathize with them, and nurture them. Servant leaders put followers first, empower them, and help them develop their full personal capacities.
- Adaptive leadership focuses on the activities of the leader in relation to the work of the followers when confronted with a technical or adaptive challenge.
- Followership accepts the influence of others to accomplish a common goal. Typically, followers comply with the directions and wishes of leaders.
- The fire officer has to be both a leader and a follower.
- Types of power may be categorized using several different methods. French and Raven refer to legitimate, reward, expert, referent, and coercive power. Yukl refers to personal and positional power.
- The fire officer works in a dynamic environment with changing conditions and evolving organizational needs. Two unique challenges include the fire station as a work location and leading a volunteer fire company.
- A company-level officer should make it a practice to walk around the fire station at various times during the workday to observe what is going on.
- The volunteer fire officer must pay attention to the satisfaction level of every member and be alert for issues.
- Just as fire companies do, the Marine Corps respond from all corners of the community and assemble to handle an emergency incident. To serve as an example, this text presents the 14 leadership traits of the Marine Corps, as qualities of thought and action which, if demonstrated in daily activities, help Marines earn the respect, confidence, and loyal cooperation of other Marines.
- Human resource management focuses on human resources planning; employee (labor) relations; staffing; human resources development; performance management; compensation and benefits; and employee health, safety, and security.
- Managing fire fighters requires physical, financial, human, and time resources.

- One of the best tools to improve time efficiency is delegation. Once a task is assigned, the fire officer must provide the fire fighter with follow-up and feedback.
- Crew resource management (CRM) is a behavioral approach to reducing human error in high-risk or high-consequence activities.
- Leading change describes the process of recommending changes (as a lieutenant) and developing a process to establish a change (as a captain).
- CRM was developed as a behavioral modification training system in a 1979 NASA workshop examining the role of human error in aviation accidents.
- The aviation industry has reduced its accident rate by 80 percent through the development, refinement, and system-wide delivery of CRM.
- Gordon Dupont determined that a "dirty dozen" of human factors contribute to tragedy:
  - Lack of communication
  - Complacency
  - Lack of knowledge
  - Distraction
  - Lack of teamwork
  - Fatigue
  - Lack of resources
  - Pressure
  - Lack of assertiveness
  - Stress
  - Lack of awareness
  - Norms
- James Reason identified two precursors to holes appearing in the layers of defense: active failures and latent conditions.
- CRM is an error management model that includes three activities: avoidance, entrapment, and mitigating consequences. Error avoidance provides the greatest opportunity for trapping errors and preventing them from evolving into full-blown catastrophes.
- A six-point CRM model serves the fire service well:
  - Communication skills
  - Teamwork
  - Task allocation
  - Critical decision making
  - Situational awareness
  - Postincident analysis
- CRM suggests that developing a standard language, maintaining a "sterile cockpit," and teaching appropriate assertive behavior are the keys to reducing errors resulting from miscommunication.
- CRM promotes the idea that members must work together for the common good. This requires developing effective teams through buy-in of all members, leaders, and followers, in the effort to be efficient and safe.
- Task overload occurs when the fire officer exceeds his or her capacity to manage the various simultaneous functions and responsibilities.
- CRM promotes the concept of team involvement in all aspects of operation. In the area of emergency scene decision making, time, experience, and training play pivotal roles in successful outcomes.
- The loss of situational awareness is frequently the first link in a chain of errors that leads to calamity.
- The CRM approach concentrates on the conditions under which individuals work and tries to build defenses to avert errors or mitigate their effects.

KEY TERMS

Active failures Unsafe acts committed by people who are in direct contact with the situation or system that create voids within the defenses, barriers, and safeguards within the safety system that could lead to a loss.

Adaptive leadership How leaders encourage people to adapt when confronted with problems, challenges and changes.

Autocratic A leadership style in which managers make decisions unilaterally, without input from subordinates.

Behavioral approach An emphasis on the personality characteristics of a leader.

Crew resource management (CRM) A program focused on improved situational awareness, sound critical decision making, effective communication, proper task allocation, and successful teamwork and leadership. (Reproduced from NFPA 1500, 2018.)

Delegation An organizational process wherein the supervisor divides work among the subordinates and give them the responsibility and authority to accomplish the respective tasks.

Democratic A leadership style in which managers reach decisions with the input of the employees but are responsible for making the final decision.

Followership The act or condition of following a leader; adherence. The characteristic that leaders can be effective only to the extent that followers are willing to accept their leadership.

Human resources development All activities to train and educate employees.

Human resources management The process of managing employees in a company. The process may include hiring, firing, training, and motivating employees.

Human resources planning The process of having the right number of people in the right place at the right time who can accomplish a task efficiently and effectively.

Laissez-faire A policy or attitude of letting things take their own course, without interfering.

Latent conditions Pre-existing conditions within the safety system defenses, barriers, and safeguards within the safety system that could lead to a loss through provoking errors or weakening system.

Leadership A complex process by which a person influences others to accomplish a mission, task, or objective and directs the organization in a way that makes it more cohesive and coherent.

Management The art of art of getting things done through and with people in formally organized groups. To forecast, plan, organize, command, coordinate, and control.

Naturalistic decision making A framework of intuitive decision making when operating under limited time, uncertainty, high stakes, team and organizational constraints, unstable conditions, and varying amounts of experience. Recognition-primed decision making is a process within this decision-making framework.

Personal power Power that reflects the effectiveness of the individual.

Positional power Power that is defined by the role the individual has within the organization.

Power The capacity of one party to influence another party.

Recognition-primed decision making (RPD) A model of how people make quick, effective decisions when faced with complex situations by generating a possible course of action based on intuition and compare it to the constraints imposed by the situation, then select the first course of action that is not rejected.

Servant leadership A leadership approach focusing on serving others, rather than accruing power or taking control.

Situational awareness The ongoing activity of assessing what is going on around you during the complex and dynamic environment of a fire incident. (Reproduced from NFPA 1410, 2015.)

Situational leadership A leadership approach where a supervisor adjusts the style of leadership to fit the developmental level of subordinates. During emergency operations, it means adjusting the style of leadership based on the changing conditions of the incident.

Skills approach An emphasis on the skills or capabilities of a leader.

Staffing The process of attracting, selecting, and maintaining an adequate supply of labor, as well as reducing the size of the labor force when required.

Trait approach An emphasis on the traits of an effective leader.

Transformational leadership A leadership approach that causes change in individuals and social systems. In its ideal form, it creates valuable and positive change in the followers with the end goal of developing followers into leaders.

REFERENCES

Bennis, Warren, and Burt Nanus. 1985. *Leaders, the Strategies for Taking Charge*. New York: Harper and Row.

Blake, Robert R., and Anne Adams McCanse. 1991. *Leadership Dilemmas-Grid® Solutions: A Visionary New Look at a Classic Tool for Defining and Attaining Leadership and Management Excellence*. Houston, TX: Gulf Professional Publishing.

Blake, Robert R., and Jane Srygley Mouton. 1982. "Theory and Research for Developing a Science of Leadership." *The Journal of Applied Behavioral Science* 18 (3): 275–291. doi:10.1177/002188638201800304

Blake, Robert R., Jane Srygley Mouton, and A. C. Bidwell. 1962. "Managerial Grid." *Advanced Management Office Executive* 1 (9):12–15.

Brunacini, Alan V. 2004. *Command Safety: The IC's Role in Protecting Firefighters*. Stillwater, OK: Fire Protection Publications.

Dupont, Gordon. 1999. "The Dirty Dozen Errors in Maintenance." Paper presented at the 11th Symposium on Human Factors in Aviation Maintenance. San Diego, CA.

Forbes, Steve. 2011. "Remembering 9/11: The Rudy Giuliani Interview." *Forbes*. Accessed June 22, 2019. https://www.forbes.com/sites/steveforbes/2011/09/09/remembering-911-the-rudy-giuliani-interview/#e52217b3a7de.

French Jr, John R. P., and Bertram Raven. 1959. "The Bases of Social Power." In *Studies in Social Power* edited by Dorwin Cartwright, 150–167. Oxford, UK: University of Michigan.

George, Bill. 2003. *Authentic Leadership: Discovering the Secrets to Creating Lasting Value*. San Francisco: Jossey-Bass.

Greenleaf, R. K. 2008. *Servant Leadership: A Journey into the Nature of Legitimate Power and Greatness*. Mawah, NJ: Paulist Press.

Heifetz, Ronald A., Marty Linsky, and Alexander Grashow. 2009. *The Practice of Adaptive Leadership: Tools and Tactics for Changing Your Organization and the World*. Boston, MA: Harvard Business Review Press.

Helmreich, Robert L., Ashleigh C. Merritt, and John A. Wilhelm. 1999. "The Evolution of Crew Resource Management Training in Commercial Aviation." *International Journal of Aviation Psychology* 9 (1): 19–21.

Hersey, Paul, and Kenneth H. Blanchard. 1969. *Management of Organizational Behavior: Utilizing Human Resources*. Englewood Cliffs, NJ: Prentice Hall.

International Association of Fire Chiefs (IAFC). 2002. *Crew Resource Management: A Positive Change for The Fire Service*. Fairfax, VA: International Association of Fire Chiefs.

Jester, James L. 2018, June 8. "Servant Leadership in Today's Fire Service." *FireRescue Magazine* 13 (5). Accessed June 22, 2019. https://www.firerescuemagazine.com/articles/print/volume-13/issue-5/departments/leaders-are-readers/servant-leadership-in-today-s-fire-service.html.

Katz, Robert L. 1955. "Skills of an Effective Administrator." *Harvard Business Review* 33 (1): 33–42.

Kern, Anthony T. 2012. *Controlling Pilot Error: Culture, Environment and CRM*. New York: McGraw-Hill.

Kotter, J. P. 1990. *Force for Change: How Leadership Differs from Management*. New York: Simon and Schuster.

Kouzes, James M., and Barry Z. Posner. 2017. *The Leadership Challenge: How to Make Extraordinary Things Happen in Organizations, Sixth edition*. New York: John Wiley & Sons.

LeSage, Paul, Jeff T. Dyar, and Bruce Evans. 2011. *Crew Resource Management: Principles and Practice. Developing a Culture for Open Communications*. Sudbury, MA: Jones and Bartlett Publishers.

Loukopoulos, Loukia D., R. Key Dismukes, and Immanuel Barshi. 2009. *The Multitasking Myth: Handling Complexity in Real-World Operations*. New York: Routledge.

Mulenberg, Jerry. 2011, March 11. "Crew Resource Management Improves Decision Making." *ASK Magazine*. Washington, DC: U.S. National Aeronautics and Space Administration.

Northouse, P.G. 2013. *Leadership: Theory and Practice, Sixth Edition*. Thousand Oaks, CA: SAGE Publications, Inc.

Powell, Michael. 2007, September 21. "In 9/11 Chaos, Giuliani Forged a Lasting Image." *The New York Times*. Accessed June 22, 2019. https://www.nytimes.com/2007/09/21/us/politics/21giuliani.html.

Reason, James. 2000. "Human Errors: Models and Management." *British Medical Journal* 320 (7237): 768–770.

Reddin, William J. 1967. "The 3-D Management Style Theory: A Typology Based On Task and Relationships Orientations." *Training and Development Journal* 21 (4): 8–17.

Reinhart, Richard O. 2007. *Basic Flight Physiology, Third edition*. New York: McGraw-Hill.

SAMattersTV. 2018, January 26. "Episode 130: Auditory Exclusion - Going Deaf Under Stress." Accessed June 22, 2019. https://www.youtube.com/watch?v=EMa62jJbHy4.

Sipe, James W., and Don M. Frick. 2015. *Seven Pillars of Servant Leadership: Practicing the Wisdom of Leading By Serving*. Mahwah, NJ: Paulist Press.

Yukl, Gary. 2013. *Leadership in Organizations, Eighth edition*. Boston, MA: Pearson.

Zsambok, Caroline E., and Gary Klein, eds. 2014. *Naturalistic Decision Making*. New York: Psychology Press.

Fire Officer in Action

The probationary lieutenant is reviewing leadership concepts.

1. Robert Katz identifies a set of skills that can be developed. Which of the following is *not* one of those skills?
   A. Conceptual
   B. Technical
   C. Transactional
   D. Human

2. Which of the following is a process whereby an individual accepts the influence of others to accomplish a common goal?
   A. Authority-compliance management
   B. Situational leadership
   C. Transformational leadership
   D. Followership

3. When a person believes that the supervisor has an obligation to comply, this is an example of which type of power?
   A. Legitimate
   B. Referent
   C. Expert
   D. Coercive

4. A democratic leadership style is found in the which of these management models?
   A. Country club
   B. Middle-of-the-road
   C. Team
   D. Authority-compliance

Fire Lieutenant Activity

NFPA Fire Officer I Job Performance Requirement 4.1.1

The organizational structure of the department incorporates many components, including, but not limited to: geographical configuration and characteristics of response districts; departmental operating procedures for administration, emergency operations, incident management system, and safety; fundamentals of leadership; departmental budget process; information management and recordkeeping; the fire prevention and building safety codes and ordinances applicable to the jurisdiction; current trends, technologies, and socioeconomic and political factors that affect the fire service; cultural diversity; methods used by supervisors to obtain cooperation within a group of subordinates; the rights of management and members; agreements in force between the organization and members; generally accepted ethical practices, including a professional code of ethics; and policies and procedures regarding the operation of the department as they involve supervisors and members.

Application of 4.1.1

Demonstrate your understanding of leadership and management theories:

1. Using Blake and Mouton's Managerial Grid, describe the type of department that would require a major effort by a company officer to improve fire fighter emergency incident competence.

2. How would a newly promoted lieutenant use situational leadership concepts when assigned to a fire company of older, experienced fire fighters?

3. Can a fire officer who is not charismatic be a successful transformational leader?

 *Access Navigate for flashcards to test your key term knowledge.*

CHAPTER 3

Fire Officer I

Leading a Team

KNOWLEDGE OBJECTIVES

After studying this chapter, you will be able to:

- List and explain the parts of the communication cycle. (**NFPA 1021: 4.2.1, 4.2.2**) (pp. 74–78)
- Identify skills for effective communication. (**NFPA 1021: 4.2.1, 4.2.2, 4.4.5**) (p. 75)
- Describe the ways to counteract environmental noise. (**NFPA 1021: 4.2.2**) (pp. 76–77)
- Identify ways to improve listening skills. (pp. 77–78)
- Identify the key points for effective emergency incident communications. (**NFPA 1021: 4.2.1**) (pp. 78–80)
- Define the term *grievance*. (**NFPA 1021: 4.2.5**) (p. 81)
- Explain general grievance procedures. (**NFPA 1021: 4.2.5**) (pp. 81–83)
- Describe the general decision-making procedure. (pp. 82–86)
- Explain considerations for planning and setting priorities. (**NFPA 1021 4.2.6**) (pp. 84–86)
- Describe a fire officer's role in supervising a single company. (**NFPA 1021: 4.2.1**) (pp. 86–88)
- Compare and contrast leadership in routine situations versus emergency situations. (**NFPA 1021: 4.2.1, 4.2.2, 4.2.6**) (pp. 88–90)

- Identify considerations for matching task assignments to individuals. (**NFPA 1021: 4.2.1, 4.2.2, 4.2.6**) (pp. 86–90)
- Describe considerations for emergency scene leadership. (**NFPA 1021: 4.2.1, 4.2.6**) (pp. 88–89)
- Describe considerations for routine situation leadership. (**NFPA 1021: 4.2.2, 4.2.6**) (pp. 89–90)
- Recognize the fire officer's responsibility for providing a safe environment for members. (**NFPA 1021: 4.2.1, 4.2.2, 4.2.3**) (p. 90)
- Identify the fire officer's role in training fire service personnel. (**NFPA 1021 4.2.3**) (pp. 90–96)
- Describe the four-step method of skill training. (**NFPA 1021 4.2.3**) (pp. 90–93)
- Describe on-the-job training and the order in which skills must be taught. (pp. 96–98)
- Describe how to develop a specific training program. (pp. 98–99)
- Describe the interrelationships among complaints, conflicts, and mistakes. (**NFPA 1021: 4.2.4, 4.2.5**) (pp. 99–100)
- Explain how to manage conflict within the department. (**NFPA 1021: 4.2.4, 4.2.5**) (pp. 100–104)
- Identify signs and symptoms of behavioral and physical health issues. (**NFPA 1021: 4.2.4**) (pp. 104–106)
- Explain member assistance policies and procedures. (**NFPA 1021: 4.2.4, 4.2.5**) (pp. 107–108)

SKILLS OBJECTIVES

After studying this chapter, you will be able to:

- Conduct clear and concise communication using the communication cycle. (**NFPA 1021: 4.2.1, 4.2.2, 4.2.3, 4.2.4, 4.2.5, 4.2.6**) (pp. 74–78)
- Provide appropriate feedback to members and supervisors on progress and assignment completion. (**NFPA 1021: 4.2.1, 4.2.2, 4.2.6**) (p. 80)
- Conduct clear and concise emergency incident communications with fire officers, fire fighters, and dispatch. (**NFPA 1021: 4.2.1, 4.2.2, 4.2.5**) (pp. 78–80)
- Communicate task assignments to unit members. (**NFPA 1021: 4.2.1, 4.2.2, 4.2.6**) (pp. 86–90)
- Supervise and account for members in emergency situations. (**NFPA 1021: 4.2.1**) (pp. 88–89)
- Employ methods of monitoring team progress. (**NFPA 1021: 4.2.1, 4.2.2, 4.2.6**) (pp. 89–90)
- Supervise and account for members in training situations. (**NFPA 1021: 4.2.3**) (pp. 90–96)
- Coordinate the completion of assigned tasks and projects using delegation. (**NFPA 1021: 4.2.6**) (p. 88)
- Create a prioritized plan for the completion of assignments or projects. (**NFPA 1021: 4.2.6**) (pp. 84–86)
- Identify actions that must be taken when a concern about safety or non-compliance arises. (**NFPA 1021: 4.2.1, 4.2.2, 4.2.3, 4.2.6**) (pp. 89–90)
- Direct members in proper completion of a prepared training evolution. (**NFPA 1021: 4.2.3**) (pp. 98–99)
- Utilize member assistance policies and procedures to address member behavioral and physical health issues. (**NFPA 1021: 4.2.4, 4.2.5**) (pp. 104–108)

You Are the Fire Officer

An early-morning response to an activated fire alarm in an industrial park became a structure fire. The crew was rocked by a fire rollover when they opened an interior door in the building. As the crews clean up from the fire, the battalion chief approaches the lieutenant.

The chief is angry that the fire crew entered the building with no tools or self-contained breathing apparatus (SCBA) and did not follow the standard operating procedure (SOP) on commercial fire responses.

1. How can you improve fire company fire-ground performance?
2. What should you do if the fire fighters point out problems with the existing SOP on commercial fires?
3. How can you improve fire fighter response to a changing incident?

 Access Navigate for more practice activities.

Introduction

Many fire officers wear a rank insignia that features bugles, representing the fire officer's speaking trumpet. At the turn of the 20th century, such a trumpet was used to shout orders on the fire ground. Today, this symbol emphasizes the fire officer's requirement to communicate. Although the technology has certainly advanced, communication skills remain critically important in the fire service. An officer must be able to communicate effectively in many different situations and contexts.

Influencing people to complete tasks requires the fire officer to be skilled as a communicator, active listener, and fire-ground commander. The fire officer meets the supervisory responsibilities by making effective decisions, properly assigning tasks, and understanding the grievance process. The fire officer develops members through effective skill training, evaluation of competence, and addressing member-related problems.

The Communication Cycle

Communication is a repetitive, circular process. Successful communication occurs whenever two people can exchange information and develop mutual understanding. When information flows from one person to another, the process is truly effective only when the person receiving the information is able to understand what the other person intended to transmit.

Effective communication does not occur unless the intended message has been received and understood. This message must make sense in the recipient's own terms, and it must convey the thought that the sender intended to communicate. The sender cannot be sure that this outcome has occurred unless the recipient sends some confirmation that the message arrived and was correctly interpreted.

The communication cycle consists of five parts:

1. Message
2. Sender
3. Medium (with noise)
4. Receiver
5. Feedback

The Message

The message represents the text of the communication. In its purest form, the message contains only the information to be conveyed. Messages do not have to be in the form of written or spoken words or verbal reports, however. For example, a stern facial expression with purposeful eye contact can convey a very clear message of disapproval, whereas a smile can convey approval. These messages are clear, yet no words accompany them.

FIRE OFFICER TIP

Accomplishing Tasks as a Manager or a Leader
A manager will use leadership to accomplish tasks, and a leader will use leadership to influence tasks.

Ensure Accuracy

A fire officer needs to have up-to-date information on the fire department's standard operating guidelines, policies, and practices. An officer should also be familiar with the personnel regulations, the approved fire department budget, and, if applicable, the current union contract.

If a fire fighter is misinterpreting a factual point or a departmental policy, the fire officer is obligated to clarify or correct the information. Ignoring an inaccurate statement may simply foster erroneous information that is transmitted along the fire service "grapevine" (discussed later in this section). If necessary, obtain the accurate information from a chief officer or headquarters staff and correct the misunderstanding.

A fire officer sometimes has to exercise control over what is discussed in the work environment. Fire station discussions can easily encroach on subjects that are intensely personal, such as politics, religion, or social values, and can quickly escalate into confrontations. To address potential problems proactively, the fire officer needs to establish some ground rules about the topics and level of intensity when discussing issues. Rumor control is useful in de-escalating the spread of inaccurate information that can harm an individual, the department, or the fire service.

The Sender

The sender is the person or entity who is sending the message. We think of the sender as a person, but it could also be a sign, a sound, or an image. The message needs to be properly targeted to the right person and formulated so that the receiver will understand the meaning. The sender is responsible for the receiver properly understanding the message.

A fire officer must be skilled if he or she is to transmit information and instructions to subordinates and co-workers effectively. The tone of voice or the look that accompanies a spoken message can profoundly influence the receiver's interpretation. Body language, mannerisms, and other nonverbal cues may all affect the interpretation.

Senders may sometimes convey messages that are not intended, not directed to anyone in particular, or not even meant to be messages. For example, a person can outwardly express personal disappointment; however, others may interpret this look as a message of disapproval aimed at them.

The Medium

The medium refers to the method used to convey the information from the sender to the receiver. The medium can consist of words that are spoken by the sender and heard directly by the receiver. Spoken words or sounds can also be transmitted as electromagnetic waves through a radio system. Written words, pictures, symbols, and gestures are all examples of messages transmitted through a visual medium. The sender should consider the circumstances, the nature of the message, and the available methods before choosing a particular medium to send a message.

When information has to be transmitted to subordinates, a fire officer can post a notice on a bulletin board, announce it at a formal beginning-of-shift meeting, or mention it during a firehouse meal. The medium that is chosen influences the importance that is attached to the message. If the information is really important, the fire officer might announce the key points at line-up, and then direct all members to read a written document and sign a sheet to acknowledge that they have read and understood it. When choosing the circumstances for transmitting a personal message, remember this guideline: Praise in public; counsel, coach, or discipline in private.

Overcoming Environmental Noise

Environmental noise is a physical or sociological condition that interferes with the message. Within this definition, "noise" includes anything that can clog or interfere with the medium that is delivering the message.

Physical noise includes background conversations, outside noises, or distracting sounds that make it difficult to hear. For example, siren noise makes it difficult to communicate over the radio. The squeal of portable radio feedback competes with the clarity of information exchange. Digital radios and cell phones can suffer from poor reception or static. A rapid flow of incoming messages can overload the receiver's ability to deal with the information, even if the words come through clearly (**FIGURE 3-1**). Darkness or bright flashing lights make it difficult to see clearly and interpret a visual message.

Another form of noise interference occurs when the receiver is distracted or thinking about something else and blocks out most or all of an incoming message. The human brain has difficulty processing more than one input source at a time. For example, you could be having a conversation on the telephone with one individual when someone else calls your name. In such a case, your attention is diverted while you try to determine who called your name and why. During that moment, you are likely to miss part of the telephone conversation. A similar situation occurs if the sender or receiver is unable to concentrate on or respond to the message due to fatigue, boredom, or fear.

Sociological environmental noise is a more subtle and difficult problem. Prejudice and bias are examples of sociological environmental noise. If the receiver does not believe that the sender is credible, the message will be ignored or result in an inadequate response. The following list provides suggestions to improve communication by minimizing this type of environmental noise:

- Do not struggle for power: Focus attention on the message, not on whether the sender has the authority to deliver it to the receiver. The situation—not the people involved—should drive the communication and the desired action.
- Avoid an offhand manner: If you want your information to be taken seriously, you must deliver it that way. Be clear and firm about matters that are important.
- Keep emotions in check: Emergency operations often deal with events that invoke intense feelings that can interfere with focus and attention to the facts of the situation at hand.
- Remember that words have meaning: Select words that clearly convey your thoughts, and be mindful of the impact of the tone of voice.
- Do not assume that the receiver understands the message: Encourage the receiver to ask questions and seek clarification if the intent of the message is unclear. A good technique of confirming understanding is for the receiver to repeat the key points of the message back to the sender.
- Immediately seek feedback: If the receiver identifies an error or has a concern about the message, encourage that individual to make the statement sooner rather than later. It is always better to solve a problem or identify resistance while there is still time to make a change.

FIGURE 3-1 Trying to talk on a cellular phone with poor reception is one example of physical noise interfering with communication.
© Jones & Bartlett Learning. Photographed by Glen E. Ellman.

- Provide an appropriate level of detail: Think about the person who is receiving the message and how much information that individual needs. Consider a fire officer telling a fire fighter to set up the annual hose pressure test. The information needs are different for a fire fighter who has 22 years of experience as a pump operator versus a rookie who has 22 days on the job.
- Watch out for conflicting orders: Make sure that your message is consistent with information coming from other sources. A fire officer should develop a network of peers to consult when dealing with unfamiliar or confusing situations. A subordinate should also be expected to inform the fire officer of a conflicting order so that the fire officer can consider the difference and then give the direction that is most appropriate at that moment.

These eight suggestions are intended to improve communications dealing with administration and supervisory activities. The fire ground or emergency scene requires different communications practices.

The Receiver

The receiver is the person who receives and interprets the message. Unfortunately, there are many opportunities for error in the reception of a message. Although it is up to the sender to formulate and transmit the message in a form that should be clearly understandable to the receiver, it is the receiver's responsibility to capture and interpret the information. The same words can convey different meanings to different individuals, so the receiver might not automatically interpret a message as it was intended. In the fire service, the accuracy of the information that is received can be vital, so both the sender and the receiver have responsibilities to ensure that messages are properly expressed and interpreted. Many messages are directed to more than one person, so there can be multiple interpretations of the same message.

Feedback

The sender should never assume that information has been successfully transferred unless some confirmation is provided that the message was received and understood. Feedback completes the communication cycle by confirming receipt and verifying the receiver's interpretation of the message. The importance that is attached to feedback depends on the nature of the message.

When relaying critical information during a stressful event, the sender should have the receiver repeat back the key points of the message in his or her own words. Without feedback, the sender cannot be confident that the message reached the receiver and was properly interpreted.

Active Listening

The ability to listen becomes increasingly important as one advances through the organization. A fire fighter must listen effectively to information that is coming from a higher level. A fire officer must also be able to listen to company members or other subordinates and accurately interpret their comments, concerns, and questions.

Listening is a skill that must be continually practiced to maintain proficiency. A typical listening situation for a fire officer could be a meeting with a fire fighter who is expressing a problem or a concern. Listening, in a face-to-face situation, is an active process that requires good eye contact, alert body posture, and frequent use of verbal engagement (**FIGURE 3-2**). The purpose of active listening is to help the fire officer understand the fire fighter's viewpoint to solve an issue or a problem.

The following techniques may help improve your listening skills:

- Do not assume anything: Do not anticipate what someone will say.
- Do not interrupt: Let the individual who is trying to express a point or position have a full say.
- Try to understand the need: Often, the initial complaint or problem is a symptom of the real underlying issue.

FIGURE 3-2 Listening is an active process that requires good eye contact, alert body posture, and frequent use of verbal engagement.
© Jones & Bartlett Learning. Photographed by Glen E. Ellman.

- Look for the real reason the person wants your attention.
- Do not react too quickly: Try not to jump to conclusions. Avoid becoming upset if the situation is poorly explained or if an inappropriate word is used. The goal is to understand the other person's viewpoint.

Stay Focused

It is easy to get sidetracked and bring unrelated issues into a conversation. Directed questioning is a good method to keep a conversation on topic. If the speaker starts to ramble, ask a specific question that moves the conversation back to the appropriate subject. For example, if you are attempting to find out why a fire fighter failed to wear a uniform shirt to the beginning of shift roll call and he starts talking about how he does not like the fire department patch, you could ask, "How does the patch affect whether you wear your uniform shirt to roll call?"

Emergency Incident Communications

A fire officer needs to communicate effectively during emergency incidents. There is no time to be elegant when communicating within the emergency environment. The direct approach requires asking precise questions, providing timely and accurate information, and giving clear and specific orders.

Under the time pressure of an emergency incident, management of the communications process can be as important as communicating effectively. The incident commander (IC) has to establish and maintain a command presence to manage the exchange of information so that the most important messages go through and lower-priority or unnecessary communications do not get in the way. The IC should be the gatekeeper for information exchange via radio communications. Additional control can be added by stating, "Unit(s) stand by." In severe situations, requesting emergency radio traffic only will reduce communications to immediate, necessary communications. This is critical in the event of a mayday situation or substantial change in fire conditions that must be communicated without delay or interference.

Key points for emergency communications are as follows:

- Be direct.
- Speak clearly (**FIGURE 3-3**).
- Use a normal tone of voice.
- If you are using a radio, hold the microphone about 2 in. (5 cm) from your mouth.

FIGURE 3-3 Emergency communications require a direct approach.
© Glen E. Ellman

- If you are using a repeater system, allow for a time delay after keying the microphone.
- Use plain English rather than "10 codes."
- Use common terminology that is recognized by the National Incident Management System (NIMS), especially when interacting with multiple agencies and disciplines.
- Try to avoid being in the proximity of other noise sources, such as running engines.

Using the Communications Order Model

The communications order model is a standard method of transmitting an order to a unit or company at the incident scene. It is designed to ensure that the message is clearly stated, heard by the proper receiver, and properly understood. It also confirms that the receiver is complying with the instruction.

Command: Ladder 2, from Command.

Ladder 2: Ladder 2, go ahead Command.

Command: Ladder 2, come in on side Charlie and conduct a primary search on the second floor. Also advise if there is any fire extension to that level.

Ladder 2: Ladder 2 going to side Charlie to do a primary search on the second floor and check for fire extension.

Command: Ladder 2, that is correct.

Initial Situation Report to Dispatch

The first-arriving company at an incident needs to describe conditions to the dispatch center concisely. The content of the initial radio report needs to meet the departmental operating procedures. In addition, this verbal picture establishes the tone of the incident. A fire officer who provides a calm and complete description of the situation on arrival demonstrates leadership and ability to control the action that will occur. An officer who gives the impression of confusion, indecision, or uncertainty fails the first test of leadership.

If a specific requirement for the initial communications has been established, a radio report should meet those criteria. Otherwise, the report should include the following information:

- Identification of the company arriving at the scene.
- A brief description of the incident situation. This may include providing information about the building size, occupancy, hazardous chemical release, or multiple-vehicle accident.
- Obvious conditions, such as a working fire, multiple victims, a hazardous materials spill, or a dangerous situation.
- Brief description of action to be taken: "Engine 2 is advancing an attack line into the first floor."
- Declaration of strategy to be used: offensive or defensive.
- Any obvious safety concerns.
- Assumption, identification, and location of command.
- Request or release resources as required.

FIRE OFFICER TIP

The C-A-N Progress Report

Many departments use the C-A-N process method to provide a progress radio report.

C: Conditions
- Your location on the fire ground
- Smoke and heat conditions
- Obstacles you encounter
- What is burning

A: Actions
- Completing a primary search
- Finishing an assigned objective
- Activities in progress, such as pulling ceilings or performing salvage
- Reporting fire is under control

N: Needs
- Urgent assistance
- Additional fire fighters
- Tools or equipment
- Ventilation

A calm, concise, and complete radio report helps the dispatch center, the chief officer, and other responding companies to understand the situation at hand. It is important to use radio terminology that meets the department's SOPs (standard operating procedures) and is clear to everyone who is listening. Close compliance with the communications SOPs is an important component of effective fire-ground command.

A visit to the dispatch center is a valuable experience for every fire officer. It is important to understand the dispatch center's operating conditions and to appreciate how dispatchers depend on fire officers at the incident scene to provide good information.

Other Responding Units

The first-arriving fire officer provides leadership and direction to responding units by implementing the incident management system. The fire officer must demonstrate the ability to take control of the situation and provide specific direction to all of the units that are operating and arriving. This requires mastery of the autocratic style of leadership.

Radio Reports

Company fire officers frequently provide verbal reports over the radio. Radio communications are essential for emergency operations because they provide an instantaneous connection and can link all of the individuals involved in the incident to share important information. During an emergency incident, both the sender and the receiver should strive to make their radio messages accurate, clear, and as brief as possible. Conditions are often stressful, with several parties competing for air time and the receiver's attention. There may only be a limited time to transmit an important message and ensure that it is received and understood.

The individual who is transmitting a radio report often feels an intense pressure to get the message out as quickly as possible. The realization that a large audience could be listening often adds to the sender's anxiety level.

When given the option, many fire officers prefer to use a telephone or face-to-face communications instead of a radio to transmit complicated or sensitive information. These options allow a more private exchange of information between two individuals, including the ability to discuss and clarify the information. A good practice is to think first, position the microphone, depress the key, take a breath, and then send a concise, specific message in a clear tone.

Project Mayday Radio Phrases

Chief Don Abbott is an incident command expert. His experience includes developing the Phoenix Fire Department Command Training Center. In 2015 he started the Project Mayday initiative to improve fire command awareness. Project Mayday examined nearly 3000 recordings from audio dispatch, dash cams, helmet cams, and body cams during a 48-month period in order to identify the events and communications that occur during operations that include a mayday (Abbott, 2019). The project found that in 87 percent of maydays, there is a major breakdown in communications, orders issued or received, missed messages, walk-over communications, and the worst of problems, missing a mayday call the first time (54 percent mayday calls are missed).

Project Mayday developed a system for tracking phrases repeated over the radio prior to a mayday being called. These phrases—which appeared in 88 percent of the mayday recordings reviewed—should serve as a trigger for an IC to reconsider their current operations. For example, hearing one of these phrases should get the IC's attention; hearing two or three of these phases could mean that a mayday is imminent (Abbott, 2018):

1. "We have zero-visibility conditions."
2. "We have fire above our heads."
3. "We have fire below us."
4. "We need more line to reach the fire; extend our line."
5. "We have not found the seat of the fire."
6. "We are running out of air" (or indications of a "low-air alarm").
7. "This is a hoarder structure."
8. "We have had a flashover."
9. "We have had a ceiling/roof collapse."
10. "We have lost multiple windows."
11. "It's really getting hot in here; we are backing out."
12. "Our exit has been blocked."
13. From a crew operating within the structure: "We are sending a fire fighter out with a problem."
14. "We have a hole in the floor" or "We have had a floor collapse."
15. "Command has lost communications with multiple crews."
16. "We have a lot of sprinkler heads going off in here."

Keep Your Supervisor Informed

The battalion chief depends on you to share information about what is happening at the fire station or in your work environment. In particular, a fire officer needs to keep the chief officer informed about three areas:

- Progress toward performance goals and project objectives: A fire officer needs to keep his or her chief apprised of work performance progress, such as training, inspections, smoke detector surveys, and target hazard documentation. It is especially important to let the chief know about anticipated problems early enough to get help and to keep the projects running on time.
- Matters that may cause controversy: The chief should be informed about conflicts with other fire officers or between shifts, or conflict that extends outside the organization. The chief also needs to know about any disciplinary issues or a controversial application of a departmental policy. When conveying these kinds of messages, sooner is better than later. Contact your supervisor to discuss a potential disciplinary issue and the best approach to enforcement and resolution.
- Attitudes and morale: A fire officer spends the workday with a group of fire fighters at a single fire station, whereas the command officer spends much of the workday in meetings or on the road. The fire officer should communicate regularly with the chief about the general level of morale and fire fighter response to specific issues.

How Quickly Do You Get Bad News?

Few things damage a new fire officer's reputation more quickly than not finding out about a situation that is going poorly until it is too late to fix it. Effective fire officers create a work environment that encourages subordinates to report bad news immediately. The sooner that an officer receives the bad news, the faster he or she can implement corrective action. Fire fighters should be encouraged to tell their supervisor whenever an event or performance is out of balance with expectations (**FIGURE 3-4**). This includes immediately reporting injuries or broken equipment, regardless of the time of day.

The same principle applies to delivery of bad news during an emergency. Fire officers who create a barrier to receiving administrative bad news can suffer catastrophic results at an emergency scene. From the fire fighter's perspective, the fire officer who does not appreciate hearing bad news in the fire station will probably not want to hear it at the emergency scene, either. In some cases, a fire fighter might even be afraid to

FIGURE 3-4 Fire fighters should be encouraged to report problems immediately.
© Jones & Bartlett Learning. Photographed by Glen E. Ellman.

point out a problem that could save a life or prevent a serious injury.

The Grapevine

Every organization has an informal communication system, often known as the "grapevine." The flow of informal and unofficial communications is inevitable in any organization that involves people. The grapevine flourishes in the vacuum created when the official organization does not provide the workforce with timely and accurate information about work-related issues. Much of the grapevine information is based on incomplete data, partial truths, and sometimes outright lies.

A fire officer can often get clues about what is going on but should never assume that grapevine information is accurate and should never use the grapevine to leak information or stir controversy. In addition, a fire officer needs to deal with grapevine rumors that are creating stress among the fire fighters by identifying the accurate information and sharing it with subordinates.

Supervisory Tasks

There are general supervisory responsibilities for first-line supervisors:

- Set the direction for the fire company work that carries out the department's mission, vision, and strategic goals.
- Ensure your fire company members deliver high-quality work.
- Manage the workload so your fire company can deliver work that is on-time and within the department's budget.
- Maintain a safe and harassment-free workplace.
- Hold your fire company members accountable for both outcomes and behavior.
- Develop your fire company members through leading, supporting, coaching and counseling. (Flaherty, 2013)

The authority having jurisdiction will have specific supervisory responsibilities, such as providing an annual evaluation of every member, that vary with location. These are described in the personnel rules of the organization as well as the fire department rules, regulations, and guidelines.

FIRE OFFICER TIP

Looking Official When Performing Formal Fire Officer Duties

Many departments provide two different uniforms that can be worn by the fire officer. The station work uniform could be a golf shirt, or "job" shirt. The more formal "Class A" uniform generally includes a long-sleeve collared shirt with insignia, badge, and tie.

One method to help fire fighters understand when the officer is performing an official supervisory task is to wear the uniform shirt. If the officer always puts on a dress shirt and tie to conduct employee evaluations or disciplinary actions, the serious and official nature of these activities is clearly visible. This increases the fire officer's effectiveness.

Grievance Procedure

A **grievance** is a dispute, claim, or complaint that any employee or group of employees may have about the interpretation, application, or alleged violation of some provision of the labor agreement or personnel regulations. A **grievance procedure** is a formal structured process that is employed within an organization to resolve a grievance. In most cases, the grievance procedure is incorporated in the personnel rules or the labor agreement and specifies a series of steps that must be followed in a particular order. If the problem cannot be resolved in a mutually acceptable manner at one level, it can be taken to the next level, and so on, up to some ultimate level, where an individual or body has the final authority to impose a binding decision.

The grievance procedure should specify a sequential process and a timeline to move through the steps. The grievance can be resolved at any point by management accepting the complaint and the corrective action requested by the grievant or by both sides reaching a negotiated settlement that is acceptable to each party. If management rejects the claim, the grievance can be taken to the next level. The timeline ensures that a grievance will not remain stalled at any level for an excessive time period.

An employee can contact a union representative at any time to discuss a situation, including how the union interprets the rule in question and whether a grievance should be submitted. The employee's union representative usually becomes formally involved at either the first or second step of the grievance process. The union representative acts as an advocate for the individual or group that submitted the grievance. The union becomes more deeply involved as the process moves through the various steps of the grievance procedure, particularly in cases in which the problem has broad impact within the organization.

The objective should always be to resolve the problem at the lowest possible level and in the shortest possible time. Grievances that must be processed through multiple steps are disruptive, time consuming, and costly to both sides. The ability to resolve problems at a low level is an indication of a healthy organization with a good labor-management relationship, whereas a steady stream of grievances moving up to the highest levels is a symptom of major relationship problems.

The following section outlines one version of the grievance procedure. The details and sequences described here may differ in other fire departments. A similar process can be established to resolve disputes in fire departments where there is no formal labor contract, even in an all-volunteer organization. The most important responsibility for a fire officer is to know and follow the procedures that apply in his or her organization.

Sample Step 1

The grievant presents his or her complaint verbally to a supervisor, shortly after the occurrence of the action that gave rise to the grievance. In some organizations, this nondocumented verbal notification is called an "informal grievance" or step zero. Even this informal step requires the grievant to provide three important pieces of information:

- The article and section of the labor agreement or personnel regulation alleged to have been violated
- A full statement of the grievance, giving facts, dates, and times of events, as well as specific violations
- A statement of the desired remedy or adjustment

Sample Step 2

The second step initiates the formal part of a grievance procedure. If the problem is not resolved at step 1, the employee may prepare and submit a written grievance (**FIGURE 3-5**). This is usually submitted on a specified grievance form. The employee, the employee's supervisor, and the personnel office all receive a copy of the grievance.

The supervisor has 10 calendar days to reach a decision and provide a written reply to the grievant. Failure to respond to the grievance within 10 days means that the supervisor has denied the grievance and the grievant can immediately go to step 3.

Sample Step 3

A step 3 grievance is written out on another specific grievance form and again specifies the article and section of the contract or personnel regulation alleged to have been violated; the dates, times, and specific violations that are alleged to have taken place; and the desired remedy or adjustment. Copies of the step 2 grievance form and the supervisor's response are attached to this document.

A step 3 grievance is submitted to a second-level supervisor, typically an administrative fire officer, who has 10 calendar workdays to respond. If the grievance is denied or the administrative fire officer does not respond within the specified time, the grievant can move to step 4 and present his or her grievance to the fire chief.

Sample Step 4

If the grievance remains unresolved, the grievant can present it to the fire chief or designee as the fourth step. The same written information must be submitted, along with all of the documentation from the previous steps. The fire chief has 10 workdays to respond to a step 4 grievance.

If the fire chief does not respond within the time frame, or if the grievance remains unsettled, the process moves out of the fire department to a mediator, personnel board, or civil service board for resolution. In this example, the grievance goes to the municipal administrator and, if not resolved at that level, goes to federal or state arbitration.

Making Decisions

A fire officer is called upon to make many different types of decisions about a wide variety of subjects. Most of the problems that arise are fairly uncomplicated, although they are not necessarily easy to solve. Moving up in the ranks means an exponential increase in decision-making situations. Often the problems become more complex at higher levels of the hierarchy and can require the participation of multiple organizations.

SAMPLE GRIEVANCE FORM

STEP TWO: TO BE COMPLETED BY UNION OR EMPLOYEE

Date of filing: _____

From: _____ _____ _____
         Employee   Rank   Assignment/shift

Grievance form must be submitted within 15 calendar days of the incident being grieved.

STATEMENT OF GRIEVANCE Must (1) contain a statement, as complete as possible under the circumstances, of the grievance and the facts upon which it is based, including dates, times, locations, names of witnesses, and other appropriate information; (2) identify the section(s) of the Contract Agreement that affect this grievance; (3) state requested remedy or corrective action.
Additional pages may be attached to the Sample Grievance Form.

(*this is where the employee enters the statement*)

Original copy of the completed Sample Grievance Form shall be delivered to the employee's immediate supervisor, with a copy to the Human Resources office and the Union representative.

_____   _____
       Employee   Date

TO BE COMPLETED BY IMMEDIATE SUPERVISOR within 10 calendar days of receipt of Sample Grievance Form

_____ _____ _____ _____
  Supervisor's Name   Rank   Work Location/Shift   Date Grievance received

(*this is where the immediate supervisor enters a response*)

_____   _____
  Supervisor Signature   Date

If employee is satisfied with Supervisor's answer, sign the original Sample Grievance Form acknowledging agreement and submit it to the Human Resources Director for placement in your employment records. If employee is NOT satisfied, shall sign the original Sample Grievance Form acknowledging disagreement and immediately notify the Union in writing. The original Sample Grievance Form shall then be submitted by the employee to the Deputy Chief within ten (10) calendar days of the decision of the immediate supervisor.

Agree_____   Do not agree:_____

FIGURE 3-5 Example of a grievance form.
© Jones & Bartlett Learning.

The following systematic approach is recommended to ensure high-quality decision making:

1. Define the problem.
2. Generate alternative solutions.
3. Select a solution.
4. Implement the solution.
5. Evaluate the result.

Although this five-step technique appears to be designed for situations where plenty of time is available, the same basic approach is used for emergency incidents. A fire officer should be able to move quickly through these steps. Training and experience will prepare you as the fire officer to identify the pertinent problem, generate realistic solutions, and select the best option quickly.

Define the Problem

The first step in solving any problem is to examine the problem closely and to define it carefully. A well-defined problem is one that is half-solved. Poorly defined problems, in contrast, waste tremendous time and effort.

Richard Gasaway, PhD, is a retired fire chief who has studied fire officer decision making. Fire officers should know what is going on within the organization and address most difficult issues before they become major problems (Gasaway, 2015). The best way to prevent major problems is to deal successfully with minor issues before they reach the crisis stage (**FIGURE 3-6**).

Peter Drucker, former professor of management at the Graduate Business School of New York University, encouraged managers to question the value of each organizational activity once per year, because what may have been vital last year may have minimal importance this year (Drucker, 1974).

Fire departments have a strong inclination to keep doing the same things in the same ways. The fire officer should identify which activities can be changed, improved, or updated. If there is a better way to do something, then give it appropriate consideration. If

FIGURE 3-6 Fire officers must pay attention to what is going on within the organization and deal with problems before they reach the crisis stage.
© Jones & Bartlett Learning. Photographed by Glen E. Ellman.

the department is doing something that is no longer worth doing, why not use the time to do something more productive?

The best people to solve a problem are usually those who are directly involved in the problem. The fire officer who is struggling with a new incident management clipboard in the rain has a valuable perspective on what might improve the clipboard's performance. The fire fighters who make roof ventilation openings have ideas about how those operations could be accomplished more efficiently. Company-level problems are most likely to be solved by involving the members of the company.

Generate Alternative Solutions

Brainstorming is a method of shared problem solving in which all members of a group spontaneously contribute ideas. A typical fire company is a good size and a natural group for brainstorming. Fire fighters are usually very adept at solving problems.

The following eight steps will assist the fire officer when brainstorming alternative solutions:

1. Using a flip chart, whiteboard, or chalkboard, write out the problem statement. Everyone should agree with the words used to describe the problem.
2. Give the group a time limit to generate ideas. Although the scope of the problem is a factor, 15 to 25 minutes seems to work well for groups of 4 to 16.
3. The fire officer should function as the scribe. The scribe writes down the ideas and keeps the group on task.
4. Tell everyone to contribute alternative solutions. At this point, there is no commenting on ideas. All suggestions are welcome.
5. Once the brainstorming time is up, have the group select the five ideas they like the best.
6. Write out five criteria for judging which solution best solves the problem. Criteria statements include the word "should"—for example, "It should be legal" or "It should be possible to complete in 6 months."
7. Have every participant rate the five alternative solutions, using a 0 to 5 scale. The value 5 means the solution meets all of the criteria, whereas 0 means the solution does not meet any of the criteria.
8. Add up the scores for each idea. The idea with the highest score is the one that provides the best solution for your problem.

Some important constraints apply to brainstorming. Most notably, the process assumes that the problem statement is accurate and the criteria are valid. On occasion, the best solution that comes out of a fire station may crash at headquarters because of a criterion or restriction that is unknown to the fire officer, such as a new federal or state regulation.

A legitimate problem-solving process has to be reasonable and based on logic and organizational values. Fire fighters should be able to anticipate that their decision will be implemented if it meets the criteria. Going through a process that results in no change or provides no feedback to the fire company members is the quickest way to destroy fire fighter participation in the decision-making process. A legitimate process reinforces the trust between the fire company and the fire administration.

Select a Solution

At this point in the decision-making process, the fire officer will have defined the problem, generated solutions, and ranked them based on criteria. Criteria are based on factors important to the department, the work group, and the fire officer. These may be found in the department's mission statement or core values. For example, if participation in local neighborhoods is one of the core values, a solution that increases the fire department's involvement in local neighborhoods would be preferred.

Implement the Solution

Once the decision has been made, the solution still has to be implemented. The implementation phase is often the most challenging aspect of problem solving, particularly if it requires the coordinated involvement of many different people. One reason for involving as many players as possible in making the decision is to capture their commitment to the plan when it is time for implementation.

Consider a plan for reducing the number of fire fatalities in residential occupancies. After careful analysis, the decision is made that the best strategy will be a community outreach program to check every smoke alarm in every multiple-family dwelling in a community over a 4-month period. This plan supports the fire department's mission statement and reflects the organization's values. To accomplish this objective, each fire company will have to invest 2 hours every evening for the next 4 months.

Before this plan can be implemented, buy-in is required. The plan has to be "sold" to the people who will perform the task, particularly to those who were not involved in the decision-making process. Rearranging schedules to do something different for 2 hours every evening involves a significant behavioral change, even if it is for a limited time and for a good cause. It would be possible simply to order everyone to "just do it"; however, this is not the best implementation strategy for a program that involves extensive public contact. Willing participation works better than involuntary compliance.

The fire officer must clearly assign tasks to individuals or teams. The work group benefits from the use of a project plan, which lists tasks, responsibilities, and due dates (**FIGURE 3-7**). Most of the projects supervised by a company-level officer do not require a sophisticated planning system. A simple project control document can be used to divide a project

Fiscal Year 2025 Capital Improvement		12/13/2024	
Fire Fighter Skills Training Station			
Goal		**Due Date**	
Fire fighters can practice forcible entry, roof ventilation, vent-enter-isolate-search (VEIS), and rapid intervention scenarios		5/1/2025	
Task	**Assigned To**	**Due Date**	**Status**
Visit and report on other skill training stations	Station commander and training center manager	8/16/2024	Done
Issue RFP for commercial training props	Logistics	9/9/2024	Issued 8/26
Review RFP responses	Resource management/station commander	10/18/2024	Done
Award RFP	Procurement	11/6/2024	Award 11/3
Remove debris and vegetation	B shift	11/13/2024	Done
Relocate storage shed	A shift	11/20/2024	Done
Build roof and pour cement foundation	Public works	12/4/2024	Done 11/27
Identify installation requirements for each prop	C shift with contractors	1/8/2025	
Lay out skill station locations	A, B, C shift, volunteers, station commander	1/15/2025	Started 12/4
Install training props	Contractors	3/5/2025	Need to confirm
Create instructions for each skill station	A, B, C shift	4/4/2025	

FIGURE 3-7 Example of a project control document for a fire station work product.
Courtesy of Mike Ward

into segments, with milestones established to identify progress. Complex and long-range tasks may require a formal project management plan and a designated coordinator, particularly if they require the coordinated activity of multiple agencies.

An implementation plan must include a schedule to ensure that the goals are met. Deadlines focus effort and help prioritize activities. If the solution requires activity by other organizations, such as changing a local ordinance or submitting a budget request, then the fire officer must determine the time it will take to accomplish that task and incorporate that time into the schedule. A schedule is valuable only if it is followed and someone ensures that the deadlines are met.

Plan B. Many problems remain unsolved long after the problem has been clearly defined and a good solution has been selected. A problem is not truly solved unless its solution is implemented. There are many reasons why good solutions are not implemented; for example, the required approvals might not have been obtained or the necessary resources might not be available. Quite simply, the organization might not have the capability to solve the problem or to implement the solution that was selected.

Fire officers should consider a "plan B" if the original solution cannot be implemented. Plan B could be an extended implementation schedule, a modified plan, or a completely different solution to the problem.

Evaluate the Results

After implementing the solution, the fire officer must assess whether it produced the desired results. Evaluation should be a standard part of the process of any problem-solving activity. The nature of the evaluation depends on the complexity of the problem and the solution; in most cases, an initial evaluation should be performed immediately after implementation, and follow-up evaluations should be performed at regular intervals.

Determining whether the solution actually solved the problem requires some type of measurement that compares the original condition with the condition after implementation. For example, do the new hose loads really result in quicker deployment of attack lines? Answering this question requires data on how long the old way took versus how long the new way takes. The evaluation should also look for situations where the original problem is solved, but another unintended and equally bad situation is created. For example, if the hose is deployed more quickly with the new technique but comes out twisted and kinked, the negative impact could outweigh the positive.

If necessary, the fire officer needs to be prepared to adjust the plan or reevaluate the original decision. Many problems are solved in stages, with gradual progress being made toward a solution. In spite of the analysis, plan B may turn out to be a better choice. Changing a plan should not be viewed as a failure.

Part of the evaluation process involves going back and listening to the people who identified the original problem and asking them for feedback. A small city adopted a big city policy of sending a single fire company to a street fire alarm pull box. In the big city, the street pull box resulted in a false alarm for 97 percent of the activations. In the small city, fire companies were frequently encountering working structure fires when responding to a fire alarm pull box. The small city fire chief listened to the fire officers and dispatchers and understood that sending a single fire company to pulled fire alarm boxes was producing undesirable results. The policy was changed, going back to the two engine and one truck response to a fire alarm pull box.

Assigning Tasks to Unit Members

Supervising a Single Company

A fire company is a basic tactical unit for emergency operations, with the fire officer in the role of a working supervisor (**FIGURE 3-8**). When the company is assigned to advance an attack line into a structure, the fire officer is alongside the crew members, directing and leading them as well as continually evaluating the environment for hazards such as backdraft, flashover, or structural collapse. The officer's personal and physical involvement in fire suppression activities should never override his or her supervisory duties.

FIGURE 3-8 Supervising emergency operations is a core fire officer task.
© Jones & Bartlett Learning. Photographed by Glen E. Ellman.

In addition to leading and participating in company-level operations, the fire officer constantly evaluates their effectiveness. Is the fire stream making progress or losing ground against the fire? What is the impact of horizontal and vertical ventilation? It is up to the fire officer to identify and mitigate fire-ground problems. For example, if the fire attack appears to have stalled, the officer determines if an additional or larger line is needed or if the attack should be made from a different direction.

The fire officer should use sound organizational management principles.(See Chapter 1, *The Fire Officer I as a Company Supervisor*.) By ensuring that each person has only one supervisor, utilizing a *unity of command* eliminates the confusion that can result when a person receives orders from multiple supervisors. Unity of command also reduces delays in solving problems as well as the potential for life and property losses. In most situations, one person can effectively supervise only three to seven people or resources (FEMA, 2013). The actual *span of control* should depend on the complexity of the incident and the nature of the work being performed. Using *division of labor*, the specific assignment of a task to an individual makes that person responsible for completing the task and prevents duplication of job assignments. *Discipline* encompasses behavioral requirements, such as always following orders from superior officers and performing up to expectations.

The fire officer relays relevant information to the branch, group, or incident commander. At a structure fire, this responsibility would include informing command when the company's assignment is completed or when the task is delayed or cannot be accomplished. It also includes informing the supervisor immediately when problems or hazards are encountered, such as signs of structural collapse. Each company officer serves as the eyes and ears for the IC.

Closeness of Supervision

The inherent risks associated with emergency operations demand close supervision at all times. Fire fighters who are assigned specific tasks, such as forcing open a door, often concentrate on the task and ignore everything else. The fire officer's role is to review the entire area of responsibility, monitor progress, coordinate with other companies, and look out for hazards.

The level of supervision should be balanced with the experience of the company members and the nature of the assignment. An inexperienced crew performing a high-risk task requires more direct supervision than an experienced crew performing a more routine task. The location of the task also affects the level of supervision. A crew that is advancing a 2½-in. (64-mm) attack line inside a large warehouse will require much closer supervision than a crew operating a master stream device 15 ft (4.6 m) away from the same burning warehouse. The company officer must know where the crew members are located and what they are doing. When performing a task in a high-risk situation, such as an immediately dangerous to life or health (IDLH) environment or below-grade rescue, the fire officer needs to see and directly communicate with all of the members.

Standardized Actions

Emergency incident operations must be conducted in a structured and consistent manner. This goal is accomplished by placing a strong emphasis on SOPs. SOPs provide a framework to allow activities to be efficiently completed through the efforts of everyone involved in the event. Just like a football team playbook, SOPs explain the standard approach that should be followed in a particular situation.

Standardized approaches are especially helpful in volunteer organizations. NFPA 1720, *Standard for the Organization and Deployment of the Fire Suppression Operations, Emergency Medical Operations, and Special Operations to the Public by Volunteer Fire Departments*, describes demand zones (urban, suburban, rural, remote, and special) and staffing response times. For example, 10 fire fighters are expected to respond within 10 minutes to 80 percent of the incidents in the suburban demand zone.

Such a standardized approach facilitates the development of an incident action plan and promotes a consistent approach to risk and safety. Utilizing an incident command system (ICS) facilitates a consistent development of efficient incident management with effective control.

Command Staff Assignments

Command staff assignments include the roles of incident safety officer, liaison officer, and public information officer. While working in one of these positions, the fire officer reports directly to the IC. The IC could assign any available and qualified officer to perform one of these roles.

The incident safety officer is responsible for overseeing the incident from a safety perspective, keeping the IC informed of safety concerns, and taking preventive action when an immediate hazard is identified. When assigned as liaison officer, the fire officer functions as the link between the IC and representatives from various agencies. At a hazardous materials incident, such representatives could include the property owner or a chemical manufacturer. When operating as the public information officer (PIO), the fire officer is responsible for gathering information to be released to the general public, developing news releases, and giving interviews or press conferences. The PIO acts as the spokesperson for the IC (FEMA, 2017).

FIRE OFFICER TIP

Do Not Forget Your Crew!
During a simulated incident exercise, single-company supervisors are assigned to command staff positions as the incident command system ramps up. Remember to identify where the rest of your fire company will be assigned. Often your company will be assigned to work with another fire company. Make sure that your crew is working under another supervisor.

Assigning Tasks During Emergency Incidents

Emergency Scene Leadership

A fire officer always has direct leadership responsibility for the company that he or she is commanding. The first-arriving fire officer has additional important responsibilities to establish command of the incident and to provide direction to the rest of the responding units. The first-arriving fire officer also has leadership responsibilities that relate to the communications center.

Methods of Assigning Tasks. The fire officer's primary responsibility is to the team of fire fighters under his or her direct supervision. The officer is responsible for their safety, their actions, and their performance at the incident scene. The fire officer should develop a consistent approach to emergency activities, in which all of the department's standard operating procedures are followed.

Most departments have some form of SOPs that specify what is expected of each company and what to do in a given situation. Some SOPs are specific and detailed, whereas others are more general. For departments with detailed procedures, the task that each fire fighter is to perform may already be determined. The SOPs may indicate which tools the fire fighter is to carry based on whether the company includes three or four members. In these cases, the fire officer is responsible for ensuring that the fire fighters know and follow the policy.

For departments that have broad SOPs or none at all, the fire officer has two choices in assigning tasks: preassigning them or assigning them as needed on the scene. The advantage of preassignment is that it relieves the officer of the burden of having to make some decisions during the emergency. The advantage of assigning tasks on the scene is that it reduces unnecessary effort.

When giving assignments at the scene, the fire officer must be clear and concise and ensure that all fire fighters understand their assigned tasks. In this situation, the fire officer uses an autocratic style of leadership because the emergency scene does not allow for participative decision making. For this method of operation to be effective, the fire fighters must clearly understand the reason for the use of different leadership styles before the incident. One method that may help a fire officer make assignments under pressure is to develop a standard method of handling situations. Making standard decisions in a consistent manner will assist in the decision-making process. If you cannot use a written checklist, then developing a mental checklist that is rapidly completed at every incident will create a predictable outcome. Because so many factors come into play when determining tactics, it is crucial that officers be trained in the critical decision-making process. For example, if you need to assign someone to rescue responsibilities, always consider first whether you have a rescue company available; if not, then a truck company and lastly an engine company can be used to handle this task. This understanding allows a sequential thought process to develop, making assignments easier.

One method of assigning tasks that allows for more participation in the process yet reduces the number of decisions the officer must make at the scene is to rely on the broad SOPs, such that the fire officer discusses with the crew the various "routine" emergencies to which they respond. The officer and crew can discuss the needs of the incident and the steps that must be taken to mitigate it. Finally, the officer and crew can decide who will be responsible for doing what.

For example, the crew may decide that at any life-threatening medical incident, the fire fighter is responsible for checking the victim's respirations, securing the airway, and then providing ventilation. The driver is responsible for checking for a pulse and uncontrolled bleeding and then providing compressions. The officer is responsible for setting up and using the automated external defibrillator (AED). This plan ensures that the crew covers all the basics of care without fumbling and waiting to be directed.

Critical Situations. Dangerous situations may develop suddenly during incident operations. A fire officer must use the autocratic leadership style when immediate action is required. Orders to evacuate a building and a fire fighter mayday are two critical events that require immediate autocratic action.

When command issues an evacuation order, all members of the fire company should immediately leave the fire building. Evacuation may mean that the company leaves a fire hose in the building and the fire fighters quickly exit the building. The fire officer is responsible for accounting for the fire fighters under his or her supervision. The fire officer provides a head count to the group command officer or the IC.

A mayday requires a complex response from the company operating within a hot zone or burning structure. The first obligation is to maintain radio discipline so that command can determine the mayday location and situation. The second obligation is to maintain company or group integrity. Incident control evaporates if every fire fighter rushes to assist the fire fighter in trouble. The fire officer needs to maintain company integrity to facilitate the mayday rescue and continue to fight the fire or control the hazard. Changes in assignment come from the group IC.

The fire officer may directly encounter a dangerous situation, such as a collapsing building, an active shooter, or a careening vehicle. In such a case, the officer must make an immediate, clear, and autocratic command to protect the fire fighters under his or her supervision by removing them from the danger area. Then, the fire officer needs to inform command.

Assistant Chief Joseph Pfeifer, when reviewing a rescue of a driver trapped in a sanitation truck that crashed through a fourth-floor exterior wall in New York City, made this observation: "As a novel incident increases in scale and complexity, it is important for incident commanders to separate management and leadership functions. Leadership is about stepping back and detaching from the management of the incident to analyze what is taking place and project future actions; then leader reconnects to guide management in adapting to novelty" (Pfeifer, 2012).

After every incident, the fire officer should briefly review the event while still at the scene or as soon as possible after returning to quarters. This is an opportunity to clarify issues and answer questions. The fire officer can reinforce good practices and immediately identify any unacceptable performance.

Assigning Tasks During Nonemergency Incidents

Most fire officer supervision is directed toward accomplishing routine organizational goals and objectives in nonemergency conditions. This includes being well

FIRE OFFICER TIP

Practice Like You Play

A key to consistent performance at emergency incidents is to "practice like you play." Every training activity and every response should be approached with the same degree of professionalism and strict adherence to SOPs. This approach leads to the company performing as expected when the situation demands maximum effort.

Most structure fire responses address relatively minor situations, such as activated fire alarms and fires that involve less than a room and contents. To prepare the fire company for larger fires and critical situations, each response should require the same consistent approach. Following are some possible examples of response actions:

- All crew members don their protective clothing, sit down, and fasten their seat belts before the apparatus moves, and they remain seated and belted as long as the apparatus is moving.
- The first-arriving engine company always establishes a water supply.
- The first-arriving ladder company positions and prepares to deploy the aerial apparatus.
- Crews entering the building wear full protective gear, including SCBA; carry their tools; and perform their assignments for a working fire.

If there is any doubt about whether a fire is present in the structure, all operations should be performed with the assumption that a fire truly exists. Without creating a customer service problem, the fire company should search the entire building, including attics and basements. This approach allows each fire fighter to develop awareness of the built environment. More importantly, when the company encounters a working fire, the standard fire-ground tasks have been practiced and are familiar.

prepared to perform in emergency situations. The fire officer accomplishes most of these goals and objectives through the efforts of fire fighters. The role of the officer involves influencing, operating, and improving to ensure that those efforts achieve the desired results.

Fifty years ago, fire officers used an autocratic style of leadership in both emergency and nonemergency duties. The officer decided what was to be done and who was to do it, and the fire fighters responded without question. Today, the workforce is very different. Employees demand to be included in the decision-making process. Consequently, effective fire officers provide a more participative form of leadership in routine activities.

Although the fire officer is commonly given specific assignments that the company must complete, much discretion may apply regarding when, how, and by whom activities are carried out. Some officers sit down with their crews at the start of the shift and cover the assignments, if any, that the administrative fire officer has indicated need to be completed for the day. The officer may also outline the status of other duties that will need to be completed over the long term but do not need to be undertaken immediately. An open question is then addressed to the crew regarding what they think should be done, including any areas that they believe should be addressed. Depending on the outcome of the discussion, the officer decides or has the group decide on the plan for the day to accomplish the activities.

Some tasks must be completed every day. In most departments, station cleaning is performed daily, making a trip to the grocery store is required, and equipment must be checked. The fire officer may decide that these duties are easiest to accomplish if they are assigned in advance. Typically, in conjunction with the crew, a determination is made about which duties will be completed every day based on the work position of the individual. The driver is usually assigned to check all equipment and clean the apparatus bay. The rookie fire fighter may be assigned to the bathroom and kitchen detail, whereas the senior fire fighter is assigned to sweeping and vacuuming floors.

Safety Considerations

Fire department operations often include high-risk situations that can occur under any weather conditions at any time of the day or night. The fire officer is responsible for ensuring that every fire fighter completes every incident or work assignment without injury, disability, or death. This idea is expressed as a fire officer's special obligation to ensure that "everyone goes home" at the end of a workday.

The responsibilities of a fire officer include identifying hazards and mitigating dangerous conditions to provide a safe work environment for fire fighters. The fire officer must also identify and correct behaviors that could lead to a fire fighter's injury or death. Safe practices must be the only acceptable behavior, and good safety habits should be incorporated into all activities. See Chapter 8, *Safety and Risk Management*, for more information.

Fire Officer Training Responsibilities

Training and coaching have been core fire officer tasks since the establishment of the first organized fire departments. **Training** is defined as the process of achieving proficiency through instruction and hands-on practice in the operation of equipment and systems that are expected to be used in the performance of assigned duties. Fire service training has evolved in complexity and sophistication at a rapid pace as new areas of expertise have been added to the list of services performed by fire departments. **Coaching** is a method of directing, instructing, and training a person or group of people with the aim of achieving some goal or developing specific skills.

NFPA 1041, *Standard for Fire and Emergency Services Instructor Professional Qualifications*, defines the standard and describes the requirements for five levels of instructor. Certification as an Fire Instructor I is a prerequisite for Fire Officer I candidates.

A basic responsibility of every fire officer is to provide training for subordinate fire department members. The specific training responsibilities assigned to fire officers vary depending on the organization and the available resources. At a minimum, a fire officer must be prepared to conduct company-level training exercises and evolutions to ensure that the company is prepared to perform its basic responsibilities effectively and efficiently. Live fire training is a high-hazard activity. Before conducting such training, the fire officer should be certified to the appropriate NFPA 1041 level as a Live Fire Instructor or Live Fire Instructor in Charge.

> **FIRE OFFICER TIP**
>
> **Four-Step Fire Fighter Development**
>
> The four-step method remains the method of choice if a candidate must describe how to improve performance with an existing procedure or explain a new procedure or device. When developing a written or oral response to such an issue, the candidate should identify each of the four steps in the training process.

The Four-Step Method of Skill Training

One of the enduring concepts of the 20th century is the four-step method of skill training. This prepare-present-apply-evaluate method originated during World War I, when the armed services were teaching farmers how to fly biplanes, drive tanks, and operate ships. The process was updated and renamed **job instruction training** when more than 1 million men and women received technical skill training during World War II.

Today, the four-step method is the foundation of the work performed at the Fire Instructor I level. Most

Voice of Experience

Leadership, including how it relates to "leading the company," is as mysterious and confrontational a topic as the firehouse conversation of smooth-bore nozzle versus fog nozzle. Leadership is getting the troops to do what we need to attain the goals of the organization or department. How do we get them to do the assigned tasks—and, better yet, how do we get them to *enjoy* those tasks? As leaders, we are tasked with getting the job done, which is more easily accomplished on the fire ground than it is around the firehouse. In our paramilitary structure, the fire ground is the battlefield. Order, unity, honor, and a strong command presence are a must. We lead by example, we lead from the front, and we listen to what the troops tell us. As command officers, we must trust our troops. As company officers, we must trust our troops. Where does this trust come from?

Many more decisions are made *away* from the fire ground than *on* the fire ground. In training, it is important to allow the troops to make a mistake (as long as it does not result in injury). If you, as their leader, make a mistake, do not be afraid to say, "I blew that." In making a mistake, you learn a very important lesson, and you can then relay that lesson to the troops. You might be surprised by the extent to which your creditability as an officer will flourish.

We ask fire fighters every day to place their lives in harm's way to potentially save someone they do not know. When was the last time we told our troops "thank you"? I firmly believe that, both on and off the fire ground, the best leaders focus on what their people are doing right and reward them for it. A simple "thank you" can have an immense impact. Treating the troops with respect and honor will pay dividends in the long run.

Good leaders know their people—both their strengths and their weaknesses. A good leader can take the most inexperienced employee and determine the positive elements that the person can bring to the table, expand on the positive traits, and motivate the employee to become a critical and vital member of the team. Our challenge as leaders is to find that desirable trait in everyone with whom we work. The true leader can develop this skill and apply it to all ranks and members of their organization.

Stephen K. Lovette
Captain
High Point Fire Department
High Point, North Carolina

fire officers use standardized curricula and training packages that are commercially prepared or developed by the local fire academy. Occasionally a company officer needs to start from scratch to develop a training program.

Step 1: Preparation

The fire officer conducts training to maintain proficiency of core competencies. Crews should be able to catch a hydrant, raise a ground ladder, and deploy attack lines so well that the task is automatic. In other words, they should be **unconsciously competent** with these tasks.

Sometimes the fire officer or the department determines the need for focused instruction. Three indicators that training is needed would be a near miss, a fire-ground problem, or an observed performance deficiency. For example, during a structure fire, the fire officer might observe that the fire fighters are having difficulty placing a 35-ft (11-m) extension ladder at an upper-floor window, or perhaps the deployment of an attack line from a standpipe connection created a tangle of hose in the stairwell. These problems need to be addressed through additional training and practice.

The fire officer begins by obtaining the necessary material and teaching aids. If needed, the fire officer writes a lesson plan (**FIGURE 3-9**). Components of a lesson plan include the following:

1. Break the topic down into simple units.
2. Show what to teach, in what order to teach it, and exactly what procedures to follow.
3. Use a guide to help accomplish the teaching objective.

If this is a new lecture or topic, the fire officer should practice delivering the lesson to make sure that the important items are covered in a timely fashion. In addition, the fire officer should preview any audiovisual items and check all of the equipment that will be used during the presentation.

The final preparation activity is to check the physical environment. Make sure you have an environment that is conducive to adult learning. This includes taking every reasonable effort to reduce distractions and student discomfort (**FIGURE 3-10**).

Step 2: Presentation

The presentation is the lecture or instructional portion of the training. The objective with step 2 is to introduce the students to the subject matter, explain the importance of the topic, and create an interest in the presentation. During this step, the fire officer could be demonstrating or showing a skill or explaining a concept.

The fire officer should present a skill one step at a time, delivering a perfect demonstration of how it should be performed (**FIGURE 3-11**). When presenting a concept or idea, recommend that others practice giving the lecture. The objective of a lecture is to provide knowledge and develop understanding so that the fire fighter will be able to perform the skills properly. The overall goal is to increase fire company efficiency.

During a lecture, use simple but appropriate language. Begin with simple concepts and move progressively to more complex information, relating the new material to old ideas. Lecture only on what is important at this time to achieve the teaching objective, and do not teach alternative methods. Teach in positive terms and avoid telling the fire fighter what *not* to do.

A lesson plan allows the fire officer to stay on topic and emphasizes the important points to be addressed. Increasing the number of senses engaged in the training session helps the fire fighter to retain more of the material. Audiovisual aids and training props should be used to enrich the presentation. Fire fighters retain information more effectively if they actually perform the skill in the process of learning it.

Step 3: Application

The fire fighter should demonstrate the task or skill under the fire officer's supervision. The objective is a correct demonstration of the task, safely performed. A good reinforcing technique is to have the fire fighter explain the task while demonstrating the skill. The fire officer should provide immediate feedback, identifying omissions and correcting errors (**FIGURE 3-12**). Success is achieved when the student can perform the task safely without input from the supervisor.

Step 4: Evaluation

At the end of the lesson or program, there should be an evaluation of the student's progress. Training that is related to a certification program always includes an end-of-class evaluation. Depending on the skill and knowledge sets involved, the evaluation may be a written or practical examination.

The fire officer can be certain that training has occurred only when there is an observable change in the fire fighter's performance when responding to a real situation where that task or skill is applied. For example, Fire Fighter Jones has not been coming to a complete stop at red traffic lights or stop signs when

FIGURE 3-9 Fire officers often write lesson plans in preparation for training activities.
© Jones & Bartlett Learning.

driving the apparatus to emergency incidents. The fire officer provides a training session to address this behavior. If, after the training session, Fire Fighter Jones always comes to a complete stop at red traffic lights and stop signs, there has been an observable change of behavior. If the behavior does not change, the training objective has not been accomplished.

Four Levels of Fire Fighter Skill Competence

Noel Burch developed a practical description of how people learn a new skill while working for Gordon Training International (Adams, 2016). It was partially based on *The Dynamics of Life Skills Coaching* (Curtiss

FIGURE 3-10 Check the physical environment to make sure it is conducive to adult learning.
© Jones & Bartlett Learning. Photographed by Glen E. Ellman.

FIGURE 3-12 The fire officer should provide immediate feedback to the fire fighter.
© Jones & Bartlett Learning. Photographed by Glen E. Ellman.

FIGURE 3-11 Fire officers should deliver a perfect demonstration of how skills are performed.
© Jones & Bartlett Learning. Photographed by Glen E. Ellman.

and Warren, 1973). The four levels of the understanding of a skill (unconscious incompetence, conscious incompetence, conscious competence, and unconscious incompetence) are described next.

Unconscious Incompetence

The fire fighter does not know what he or she does not know. These are fire fighters who fail an evolution or skill performance for one or more of the following reasons:

- Exceeded time limit
- Missed a critical step in the procedure
- Created a safety hazard
- Unable to perform the task

The fire fighter is incompetent in this evolution or skill performance. They require remedial training and practice and a mandatory reevaluation.

Conscious Incompetence

The fire fighter knows what he or she does not know. Learning starts at this level when there is a sudden awareness of how poorly the evolution or skill is performed. These are fire fighters that completed the evolution or skill performance on a first attempt and may meet the minimum requirements from recruit training. The following statements are true for fire fighters at this level:

- There will be improved performance with additional practice.
- Delivery of the evolution or skill may require supervision in the field to assure adequate performance.
- The fire fighter takes all appropriate safety precautions.

The fire fighter is capable to perform the task on the training ground but needs oversight to assure an adequate and complete delivery of the evolution or skill set on the fire ground.

Conscious Competence

The fire fighter knows how to perform the evolution or skill performance the correct and complete way. The following statements are true for fire fighters at this level:

- The evolution or skill performance is completed well within the time requirements.
- The evolution or skill can be performed on the fire ground *without* supervision.
- The fire fighter makes no serious or critical errors.
- The fire fighter observes and performs all appropriate safety practices.

This is an experienced fire fighter who has performed enough repetitions and sets of the evolution to be a competent on the fire ground. The terms *sets* and *reps* come from physical training practice. A **rep (repetition)** is one complete motion of an exercise. A **set** is a group of consecutive repetitions. For example, "I did two sets of ten reps on the chest press." This means that you did ten consecutive chest presses, rested, and then did another ten chest presses.

Unconscious Competence

The fire fighter is experienced with this evolution or skill performance to the point that the activity appears natural and easy. These are the senior fire fighters who are the pillar of that fire company who are admired and closely observed. The following statements are true for fire fighters at this level:

- Evolutions or skill sets are error-free, and the fire fighter can easily adopt to any problem or barrier to completing the task.
- Evolutions or skill sets are accomplished in the least amount of time.
- The fire fighter could supervise others doing the same evolution or skill performance and identify errors or suggest improvements.
- The fire fighter knows the role and importance of this task in relation to other fire-ground operations.

This is the performance result from a competent and confident fire fighter.

Mentoring

Mentoring is a developmental relationship in which a more experienced person, or mentor, helps a less experienced person, referred to as a protégé. The unique organization of fire departments, where company officers function as working supervisors within a high-hazard environment, makes mentoring easy.

Mentoring is a one-on-one process in which the more experienced person provides a deliberate learning environment through instructing, coaching, providing experiences, modeling, and advising. Mentoring may be provided through feedback while picking up after an emergency incident. Failures as well as successes make up the mentoring experience.

The mentoring process extends beyond a particular fire company assignment and rank, usually lasting for a long period of time. Boston Fire Commissioner Leo Stapleton described how he would still get feedback from a senior fire fighter, decades after Stapleton was his rookie (Stapleton, 1983).

The following qualities make for an effective mentor:

- A desire to help
- Current knowledge
- Effective coaching, counseling, facilitating, and networking skills

Provide New or Revised Skill Sets

On occasion, the fire officer is required to provide the initial training for a new or revised skill set. This is often related to a new device that has been acquired, such as a new type of breathing apparatus or a thermal imaging device. The fire fighters need to become familiar with the new equipment and proficient in its use. In the case of a thermal imaging device, the fire fighters have to learn how it works and how to maintain it, as well as how to incorporate its use into fire-ground operations. This type of training could also be required to introduce a change in standard operating procedures. Sometimes procedures are changed or additional training is mandated in response to a near-miss incident.

Teaching new skills takes more time than maintaining proficiency of existing skills. The fire officer should obtain as much information as possible about the device or procedure, especially identifying any fire fighter safety issues. The emphasis of fire station-based training should be on the safe and effective use of the device or procedure. The fire officer should plan to spend a couple of training periods developing competency and showing how the new skill relates to existing procedures. Active learning research recommends spending no more than 10 minutes presenting a formal lecture or video presentation; the student loses focus after 15 to 20 minutes of passive learning (Kiewra, 2002). Practicing new skills is facilitated by encouraging adventure, challenge, and competition. For example, using a thermal imaging device to complete a scavenger hunt for heat sources around the fire station is one way to reinforce the capability of the device.

Ensure Competence and Confidence

The fire officer works as a coach when providing training to an individual or a team. After team members have learned the basic skills and can appropriately demonstrate them, the coach has to work with them to build competence and confidence. The coach must provide the guidance that advances them from being capable of performing the basic required skills to being able to perform those skills effectively, efficiently, and consistently.

FIGURE 3-13 Skills practice may require training props like the ones shown in these photos.
© Jones & Bartlett Learning. Photographed by Glen E. Ellman.

Many fire fighter tasks involve psychomotor skill sets. Psychomotor skill levels can be classified into four categories. These levels may be described using the following example of a new driver/operator who is required to know, from memory, every address in the fire district.

- Initial: The driver/operator knows the main streets and has a basic understanding of how the street grid and numbering system work. The fire officer has to help by reading the map when responding in subdivisions and office parks.
- Plateau: The driver/operator can drive to more than 85 percent of the streets in the company's response area without assistance from the officer. At this level, the driver/operator is competent.
- Latency: The driver/operator can remember the route to an area of the district where the engine company has not had a response for several months.
- Mastery: The driver/operator can easily drive to any address in the district and knows at least two or three alternative routes to each area. The driver knows all of the subdivisions, the layout of every office park, and the locations of all hydrants and fire protection connections. At this level, the driver/operator is confident in his or her knowledge of the district.

To bring fire fighters up to the mastery level, the fire officer must work every day to reinforce their skills. Many fire fighter skill sets are used infrequently, yet when that task is needed, the fire fighter must deliver a near-perfect performance under urgent or critical conditions. It is dangerous to ask a fire fighter to demonstrate a rusty skill set in a critical emergency situation.

The fire officer needs to provide enough repetition and simulations to maintain fire fighter confidence in seldom-used skill sets. The continuing expansion of fire fighters' responsibilities as all-hazard mitigation specialists increases the fire officer challenge.

For best performance, many of these practice sessions should be performed while the fire fighters are wearing full personal protective clothing, operating within a realistic fire-ground situation. This may require construction of training props, such as an assembly that allows the fire fighters to practice opening a roof using a power saw (**FIGURE 3-13**). The fire officer should always be on the alert for opportunities to acquire abandoned structures where realistic fire-ground skills can be practiced.

When New Member Training Is On-the-Job

Fire officers have special responsibilities when operating with inexperienced fire fighters and fire fighters in training. At the very first meeting with the new fire fighter trainee, the fire officer should explain the procedures in the fire station when the company receives an alarm, assign a senior fire fighter to function as a mentor, and describe any restrictions that are placed on fire fighters in training. Many departments require that a new member obtain NFPA Fire Fighter I certification before responding in the field.

Skills That Must Be Learned Immediately

Federal regulations mandate that four topics are covered as part of any emergency service training program:

- Bloodborne pathogens: Even "suppression only" fire fighters are at risk of being exposed to blood and other bodily fluids. The Occupational

Safety and Health Administration (OSHA) has issued regulation 29 CFR 1910.1030, *Bloodborne Pathogens*, which requires all fire fighters to be trained regarding the department's exposure control plan, the personal protective equipment used by the fire fighter, and the reporting requirements if an exposure occurs. Usually, this training takes about 4 hours.

- Hazardous materials awareness and operations: Every public safety member is required to have training at the hazardous materials awareness and operations levels. OSHA regulation 29 CFR 1910.20, Hazardous Waste Operations and Emergency Response (HAZWOPER) training, describes these requirements. At the awareness level, first responders are able to recognize a potential hazardous materials emergency, isolate the area, and call for assistance. At the operations level, first responders are able to recognize a potential hazardous materials incident, isolate and deny entry to other responders and the public, evacuate persons in danger, and take defensive actions such as shutting off valves and protecting drains without having contact with the product.

- SCBA fit testing: Training regarding OSHA regulation 29 CFR 1910.134, *Respiratory Protection*, requires that anyone who uses respiratory protection during job tasks be provided with appropriate training, be fit-tested for a mask, and be subject to a health monitoring program. This is usually covered in the Fire Fighter I training program.

- National Incident Management System: Homeland Security Presidential Directive 5 (HSPD-5), *Managing of Domestic Incidents*, requires incident management training. The NIMS provides a consistent framework that operates at all jurisdictional levels regardless of the cause, size, or complexity of the incident. NIMS creates an all-hazard template for fire fighters to operate within a multiple-jurisdiction domestic incident within the federal national response plan.

In addition to the federally required training, the fire department needs to provide emergency scene awareness training to reduce the risk that a probationary fire fighter will be injured at the emergency scene. This training would include such items as how to avoid being struck by a car when operating on an interstate highway. The fire officer should spell out the expected behavior of the trainee when operating at emergencies. This effort may include assigning the trainee to shadow, or work with, a senior fire fighter on the team.

The expected behavior falls into three areas: responding to alarms, on-scene activity, and emergency procedures. Responding to alarms includes the trainee behavior that should occur when an alarm is received at the fire station. Be specific about the appropriate actions necessary when a fire fighter is preparing to respond to an alarm. Emphasize the importance of not running in the station, donning protective clothing correctly, and always wearing a seat belt when riding in a fire vehicle.

On-scene activity includes clearly defining the expected location, activities, and behavior of the trainee at an emergency incident. Make sure that the trainee knows the safe locations when working at an incident on a roadway. Identify off-limits activity. The fire officer should know all the restrictions that apply to each individual and ensure that the trainee is equally aware of those limitations. A trainee who has not completed SCBA training and fit testing must not wear breathing apparatus or enter an IDLH environment. If the trainee is a teenager, additional restrictions might apply; allowing a 15-year-old to operate inside a burning building is prohibited in many jurisdictions.

Finally, the officer should clearly identify the trainee's expected behavior when operating during an emergency situation, such as a mayday (fire fighter down or lost) or a potentially violent situation. The fire officer should ensure that the trainee will be in a safe location and stay out of the way until the emergency situation is controlled.

Skills Necessary for Staying Alive

Once the skills that must be learned immediately are covered, the fire officer should concentrate on the skills the fire fighter in training needs to know to stay alive. Most states and commonwealths require the trainee to pass a knowledge exam and a skills test for Fire Fighter I before engaging in interior structural firefighting. Some communities start with the trainee being limited to "outside only" activities at fire scenes. Examples of skills and topics that should be taught first in a training program include the following:

- Fire-ground tasks that emphasize teamwork and require mastering the location of all of the equipment on the apparatus:
  - Supply line evolutions
  - Ropes and knots
  - Laddering the fire building
  - Lights, fan, and power deployment

- Motor vehicle crashes and medical emergencies
  - Cardiopulmonary resuscitation (CPR) and AED training
  - Outside-circle activities on a crash extrication
  - Helicopter landing zone procedure
  - Assisting the paramedics on a medical emergency

The trainee should have received enough training and demonstrated adequate stay-alive skills to begin responding to emergencies. This allows the student to gain experience by functioning in an "outside only" role while completing the Fire Fighter I certification training. Some fire departments do not allow trainees to respond to any emergency incidents until they have achieved the initial Fire Fighter I certification level.

Developing a Specific Training Program

On occasion, the fire officer may need to develop a specific training program that is not covered by an existing certification training program or prepared lesson plan. This may be a work improvement plan or training related to a new device or procedure. NFPA Fire Instructor II covers the development of a training program in more detail. Here is an overview of the five steps for developing a training program.

Assess Needs

The fire officer must first confirm that there is a need for a training program. Some performance problems may be better solved through an engineering solution. For example, the effective use of personal alert safety system (PASS) devices was a problem with the first generation of devices. Many fire fighters were operating in an IDLH environment while their PASS device was not armed. Despite training programs and, in some departments, progressive discipline, the devices remained unarmed. An engineering solution has resolved this performance issue. The PASS device is now integrated into the SCBA and is armed every time the high-pressure hose is charged.

Establish Objectives

Training has occurred when there is an observable change in behavior. Prior to undertaking the training program, you should identify the specific behavior you want the fire fighters to exhibit after the training. The desired behavior could be that fire fighters will demonstrate the proper procedures for deploying a ground ladder. The description must include the conditions under which the behavior will be demonstrated. For example, if an expectation is that a ground ladder will be deployed while fire fighters are wearing full protective clothing and SCBA, then that condition must be part of the expected fire fighter behavior.

The final part of the objective is the measure of performance. Fire fighter evolutions are often timed events, so the measure of performance is usually described in terms of the time required to complete a task properly.

The completed behavioral objective could look like this:

> Given a fire department pumper and a building two or more stories high, a crew of two fire fighters will exit the pumper in full personal protective clothing and SCBA, select and remove the appropriate ground ladder from the pumper, and properly deploy the ground ladder to the assigned window within 2 minutes.

Develop the Training Program

Various methods exist for developing the training program. If training is needed for a new device, the manufacturer or vendor may have a training package available. If training is needed for a new procedure or company evolution, the department may have a template the fire officer can use. Other departments may have developed training programs, props, or resources that they will share. Organizations such as IAFC's Near Miss, Responder Safety Learning Network, and special-interest websites may post resources available for downloading. A few minutes of research may provide a rich response.

The fire officer also needs to consider how the training will be delivered. Will it require a skill drill using multiple companies, or can the skill be practiced by an individual fire fighter?

Deliver the Training

New programs should be subjected to a pilot class or trial run before finalization. This sort of testing provides the opportunity to tweak the program, identify any problems, and correct any unforeseen issues.

Although the use of a lesson plan is part of the four-step method, it is important that the fire officer develop lesson plans for every training program. A good lesson plan satisfies four criteria:

1. Organizes the lesson
2. Identifies key points
3. Can be reused
4. Allows others to teach the program

Evaluate the Impact

Have you accomplished the change of behavior? Was the training program worth the fire fighters' or instructor's time? Was the instructional method appropriate for the learning objective? What can the instructor-developer do to improve the training program?

Many successful national fire service training programs, such as the "Reading Smoke" size-up technique and the "Rapid Intervention" program, have been developed by fire officers and fire fighters. They have the perspective, the understanding, and the need to develop and share vital emergency activity training.

Addressing Member-Related Problems

A problem may be defined as the difference between the current situation and the desired situation. If the way the apparatus driver parks at a highway incident places the fire crew at greater risk than is necessary, it is a problem. Forcing a commercial door and destroying a display window when a rapid-entry key safe is available is a problem. In each of these cases, there is a discrepancy between the current and the desired situations.

Fires and emergency incidents represent a unique category of problems that call for specialized problem-solving skills. Nonemergency situations require the application of conventional problem-solving skills and techniques. These situations include supervisory, management, and administrative activities in which the fire officer is directly responsible for solving the problem as well as initially processing situations that require resolution from an administrative or executive fire officer. These circumstances include situations that involve individuals or organizations outside the fire department.

Decision-making skills are used whenever the fire officer faces a problem or situation that requires a response. Promotional examinations evaluate the ability of fire officer candidates to exercise good judgment and make sound decisions. Decisions should always be guided by organizational values, guidelines, policies, and procedures. This is true even when the decision made does not agree with the fire officer's opinion or personal preferences.

Requiring the fire officer to act in the best interest of the department in solving problems and making decisions does not mean that other values and considerations are ignored, however. Several solutions to a problem may exist—some better than others and some more desirable to one set of interests than another. In most cases, there is a reasonable solution that serves multiple interests and concerns; in other cases, however, one concern must prevail over all others. Problem-solving techniques are designed to identify and evaluate the realistic potential solutions to a problem and determine the best decision.

Complaints, Conflicts, and Mistakes

Complaints, conflicts, and mistakes are special categories of problems. One of the key factors in decision making is how to deal with situations that involve conflicts or complaints. These three terms are defined as follows:

- A **complaint** is an expression of grief, regret, pain, censure, or resentment; a lamentation; an accusation; or fault finding.
- A **conflict** is a state of opposition between two parties. A complaint is often a manifestation of a conflict.
- A **mistake** is an error or fault resulting from bad judgment, deficient knowledge, or carelessness. It can also be a misconception or misunderstanding. Mistakes happen; the issue is how to deal with a mistake, or the perception of a mistake, when someone complains to the fire officer about it.

Sometimes a fire officer has to make a decision or enforce a policy that is not popular with the crew members. A citizen could be frustrated with a fire department action or may be unhappy with a situation. People misbehave and make mistakes. Disagreements and differences of opinion occur. It is not possible to make everyone happy all of the time, but a fire officer must deal with all of these situations in a professional manner. Dealing appropriately with problems and conflicts requires maturity, patience, determination, and courage (**FIGURE 3-14**).

The types of problems that a fire officer could be expected to encounter can be classified into four broad categories:

1. In-house issues: Situations or decisions occurring at the work location that are within the direct scope of supervisory responsibilities. An example might be a complaint about the assignment of duties to different individuals within a fire station. Most of these conflicts begin and end at the company officer level.
2. Internal departmental issues: Operational policies, decisions, or activities that go beyond the scope of the local fire station. An example is a conflict over where a reserve ladder truck

FIGURE 3-14 Dealing appropriately with problems and conflicts requires maturity, patience, determination, and courage.
© Glen E. Ellman

will be housed and which company will be responsible for maintaining it. Another example is a dispute between companies over the tasks of a rapid intervention team in a high-rise fire. The resolution usually requires action by command officers at a higher level on the organizational chart.

3. External issues: Fire department activities that involve private citizens or another organization. An example is a citizen making a complaint about an inappropriate remark uttered during an emergency medical services (EMS) incident. External issues require the fire officer to perform one additional task early in the conflict resolution process—making sure that the fire officer's supervisor is not surprised. It is poor form for the battalion chief to learn about a fire department incident from the local media.

4. High-profile incidents: Any issues that are likely to become major events. An example might be a fire fighter who is arrested while on duty. The department must take immediate actions to respond to these events. Senior fire administrators often become directly involved in these situations or keep a close watch on how they are handled.

As discussed earlier, a problem should be solved at the lowest possible level within an organization. A fire officer is expected to manage problems within the level of authority for a supervising or managing fire officer. At the same time, the fire officer should recognize those problems that need to be handled at a higher level and make the appropriate notifications without delay. If there is any doubt, it is wise to discuss the situation with the officer at the next higher level in the chain of command.

Managing Conflict

One factor that distinguishes a fire officer from a fire fighter is the responsibility to act as an agent of the formal organization. A fire officer is the official first-level representative of the fire department administration when dealing with subordinates and enforcing policies and procedures. This responsibility places the fire officer in a position to be the initial contact in dealing with a wide variety of problems, including situations that potentially involve conflict, emotions, or serious differences of opinion.

Situations that involve conflicts and grievances require an additional set of skills that go beyond the general problem-solving model. The general model is designed to focus on solving the problem itself. In conflict situations, the issues are often much more complicated and sensitive. A relatively simple problem, for example, may become complicated by the ways that different individuals react to it or to one another. In some cases, the problem centers on the relationship between individuals or groups and plays out in relation to other issues.

Personnel Conflicts and Grievances

The close living relationships within a fire company can produce a variety of tensions, anxieties, and interpersonal conflicts. This friction can occur in addition to the types of conflicts commonly experienced in most workplaces. One of the most difficult situations for a fire officer is an interpersonal conflict or grievance within the company or directly involving a company member.

A fire officer may face four different types of internal conflict situations. A fire fighter might come to an officer with a complaint about:

- A co-worker (or co-workers)
- The work environment, including the fire station, apparatus, or equipment
- A fire department policy or procedure
- The fire officer's own behavior, decisions, or actions

The fire officer is the individual's first point of contact with the formal organization. The official response to the problem begins when the officer becomes aware that a problem exists. The relationship of the fire officer to the conflict and the complainant makes a significant difference in the role the officer can play in resolving the conflict.

Fire officers with staff assignments must also be prepared to deal with problems that involve conflict. Their relationship to the individuals involved is likely to be different, but their responsibility to officially represent the formal organization is the same.

Conflict Resolution Model

The conflict resolution model is a basic approach that can be used in situations where interpersonal conflict is the primary problem or a complicating factor.

Listen and Take Detailed Notes. The first phase of the conflict management template is to obtain as much information as possible about the problem. The fire officer should encourage the complainant to explain the situation completely. If the details are even slightly complicated, the fire officer should take notes. The person who is making a complaint has a certain perspective on the situation. Whether you agree or disagree with that person's perspective, an important starting point is to find out what the complainant thinks about the situation.

FIGURE 3-15 Give the speaker your full attention.
© Jones & Bartlett Learning. Photographed by Glen E. Ellman.

FIRE OFFICER TIP

Conflict Resolution

You should anticipate handling some type of conflict issue. The four-step conflict resolution model provides an effective framework for managing internal and external conflicts. Active listening may provide the candidate with additional information that the role player may hold back if not questioned.

FIRE OFFICER TIP

Fire Station Communication and Conversations

A fire officer has a different relationship with the company members than most supervisors have with their subordinates. Spending 10- to 48-hour shifts together in a fire station and engaging in emergency operations tend to create close relationships. As a consequence, the fire officer becomes aware of most internal problems early, when there is an opportunity to take preventive action. Unfortunately, the fire officer can become too close to some problems to deal with them effectively.

The kitchen is the most important room in the fire station. Given enough coffee and time, the members will discuss and solve all of the department's problems. The kitchen is a great place for a fire officer to learn about the company's issues and concerns. It also provides a good opportunity for the fire officer to informally release information and measure reactions to different issues. Conversely, if the fire officer has important official information to share with the company, it should not be slipped into casual conversation.

Although much information is exchanged in the kitchen, it should not substitute for official communication mechanisms. Although a fire officer is entitled to have a personal opinion, it is confusing to express mixed messages when presenting official communications.

When dealing with an individual who is expressing a concern or a problem, the fire officer should focus on active listening. Engaged, or active listening is the conscious process of securing all kinds of information through a combination of listening and observing. The listener gives the speaker his or her full attention, staying alert to any clues of unspoken meaning while also listening intently to every word that is spoken. The fire officer should be aware of nonverbal clues that may indicate agreement, dissatisfaction, anger, or other emotions. Often, these nonverbal clues provide great insight into the disposition of the speaker. The fire officer actively seeks to keep the conversation open and satisfying to the speaker, showing an interest in feelings and emotions as well as raw information (**FIGURE 3-15**).

Paraphrase and Receive Feedback. The first objective should be to understand the issue and why the individual is complaining. After listening, the fire officer should be able to paraphrase the complaint and recite it back to the complainant. Paraphrasing the issue and receiving feedback from the complainant accomplishes two goals: The fire officer finishes this phase with a good understanding of the issue from the complainant's perspective, and the complainant feels that the fire officer really listened.

Do Not Explain or Excuse. In situations where the complaint is directly related to actions taken or policies enforced by the fire officer, it is understandable that the fire officer would want to respond immediately to the complaint. In this situation, it is important to listen and to process the information before deciding on an appropriate response. A reflexive explanation or excuse gives the individual an additional reason to complain. If the complainant feels strongly enough to complain about something the officer has done, that officer's explanation probably will not solve the problem.

Investigate

An **investigation** is a detailed inquiry or systematic examination. All complaints should be investigated, even if the foundation for the complaint appears to be weak or nonexistent. Fire department procedures should determine who will conduct the investigation, depending on the nature of the complaint and the relationship of the individuals who are involved. Sometimes the fire officer who received the complaint is assigned to conduct the investigation; however, a fire officer who is directly or personally involved in the problem should never be involved in conducting the investigation. Sensitive matters require an appropriate level of investigator.

The purpose of the investigation is to obtain additional information beyond the original complaint. The investigator must be impartial in gathering and documenting information. The information could come from other individuals, reference documents, or incident-specific data. When investigating a human resources conflict, such as a payroll or work assignment issue, the fire officer might have to refer to specific departmental directives and regulations.

The product of an investigation is a report, which is provided in an appropriate format for the fire officer's immediate supervisor (**FIGURE 3-16**). A complete investigative report has three objectives:

1. The report must first identify and clearly explain the issues.
2. The report should then provide a complete, impartial, and factual presentation of the background information and relevant facts.
3. The conclusion should be a recommended action plan, which is based on and supported by the information.

Take Action

Once the investigation is completed, the fire officer presents the findings and recommended action to a supervisor at a higher level. There are four possible responses:

1. Take no further action: The investigation may conclude that the complaint was unfounded or requires no further action. If the complaint was related to an earlier decision or action, that original decision is affirmed. The response should include the reasons why no further action is recommended.
2. Recommend the action requested by the complainant: The investigation may conclude that the complaint was justified and that the requested action is the best solution to the problem.
3. Suggest an alternative solution: The investigation may conclude that some alternative action or policy is the best solution to address the complainant's concerns. For example, a citizen might complain that the fire truck blocks several spaces in the parking lot at a local gym and does not want the fire fighters to go there. The fire officer meets with the citizen and proposes a more appropriate parking space for the fire truck. The compromise is acceptable to the citizen and to the department.
4. Refer the issue to the office or person who can provide a remedy: Other members of the fire department or some other municipal agency may be able to provide the relief the complainant seeks. Grievance procedures require that the employee start with the immediate supervisor for all complaints. If the employee is not satisfied with the response at that level, then he or she can take the grievance to a higher level. If the problem involves a paycheck deduction issue, it will probably have to be resolved by the payroll clerk or human resources office. The fire officer's duty is to refer the complaint to the appropriate person.

Follow-Up

For many of the conflicts, the fire officer needs to follow up with the complainant to see whether the problem is resolved.

Understanding Emotional Confrontations

Emotional confrontation between the fire officer and a fire fighter may be an example of clashing psychological contracts. **Psychological contract** refers to mutual unwritten expectations that exist between an

> **Municipal City Fire and Rescue Department**
> **Internal Memorandum**
>
> Date: May 08, 2019
> To: Assistant Chief James Arrow, A Platoon Commander
> Thru: Battalion Chief Frank Johnson, 3rd Battalion
> Thru: Captain Jean Davis, Fire and Rescue Station 100
> From: Lieutenant Taylor Williams, Quint 100
> Subj: Civilian property damage
> Ref: Incident #201705051473, 437 Western Way
>
> On Friday, May 03 Quint 100 was dispatched for "smoke in the building" at Exotic Food Emporium. Municipal City was experiencing a severe thunderstorm and there was no electrical power in the neighborhood.
>
> I observed a white haze in the store. While completing my size-up, Firefighter James Grynski started to force open the front door of the store. Swinging the flathead axe shattered the storefront window.
>
> The white haze was from fumigation. The notice was posted on the front and back doors of the store. With the release of the fumigation gas, requested a haz-mat, EMS, and chief response to our location.
>
> There was no notation in the dispatch that the store was undergoing fumigation. Quint 100 staged at the A/B corner on arrival. I headed to Side C to complete the size-up while Grynski and Kinders headed to the Side A front door. My goal was to determine if there was a working fire in the store.
>
> Investigation
> Neither Grynski nor Kinders recall seeing the bright orange fumigation sign that was hanging on the front door.
> After the window shattered Kinders located the rapid entry keybox that was located to the right of the door.
>
> Action Taken
> Once the event was stabilized, took statements from Grynski, Kinders, and Rollo.
> Took pictures of the damage.
> Acquired contact information for the store owner and shopping center representative.
> Placed plywood over the window and secured the front door.
>
> Follow-up
> Referred to Chief Johnson

FIGURE 3-16 Example of a fire officer investigation report.
Courtesy of Mike Ward

employee and his or her employer regarding policies and practices in their organization. The psychological contract influences the attitude and job performance of the employee (Patrick, 2008).

Each of us has a system of beliefs that define our unwritten expectations with the fire department. Our beliefs are shaped by our values, on-the-job experiences, and broader societal norms with which we agree (Dabos and Rousseau, 2004). When a decision or situation occurs that is in conflict with our unwritten expectations, we often respond emotionally.

When we get emotional (scared, angry, or euphoric), adrenaline fills up the prefrontal lobes of the neocortex of the brain, creating an "emotional bubble" that interferes with the ability to hear or consider any response to the issues that make up this confrontation.

Michael Taigman, an emergency service performance consultant, provides a conflict resolution model that is especially effective when emotions are high (Taigman and Dean, 1999). This model was effective for Taigman when working with employees during the stressful creation of a paramedic ambulance service under a tight schedule. He continued to refine the model while working as a consultant. The model follows four steps:

1. Drain the emotional bubble.
2. Understand the complainant's viewpoint.
3. Help the complainant feel understood.
4. Identify the complainant's expectation for resolution.

Step 1: Drain the Emotional Bubble. The body reacts to emotional conflict or stress the same way it does when you are a member of the first-arriving company at a working structure fire. Both situations result in dumping of adrenaline to prepare the body to fight or run away. The same adrenaline-induced "red haze" that reduces fire fighter effectiveness at emergencies also happens to ordinary citizens. It tends to bring complaints to the surface and impedes resolution of conflicts. Adrenaline fills up the prefrontal lobes of the neocortex of the brain, creating an emotional bubble that interferes with the ability of the complainant to hear the fire officer or consider any response to the issue.

Taigman recommends listening deeply, actively, and empathetically to drain this emotional bubble. This is not an easy task for the fire officer, but writing detailed notes and not explaining or excusing can facilitate this process. The fire officer asks questions and encourages responses, draining the emotional bubble by allowing the complainant to express grief, regret, pain, censure, or resentment completely.

This discussion should be held in private and should adhere to a few ground rules. Most importantly, there should be no physical contact. If the discussion is between a fire fighter and a fire officer, avoid personal attacks and concentrate on the work issues.

Step 2: Understand the Complainant's Viewpoint. The initial complaint or behavior may be a sign or symptom of a larger problem. By draining the emotional bubble through active listening, the fire officer may identify the root cause or issue of the complaint.

Internal conflicts, grievances, or issues occasionally suffer from long memories. It may require some investigating to understand the current issue, which may be related to something that happened months ago or may be wrapped up in history and tradition.

Step 3: Help the Complainant Feel Understood. The "listen and take detailed notes" part of the basic conflict management template includes the recommendation to paraphrase and repeat back what you heard from the complainant. Some issues may be more readily resolved if the complainant feels that the fire officer understands the issue, conflict, or problem.

Step 4: Identify the Complainant's Expectation for Resolution. By this final step, the complainant has drained the emotional bubble, has been able to describe what is going on, and feels that the fire officer understands the issue, problem, conflict, or grievance. The fire officer should now ask what the complainant expects the department to do to resolve this issue. If the problem is an internal grievance, this is where the employee should be asked to describe the desired action.

Behavioral and Physical Health Issues

A rising leadership challenge is fire fighter behavioral and physical health. Since 2017, the rate of fire fighter suicide has exceeded the rate of line-of-duty deaths (Heyman, Dill, and Douglas, 2018). A 2015 study by Firefighter Behavioral Health Alliance (FFBHA) reports that 37 percent of those surveyed contemplated suicide and almost 7 percent had attempted it. That is more than 10 times the rate of the general population (Fisher, 2018).

Awareness

The fire service has several factors that promote resiliency, such as a sense of belonging and support from one another, an enduring sense of purpose, and often a strong sense of gratitude and respect from the public (NFFF, 2017). Fire officers should consider the following warning signs of stress-related issues in an individual:

1. Isolation from others
2. Disturbed sleep
3. Increased irritability
4. Decreased interest in significant activities
5. Self-destructive or reckless behavior (IAFF, 2017)

Substance Abuse

Fire fighters are heavier users of alcohol than the general population. One study found 58 percent of the fire fighters reporting binge drinking (Carey et al., 2011), in comparison to the 23 percent of males in the general population who report binge drinking (CDC, 2016).

Less data exists regarding fire fighter drug abuse. The following are signs that a fire fighter may be grappling with substance abuse:

- Mood swings and desire to spend time alone
- Loss of interest or lack of socialization
- Increased anxiety and crankiness
- Missing work or continuous issues at work
- Change in diet
- Lack of care in physical appearance (Tagliareni, 2018)

Family and Marital Problems

Three things make firefighting different from other professions: exposure to danger, exposure to trauma, and the 24-hour shift (Gagliano and Gagliano, 2018). Gagliano states that a successful fire fighter marriage requires five essential conversations:

1. Reentry time: Allowing the returning spouse some time to enter the home without hassle or massive conversation. Sleep deprivation, adrenaline backlash, and the impact of trauma require transition time from fire-ground warrior to loving spouse and parent.
2. Harshness and gallows humor: Anger or dark humor from trauma, combined with aggression from adrenaline, results in a defense mechanism by the fire fighter that does not travel well outside the fire station. Develop key phrases ahead of time so the spouse can lovingly point out when the fire fighter is acting inappropriately at home.
3. Handling the tough runs: Generally, the stressful things a fire fighter encounters at work remain at work. If it has been a particularly bad incident or work day, the fire fighter should tell their spouse their real feelings and fears after a tough run or shift.
4. Dealing with the fix-it mentality: Fire fighters are "fixers" on steroids. When the spouse is venting about an issue, it is difficult for the fire fighter to empathically listen and not respond to it as a situation requiring immediate fixing.
5. Keep your first family first: Set boundaries so that the second family, the fire service, does not push away the first family.

One of the deadliest enemies of marriage is the pursuit of wealth at the cost of family time. Annual reports that identify the highest paid municipal workers often include fire fighters and fire officers who will occasionally earn more in overtime than what they received for their regular salary. A West Coast fire fighter made $300,000 in overtime pay on top of his $92,000 annual pay, telling a reporter that he "basically lived at the station" and did not go home often (Boehm, 2018).

Financial Problems

ABC's *20/20* television program produced a series of episodes called "My Reality: Hidden America" that looked at working Americans and the realities they face in making ends meet. The interviewees included a fire fighter/paramedic named Chris Smith who works in three different towns and spends up to 104 hours a week at work to make ends meet. He slept in his home for a total of three nights in February 2017. A 30 percent jump in his health insurance premium and a flooded backyard wiped out the family budget (Sawyer, 2017).

There are three reasons for this struggle to maintain adequate income:

- Greater personal debt: Student tuition loans are the fastest growing segment of household debt, increasing by almost 157 percent since 2007. Auto loan debt has increased by over 50 percent since 2007, with the average auto loan stretched to 6 years (Griffin, 2018).
- Stagnant purchasing power: While paychecks are bigger, purchasing power has hardly budged. "After adjusting for inflation, however, today's average hourly wage has just about the same purchasing power it did in 1978, following a long slide in the 1980s and early 1990s and bumpy, inconsistent growth since then. In fact, in real terms average hourly earnings peaked more than 45 years ago: The $4.03-an-hour rate recorded in January 1973 had the same purchasing power that $23.68 would today" (Desilver, 2018).
- Slow recovery from recent recessions: Local governments have not returned to the financial health they had in the late 1990s. The recession of 2001 and the "Great Recession" of 2007–2009 were different than earlier recessions. The decline in local and state tax revenue was steep in both periods, and the recovery has been slower than any earlier recession. The Urban Institute pointed out: "Even though 42 states had recovered their nominal 2008 tax revenue levels by 2013, only 24 experienced real revenue growth" (Francis and Sammartino, 2015).

Jason Hoschouer provides a vivid example of this situation. "In 2009, I made more than $100,000. Mind you, that included an average of about 40 hours of overtime a month. Sounds great, right? The problem? I couldn't afford to pay a $300 electric bill. I had a wife, a daughter, and another kid on the way." (Hoschouer, 2016).

Resiliency to Stress

The American Psychological Association describes **resiliency** as the process of adapting well in the face of adversity, trauma, tragedy, threats, or significant sources of stress, such as family relationship problems, serious health problems, or workplace and financial stressors. It means "bouncing back" from difficult experiences (APA, n.d.). The nature of firefighting and emergency medical responses can create fear or horror in the responder. Fear is encountering a situation that is perceived as a threat to you, such as encountering gunfire as the fire company is arriving at an incident location. Horror is a reaction to a situation experienced by the responder, such as finding an infant burned to death in a structure fire. In addition to these specific sentinel events, fire fighters experience an adrenaline dump and a "fight-or-flight response" every time they respond to a 911 call (Hall et al., 2016).

Fire fighters encounter the acute stress of a sentinel event as well as the accumulated stress of everyday 911 responses. Both situations create physical and emotional reactions that will be different for each fire fighter.

Moral Injury

Moral injury was first used to describe soldiers' response to their actions in war. It represents "perpetrating, failing to prevent, bearing witness to, or learning about acts that transgress deeply held moral beliefs and expectations" (Litz et al., 2009). Physicians adopted the concept when describing the frustration of being unable to provide high quality care and healing in the context of health care (Talbot and Dean, 2018). Like public safety workers and soldiers, physicians are experiencing a high level of suicide (Price and Norbeck, 2018).

Moral injury occurs for fire fighters when there is a gap between what they want to do and the conditions that they are confronting. The opioid epidemic, where fire company first responders are finding themselves resuscitating the same person a couple of times a year with no indication of improvement, is one example. In some communities the number of fire fighters on duty has shrunk while the workload has increased. Disasters and local incidents occur in which fire fighters are faced with devastating human conditions that they cannot reverse or mitigate, creating a situation described by Litz as "a deep soul wound that pierces a person's identity, sense of morality, and relationship to society" (Litz et al., 2009).

Acute Stress Disorder

In **acute stress disorder (ASD)**, a person who experienced, witnessed, or was confronted with a traumatic event may experience "numbing, reduced awareness, depersonalization, derealization, or amnesia." The person feels disconnected and has significant impairment in social, occupational, or other important areas of functioning. To get an ASD diagnosis, the disorder must last for a minimum of 3 days and a maximum of 4 weeks. Onset must occur within 4 weeks of the event. ASD may be the precursor to PTSD (APA, 2013).

Posttraumatic Stress Disorder

Posttraumatic stress disorder (PTSD) is characterized by symptoms of avoidance and nervous system arousal after experiencing or witnessing a traumatic event. Specific criteria to identify PTSD are as follows:

1. Traumatic event: A person who experienced, witnessed, or was confronted with a traumatic event.
2. Intrusion or reexperiencing: Describing ways that someone reexperiences the event. This could look like:
   a. Intrusive thoughts or memories
   b. Nightmares related to the traumatic event
   c. Flashbacks, feeling like the event is happening again
   d. Psychological and physical reactivity to reminders of the traumatic event, such as an anniversary
3. Avoidant symptoms: Ways that someone may try to avoid any memory of the event, and must include one of the following:
   a. Avoiding thoughts or feelings connected to the traumatic event
   b. Avoiding people or situations connected to the traumatic event
4. Negative alterations in mood or cognitions. This newer criterion captures many symptoms that have long been observed. There is a decline in someone's mood or thought patterns, which can include:
   a. Memory problems that are exclusive to the event
   b. Negative thoughts or beliefs about one's self or the world

c. Distorted sense of blame for one's self or others, related to the event
  d. Being stuck in severe emotions related to the trauma (e.g., horror, shame, sadness)
  e. Severely reduced interest in pre-trauma activities
  f. Feeling detached, isolated or disconnected from other people
5. Increased arousal symptoms: Person remains hypervigilant to threats.
  a. Difficulty concentrating
  b. Irritability
  c. Increased temper or anger
  d. Difficulty falling or staying asleep
  e. Hypervigilance
  f. Being easily startled (Staggs, 2019)

Behavioral Health Resources for Fire Officers

Fire service organizations have been mobilizing to provide resources and tools to reduce the impacts of behavioral health issues for fire fighters. This remains an evolving issue but, as of this publication, resources are available from the following sources:

- National Fallen Firefighters Foundation
- Firefighter Behavioral Health Alliance
- International Association of Fire Chiefs Safety, Health and Survival Section
- International Association of Fire Fighters, Behavioral Health Program

Employee Assistance Programs

An **employee assistance program (EAP)** is designed to deal with issues such as substance abuse, emotional or mental health issues, marital and family difficulties, or other difficulties that affect job performance. EAPs help the employees cope with underlying issues that might be affecting workplace performance. Fire department EAPs are comprehensive programs that deal with a wide range of issues that can affect fire fighters. When an EAP is available, it is a resource that fire fighters can turn to when in crisis.

One important characteristic of an EAP is its ability to maintain the value to the organization of highly trained emergency service professionals. Consider a fire fighter with chronic tardiness. If the fire fighter is unable to correct this behavioral problem, and the progressive discipline process is followed, then termination is inevitable. The underlying problem, however, could be an off-the-job issue that the individual cannot solve without assistance. Termination of an otherwise good employee would be a tremendous waste in resources because it could cost the department as much as $100,000 to find and train a replacement. If EAP involvement can help solve the problem, it is worth the effort.

Consider some of the reasons a fire fighter might continue to report late for work:

- Child or elder care issues: The employee is late due to unanticipated coverage problems.
- A family crisis: Examples can include a divorce proceeding or a dying family member.
- Alcoholism or substance abuse.
- Coming from a second job: This job may be needed to handle a financial crisis, such as a child with a significant health problem not covered by insurance, a crushing debt, or gambling losses.
- A psychological condition or chemical imbalance.

Often, these situations cause stress for the employee. A referral to the EAP may assist in addressing the issues. For an EAP to be successful, however, the fire officer must be able to recognize stress in an employee. Signs that may be noticed at work could include absenteeism, unexplained fatigue, memory problems, irritability, insomnia, increased use of products with caffeine/nicotine, withdrawal from the crew, resentment toward management/co-workers, stress-related illnesses, moodiness, or weight gain or loss. Although any of these points individually may not indicate the employee is stressed, multiple signs may indicate that the fire officer should ensure that the employee is aware of the EAP.

If the stressful situation goes unresolved, it can seriously affect performance on the job. Obviously, an employee who is absent reduces the efficiency of a fire company. More subtle effects are produced by symptoms of stress such as physical exhaustion from stress-related fatigue or lack of sleep. The fatigue or indecision caused by this condition can have serious safety consequences for every person on scene. It may also affect the group dynamics of the crew. A crew member who is stressed may not intend on taking it out on the crew at work, but the tension created is often evident to the rest of the company.

The goal of the EAP is to provide counseling and rehabilitation services to get the employee back to full productive duty as soon as possible. Fire department EAPs have been successful in lowering employee turnover and reducing absenteeism, tardiness, accidents, and injuries. In addition, fewer employee grievances and severe disciplinary actions are encountered when an EAP is in place.

Successful EAPs place a high value on confidentiality and require that fire fighters enter the program voluntarily. Although a fire officer can recommend or suggest that a fire fighter consider seeking assistance from an EAP, the fire officer cannot know the details of any fire fighter/EAP interaction. The fire officer's focus is on the fire fighter's job performance.

You Are the Fire Officer Conclusion

The lieutenant consults with other company officers to brainstorm on how to resolve the issue of poor fire company performance on commercial fires. A captain points out that the department runs very few working commercial fires, about two per year. A lieutenant from an adjacent fire station suggests that they schedule some commercial occupancy walkthroughs, followed by a multiple-company drill where they will practice on the commercial fire standard operating procedures. This will ensure the department is familiar with the commercial occupancies in the area and that the SOPs for commercial occupancies are effective.

After-Action REVIEW

IN SUMMARY

- A fire officer must be able to process several types of information to supervise and support the fire company members effectively.
- Successful communication occurs when two people can exchange information and develop mutual understanding.
- The communication cycle includes five components: message, sender, medium, receiver, and feedback.
- To improve your listening skills, do not assume, do not interrupt, try to understand the need, and do not react too quickly.
- Fire officers should keep their superior officers informed about progress toward goals and projects, potential controversial issues, fire fighter attitude, and morale.
- A flow of informal and unofficial communications is inevitable in any organization that involves people. The grapevine flourishes in the vacuum created when the official organization does not provide the workforce with timely and accurate information about work-related issues.
- The direct approach to emergency communications entails asking precise questions, providing timely and accurate information, and giving clear and specific orders.
- Radio communications are essential for emergency operations because they provide an instantaneous connection and can link all of the individuals involved in the incident to share important information.
- An officer should be as consistent as possible when sending verbal messages over the radio. The performance goal should be to sound the same and communicate just as effectively when reporting a minor incident as when communicating under intense stress.
- A systematic approach to high-quality decision making is recommended:
  - Define the problem.
  - Generate alternative solutions.
  - Select a solution.
  - Implement the solution.
  - Evaluate the result.
- The first step in solving any problem is to examine the problem closely and to define the problem carefully. A well-defined problem is one that is half-solved.

- The implementation phase is often the most challenging aspect of problem solving, particularly if it requires the coordinated involvement of many different people.
- Determining whether the solution actually solved the problem requires some type of measurement that compares the original condition with the condition after implementation.
- The inherent risks associated with emergency operations demand close supervision at all times.
- Command staff assignments include the safety officer, liaison officer, and public information officer positions.
- The fire officer's primary responsibility is to the team of fire fighters under his or her direct supervision.
- The first-arriving fire officer must demonstrate the ability to take control of the situation and provide specific direction to all of the units that are operating and arriving.
- Most fire officer leadership activity is directed toward accomplishing routine organizational goals and objectives in nonemergency conditions.
- Training and coaching have been core fire officer tasks since the establishment of the first organized fire departments.
- A fire officer must be prepared to conduct company-level training exercises and evolutions to ensure that the company is prepared to perform its basic responsibilities effectively and efficiently.
- The four-step method is a core part of most Fire Instructor I certification programs. It includes the following components:
    - Preparation
    - Presentation
    - Application
    - Evaluation
- The four levels of fire fighter skill competence are unconscious incompetence, conscious incompetence, conscious competence, and unconscious incompetence.
- Mentoring is a one-on-one process in which the more experienced person provides a deliberate learning environment through instructing, coaching, providing experiences, modeling, and advising.
- Teaching new skills takes more time than maintaining proficiency of existing skills. The fire officer should obtain as much information as possible about the device or procedure, especially identifying any fire fighter safety issues.
- Psychomotor skill levels are divided into four categories:
    - Initial
    - Plateau
    - Latency
    - Mastery
- At the very first meeting with a new fire fighter trainee, the fire officer should explain the procedures in the fire station when the company receives an alarm, assign a senior fire fighter to function as a mentor, and describe any restrictions that are placed on fire fighters in training.
- Four federal regulations govern fire fighter training:
    - OSHA regulation 29 CFR 1910.1030, *Occupational Exposure to Bloodborne Pathogens*
    - OSHA regulation 29 CFR 1910.20, *Hazardous Waste Operations and Emergency Response (HAZWOPER)* training
    - OSHA regulation 29 CFR 1910.134, *Respiratory Protection* training
    - Homeland Security Presidential Directive 5 (HSPD-5), *Managing Domestic Incidents*
- The fire officer may need to develop a specific training program that is not covered by an existing certification training program or prepared lesson plan. Five steps can help accomplish this goal:
    - Assess needs.
    - Establish objectives.
    - Develop the training program.

- Deliver the training.
- Evaluate the impact.
- Complaints, conflicts, and mistakes are special categories of problems. One of the key factors in decision making is knowledge about how to deal with situations that involve conflicts or complaints.
- One of the most difficult situations for a fire officer is an interpersonal conflict or grievance within the company or directly involving a company member.
- The conflict resolution model is a basic approach that can be used in situations where interpersonal conflict is the primary problem or a complicating factor.
- All complaints should be investigated, even if the foundation for the complaint appears to be weak or nonexistent.
- The fire officer may take or recommend four actions after completing an investigation:
  - Take no further action.
  - Recommend the action requested by the complainant.
  - Suggest an alternative solution.
  - Refer the issue to the office or person who can provide a remedy.
- With many conflicts, the fire officer needs to follow up with the complainant to see whether the problem is resolved.
- A rising leadership challenge is fire fighter behavioral and physical health issues such as substance abuse, financial problems, acute stress disorder, and posttraumatic stress disorder.
- Fire service organizations have been mobilizing to provide resources and tools to reduce impact of physical and behavioral health issues for fire fighters.

KEY TERMS

Acute stress disorder (ASD) An intense, unpleasant, and dysfunctional reaction beginning shortly after an overwhelming traumatic event and lasting less than a month.

Brainstorming A method of shared problem solving in which all members of a group spontaneously contribute ideas.

Coaching A method of directing, instructing, and training a person or group of people with the aim to achieve some goal or develop specific skills.

Complaint Expression of grief, regret, pain, censure, or resentment; lamentation; accusation; or fault finding.

Conflict A state of opposition between two parties. This is often the manifestation of a complaint.

Employee assistance program (EAP) An employee benefit that covers all or part of the cost for employees to receive counseling, referrals, and advice in dealing with stressful issues in their lives. These problems may include substance abuse, bereavement, marital problems, weight issues, or general wellness issues.

Environmental noise A physical or sociological condition that interferes with the message in the communication process.

Grievance A dispute, claim, or complaint that any employee or group of employees may have in relation to the interpretation, application, and/or alleged violation of some provision of the labor agreement or personnel regulations.

Grievance procedure A formal, structured process that is employed within an organization to resolve a grievance. In most cases, the grievance procedure is incorporated in the personnel rules or the labor agreement and specifies a series of steps that must be followed in a particular order.

Investigation A systematic inquiry or examination.

Job instruction training A systematic four-step approach to training fire fighters in a basic job skill: (1) prepare the fire fighters to learn, (2) demonstrate how the job is done, (3) try them out by letting them do the job, and (4) gradually put them on their own.

Mentoring A developmental relationship between a more experienced person and a less experienced person (a protégé).

Mistake An error or fault resulting from defective judgment, deficient knowledge, or carelessness; a misconception or misunderstanding.

Moral injury The damage done to one's conscience or moral compass when that person perpetrates, witnesses, or fails to prevent acts that transgress one's own moral beliefs, values, or ethical codes of conduct.

Posttraumatic stress disorder (PTSD) A condition characterized by symptoms of avoidance and nervous system arousal after experiencing or witnessing a traumatic event.

Psychological contract Refers to mutual unwritten expectations that exist between an employee and his or her employer regarding policies and practices in their organization.

Rep (repetition) One complete motion of an exercise.

Resiliency The process of adapting well in the face of adversity, trauma, tragedy, threats or significant sources of stress, such as family relationship problems, serious health problems, or workplace and financial stressors.

Set A group of consecutive repetitions.

Training The process of achieving proficiency through instruction and hands-on practice in the operation of equipment and systems that are expected to be used in the performance of assigned duties.

Unconsciously competent The highest level of the Conscious Competence Learning Matrix developed by Dr. Thomas Gordon in the 1970s. At this level, the skill becomes so practiced that it enters the unconscious parts of the brain—it becomes second nature.

REFERENCES

Abbott, Donald. 2018, February 1. "Communications Clues: Don Abbott Offers 16 Phrases That Should Alert the IC of a Possible Mayday Event." *Firehouse Magazine.* Accessed June 25, 2019. https://www.firehouse.com/tech-comm/radios-pagers-accessories/article/12387578/communications-clues.

Abbott, Donald. 2019. "Project Mayday: Career 2015-2018 48 months." Glendale AZ, Command Emergency Response Training. http://projectmayday.net. Accessed June 25, 2019.

Adams, Linda. 2016. "Learning a New Skill Is Easier Said Than Done." Accessed June 25, 2019. https://www.gordontraining.com/free-workplace-articles/learning-a-new-skill-is-easier-said-than-done/.

American Psychiatric Association (APA). n.d. "Understanding Chronic Stress." Accessed June 25, 2019. https://www.apa.org/helpcenter/understanding-chronic-stress.

American Psychiatric Association (APA). 2013. *Diagnostic and Statistical Manual of Mental Disorders, Fifth edition.* Philadelphia: American Psychiatric Publishing.

Boehm, Eric. 2018, June 21. "Firefighter Earned $300K in Overtime by Working More Hours Than Actually Exist." *Reason.* Accessed June 25, 2019. https://reason.com/2018/05/21/firefighter-earned-300k-in-overtime-by-w/.

Carey, Mary G., Salah S. Al-Zaiti, Grace E. Dean, Loralee Sessanna, and Deborah S. Finnell. 2011. "Sleep Problems, Depression, Substance Use, Social Bonding, and Quality of Life in Professional Firefighters." *Journal of Occupational and Environmental Medicine* 53 (8): 928–933.

Centers for Disease Control and Prevention (CDC). 2016. "Alcohol and Public Health: Excessive Alcohol Use and Risks to Men's Health." Accessed June 25, 2019. https://www.cdc.gov/alcohol/fact-sheets/mens-health.htm.

Curtiss, Paul R., and Phillip W. Warren. 1973. *The Dynamics of Life Skills Coaching: Life Skills Series.* Prince Albert, Saskatchewan: Training Research and Development Station, Department of Manpower and Immigration.

Dabos, Guillermo E., and Denise M. Rousseau. 2004. "Mutuality and Reciprocity in the Psychological Contracts of Employees and Employers." *The Journal of Applied Psychology* 89 (1): 52–72.

Desilver, Drew. 2018, August 7. "For Most U.S. Workers, Real Wages Have Barely Budged in Decades." FactTank: News in the Numbers. Pew Research Center. Accessed June 25, 2019. https://www.pewresearch.org/fact-tank/2018/08/07/for-most-us-workers-real-wages-have-barely-budged-for-decades/.

Drucker, Peter F. 1974. *Management: Tasks, Responsibilities, Practices.* New York: Harper and Row.

Federal Emergency Management Agency (FEMA). 2017. *National Incident Management System, Third edition.* Washington, DC: Department of Homeland Security.

Flaherty, Michelle Poche, editor. 2013. *Effective Supervisory Practices, Fifth edition.* Washington, DC: International City/County Management Association.

Fisher, Nicole. 2018, August 23. "More Firefighters Committed Suicide in 2017 Than Died in Line of Duty." *Forbes.* Accessed June 25, 2019. https://www.forbes.com/sites/nicolefisher/2018/08/23/haunted-heroes-more-firemen-committed-suicide-in-2017-than-died-in-line-of-duty/#4d4ed312a243.

Francis, Norton, and Frank Sammartino. 2015. "Governing with Tight Budgets: Long-Term Trends in State Finances State and Local Finance Initiative." Washington, DC: The Urban Institute.

Gagliano, Anne, and Mike Gagliano. 2018. "Challenges of the Firefighter Marriage." Tulsa, OK: Pennwell Corporation.

Gasaway, Richard. 2015. *It's More Than Paying Attention: Situational Awareness Matters.* St. Paul, MN: Gasaway Consulting Group.

Griffin, Riley. 2018, October 17. "The Student Loan Debt Crisis Is About to Get Worse: The Next Generation of Graduates Will Include More Borrowers Who May Never be Able to Repay." *Bloomberg.* Accessed June 25, 2019. https://www.bloomberg.com/news/articles/2018-10-17/the-student-loan-debt-crisis-is-about-to-get-worse.

Hall, Sarah J., Brad Aisbett, Jamie L. Tait, Anne I. Turner, Sally A. Ferguson, and Luana C. Main. 2016. "The Acute Physiological Stress Response to an Emergency Alarm and Mobilization during the Day and at Night." *Noise and Health* 18 (82): 150–156.

Hayman, Miriam, Jeff Dill, and Robert Douglas. 2018. "The Ruderman White Paper on Mental Health and Suicide of First Responders." Boston: Ruderman Family Foundation.

Hoschouer, Jason. 2016. *Badges and Budgets: Personal Finance From a Law Enforcement Perspective*. Clayton, CA: The MotorCop Mindset.

IAFF staff. 2017. "Recognizing PTSD in Fire Fighters: 5 Warning Signs." Accessed June 25, 2019. https://www.iaffrecoverycenter.com/blog/recognizing-ptsd-fire-fighters-5-warning-signs/.

Kiewra, Kenneth A. 2002. "How Classroom Teachers Can Help Students Learn and Teach Them How to Learn." *Theory into Practice* 41 (2): 72–80.

Litz, Brett T., Nathan Stein, Eileen Delaney, Leslie Lebowitz, William P. Nash, Caroline Silva, and Shira Maguen. 2009. "Moral Injury and Moral Repair in War Veterans: A Preliminary Model and Intervention Strategy." *Clinical Psychology Review* 29 (8): 695–706.

National Fallen Firefighters Foundation (NFFF). 2017. *Fire Service Behavioral Health Management Guide*. Emmitsburg, MD: National Fallen Firefighters Foundation.

Patrick, Harold A. 2008. "Psychological Contract and Employment Relationship." *The ICFAI University Journal of Organizational Behavior* 7 (4): 7–24.

Pfeifer, Joseph W. 2012. "Adapting to Novelty: Recognizing the Need for Innovation and Leadership." *With New York Firefighters* 72 (1): 20–23.

Price, Gary, and Tim Norbeck. 2018. "Physicians are Human Too." The Physicians Foundation. Accessed June 25, 2019. https://www.forbes.com/sites/physiciansfoundation/2018/07/18/physicians-are-human-too/#46cf8ca34a29.

Sawyer, Diane. 2017. "My Reality: Hidden America: Chris Smith: A Firefighter Searches for Time." 20/20, ABC News Corp. Accessed June 25, 2019. https://abcnews.go.com/US/deepdive/diane-sawyer-income-inequality-my-reality-hidden-america-44770807.

Staggs, Sarah. 2019. "Posttraumatic Stress Disorder (PTSD) Symptoms." Copyright 2019 PsychCentral.com. All rights reserved. Reprinted with permission. Accessed June 25, 2019. https:// psychcentral.com/ptsd/posttraumatic-stress-disorder-ptsd-symptoms/.

Stapleton, Leo. 1983. *Thirty Years on the Line*. Dover, NH: DMC Associates.

Tagliareni, Sonia. 2018. "Firefighters and Addiction." Accessed June 25, 2019. https://www.drugrehab.com/addiction/firefighters/.

Taigman, Michael., and S. Dean. 1999. "Complaints" In *Secrets of Successful EMS Leaders: How to Get Results, Advance Your Career, and Improve Your Service*. Midlothian, VA: Sempai-Do.

Talbot, Simon G., and Wendy Dean. 2018, July 26. "Physicians Aren't 'Burning Out.' They're Suffering from Moral Injury." *STAT*. Accessed June 25, 2019. https://www.statnews.com/2018/07/26/physicians-not-burning-out-they-are-suffering-moral-injury/.

Fire Officer in Action

The lieutenant is reviewing some of the factors of poor communication.

1. What should the sender do when relaying critical information during a stressful event?
   A. Send message in voice and text.
   B. Repeat the message twice.
   C. Have the receiver repeat key points of the message in their own words.
   D. Print out the message on paper and have a runner deliver it to the receiver.

2. Best practices in active listening do *not* include:
   A. interrupting the speaker to summarize the issue.
   B. looking for the real reason the person wants your attention.
   C. understanding the other person's viewpoint.
   D. trying to understand the speaker's need.

3. Prejudice and bias are examples of:
   A. digital noise.
   B. physical noise.
   C. distracted noise.
   D. sociological noise.

4. Project Mayday points out that hearing _____ of the 16 identified phrases should serve as a trigger for the incident commander to reconsider current operations.
   A. 1
   B. 3
   C. 6
   D. 8

Fire Lieutenant Activity

NFPA Fire Officer I Job Performance Requirement 4.2.1

Assign tasks or responsibilities to unit members, given an assignment at an emergency incident, so that the instructions are complete, clear, and concise; safety considerations are addressed; and the desired outcomes are conveyed.

Application of 4.2.1

Describe how a fire officer can improve communications on the fire ground:

1. Describe how a fire officer can improve personal effectiveness in radio transmissions.
2. Identify two ways that fire officer to fire fighter communication can be improved.
3. Develop a procedure that uses the information from Project Mayday to identify an impending mayday situation.
4. How can a fire officer address safety considerations when making a fire-ground assignment?

Access Navigate for flashcards to test your key term knowledge.

CHAPTER 4

Fire Officer I

Community Relations and Risk Reduction

KNOWLEDGE OBJECTIVES

After studying this chapter, you will be able to:
- Explain how demographics can improve community risk reduction and fire department community relations. (**NFPA 1021: 4.3.1**) (pp. 116–117)
- Define *community risk reduction*. (**NFPA 1021: 4.3.1**) (pp. 117–118)
- Identify the 5 Es of community risk reduction. (**NFPA 1021: 4.3.1**) (pp. 118–119)
- Identify obstacles to implementing community risk reduction plans and explain how to overcome them. (**NFPA 1021: 4.3.1**) (pp. 118–119)
- Explain the six steps of implementing a community risk reduction plan. (**NFPA 1021: 4.3.1**) (p. 119)
- List and describe opportunities for public education. (pp. 119–120)
- Describe how to gather information from a citizen with a concern. (**NFPA 1021: 4.3.2**) (p. 122)
- Describe how to resolve citizen complaints. (**NFPA 1021: 4.3.2**) (pp. 120–121)
- Explain why it is important to respond to citizen concerns and inquiries. (**NFPA 1021: 4.3.2, 4.3.3**) (pp. 121–122)
- Describe the difference between customer service and customer satisfaction. (pp. 122–124)

SKILLS OBJECTIVES

After studying this chapter, you will be able to:
- Identify and allocate needed resources to implement a community risk reduction plan. (**NFPA 1021: 4.3.1**) (p. 119)
- Create a timeline with milestones for community risk reduction plan implementation. (**NFPA 1021: 4.3.1**) (p. 119)
- Assign tasks and responsibilities for carrying out the community risk reduction plan. (**NFPA 1021: 4.3.1**) (p. 119)
- Communicate goals expectations for community risk reduction plan activity. (**NFPA 1021: 4.3.1**) (p. 119)
- Gather information about a citizen concern and respond to the concern. (**NFPA 1021: 4.3.2**) (p. 122)
- Answer a public inquiry in a professional manner. (**NFPA 1021: 4.3.3**) (pp. 121–124)

ADDITIONAL NFPA STANDARDS

- **NFPA 1452**, *Guide for Training Fire Service Personnel to Conduct Community Risk Reduction for Residential Occupancies*
- **NFPA 1035**, *Standard on Fire and Life Safety Educator, Public Information Officer, Youth Firesetter Intervention Specialist, and Youth Firesetter Program Manager Professional Qualifications*

- **NFPA 1300**, *Standard on Community Risk Assessment and Community Risk Reduction Plan Development*
- **NFPA 1600**, *Standard on Continuity, Emergency, and Crisis Management*

You Are the Fire Officer

The battalion chief is reviewing target hazards with the lieutenant. The talk focuses on the frequent responses to the Restful Vista senior living residential high-rise. Almost every week there is a "food-on-the-stove" incident that generates a 911 response.

1. How can a fire officer identify a community risk?
2. What can be measured to document the impact of a community risk reduction program?
3. What options does the fire officer have when addressing community risks?

Access Navigate for more practice activities.

Introduction

Fire fighters are important members of their community. They quickly respond to calls for help and are known as creative and compassionate problem solvers. One of the most powerful contributions fire fighters can make to their community is identifying and correcting situations that may cause a fire, hazardous incident, or serious injury.

The fire station is an important part of the community. Citizens tend to think of the fire department in relation to their local fire station. The financial crises showed the value of strong community ties, as residents rallied to keep their local fire stations open in the face of budget cuts, measuring their personal perceptions of safety by the proximity of the closest fire station.

Jurisdictions are moving toward more community-based local government. Next to schools, the fire department is the most decentralized and community-based function of local government. Some cities use their neighborhood fire stations as a primary point of contact for local government services.

The fire officer is the official fire department representative and also ensures that the community's needs are being addressed by the department. This chapter provides the fire officer with information about how to learn more about the community; how to identify risks that could lead to a loss of life, property, or community; how to develop a community risk reduction program; and how to effectively respond to a citizen concern.

Understanding the Community

Each community has special needs and different characteristics that should be considered in relation to every service and program provided by the fire department. The most significant information comes from understanding which types of people live and work in the community. The fire officer should develop a good understanding of the population and demographics of the particular areas where the company responds.

The federal government undertakes a nationwide census once every decade. The data gathered via the census are readily available and provide an excellent starting point to begin an analysis of the local community. The census collects and identifies a massive amount of information about the **demographics** of the nation. Demographic data describe the characteristics of human populations and population segments. The census data can be analyzed with sophisticated database software tools and digitized mapping to develop a profile of the many characteristics of local populations down to the neighborhood level.

A variety of demographic analysis techniques may be applied. Demographic data are often used to identify and analyze consumer markets, to help retailers predict what consumers will buy, and to give advertisers insight into the messages that will be most effective in particular markets. Politicians make extensive use of demographic data to predict voters' response to policies and messages in each area. Both nonprofit and for-profit programs may be fine-tuned to reach certain segments of the population based on detailed analysis of what different types of people like, what they value, what they believe, and where they live (Gronbach and Moye, 2017). The same approach can be used to ensure that the fire department is delivering the appropriate services and information to the local community.

> **FIRE OFFICER TIP**
>
> **Se Habla Espanol**
>
> In communities where English is not the primary language, it is important to be able to communicate in the language of the neighborhood. Many departments encourage their members to learn additional languages, and some even provide special classes or offer pay incentives for multilingual fire fighters. A few departments have experimented with language immersion programs, making the primary language of the neighborhood (e.g., Spanish) the primary language at a designated fire station. The fire fighters assigned to that station answer the telephone, speak to one another, and handle emergencies while speaking that language.

Understanding the cultural factors that influence particular behaviors will increase the effectiveness of fire department messages. As a fire officer, you can identify groups with special needs so as to improve the delivery of emergency services. An effective program should be designed to meet the needs of the particular community, and the message must be formulated to reach and be understood by the target audience. With each census, it has become apparent that the United States is growing increasingly multiethnic. The country is experiencing a continual flow of immigrants who bring a variety of languages, cultures, religions, traditions, and beliefs into their new communities (Dorling and Geitel-Basten, 2018).

One dimension of demographic analysis is the identification of communities where people share the same diversity classifications, such as cultural background, language, religion, and age. For example, one neighborhood may have residents who speak one dialect and share a culture that is totally different from that of the adjoining neighborhood. Various diversity classifications often challenge the fire service's ability to meet the needs of different groups.

Safety information shared through risk reduction campaigns should be applicable to the particular community and delivered in a format that the community can understand and act on. The format and the method of delivery should be tailored to meet the language, cultural, and community history of the community (Kirtley, 2008).

Both emergency services and community risk reduction efforts must be fine-tuned to identify and meet those needs in particular communities. The fast, flashy, and loud aspects of the typical U.S. response to a 911 call could be overwhelming, terrifying, or embarrassing to some who called for assistance. The customs and traditions of another culture may clash with the aggressive "pit crew"-style response to a cardiac arrest. A shop owner may be angered by the arrival of a fire company to perform a fire safety inspection. Cultural sensitivity is required to help the customers appreciate the services the fire department is providing.

> **FIRE OFFICER TIP**
>
> **Constructing an Effective Message**
>
> The Vision 20/20 Project is guided by a coalition of national organizations and experts exemplifying how collaboration, communication, and commitment to data-based solutions save lives and property. They have developed an online Fire Safety Materials Generator that allows the user to choose messages, pictures, and designs that fit the needs of the community:
>
> 1. Choose your audience.
> 2. Choose your materials.
> 3. Customize your message.
> 4. Save and share.
>
> Go to https://materialsgenerator.strategicfire.org/ and make your own media.

Community Risk Reduction

Community risk reduction (CRR) is a process to identify and prioritize local risks, followed by the integrated and strategic investment of resources to reduce their occurrence and impact (NFPA 1300, 2020). CRR is different from the traditional ideas of fire prevention because it is concerned with a comprehensive approach to reducing the overall incidence and impact of emergencies within the community. The fire service has a rich history of striving to reduce the risk of fire in the community. This has been achieved through emergency response, fire safety education, adoption of fire codes, and enforcement of those codes. Today, fire departments recognize that their primary goal is to save lives and property from more than just fires. This broader, "all hazards" vision has become evident as the fire service has responded to extrication, drowning, heart attack, stroke, hazardous materials release, trench and building collapse, high-angle rescue, swiftwater rescue, underwater and ice rescue, active shooter, and weapons of mass destruction incidents.

The fire service has taken on the role of prevention and mitigation of these types of incidents. The best method of preventing fire injuries and deaths, of course, is to prevent the fire from ever starting or to reduce the severity of the fire if it does start. The same is true for the other types of incidents to which the fire department is called.

This shift in thinking often requires a transformation of the culture of the fire department. Members, after all, joined to fight fires—not to identify trip hazards. The company officer's challenge is to promote the concept that the fire fighter role is to save lives and property from every type of incident to which department members might be called. A link must be made between the incidents to which the fire department responds and the prevention of those types of incidents.

A basic overview of CRR is provided here for insight into the methods of risk reduction (Stouffer, 2016a):

1. Identify risks.
2. Prioritize risks.
3. Develop strategies and tactics to mitigate risks.
4. Prepare the CRR plan.
5. Implement the CRR plan.
6. Monitor, evaluate, and modify the plan.

Step One: Identify Risks

The first step in developing a CRR program is to identify risks. A risk can be defined as any factor, human or otherwise, that could lead to an emergency, negatively affecting people, places, or resources of that community. A logical place to begin identifying community risks is to look at what the department experiences in terms of the types of incidents, call volume, locations, and causes. These data are available in many departments through the records management system (RMS). Another good source for data concerning risks in the community is likely available in a geographic information system (GIS). A GIS understands location and is particularly well suited to pinpoint risks. It can analyze preparedness capabilities and build layers of data, which can help members of the department visualize the relationships between variables impacting risk and preparedness. Your GIS might be located in the fire department or in another government agency.

Step Two: Prioritize Risks

The second step in developing a CRR program is to prioritize risks. When building a model for risk reduction, evaluate each of the identified risks based on factors such as the severity, frequency, and duration of an event. It is also important to consider the likelihood of occurrence and the department's capacity to respond. A special consideration would be the economic impact of a loss. One particularly important risk factor to consider is people. Often, identifying the relationship between the most vulnerable populations and high call volume areas will help you to identify risk reduction priorities.

FIRE OFFICER TIP

Economic Impact of Successful Commercial Fire Interventions

Anthony Evans, PhD, evaluated the impact of successful commercial fire interventions of the Phoenix, Arizona, Fire Department for a 1-year period. The report from the L. William Seidman Research Institute showed that 7446 private, farm, and government jobs could have been lost if the fire department had not successfully intervened at 42 commercial fires between June 1, 2012, and May 31, 2013. The consequences of not successfully intervening mean that a company would have to temporarily or permanently close due to the fire. There would also be a loss of $650 million in gross state product, $295.6 million in real disposable income, and $35 million in state tax revenue.

This study was triggered by an earlier study that examined at the impact of suppressing a fire at a furniture manufacturer in 2011. The successful suppression saved up to 203 jobs, preserved $20 million gross state product, and $9 million in real disposable personal income (Evans, 2014).

Step Three: Develop Strategies and Tactics to Mitigate Risks

The third step in developing a CRR program is to develop strategies and tactics to mitigate risks or reduce the occurrence of preventable incidents. To do this effectively, one must utilize all the available resources within the community (Stouffer, 2016b). The Five Es of prevention—education, engineering, enforcement, economic incentives, and emergency response—provide the basic strategies and tactics to address this third step (Sawyer, Phillips, Catts, and Sawyer, 2016). All risks should be examined through the lens of each of the following strategies:

- Education: Changing behavior by teaching people about fire and emergency prevention and response. This is the traditional approach to fire safety.
- Engineering: Using technology to make buildings and products safer. Smoke alarms,

child-resistant medication caps, and car seats are all examples of using technology to improve safety.

- Enforcement: Fire and building codes are used to create and maintain safety in the built environment. Enforcing these codes reduces the risk in the community in both the short and long term.
- Economic incentives: Providing financial motivation can encourage beneficial behaviors and choices. Financial motivation can be in the form of reward or punishment. For example, providing tax credits for sprinkler systems or imposing fines for the absence or removal of smoke alarms.
- Emergency response: A well-trained and prepared emergency response force is essential. Deployment should be responsive to the unique aspects of the communities served.

These first three steps of developing a CRR program comprise a risk assessment. A **risk assessment** is the identification and prioritization of potential and likely risks within a particular community. More information on conducting a risk assessment is available in NFPA 1300, *Standard on Community Risk Assessment and Community Risk Reduction Plan Development,* as well as the Vision 20/20 Community Risk Assessment Guide (Stouffer, 2016b).

Steps four through six of developing a CRR program are related to the development of a CRR plan.

Step Four: Prepare the CRR Plan

The fourth step is to prepare the CRR plan. This plan should be developed with input from fire fighters working with the fire officer and other public safety personnel, as well as suggestions from community groups and people outside the fire department. Team members should examine each strategy to begin outlining goals, benchmarks, and time frames. Once finished, the plan should be communicated broadly to build understanding and buy-in.

Step Five: Implement the CRR Plan

The fifth step is the implementation of the CRR plan. There are six strategies of implementation.

1. Identify and allocate needed resources: The fire officer in charge of the CRR efforts will need to identify and justify resources necessary for performing the services.
2. Prepare a timeline with milestones: Be realistic in developing objectives and timelines. Include benchmarks for program development, training, implementation, delivery, and evaluation.
3. Assign responsibilities: Some people will volunteer to handle tasks and others will be assigned. The most difficult challenge is to hold members accountable for their area of responsibility.
4. Communicate goals and expectations: Goals and expectations should be clearly defined and communicated from the start of the work. Revisit the guidelines periodically.
5. Monitor progress: Monitoring is an ongoing process. Check in to ensure consistency and accountability.
6. Make adjustments as necessary: The proposed strategies may bring unintended results. The leader should be skillful in making decisions and willing to accommodate necessary changes.

Step Six: Monitor, Evaluate, and Modify the Plan

The sixth step is monitoring, evaluating, and modifying the plan as needed. Members will provide progress reports from their areas of activity to the leader. The leader will also be providing updates and progress reports to the administration and partners.

Public Education

An integral part of community risk reduction is fire and life safety education. Often referred to as *public education*, the goals of fire and life safety education are to help people understand how to prevent the loss of life, injuries, and property damage from occurring and to teach them how to react in an appropriate manner if an emergency occurs. Fire companies work to create a change in behavior that results in greater safety for their community. For example, making people aware of common fire risks and hazards and providing information about reducing or eliminating those dangers can prevent many fires, injuries, and deaths from occurring. Public education programs teach techniques to reduce the risks of death or injury in the event of a fire and prepare for other hazards such as severe weather. The programs are designed to prevent all types of accidents and injury to which the fire service commonly responds. Public fire safety education programs include, but are not limited to, those focused on the following topics:

- Stop, Drop, and Roll
- Close your door safety initiative
- Crawl low under smoke
- Exit Drills In The Home (E.D.I.T.H.)
- Emergency notification
- Installation and maintenance of smoke alarms
- Advantages of residential sprinkler systems
- Selection and use of portable fire extinguishers
- Learn Not to Burn® preschool program
- Fire safety for special populations
- Fall prevention
- Wildland fire prevention programs

Most fire and life safety programs are presented to groups such as school classes, scout troops, church groups, senior citizen groups, civic organizations, hospital staff, and business employees. Presentations are often made at community events and celebrations, as well as during Fire Prevention Week activities. Another popular fire safety education activity is a fire station tour. Both children and adults enjoy the opportunity to tour the local fire station, and a fire station tour is an excellent opportunity to promote fire prevention. Leadership in public education activities is a critical component of serving a company and a community. Officers responsible for community risk reduction implementation at the company level should be well versed in public education.

Community Emergency Response Team

The **Community Emergency Response Team (CERT)** concept was developed and first implemented by the Los Angeles Fire Department (LAFD) in 1985. The goal is to provide basic training to local community residents that would allow them to function effectively during the first hours after a catastrophic event. It is a hands-on community risk reduction activity.

CERT groups can provide immediate assistance to victims in their area and collect disaster intelligence that assists professional responders with prioritization and allocation of resources after a disaster (**FIGURE 4-1**). The training also teaches the CERT participants how to organize spontaneous volunteers who have not undergone the training.

CERT was moved to the Citizen Corps section of the federal government in 2004. In 2019, Citizen Corps listed 2700 CERT programs.

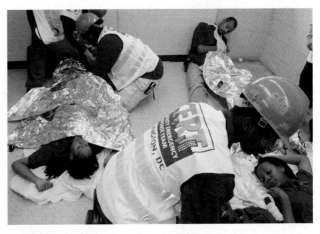

FIGURE 4-1 CERT groups can provide immediate assistance to victims and help professional responders with prioritization and allocation of resources.
© Alex Wong/ Getty Images News/ Getty Images

CERT Course Schedule

The CERT course is delivered in the community by a team of first responders trained as CERT instructors. The organization of the training varies from program to program and covers the following topics:

- Session 1, Disaster Preparedness: Addresses hazards specific to the community. Materials cover actions that participants and their families take before, during, and after a disaster as well as an overview of CERT and local laws governing volunteers.
- Session 2, Fire Suppression: Covers fire chemistry, hazardous materials, fire hazards, and fire suppression strategies. The thrust of this session is safe use of fire extinguishers, size-up of the situation, control of utilities, and extinguishment of a small fire.
- Session 3, Medical Operations, Part I: Participants practice diagnosing and treating airway obstruction, bleeding, and shock by using simple triage and rapid treatment techniques.
- Session 4, Medical Operations, Part II: Covers evaluating patients by performing a head-to-toe assessment, establishing a medical treatment area, performing basic first aid, and demonstrating medical operations in a safe and sanitary manner.
- Session 5, Light Search-and-Rescue Operations: Participants learn about search-and-rescue planning, size-up, search techniques, rescue techniques, and—most important—rescuer safety.

- Session 6, Psychology and Team Organization: Covers signs and symptoms that might be experienced by the disaster victim and workers, and addresses CERT organization and management.
- Session 7, Case Review and Disaster Simulation: Participants review and practice the skills that they have learned during the previous six sessions in a disaster activity.

Since the CERT program was moved to the Citizen Corps, there has been an increased emphasis on assessing community needs and developing CERT response goals that address local needs. The CERT Participant Hazard Annex includes lesson plans for 13 hazardous situations.

FIRE OFFICER TIP

Catastrophe Planning

Some fire departments have developed media communication/public education plans to assist their communities in responding to natural and human-made catastrophes. For example, the San Francisco Fire Department accepted the services of a retired local television anchorperson to develop preassembled public safety messages and citizen instructions that may be issued in a wide variety of major emergencies.

Fire departments have to be able to respond to dynamically changing conditions. For example, national concerns, such as the Ebola virus disease (EVD) outbreaks, have required fire departments to develop specific public information messages quickly, using information updates from the Centers for Disease Control and Prevention (CDC). Hurricane Michael made landfall as a Category 5 storm on October 10, 2018, destroying Mexico Beach, Florida. More than a dozen federal, state, and local public safety agencies coordinated their efforts to provide a consistent message to the general public as the hurricane moved through Florida, Georgia, North Carolina, and Virginia.

Responding to Public Inquiries

Sometimes, the public makes an inquiry to the fire officer for general information, such as a request for a description of the services the fire department or city provides or a request to participate in a community event. Both of these situations have the potential to leave the citizen satisfied or for the citizen to go away with a negative view of the fire department.

The fire officer must treat all requests professionally and with respect. Even though you may find the request less than critical, the citizen believes that it is valid and important. Failure to approach the request with sincerity can have a lasting negative impact.

Every effort should be made to answer each question fully and accurately. The fire officer may not have all the information that the individual is seeking. For example, the fire officer may not know where the closest flu shot clinic is located. When this occurs, the fire officer should seek out the information immediately rather than just expressing that he or she does not know. A phone call or two can usually lead to the answer and leaves the citizen with a positive image of the fire department.

Some requests that citizens make may not be within the fire officer's authority. In these situations, the fire officer should provide a method of moving the request to the level where it can be resolved. If time allows, the best method is to get the citizen's contact information, write up a summary of the discussion, and forward it to the appropriate administrative fire officer at headquarters. The fire officer should also follow up to ensure that the citizen is contacted in a reasonable amount of time. Some citizens may prefer to contact fire administration themselves. In these situations, the fire officer should give the citizen specific information on whom to contact. The fire officer should also contact the administrative or executive fire officer before the citizen does to provide information on the situation.

FIRE OFFICER TIP

Reducing Cooktop Fires: The Worcester Project

Lieutenant Annie Pickett started working with the Worcester Housing Authority after learning that 759 apartments in four buildings were generating about 12 stovetop fires per month, causing about $223,000 in direct damages and displacing one or two families per month (Pickett, 2018). The residents were low income elderly, 86 percent of whom had a physical mobility or mental disability issue.

The fire department and housing authority took a two-pronged education and engineering approach to manage this risk. They provided person-to-person training on fire-safe cooking techniques while installing a temperature limiting control (TLC) heating element on the stove. The TLC would never get hot enough to ignite oil while cooking. A big part of the training was learning new cooking techniques with the TLC burner elements. There have been no stovetop fires in the 800 apartments that received the TLC burners and training. That success lead to grant funding to complete the installations of these burner elements in the rest of the complex (Crawford, 2018).

Many departments have specific policies that address how citizen inquiries are to be handled. The fire officer should understand and follow these policies. Failure to do so may lead the citizen to believe that he or she is being treated unfairly. It can also leave the fire officer open to disciplinary action.

Citizen Concerns

The fire officer represents the department in dealing with citizens, public or private organizations, and other governmental agencies. A fire officer could be faced with three different types of citizen concerns. A citizen might have concerns about:

1. The conduct or behavior of a fire fighter (or a group of fire fighters)
2. The fire company's performance or service delivery
3. Fire department policy

Sometimes, a citizen may want to express an alternative viewpoint on an issue and try to see whether some resolution might be mutually agreeable to the department and the citizen. At other times, the citizen is making a formal complaint. When this occurs, the fire officer is functioning as the official recipient of the concern. On other occasions, the fire officer may be the subject of the concern. The first role of the fire officer is to respond in a professional manner that effectively obtains the needed information. The methods outlined for resolving conflict within the company also apply to a citizen concern.

The fire officer must take notes and function as an active listener. By listening attentively and taking detailed notes, the fire officer demonstrates that the concern is officially considered to be important and is receiving the fire officer's full attention. If the immediate response to a concern is an explanation or excuse, the complainant will feel that the fire officer is not paying attention, does not care about the issue, or has something to hide. The person who had a concern will likely stop providing information and feel even more strongly that the concern was justified.

Be empathetic and listen to the complainant's statement of the problem without interrupting. Frequently, these kinds of complaints are a method of venting frustration by a citizen rather than a real expectation that the situation will change; the resolution of the problem may be as simple as acknowledging that the complainant has had a bad experience with the department. Allowing the citizen to express the frustration is important.

Unless it is within the scope of the fire officer's authority, the fire officer should make no promises or imply that certain actions will be taken in the discussion with the complainant. If the fire officer does not have the authority to make a decision on the issue or the citizen is dissatisfied with the officer's decision, the officer should ask whether the citizen would like the issue forwarded to the next level within the organization. If the citizen would like further action, the fire officer should determine the appropriate organizational level where the decision can be made. A preferred method is to consider the scope of the concern. If it deals with an operational issue, it should be sent to the chief of operations. If it is an issue about codes, it would be sent to the fire marshal. If the issue affects all areas of the department, it would be forwarded to the chief of the department. Concerns or complaints about personnel should be forwarded for action to the supervisor of the individual who is involved. If the proposed resolution involves discipline, the fire officer is obligated to protect employee privacy and civil service due process (**FIGURE 4-2**).

All relevant facts should be identified and forwarded, along with the details of the concern. In some departments, the proper procedure is to follow the chain of command. In other departments, fire officers are encouraged to forward the information directly to the decision maker to ensure prompt attention to the matter. Even in these circumstances, the fire officer should inform his or her supervisor about the situation. If any doubt arises about the appropriate response, discuss the issue with your supervisor before taking any action.

Customer Service versus Customer Satisfaction

Customer service is a term that public safety has borrowed from the retail business world. A focus on customer service fixes problems, straightens out procedural glitches, corrects errors of omission (or commission), and provides information.

Customer satisfaction focuses on meeting the customer's expectations. The Center for Public Safety Excellence describes seven types of expectations (CPSE, 2015):

1. Explicit: "I called for an ambulance, why did a fire truck respond?"
2. Implicit: "I expect that the fire will quickly be extinguished after the fire company arrives."
3. Static performance: The jurisdiction's annual report states an average response time of 6 minutes and 30 seconds. "I expect it will take that long to get to my location after calling 911."

> Municipal City Fire and Rescue Department
> Headquarters
>
> May 24, 2019
>
> Mrs. Caroline Marks
> Exotic Food Emporium
> 437 Western Way
> Municipal City
>
> Mrs. Marks
>
> We have completed our investigation into the fire and rescue department response to your store on May 5th. During an investigation for smoke in the building, the fire department damaged the front door, interrupted a fumigation service, and destroyed a storefront window.
>
> The initial actions taken by our members were not consistent with departmental policies and procedures. The city will pay for the repairs to the storefront and a second fumigation. Cynthia Bowers from the Mayor's office will continue to be your point of contact during this process.
>
> I apologize for the actions taken by our members. The members involved in this incident are receiving appropriate training and disciplinary actions. We are making sure that all of our fire fighters know how to properly respond to a building with a rapid entry key lock-box that is undergoing fumigation.
>
> Sincerely,
>
> Fire Chief
> Municipal City Fire and Rescue Department

FIGURE 4-2 Example of a letter to a civilian in response to a complaint.
Courtesy of Mike Ward.

4. Dynamic: How the service evolves over time/experience. "After my last experience, drive me to the hospital instead of calling 911."
5. Technological: "Why can't fire department A speak to fire department B on the radio?"
6. Interpersonal: "Those fire fighters were fantastic. They took care of the problem and were great at keeping my kids calm."
7. Situational: "I expected the fire department to restore the sprinkler heads that were activated in the fire."

Fire departments often meet customers during one of the worst days of their life. They did not start the day planning to crash the car, melt the stove, or have trouble breathing. Good customer service requires sensitivity on the part of every person involved. Although there is little competition for others to provide public safety services, having satisfied customers is important to the jurisdiction (**FIGURE 4-3**).

FIGURE 4-3 Creating satisfied customers is one of the fire officer's most important activities.
© Jones & Bartlett Learning. Photographed by Glen E. Ellman.

Keep the Complainant Informed

If the fire officer needs to do research, consult with others, or obtain direction from supervisors, he or she should keep the complainant informed during this

FIRE OFFICER TIP

Use All of Your Tools to Solve the Problem at Your Level

You need to be proficient in solving problems. There is a natural tendency to push complaints and conflicts up the chain of command. In some departments, this practice can lead to the administrative fire officer handling an issue that should have remained at the supervising fire officer level. The further up the chain of command you go with an issue or problem, the more likely the resolution will not work to your advantage.

If you wish to be promoted to higher positions within the department, developing mastery in handling problems, conflicts, and mistakes increases your value to the formal organization. Even if you just want to improve the situation where you work, the skills described in this chapter will help you meet that goal.

process. Such communication demonstrates that the fire officer is not ignoring the issue, and it educates the complainant on the process.

Follow Up

Citizens frequently complain about local government unresponsiveness. By following up with the complainant, the fire officer reinforces the impression that the complainant's issue is important. This consideration is especially critical if the fire officer has referred the issue to another individual, agency, or organization.

Follow up by the fire officer may be inappropriate, however, if the fire officer was the subject of the concern or if the conflict was handled at a higher supervisory level. The fire officer may consider consulting with a supervisor before conducting a follow up in some cases.

You Are the Fire Officer Conclusion

A fire officer can identify a community risk by the frequency of times it generates a 911 response or the potential severity of a single incident. The impact, or effectiveness, of a CRR can be measured by the reduction of the frequency of 911 responses, the reduced level of severity if an incident occurs, and the increase in the quality of life within that community. The fire officer has five options for resolving a community risk: education, engineering, enforcement, economic incentives, and emergency response.

After-Action REVIEW

IN SUMMARY

- The fire officer is the official fire department representative and also ensures that the community's needs are being addressed by the department.
- The fire officer should develop a good understanding of the population and demographics of the particular areas where the company responds.
- Today, fire departments recognize that their primary goal is to save lives and property from more than just fires.
- Community risk reduction is concerned with a comprehensive approach to reducing the overall incidence and impact of emergencies within the community.
- Developing a community risk reduction program includes the following steps:
  - Step 1: Identify risks.
  - Step 2: Prioritize risks.
  - Step 3: Develop strategies and tactics to mitigate risks.
  - Step 4: Prepare the CRR plan.
  - Step 5: Implement the CRR plan.
  - Step 6: Monitor, evaluate, and modify the plan.

- An integral part of community risk reduction is fire and life safety education. The goals of fire and life safety education are to help people understand how to prevent the loss of life, injuries, and property damage from occurring and to teach them how to react if an emergency occurs. Fire and life safety education is often referred to as *public education.*
- Sometimes, the public makes an inquiry to the fire officer for general information.
- The fire officer must treat all requests professionally and with respect. Every effort should be made to answer each question fully and accurately.
- If the fire officer does not have all the information that the individual is seeking, the fire officer should seek out the information immediately rather than just expressing that he or she does not know.
- The fire officer should also follow up to ensure that the citizen is contacted in a reasonable amount of time.

KEY TERMS

Community Emergency Response Team (CERT) A fire department training program to help citizens understand their responsibilities in preparing for disaster and increase their ability to safely help themselves, their families, and their neighbors in the immediate hours after a catastrophe.

Community risk reduction (CRR) A process to identify and prioritize local risks, followed by the integrated and strategic investment of resources to reduce their occurrence and impact. (Reproduced from National Fire Protection Association, Community Risk Reduction (CRR), NFPA 1035. Retrived from https://community.nfpa.org/community/nfpa-today/blog/2014/05/13/community-risk-reduction-as-defined-by-nfpa-correlating-committee-on-professional-qualifications).

Demographics The characteristics of human populations and population segments, especially when used to identify consumer markets; generally includes age, race, sex, income, education, and family status.

Risk assessment Identification of potential and likely risks within a particular community records management system (RMS).

REFERENCES

Center for Public Safety Excellence (CPSE). 2015. *Exceeding Customer Expectations: Quality Concepts for the Fire Service.* Chantilly, VA: Center for Public Safety Excellence. Accessed July 12, 2019. www.cpse.org.

Crawford, Jim. 2018, December 26. "Taking the Fire Incident Rate to Zero: The Worcester Fire Department Has Shown Us a Clear Path Forward." *Fire Rescue Magazine.* Accessed June 25, 2015. https://www.firerescuemagazine.com/articles/print/volume-13/issue-12/departments/fire-prevention-and-education/taking-the-fire-incident-rate-to-zero.html.

Dorling, Danny, and Stuart Geitel-Basten. 2018. *Why Demography Matters.* Cambridge, UK: Polity Press.

Evans, Anthony. 2014. "The Economic Impact of Successful Commercial Fire Interventions: Phoenix Fire Department June 1, 2012 – May 13, 2013." Accessed June 25, 2015. http://azsmart-dev.wpcarey.asu.edu/wp-content/uploads/2016/03/PFD-Final-Report.pdf.

Gronbach, Kenneth W., and M. J. Moye. 2017. *Upside: Profiting from the Profound Demographic Shifts Ahead.* New York: AMACOM.

Kirtley, Edward. 2008. "Principles and Techniques of Fire and Life Safety Education." In *Fire Protection Handbook*, edited by Arthur E. Cote. Quincy, MA: National Fire Protection Association.

National Fire Protection Association. 2015. NFPA 1035, *Standard on Fire and Life Safety Educator, Public Information Officer, Youth Firesetter Intervention Specialist and Youth Firesetter Program Manager Professional Qualification.* Accessed March 25, 2019. www.nfpa.org.

National Fire Protection Association, 2020. NFPA 1300, *Standard on Community Risk Assessment and Community Risk Reduction Plan Development.* Accessed July 12, 2019. www.nfpa.org.

Pickett, Annie. 2018. "Teaching and Technology: A Recipe for Fire Safe Cooking." Vision 20/20 Symposium, Reston, VA.

Sawyer, Derrick, Derrick Phillips, Dana Catts, and David Sawyer. 2016. *Community Risk Reduction: Doing More with More.* Quincy, MA: NFPA Urban Fire and Life Safety Task Force.

Stouffer, John A. 2016a. *Community Risk Assessment: A Guide for Conducting a Community Risk Assessment.* Washington, DC: Vision 20/20.

Stouffer, John A. 2016b. *Community Risk Reduction Planning: A Guide for Developing a Community Risk Reduction Plan.* Washington, DC: Vision 20/20.

Fire Officer in Action

The lieutenant is reviewing community risk reduction practices.

1. Which of these is the second step in developing a community risk reduction plan?
   A. Identify the population at risk.
   B. Determine the budget.
   C. Create multiple strategies.
   D. Prioritize risks.

2. An annoyed citizen's statement, "I called for an ambulance, why did you show up in a fire truck?" is an example of which type of expectation?
   A. Interpersonal
   B. Technological
   C. Explicit
   D. Static performance

3. Which of these are involved in the preparation of a community risk reduction plan?
   A. Journalists
   B. Battalion and deputy chiefs
   C. University professors
   D. Community groups

4. A community risk reduction plan has been implemented, but the initial results are not what was anticipated. The fire officer should:
   A. make no changes until the plan was been in effect for 12 months.
   B. modify the plan.
   C. double the assigned resources.
   D. close the program.

Fire Lieutenant Activity

NFPA Fire Officer I Job Performance Requirement 4.3.1

Implement a community risk reduction (CRR) plan at the unit level, given an authority having jurisdiction (AHJ) CRR plan, and policies and procedures, so that a community need is addressed.

Application of 4.3.1

Develop a community risk reduction plan for one of the following scenarios:

1. The number of fire company "lift assist" responses is increasing, especially to the homes of older adults who live alone. Some take hours to crawl to the phone after they have fallen. Others are calling multiple times per month.

2. The winter brings a surge of single-family dwelling fires due to improperly disposed fireplace ashes, resulting in one or two large-loss fires each year.

3. Complete a community risk reduction plan to eliminate stovetop fires.

4. The summer brings pediatric drownings at private pools. Describe a community risk reduction plan to eliminate such events.

 Access Navigate for flashcards to test your key term knowledge.

CHAPTER 5

Fire Officer I

Fire Department Administration

KNOWLEDGE OBJECTIVES

After studying this chapter, you will be able to:

- Describe the process for recommending changes to policies. (**NFPA 1021: 4.4.1**) (p. 129)
- Describe the process for implementing a new or changed policy. (**NFPA 1021: 4.4.1**) (pp. 129–130)
- Describe documentation considerations for policy change recommendations or new policy implementation. (**NFPA 1021: 4.4.1**) (pp. 129–130)
- Describe the budget cycle. (**NFPA 1021: 4.4.3**) (pp. 130–134)
- Compare and contrast an operating budget and a capital budget. (p. 130)
- Describe information that should be included in a budget request. (**NFPA 1021: 4.4.3**) (pp. 132–134)
- Justify a budget request. (**NFPA 1021: 4.4.3**) (p. 134)
- Identify revenue sources. (pp. 134–136)
- Describe options used to address declining revenues. (pp. 136–138)
- Identify types of management actions governed by human resource policy and procedures. (**NFPA 1021: 4.2.5**) (pp. 138–144)
- Explain how a chief officer can determine appropriate steps to take for an administrative action. (**NFPA 1021: 4.2.5**) (pp. 140–144)
- Explain the purpose of regular fire fighter evaluations. (**NFPA 1021: 4.2.5**) (pp. 139–140)
- Describe methods of positive discipline. (**NFPA 1021: 4.2.4**, 4.2.5) (pp. 141–142)
- Describe methods of corrective discipline. (**NFPA 1021: 4.2.4, 4.2.5**) (pp. 142–144)
- Explain the role and importance of employee documentation and record keeping. (**NFPA 1021: 4.2.4, 4.2.5**) (p. 145)
- Describe the features of an effective records management system. (pp. 145–147)
- Describe the types of information that should be documented. (p. 145)
- Explain considerations for the following types of record keeping:
  - Inventory record keeping
  - Inspection and maintenance record keeping
  - Training and credentialing record keeping
  - Individual exposure history and medical records record keeping
- Explain the importance of properly made reports and logs to the organization. (**NFPA 1021: 4.1.2, 4.4.2**) (p. 147)
- Identify common types of reports used in the fire service. (**NFPA 1021: 4.4.2**) (pp. 147–151)
- Identify reasons fire departments collect data and conduct data analysis. (**NFPA 1021: 4.4.5**) (pp. 147–149)
- Prepare a written report, given incident reporting data from the jurisdiction, so that the major causes for service demands are identified for various planning areas within the service area of the organization. (**NFPA 1021: 5.6.3**) (pp. 147–151)

- Explain the importance of gathering incident data in a standardized format. (**NFPA 1021: 4.1.2, 4.4.5**) (pp. 148–149)
- Identify situations that may require an expanded incident report. (**NFPA 1021: 4.4.5**) (p. 150)
- Present a proposal to change a policy. (**NFPA 1021: 4.4.1**) (p. 129)
- Communicate new or changed policies to members. (**NFPA 1021: 4.4.1**) (pp. 129–130)
- Complete forms and records. (**NFPA 1021: 4.4.2**) (pp. 145–146)
- Prepare a budget request. (**NFPA 1021: 4.4.3**) (pp. 132–134)
- Complete a postincident report. (**NFPA 1021: 4.4.5**) (pp. 148–149)

SKILLS OBJECTIVES

After studying this chapter, you will be able to:

- Identify the need for a policy change. (**NFPA 1021: 4.4.1**) (p. 129)
- Create a written proposal to change a policy. (**NFPA 1021: 4.4.1**) (p. 129)

You Are the Fire Officer

The lieutenant notices that a fire fighter on the shift seems preoccupied and not focused on the job tasks at hand. Recently the fire fighter has been 1 to 2 hours late to work on four out of the past 12 workdays. While coaching the fire fighter about the second tardiness, the response was "I will take care of it." Unfortunately, the frequency of tardiness has increased. Today the fire fighter did not show up for work and did not call the fire station. When the lieutenant called the fire fighter, he apologized and said that he could not make it into work today. Today's no-call/no-show and the four earlier late arrivals puts the fire fighter into corrective discipline.

1. How do you approach a deteriorating performance issue that has not responded to coaching?
2. What options does the lieutenant have regarding the no-call/no-show?
3. What is the fire officer's role when off-duty situations impact on-duty fire fighter job performance?

Access Navigate for more practice activities.

Introduction

In terms of time-on-task, the fire officer spends more time performing personnel-related functions than emergency response and mitigation. The authority having jurisdiction (AHJ) expects the fire officer to be a caretaker of municipal resources and the first official to assure that policies are followed, and precious human and fiscal resources are protected and effectively used.

As defined by NFPA 1021, *Standard for Fire Officer Professional Qualifications*, administrative activities include the following:

- Implement, evaluate, and change department policies.
- Develop and manage a budget process.
- Identify and monitor revenue sources.
- Be the first stop for subordinate human resource issues, including positive and corrective discipline.
- Function as a witness and documenter, and maintain appropriate record keeping.

Policy Recommendations and Implementation

The fire officer is responsible for subordinate compliance with existing policies and practices. Because the fire officer is in direct contact with fire fighters and citizens, the officer gets a vivid view of the opportunities, challenges, and barriers. This places the officer in the best position to evaluate the effectiveness of existing policies and to recommend a revision of existing policy or the development of a new departmental policy (Yukl, 2013). This change could be the result of a citizen or employee complaint, or it could be an officer- or fire fighter-identified problem. The problem could be anything that creates a disparity between the actual state and the desired state. For example, not having a standardized location for all equipment on the apparatus reduces efficiency on incident scenes; therefore, it is a problem.

Recommending Policies and Policy Changes

The fire officer must understand the procedure for adopting new policies within the department. Although many fire fighters may believe that this procedure simply consists of getting the chief to put his or her name on a piece of paper, usually it entails much more. Some departments have a policy on how to implement a new policy. In discussing the challenge of implementing the National Fallen Firefighters Foundation's Life Safety Initiatives, Bill Manning (Manning, 2007) describes five "disconnects":

1. Culture change is viewed by some as a threat.
2. Bad (unsafe) behaviors and attitudes are allowed to leach into what the membership see as part of "tradition."
3. Safety and mission within organizational cultures are imbalanced.
4. The voices (and actions) of safety leadership are either subconsciously muffled or consciously subdued.
5. The lessons from behavioral safety science have not been embraced by fire service leaders, much less blended into everyday operations.

The most common method to attempt to get new policies adopted or to change existing policies is to outline the problem one or more fire fighters have identified to their supervisor with the recommendation that "someone ought to do something about this." This places all responsibility on the supervisor to come up with a solution and get the department to adopt it. The fact that this method is easy for the officer, however, is far outweighed by its ineffectiveness. Most of the time, nothing changes as a result of this process, and if it does, the change may create more problems than it solved.

To recommend a new policy or a change to an old policy successfully, the fire officer should carefully identify the problem and develop documentation to support the need for a change. This support could consist of statistical measures, anecdotal evidence, or both (Daly, Watkins, and Reavis, 2006). Many times the results of a postincident briefing provide a vivid opportunity to illustrate the need. Debriefings should capitalize on the positives, identify shortcomings, discuss strategies for improvement, and allow participants an opportunity to see and hear the incident through different sets of eyes and ears. The five-step model (Okray and Lubnau, 2004) can provide the granular information.

The fire officer's opinion seldom carries enough weight to prompt a change, because other officers may have different opinions. One effective technique is to discuss the situation with other fire officers initially to determine how widespread the perceived problem is.

Once evidence supporting the existence of a widespread problem has been obtained, the conflict resolution model (outlined in Chapter 3, *Leading a Team*) should be used to develop and choose the best alternative. At this point, the fire officer is ready to write out a proposal to administration. The problem should be carefully outlined, along with the proposed policy change that will resolve it. Resources that will be needed for the solution must also be identified, including financial and time commitments. The benefits of the solution should be identified, as should any potential negative effects and the means by which they will be addressed.

Once a written proposal has been developed, the fire officer should approach his or her superior if that person has not already been part of the development. Along with the recommendation, the policy should be presented to the appropriate officer whose scope is to oversee the area affected by the policy. It is critical to get the agreement of this fire officer, because without this officer's recommendation, the policy will likely not be implemented (Yukl, 2013).

With this officer's recommendation, the proposal is usually forwarded to the chief of the department for review. Usually, a review committee comprising senior staff officers evaluates the proposal and makes changes and accepts or rejects the proposal. Once it is accepted, a draft policy is usually developed in the proper format. The draft policy is distributed to company officers to share with all personnel for review and comment. After a comment period, all comments are addressed, and a final policy is signed by the chief and distributed to the department.

Implementing Policies

Like the recommendation process, the process for implementing new policies varies widely. The fire officer should follow departmental procedures. In their absence, the officer may use the following methods to improve his or her communication and understanding.

Because policies are the backbone of order for the fire department, every individual should understand the policies that affect him or her. For this to occur, the fire officer must take responsibility for ensuring that the fire fighters are informed about the policy and that they fully understand it. The fire officer must also ensure that he or she follows all policies. Failure of the officer to follow all policies undermines their importance, and fire fighters will develop the attitude that they can choose which policies they will follow.

When a new or amended policy is distributed, the fire officer should make sure that it is communicated to the subordinates. At the beginning of each work shift, any new policies should be discussed. The fire officer should identify the points that are most relevant, particularly those areas that change the current practice. For example, if a revised driving policy requires every apparatus responding to an emergency to come to a complete stop at all red signals and stop signs rather than just slow to a speed necessary to avoid an accident, the fire officer should specifically point out this change (Smeby, 2013). Company officers should share changes and show ownership. This communicates buy-in and expressed value to the organization.

The fire officer should require that all fire fighters read the policy and sign off that they understand it. This practice helps ensure accountability. Employees tend to follow policies more closely if they believe they will be held accountable for them. To make sure that employees understand the policy, the fire officer should provide fire fighters with situations covered by the policy and ask them to apply the policy. Many departments also require that a copy of the policy be posted on the bulletin board for a period of time. In addition, some require a notation in the station log book that the policy was distributed.

The fire officer should evaluate the employees' actions against the policy. If a policy is violated, the fire officer should review the policy with the employee involved. For repeated violations or a violation of a safety policy, disciplinary action should be considered. Employees should never get the feeling that it is acceptable to ignore policies.

On a long-term basis, a regular review of policies should occur. Selecting policies that are not routinely encountered and testing fire fighters' knowledge about them will help keep everyone up-to-date.

The Budget Cycle

NFPA 1021 starts fire officer participation in the budget process at the Fire Officer I level, with preparing a budget request given a unit level need, so that the request is in the proper format and supported by data. Here we describe the budget process with the focus on the Fire Officer I making a request for a new or replacement item that would come from the capital funds, like a new hydraulic rescue tool or replacement of a fire station appliance.

A **budget** is an itemized summary of estimated or intended revenues and expenditures. **Revenues** are the income of an organization from all sources; **expenditure** is the money spent for goods or services. Every fire department has a budget that defines the funds that are available to operate the organization for 1 year. The budget process is a cycle consisting of six steps:

1. Identification of needs and required resources
2. Preparation of a budget request
3. Local government and public review of requested budget
4. Adoption of an approved budget
5. Administration of approved budget, with periodic review and revision
6. Close-out of budget year (Truchil, 2018)

The budget document describes where the revenue comes from (input) and where it goes (output) in terms of operating and capital expenditures. The **operating budget** is an estimate of the income and expenditures of the jurisdiction. The **capital budget** is an allocation of money for the acquisition or maintenance of fixed assets such as land, buildings, and durable equipment. Fire apparatus, powered ambulance stretchers, and self-contained breathing apparatus (SCBA) units are examples of a capital budget item as they last more than one year and exceed a monetary threshold established by the AHJ.

Annual budgets sometimes apply to a **fiscal year**, such as the year starting on July 1 and ending on June 30 of the following year. Using these dates, the budget for fiscal year 2025, referred to as FY25, would start on July 1, 2024, and end on June 30, 2025. The process for developing the FY25 budget would start in 2023, a full year before the beginning of the fiscal year.

Consider the timeline for a replacement pumper. If the initial request is submitted in August 2024 and it is approved at every stage of the process, the funding will not be authorized until July 2025. **TABLE 5-1** provides a description of the process (Menifield, 2017).

TABLE 5-1 Fiscal Year 2025 Timeline

August–September 2023	Fire station commanders, section leaders, and program managers submit their FY25 requests to the fire chief. These are for proposed new programs, expensive new or replacement capital equipment, and physical plant repairs. This is the point at which a fire officer would request funds for a new storage shed or a replacement dishwasher. Documentation to support replacing a fire truck or purchasing new equipment, such as a thermal imaging device or a hydraulic rescue tool, would be submitted, and a proposal for a special training program would be prepared. These documents are prepared by fire officers throughout the organization.
September 2023	The fire chief reviews the requests and develops a prioritized budget, including new items and programs, for FY25. This is the point at which department-wide initiatives or new programs are developed. Purchasing a new radio system would be a new item.
October 2023	The budget office distributes FY25 preparation packages. Included are the forms for personnel, operating, and capital budgets. The senior local administrative official (mayor, city manager, county executive) provides guidelines. For example, if the economic indicators predict that tax revenues will not increase during FY25, the guidelines could restrict increases in the operating budget to 0.5 percent or freeze the number of full-time equivalent positions.
November 2023	All agency heads must submit their FY25 budget requests to the budget director.
December 2023	The budget director assembles all of the proposals. Any request that exceeds the amounts specified in the guidelines must be the result of a legally binding regulation, agreement, or settlement. Other submissions above the limit will need the support of elected officials.
January 2024	Local officials receive the preliminary budget from the budget director. There may be two or three alternative proposals that require a decision from the officials. Those decisions are made, and the proposal is completed. This is the "proposed fiscal year 2025 budget."
February or March 2024	The proposed budget is made available to the public for comment. Many cities make the proposed budget available on an Internet site, inviting public comment to the elected officials. In smaller communities, the proposed budget may be published in the local newspaper.
March or April 2024	Local officials hold a hearing or town meeting to receive input on the FY25 proposed budget. Based on the hearings and on additional information from staff and local government employees, the elected officials debate, revise, and amend the budget. During this process, the fire chief may be called on to make a presentation to explain the department's budget requests, particularly if large expenditures have been proposed. The department may be required to provide additional information or submit alternatives as a result of the public budget review process. At the completion of this step, the AHJ has what is called an "amended budget," reflecting all of the changes made by local officials. The amended budget is returned to the local leadership for review.
May 2024	Local leaders approve the amended budget. This becomes the "approved" or "adopted" FY25 budget for the municipality.
July 1, 2024	FY25 begins. Some fire departments immediately begin the process of ordering expensive and durable capital equipment, particularly items that have long lead times for delivery.
August–September 2024	The budget process for 2026 starts.
October 2024	Informal first-quarter budget review, looking for trends in expenditures to identify any problems. For example, if the cost of diesel fuel has increased and 60 percent of the motor fuel budget has been exhausted by the end of September, there will be no funds remaining to buy fuel in January. Amounts that have been approved for one purpose may have to be diverted to a higher priority.

(continues)

TABLE 5-1 Fiscal Year 2025 Timeline (continued)

January 2025	Formal mid-year budget review. The approved budget may be revised to cover unplanned expenses or shortages. In some cases, these changes come in response to a decrease in revenue, such as an unanticipated decrease in sales tax revenue. Expenditures for the remainder of the year have to be reduced.
April 2025	Informal fourth-quarter review. Final adjustments in the budget are considered after 9 months of experience. Year-end figures can be projected with a high degree of accuracy. Some projects and activities may have to stop if they have exceeded their budget, unless unexpended funds from another account can be reallocated.
May 2025	The finance office begins looking at the end-of-year budget projections. Departments with approved funding need to finalize spending funds allocated.
June 2025	All expenditures for fiscal year 2025 are complete. No new items can be purchased after June 30 on the FY25 budget.
July 2025	The FY25 budget is closed out.

Data from Menifield, C. E. (2017) The Basics of Public Budgeting and Financial Management: A Handbook for Academics and Practitioners, 3rd edition. Lanham, MD: Hamilton Books.

Base and Supplemental Budgets

Most municipal governments use a base budget in the financial planning process. The **base budget** is the level of funding that would be required to maintain all services at the currently authorized levels, including adjustments for inflation and salary increases. A built-in process assumes that the current level of service has already been justified and that the starting point for any changes in the budget should be the cost of providing the same services next year.

Budgets can increase or decrease from the base level. Proposed increases in spending to provide additional services are classified as **supplemental budget** requests. When budgets have to be reduced, the change is calculated as a percentage of the base budget. For example, a 15 percent budget cut means that only 85 percent of the base budget expenditures for the current year will be approved for the next fiscal year. The department would need to reduce its expenditures by more than 15 percent, however, due to inflation and already negotiated salary increases. The department may have to prioritize programs and activities to achieve the budget reductions. Instead of an across-the-board expenditure reduction, there may be elimination of an existing activity, such as the staffing of one engine, to accomplish the budget goal.

Increases in the fire department budget require early notification and support of elected officials. A new program that involves additional employees or a major capital expenditure would require funds that would have to be obtained by increasing revenues or decreasing expenditures from some other part of the budget. These decisions are made by the elected officials. As part of the effort to present a balanced proposed budget, the budget director usually has a good idea of what the elected officials are likely to approve.

Elected officials are both advocates and gatekeepers in developing the local budget. They want to avoid raising taxes and, at the same time, deliver the services that the voters expect. They do this through direct involvement in the budget preparation process and participation in public hearings. Elected officials often request private, face-to-face meetings with fire department leadership to discuss issues that are expected to be controversial (Fischer, 2016). Chapter 12, *Administrative Communications*, provides information on levys, grants, and other revenue sources.

Making a Budget Request

Understanding the budget cycle will help the fire officer develop a budget request. At the lieutenant level, we will look at two types of requests: replacing an existing capital item and requesting a new capital item. The AHJ will require that a capital budget request form be completed, along with appropriate documentation (**FIGURE 5-1**). The information on that form will include:

- Type of project
  - Replace or repair existing facilities or equipment.
  - Improve quality of existing facilities or equipment.

Type of Project:	☐ Replace or repair existing facilities or equipment
	☐ Improve quality of existing facilities or equipment
	☐ Expand capacity of existing service level/facility
	☐ Provide new facility or service capacity

Department Priority: Request number _____ of _____ requests

Type of Priority: ☐ Urgent
☐ Safety/Health
☐ Preserve Asset

Project Title: _____

Work Location: _____

Contact Person
Name: _____ Phone: _____ Email: _____

Project Description and Justification:

[   ]

Total Cost of Project: FY Year 2021 $ _____
FY Year 2022 $ _____
FY Year 2023 $ _____

Funding: Currently allocated funding $ _____
Grant funding $ _____
Other sources _____ $ _____ (describe sources)

Annual Operating Cost: Maintenance $ _____
Energy $ _____
Salary $ _____
Other $ _____

FIGURE 5-1 Example of a capital request form.

- Expand capacity of existing service level/facility.
- Provide new facility or service capacity.
- Department priority (where does this request rank on a list of submitted capital items?)
- Type of priority
  - Urgent
  - Safety/health
  - Preserve asset
- Project title
- Work location (fire station)
- Contact person (name, e-mail, and work phone number)
- Project description and justification
- For the purchase of a capital item: the exact information for the item to be purchased, including purchase price

- For construction, painting, or remodeling: an official cost estimate from the facilities or maintenance department or three estimates from contractors.
- Total cost of project. If spread over multiple fiscal years, how much would be allocated each year?
- Details of funding available:
  - What is currently allocated in the fiscal year budget for this item?
  - Are there other sources of funding (grants, transfer of existing funds, portions covered by existing allocations in other accounts)?
- Annual operating cost impact. The additional cost or savings in terms of maintenance, energy, salary, or other factors

Project Description and Justification

The process of getting a budget request approved will require accurately answering the question: "What happens if we do not purchase the item or do the construction, painting, or remodeling?" There are always more requests than available resources, so the decision makers have to triage the requests.

For a replacement item, focus on the reliability, quality of output, and repair costs. For example, a 21-year-old industrial washing machine has been broken for four months because there are no parts available to repair the machine. Failure to have a working industrial washing machine means that fire fighters are unable to wash work clothes contaminated with body fluids, soot, and harmful materials. This creates a health hazard within the workplace.

Motor vehicles get replaced when the cost of keeping the current vehicle on the road exceeds the cost of a replacement vehicle. You may be able to calculate the cost per mile by adding up the scheduled maintenance, fuel, and repairs and dividing that by the mileage. If that cost is higher that the anticipated cost per mile for a new vehicle's purchase price, fuel, and scheduled maintenance, it strengthens the case for a replacement vehicle (Anderson, 1998).

It is a challenge to get new capital equipment using existing funding. The justification will have to vividly describe the value of the purchase in terms of quality of service or improved capacity of service. Consider a request to purchase a hydraulic rescue tool for a fire company. The justification should contrast the current situation with the improved condition if the tool is purchased. For example: "Engine 7 is serving a geographically isolated area with narrow roadways with frequent motor vehicle crashes. Engine 7 handles an average of 27 vehicle crashes a year that require patient extrication using a hydraulic rescue tool. The nearest tool is 35 minutes away, on Rescue 30 at the Central Fire Station. Due to the workload of Rescue 30, the tool was unavailable 40% of the time for crashes with extrication that Engine 7 handles. Adding a hydraulic rescue tool to Engine 7 will reduce the time to extricate from 40 minutes to 15 minutes."

Purchase Information

The AHJ maintains a catalog of capital items that are described in a specific way that needs to be used on the budget request form. Some items may come from state or regional purchase agreements, others were developed by the purchasing department. The fire officer should meet with the purchasing officials to understand the process and the terminology used in budget requests.

The fire officer may be required to get a quote from a vendor for the capital item. For more common items, the officer may need to get a quote from three different vendors. If the AHJ has existing purchase agreements, use their existing agreement. For example, many states have negotiated a specific price for motor vehicles that can also be used by counties, cities, and towns (NASPO, 2019).

Total Cost of Project

Adding a resource will generate additional costs beyond the initial purchase. These will be added to the operating budget as maintenance costs, energy costs, and other items, like an annual license or maintenance fee. The budget submission form may require you to identify those costs with the new capital item as well as any savings. For example, an 11-year old Type 1 ambulance with a diesel engine, 360,000 miles, and significant rust issues is costing the fire department $23.50 per mile for operating, maintenance, and repair costs. A new Type II ambulance with a 5-cylinder gasoline engine will cost $10.75 per mile in operating, maintenance, repair, and financing costs.

Revenue Sources

The revenue stream depends on the type of organization that operates the fire department and the formal relationship between the organization and the local community. There is a wide diversity both in organizations and in revenue stream sources. Each type of organization has a different process for obtaining revenue and authorizing expenditures.

Most municipal fire departments operate as components of local governments, such as towns, cities, and counties. A fire district is a separate local government unit that is specifically organized to collect taxes that support fire protection. In some areas, fire departments are operated as regional authorities or an equivalent structure.

Local Government Sources

The mix of revenues available to local governments varies considerably because state governments set the rules for local governments. The fire officer needs to know the rules that apply to his or her jurisdiction. General tax revenues can be spent for any purpose that is within the authorized powers of the local government.

Some funds are restricted and can be used only for certain purposes. A **fire tax district** is created to provide fire protection within a designated area. A special fire protection tax is charged to properties within that service area, in addition to any other county or municipal taxes. For example, the fire tax could be $0.07 per $100 of value for property within the defined area. All of the revenue from the fire tax is dedicated to pay for the provision of fire protection services in the fire tax district.

The use of other revenues can also be restricted. For example, a fuel tax can be used only to build or maintain roads. Likewise, taxes that are levied to repay capital improvement bonds can be used only for that purpose. As demonstrated during the Great Recession of 2007–2009, however, funding sources that are not restricted by federal legislation may be taken from one part of the budget to cover another part of the budget (Gordon, 2012).

The United States census summarizes the sources of state and local government tax revenues in nine general areas. The top three local tax revenue sources represent about half of the total revenues collected by local governments:

- General sales and gross receipts taxes
- Property taxes
- Individual income taxes

Other sources of revenues for local governments include federal support and charges levied on consumers, such as user fees.

Many fire departments also obtain revenue through direct fees for service. Fire departments that operate ambulances obtain revenue by charging for patient treatment and transportation to offset the operating costs of the service. Direct revenue might also come from administrative fees for fire prevention permits, special inspections, and other services that are provided to individual property owners or contractors. Some fire departments have tried to obtain reimbursement for routine activities, such as responding to automobile accidents or alerts generated by fire alarm systems. Depending on the jurisdiction, these fees may be deposited into the general fund or a designated fire department fund.

Volunteer Fire Departments

Some volunteer departments are operated by municipal governments and are completely supported by local tax revenues, with all capital and operating expenses being handled by the jurisdiction. Such a budget process treats the volunteer fire department as a local government agency.

Other volunteer fire departments are organized as independent 501(c) nonprofit corporations. Some of these volunteer organizations raise their own operating funds through public donations or subscriptions and are entirely independent of local government. In other areas, local tax revenues are allocated to the volunteer corporations. Their relationship with a local jurisdiction may be defined through a memorandum of understanding, a contract for services, or a regional association of independent volunteer fire departments. Some jurisdictions pass local legislation or approve a charter that details the relationship, duties, services, and compensation that the volunteer fire department will provide to the community in return for the tax revenues.

Tax revenues support the entire operation of some volunteer fire departments, whereas others supplement their tax allocation with fund-raising activities. Sometimes the volunteer corporations own and operate the fire stations and apparatus, whereas other volunteer organizations occupy buildings and operate vehicles that are owned by the local government.

Volunteer organizations use a wide variety of fund-raising methods. Direct fund-raising often includes door-to-door solicitations or direct mail campaigns.

Bingo and Other Activities

Many volunteer departments sponsor activities to generate revenue, such as dinners, bake sales, car washes, raffles, and bingo nights or other gaming activities. These activities are state or regionally regulated, with specific procedures for conducting the games and handling the money. Volunteer fire departments have a revenue advantage over other operators because they do not have to pay their members to run the games. Some volunteer departments generate revenue by renting out social halls or meeting rooms for private events (Carter, 2015).

Some departments hold an annual carnival or other entertainment event that provides revenue to the department. Holding the event on fire department property and using unpaid staff allows the fire department to minimize operating costs. Many of these fund-raisers are major events in the community, creating goodwill and strong citizen awareness and support.

Real Estate and Portfolio Management

As a corporation, some fire departments have invested in real estate to generate revenue for fire department operations. Others have developed a financial portfolio that provides interest income to the department.

Lower Revenue Options

Planned expenditures must be balanced against anticipated revenues a year or more in advance. If the revenues do not meet expectations, adjustments must be made to reduce spending. If revenues exceed expectations, some extra funds could be available during the year. Local revenue has a history of cyclical behavior, driven by external and internal forces that affect the delivery of services.

Six of the seven recessions that have occurred in the United States since 1969 lasted an average of 10.8 months. The most recent, called the "Great Recession," spanned from December 2007 through June 2009. In the first six recessions, the economy sprang back within a year, restoring municipal revenue within 2 years. As the U.S. Bureau of Labor Statistics noted in its report *The Recession of 2007–2009*, however, 3 years after the most recent recession, many parts of the U.S. economy had not shown signs of recovery (BLS, 2012). Changes in the local economy often result in major changes in the amount and types of revenues collected by local government. If property values fall, property tax revenue decreases. Where local governments receive a percentage of sales taxes, a healthy economy provides increasing revenues, whereas a weak economy results in a revenue drop. A business that shuts down or reduces the number of employees can have a major impact on the local government budget.

Consider the impact of an empty office, retail, or manufacturing building. When the building is occupied and operating, it provides revenue to local government through property taxes and sales or gross receipts taxes. When the building is vacant, there are no sales or gross receipts to be taxed. Thus an empty building generates dramatically lower revenues to local government. For this reason, local officials closely monitor the vacancy rate for office, business, and mercantile space.

When a new business moves into a community, it brings additional revenue to the local government. When a business moves to another jurisdiction, its sales tax revenue also moves to the new jurisdiction. The local jurisdiction could also lose the income tax from people who move with the business. If a property owner goes into bankruptcy, the building could be abandoned and the property taxes may not be paid.

FIRE OFFICER TIP

Verify Your Sources

One of the best ways to ensure approval of a budget proposal is to identify a federal regulation, state law, or local ordinance that mandates equipment replacement or delivery of a program. Budget officials know that they must comply with all legally binding requirements, but they are not always eager to spend limited resources on recommended practices, nonbinding consensus standards, or suggested best practices. Thus, if the state requires replacement of Level A chemical suits after 5 years, the budget request should identify the specific regulation to ensure funding for this item.

Budget reductions may drastically affect fire department operations. When faced with declining revenues, fire departments have to make difficult choices. Five such options, provided in increasing order of difficulty, are the following:

1. Defer scheduled expenditures, such as apparatus replacement and station maintenance.
2. Privatize or contract out elements of the service provided by the department.
3. Regionalize or consolidate services.
4. Reduce the workforce.
5. Reduce the size of the department.

Defer Scheduled Expenditures

Deferring scheduled expenditures means delaying the purchase of replacement fire apparatus or other expensive equipment. For example, a department that normally replaces pumpers on a 15-year schedule might decide to stretch the life of the rigs to 18 or 20 years. During the recession of the 1980s, Los Angeles and Chicago could not afford to replace aging ladder trucks. Instead, they obtained funding to rehabilitate existing rigs. The equipment was updated, rewired, repainted, and recertified as aerial apparatus. The Los Angeles Service Life Extension Program (SLEP) replaced gasoline motors and manual transmissions with diesel engines and automatic transmissions—a rehabilitation program that squeezed an additional 5 to 7 years of operational service from the organization's rigs.

Another method of extending the life of heavy apparatus is the use of smaller vehicles to handle emergency medical services (EMS) first responder responses. This means an engine company crew will respond to a medical emergency in a four-door sport utility vehicle (SUV) instead of the fire apparatus.

Privatize or Contract Out Elements of the Service Provided

Privatization means replacing municipal employees with contract employees. The concept is that the cost to the municipality to provide the service should be lower because a private company can operate more efficiently than a local government agency. Trash pickup, vehicle fleet maintenance, and school bus operations are three examples of services that are often contracted out by local government.

Some fire departments privatize or contract out services that are so specialized that a private company may be able to provide an economic advantage to the municipality. Paramedic training, special operations classes, hazardous materials site clean-up, ambulance billing, apparatus maintenance, and communications system maintenance are popular candidates for privatization.

Regionalize or Consolidate Services

Regional or consolidated fire departments are established to increase efficiency by reducing duplication in staff and services. Miami–Dade Fire Rescue Department in Florida started in 1935 as a fire patrol with one employee and one truck reporting to the Agriculture Department. Today this department provides fire suppression, EMS, and specialized operations services in an 1883-square-mile (4877-km^2) area that includes the unincorporated portions of Dade County and 30 municipalities.

California provides many examples of consolidation. Los Angeles County provides fire protection and paramedic services to 59 cities and towns, as well as the Lifeguard Division. The Orange County Fire Authority contracts with 23 cities.

Regionalization of specialized and infrequently used resources is a way of reducing expenditures. The Mutual Aid Box Alarm System (MABAS) provides a resource response plan for Illinois, Wisconsin, and parts of Indiana, Iowa, and Missouri. MABAS encompasses 42 hazardous materials teams, 26 underwater rescue/recovery teams, and 41 technical rescue teams.

Metrofire is an association of 34 metropolitan Boston fire departments that includes a regional hazardous materials response team as well as evacuation/rehabilitation buses.

Reduce the Workforce

Sometimes cities have to reduce their workforces. Fire departments try to protect staffing on emergency response vehicles by reducing positions in administrative and support areas. This approach has limited effectiveness, however, both because these personnel represent a small proportion of the workforce and because they perform important tasks that support the emergency responders.

One option is to maintain the same number of companies but to reduce the staffing per vehicle. Another option is to limit the number of units in service. Neither option is popular or attractive; however, budget realities sometimes leave no other choices.

Some fire departments temporarily close fire companies for one day or work shift at a time, reassigning the on-duty fire fighters to fill vacancies in other fire stations. This practice may create huge political repercussions if a civilian dies or suffers serious injury in an incident where the nearest fire company was closed for the day.

Reduce the Size of the Fire Department

Departments have been asked to reduce their workforce in response to reduced local revenues. The Baltimore Fire Department disbanded 23 engine and ladder companies between 1990 and 2012, reducing its fire fighter workforce by 26 percent. The seven fire companies that were closed in 2000 did not reduce the workforce, because the department instead shifted the staff from these companies to other positions, where they were used to reduce fire fighter overtime and staff additional ambulances. While additional emergency medical technician (EMT)-ambulances have been staffed in response to EMS incident workload, 2019 often showed a loss of up to three engine companies that were reassigned to staff "critical alert" ambulances in the afternoon and evening.

Closing a fire station or eliminating a fire company often creates a significant and high-profile political issue that mobilizes the citizens as well as the fire fighters. No community wants to lose its neighborhood fire company. The Los Angeles Fire Department's

2011 deployment plan stopped 3 years of rotating closures and permanently closed 11 engine companies, 7 "light forces" (pumper and tiller truck), and 13 command positions. Compared to 2007, in 2013 there were 228 fewer fire fighters on duty each day (Ward, 2018). In fiscal year 2019, Los Angeles used a federal Staffing for Adequate Fire and Emergency Response (SAFER) grant to initially fund the restoration of four engine companies that were closed in fiscal year 2012 (LAFD, 2018).

FIRE OFFICER TIP

The Bicentennial Layoff

New York started its fiscal year 1976 by laying off more than 40,000 city employees, including 1600 fire fighters. Although the city hired some workers back, 900 people lost their New York City Fire Department (FDNY) jobs. Some of the laid-off individuals became temporary employees under a federal Department of Housing and Urban Development grant. They were assigned as the fourth or fifth member of an FDNY ladder company, with the job of boarding up roofs and windows of fire-damaged buildings. It would take two years before the city could rehire all the laid-off fire fighters.

Human Resources

The municipality that oversees the fire department has a human resources department that issues rules and regulations as well as provides a range of services. The proper administration of the human resources function is one of the most important and sensitive areas of jurisdictional management. Because of liability factors and potential labor relations problems, renewed emphasis is being placed on standardized testing, uniform job descriptions, position classification and compensation systems, comprehensive pre-employment physicals, drug testing, reference checking, and background investigations (Kemp, 1998).

Human resources departments (HR) also provide guidance for compliance with state and federal laws and regulations. They provide the interface between the municipality and the regulators. These include hiring; coordinating discipline; and managing Family and Medical Leave Act (FMLA), disability, payroll, and retirement services. Behind the fire department's rules and regulations, this is the second most important obligation of a first-line supervisor (Belker, McCormick, and Topchik, 2018). The following are areas that should be closely examined by the fire officer:

- Employee behavior (coaching and discipline) process. There are procedures, timelines, and forms used in rewarding great performance and coaching for performance improvement, and there is a process to correct performance deficits.
- Family Medical Leave Act. Employees with emergent, planned, or intermittent conditions need to follow a process that requires physician documentation.
- Grievance or problem-solving procedure. Company-level fire officers are the first person a fire fighter approaches to get a problem solved or a grievance addressed. There are forms and a timeline.
- Transfers. Often this is a process that is combined between the fire department standard operating procedures (SOPs) and HR. The fire department makes the decision and HR performs the record changes.
- Promotions. There is an extensive process by both HR and the fire department to determine eligibility, create an assessment tool, and administer a promotional exam.
- Code of conduct. Either within the personnel regulations or as a stand-alone document, the municipality has published a code of conduct describing their expectations of employee behavior. Formal corrective actions in response to social media posts or citizen videos of employee behavior are often referenced to the municipality code of conduct.
- Leave management. This includes the process of obtaining vacation leave, documenting sick leave, and processing special leave categories.
- Payroll. This includes recording overtime, acting-out-of-class pay, proficiency pay, leave without pay, and other issues.
- Benefit administration. Health benefits, wellness programs, employee assistance, and professional development are part of benefit administration.
- Retirement. Includes administration of retirement programs, preretirement education, and information about Social Security, Medicare, and Medicaid.

The fire officer should invest some time with a human resources representative to understand the specific process for assisting a fire fighter with these issues. There are unique procedures, forms, and timelines that need to be followed for each area.

There often is a civil service board or commission that advises municipal leadership and HR on personnel policies for merit-based employees. They also conduct the final level of employee grievance appeal hearings.

Evaluation

Regular evaluations are performed to ensure that each fire fighter knows which type of performance is expected while on the job and where he or she stands in relation to those expectations. This process helps the fire fighter set personal goals for professional development and performance improvement and provides positive motivation to perform at the highest possible level.

Most career fire departments require a supervisor to conduct an annual performance evaluation for each assigned employee. The annual performance evaluation is a formal written documentation of the fire fighter's performance during the rating period. This permanent personnel record is important to both management and the fire fighter, but it cannot replace the ongoing supervisory actions that maintain and improve employee performance throughout the year.

Volunteer fire departments provide an equivalent form of periodic evaluation for each member by a higher-ranking individual. Although these documentation procedures are often less structured, every member deserves evaluation. The responsibilities of a supervisor are just as important in a volunteer fire department as in a full-time career organization.

Starting the Evaluation Process with a New Fire Fighter

Fire officers have a special responsibility when starting an evaluation process with a probationary fire fighter, because the fire officer is helping to shape that individual's fire department career. The officer will be shepherding the probationary fire fighter through his or her first real-life emergency experiences, providing feedback and guidance. Most fire fighters have vivid memories of their first supervisor and continue to respond to the expectations established by that officer.

New fire fighters start with wide variations in the range and depth of their skills. In some departments, probationary members can start to ride the apparatus after receiving a few hours of orientation. Many fire departments require new members to achieve NFPA 1001, *Standard for Fire Fighter Professional Qualifications*, and EMT certification before they are authorized to respond to any alarms. It is the fire officer's responsibility to determine each individual's skills, knowledge, aptitudes, strengths, and weaknesses and then to set specific expectations for each new fire fighter.

Recruit Probationary Period. Most fire departments include structured in-station training as part of the recruit fire fighter's probationary period. Regardless of the new fire fighter's level of certification or pre-employment experience, the fire officer is responsible for evaluating each individual during the probationary period. Often a classified job description specifies all of the required knowledge, skills, and abilities that a fire fighter is expected to master within a specified time period to complete the probationary requirements. The fire fighter should be provided with a copy of these specific requirements to use as a checklist (**FIGURE 5-2**).

The recruit fire fighter is expected to obtain experiences and demonstrate competencies related to the classified job description during the probationary period. This period often includes assignments on different types of companies and specialty units operated by the department, such as an engine company, a ladder company, an ambulance, and the telecommunications center. Competencies may include demonstrating all of the skill sets required for NFPA 1001 Fire Fighter II certification. Structured probationary programs require the fire officer to complete a monthly evaluation of each probationary fire fighter. This evaluation typically assesses the probationary fire fighter's progress in four areas:

- Fire fighter skill competency, including proficiency as an apparatus operator
- Progress in learning job-specific information not covered in basic training, such as the local fire prevention code and the department's rules, regulations, and SOPs
- Progress in learning the fire company district, including streets and target hazards
- Performance of other job tasks, such as recording deliveries, performing housework, conducting in-station tours, and completing reports

In volunteer fire departments, it is common for different individuals to certify that a probationary fire fighter has met the requirements for different components of the program. One officer should be specifically assigned to oversee the progress of each

City of Charlottesville Fire Department

Fire Fighter Job Description

General Definition of Work
The fire fighter performs responsible service work in fire suppression and prevention; does related work as required. Work is performed under the regular supervision of a company and/or shift commander.

Typical Tasks
Responds to alarms, drives and operates equipment and related apparatus, and assists in the suppression of fires, including rescue, advancing lines, entry, ventilation and salvage work, extrication, and emergency medical care of victims.
Performs cleanup and overhaul work, establishes temporary utility services.
Assists in maintaining and repairing fire apparatus and equipment, and cleaning fire stations and grounds.
Checks fire hydrant flows.
Makes fire code inspections of business establishments and prepares pre-fire plans.
Responds to emergency and nonemergency calls, pumps out basements, inspects for gas leaks, secures vehicle accidents, inspects chimneys, etc.
Participates in continuing training and instruction programs by individual study of technical material and attendance at scheduled drills and classes.
Conducts station tours for the public, school, and community demonstrations and programs.
Backs up for dispatching personnel, monitors alarm boards, receives and transmits radio and telephone messages.
Performs related tasks as necessary.

Knowledge, Skills, and Abilities
General knowledge of elementary physics, chemistry, and mechanics; general knowledge of technical firefighting principles and techniques, and principles of hydraulics applied to fire suppression; general knowledge of the street system and physical layout of the city; general knowledge of emergency care methods, techniques, and equipment; ability to understand and follow written and oral instructions; ability to establish and maintain cooperative relationships with fellow employees and the public; ability to keep simple records and prepare reports; possess a strong mechanical aptitude; ability to perform heavy manual labor; skill in operation of heavy emergency equipment.

Education and Experience
Any combination of experience and training equivalent to graduation from high school.

Special Requirements
Possession of a valid driver's permit issued by the Commonwealth of Virginia.

Future Requirements at Three Years of Service
NFPA 1001   Fire Fighter Level II
NFPA 1002   Driver/Operator Certification
Commonwealth of Virginia Emergency Medical Technician or greater

FIGURE 5-2 Sample fire fighter job description.
© Jones & Bartlett Learning.

individual probationary member, thereby ensuring that the overall program requirements are being accomplished successfully.

The fire fighter probationary period lays the foundation for a long career. The fire officer has a special opportunity to prepare future departmental leaders by providing a comprehensive and effective probationary period.

Providing Feedback after an Incident or Activity

Performance evaluation should be a continual supervisory process, not a special event that is performed only when a scheduled rating has to be submitted. Frequent feedback from the fire officer should keep fire fighters aware of how they are doing, particularly after incidents or activities that present a special challenge.

Feedback on individual performance is most effective when delivered as soon as possible after an action or incident. That means providing essential feedback before the ashes get cold after a structure fire. At this point, the fire fighter is intensely aware of the event and wants to know how well he or she performed. The fire officer needs to be ready to provide specific information to recognize or improve fire fighter performance (**FIGURE 5-3**).

FIGURE 5-3 Performance feedback should be delivered as soon as possible after an action or incident.
© Jones & Bartlett Learning. Photographed by Glen E. Ellman.

FIRE OFFICER TIP

Immediately Correct Unsafe Conditions

Although negative feedback should be issued to an individual in private, the fire officer must correct unsafe conditions as soon as they are noticed. There is no excuse for allowing an unsafe action or situation to occur without taking corrective action, even if that means shouting an order at a crowded fire ground or issuing a direct instruction over the radio. Once the incident is under control, the fire officer needs to follow up with a private face-to-face meeting with the fire fighter or fire fighters who created or ignored an unsafe condition.

Positive and Corrective Disciplinary Actions

Discipline is a moral, mental, and physical state in which all ranks respond to the will of the leader. The fire officer builds discipline by training to meet performance standards, using rewards and punishments judiciously, instilling confidence in and building trust among team leaders, and creating a knowledgeable collective will.

Within the fire department, discipline is divided into positive and negative or corrective sides. Positive discipline is based on encouraging and reinforcing appropriate behavior and desirable performance. Corrective discipline is based on punishing inappropriate behavior or unacceptable performance. Both positive and corrective discipline can be applied to a full range of activities, including emergency incidents and administrative functions.

Positive discipline should be used before corrective discipline is applied. Progressive discipline refers to starting out to correct a problem with positive discipline and then increasing the intensity of the discipline if the individual fails to respond to the positive form, perhaps by using mild corrective discipline. Corrective discipline might have to be used to an increasingly greater extent in a situation in which an individual fails to respond in an appropriate manner to correct the problem. There are exceptions to this rule: Some actions or behaviors are so unacceptable that they must result in immediate corrective discipline.

Positive Discipline

Positive discipline is directed toward motivating individuals and groups to meet or exceed expectations. The key to positive discipline is to convince these parties that they want to do better and are capable of and willing to make the effort. A fire officer provides positive discipline by identifying weaknesses, setting goals and objectives to improve performance, and providing the capability to meet those targets.

The starting point for positive discipline is to establish a set of expectations for behavior and performance. Once these expectations are known, there must be a consistent and conscientious effort to meet them. The expectations have to apply to the entire team as well as to each individual team member.

Positive discipline is reinforced by recognizing improved performance and rewarding excellent performance. Due to the nature of firefighting work, outstanding performance is recognized through the use of awards, ribbons, and bonuses. Departments often have an awards ceremony to celebrate these achievements.

Teamwork is a key factor in ensuring success for fire companies. In addition to each individual having the required knowledge and skills, the company must be able to work efficiently and effectively as a team. All of the company members have to work, learn, and practice together to become capable and confident. The fire officer has to provide the leadership to make this cohesiveness happen.

The officer sets the stage for positive discipline by setting clear expectations and by "walking the talk." Fire officers are working supervisors; that is, they supervise while directly participating in firefighting activities and performing nonemergency duties. The officer should demonstrate a personal commitment to the department's goals, objectives, programs, rules, and regulations by participating in all of the activities that are expected of fire fighters, such as physical fitness training and regular company drills. Fire fighters can gauge an officer's level of commitment by observing his or her own self-discipline (**FIGURE 5-4**).

Competitiveness can also be used as a stimulant in positive discipline, particularly at the company level. Fire fighters are naturally competitive, and most fire companies work hard to prove that they can be better, faster, more skillful, or more impressive than a rival company. The officer has to point that competitive energy in a positive direction, making sure that the ultimate objective is high performance.

Empowerment. Empowerment is one of the most effective strategies within the realm of positive discipline. Fire fighters often complain that they have little control over their work environment; they are told where to go, what to do, and when to do it. Fire officers can help make fire fighters feel stronger by learning how to control their own destiny. An officer who identifies an area where improvement is needed can often empower fire fighters to correct the problem on their own. It is important for the officer to identify the target and provide the resources, but doing the work on their own and demonstrating their capabilities can be a very positive motivator for the fire fighters.

Providing information to help fire fighters learn more about the job can support the empowerment process. The first component could be described as "Local Government 101." This information helps fire fighters learn more about how the fire department and local government work. It could include reviewing the approved budget for the next fiscal year or discussing which functions are performed by each part of the organization. The more fire fighters understand about how the organization and local government work, the more they can feel a sense of participation.

The second component could be described as "Success 102." This information identifies the tools that others have used to achieve success within the fire department. If some companies are consistently recognized for positive performance, what are they doing right? When a few individuals achieve high scores on every promotional examination, what is their secret method? Do fire fighter self-study groups improve promotional examination results? Answering these questions may take some research by the fire officer, such as identifying the individuals or groups that have performed well in the promotion process and asking them how they prepared. An officer can often help fire fighters succeed by identifying successful practices.

FIRE OFFICER TIP

Hands-On Skill Drills

A commitment to in-station training produces competent and confident fire fighters and fire companies. A fire officer should provide frequent hands-on opportunities for fire fighters to practice their emergency service delivery skills, such as laying out hose, throwing ladders, forcing doors, and cutting roofs. Some type of activity should be scheduled every day, as skills have been shown to deteriorate if not used within 90 days of training.

Many fire departments have developed training props for practicing with hand and power tools. As a confidence builder, fire fighters should practice working with hose lines and tools in simulated fire-ground scenarios while wearing full personal protective clothing (**FIGURE 5-5**). Fire officers should also keep an eye out for opportunities to conduct realistic exercises in buildings scheduled for demolition.

Corrective Discipline

If an individual fire fighter's performance needs improvement in a particular area, the fire officer should coach that person, providing guidance and extra opportunities to correct the problem. Often, this step is needed when the fire fighter is unable to perform the task or skill because the task or skill was never learned. Sometimes, simply pointing the individual in the right direction and offering encouragement can achieve the objective. In many

FIGURE 5-4 Fire officers set the stage for positive discipline by "walking the talk."
© Glen E. Ellman.

FIGURE 5-5 Fire fighters should practice working with hose lines and tools in simulated fire-ground scenarios while wearing full personal protective clothing.
© Jones & Bartlett Learning. Courtesy of MIEMSS.

cases, the officer can arrange for another fire fighter to work with the individual to correct the problem.

Oral Reprimand, Warning, or Admonishment. An oral reprimand, warning, or admonishment is the first level of corrective discipline, considered "informal" by many organizations. An informal discipline action stays with the fire officer and does not become part of the employee's official record.

For example, a new fire officer observes that the apparatus driver does not comply with the department regulation requiring that units responding to an emergency come to a complete stop when encountering a red traffic light. When this occurs, the fire officer should have a private face-to-face meeting with the apparatus driver. In this meeting, the fire officer would determine why the driver did not stop at a red traffic light and would clearly state the policy and expectation for all subsequent responses. If the fire officer determines the reason was willful noncompliance with the department regulation, he or she would take the following steps:

1. Tell the driver that the officer expects compliance with the regulation requiring that rigs stop at red lights before proceeding through an intersection. Provide the driver with a printed copy of the regulation.
2. Inform the driver that this is a verbal reprimand and that the fire officer will maintain a written record of the reprimand for whatever time period is required by the AHJ. (A common practice is to maintain the written record for 366 days.) At the end of that period, if there have been no further problems related to the same individual or issue, the paper record is removed and destroyed.)
3. Inform the driver that continued failure to comply with the regulation will result in more severe discipline.

In most situations, these actions will suffice to correct the fire fighter's behavior. If the fire fighter continues to have difficulty complying with the regulation, however, this is the first step in progressive corrective discipline. As part of this process, determine whether the fire fighter is unable to meet the required performance, which will require additional focused training, or unwilling to perform, which will start a fact-finding process.

FIRE OFFICER TIP

Evaluation and Discipline Procedures
Evaluation and discipline procedures are part of almost every promotional exam. The candidate must be completely familiar with the local procedures in these areas. For example, preparing for a captain exam will require knowledge of evaluation and discipline activities at the battalion chief level. Frequently, the candidate is asked to write a document pertaining to these issues for a superior fire officer's signature as part of the promotional process.

One group of captain candidates struggled with a promotional writing exercise that required them to document why one of their subordinates should receive a "Fire Fighter of the Year" award from a service club. It was a struggle for them to issue positive discipline because all of their promotional preparation to that point had focused on issuing only corrective discipline.

Informal Written Reprimand. Some fire departments require the fire officer to use a standard form when issuing an oral reprimand, warning, or admonishment. This type of form covers the details of the informal discipline and provides a space for the employee to respond to the reprimand. The form ensures that the fire officer covers all of the requirements of an informal reprimand. In addition, the form allows the fire fighter to understand clearly that this is a corrective disciplinary issue and to have an opportunity to respond. This type of record is valued by the personnel office if the issue becomes a grievance or results in a civil service hearing.

Voice of Experience

In the fire service we discuss many aspects of the job that impact us on a daily basis. Perhaps one of the most misunderstood ones, although it is at the core of the fire department's existence, is an understanding of the fire department's budget. Fire officers work with men and women who are truly enthused and energized with ideas, plans, and goals. One of the responsibilities we have as fire officers is to make sure that members within the organization have a basic understanding of budgeting and the components that are included within budgets.

The questions that are asked are: Where does funding it come from? How much money do we receive? What can we do with the funds, and how? Budgeting is a complete process that starts with understanding your jurisdiction, be it a municipal setting or a fire district. How do you develop a budget based on revenues that include taxes, grants, EMS charges, inspections, and plan reviews?

Today's fire service leader must also understand investments and how they work to our advantage or perhaps disadvantage. How do we spend our funds? What are the components of budget preparation based on past practice, bidding for services, bidding for equipment, and economic constraints based on what we can afford or what we can sustain?

The budgeting process must cover purchase orders based on competitive bids, competent vendors, and service agreements that guarantee the necessities of operations.

Recently, my focus with the budget process has been centered on balance. I have been dealing with increasing pension contributions, increasing insurance costs, maintaining the framework of a collective bargaining agreement, and assuring that the fire department has an adequate reserve fund for future projects and purchases. Striving to achieve balance and sustainability without overburdening the taxpayers has led to a period of looking at all sources of revenue—places where we can increase our budgets so that we can continue to provide essential service.

Over the past few years, I have taken a more hands-on approach, with assistance from all officers when we review project requests and how they apply to our budget, mindful of our revenue flow. The analogy I use is that of a wheel where the center hub holds spokes that attach to the wheel. The wheel is our fire department and the spokes are the functions and services we provide, along with the costs of providing these services. The hub is the budget. Without a solid budget we will be stressed to provide extended services, let along essential ones.

Jim Grady III
Chief
Frankfort Fire Protection District
Illinois Fire Chiefs Association
Illinois Fire Service Institute
Frankfort, Illinois

Employee Documentation and Record Keeping

Municipal personnel rules usually require that all of the official records of an employee's work history be secured in a central repository. This repository includes all of the official documents accumulated during employment, including the following items:

- Hiring packet (application, candidate physical aptitude test score, and medical examination results)
- Tax withholding, I-9 status
- Benefits
- Personnel actions (changes in rank and pay)
- Evaluation reports (probationary, annual, and special)
- Grievances
- Formal discipline

Some fire departments maintain a second personnel file at fire headquarters. That file may include letters of commendation, transfer requests, protective clothing record, work history, copies of certifications, and other fire department–specific information that does not need to be kept in the secured personnel file.

Record-Keeping Systems

The fire officer is expected to maintain reports, logs, and files within the records management system in the manner outlined in the AHJ's policies and procedures documentation. A secure, accessible, and protected records management system is essential so that the business of the jurisdiction can be legally and effectively completed (Reese, 2005). Features of effective records management include the following:

- Preserve the right information for the correct length of time.
- Meet legal requirements for records retainment.
- Archive vital information for business continuity and disaster recovery.
- Respond to information requests in a timely manner.
- Be resistant to physical or digital destruction or theft.

The fire department is funded through local tax revenue, state funds, federal grants, and other sources. As a local government entity, the department must be transparent and responsive in documenting the

FIRE OFFICER TIP

Beware Ransomware Attacks

One of the casualties in the international cyberwars are local government computer systems. These systems may be infiltrated and taken over, or hijacked, by an outside entity using "ransomware." The hijacking entity then demands a ransom to restore access to the network. A 2017 attack in Bingham County, Idaho, disrupted the emergency communications center, interfering with incoming 911 calls and dispatching. A 2019 attack put the City of Baltimore, Maryland's network down for over a month. Every year, more than 50 jurisdictions experience a ransomware attack. Despite this threat, many local governments are reluctant to upgrade information management system software due to the expense; this increases their vulnerability to a ransomware attack (Farmer, 2019).

utilization of these resources. Effective documentation covers a range of records.

Inventory

Every capital item, from the fire station to the industrial dryer, needs to be documented. The document includes information on acquisition, purchase price, model number, amount purchased, and serial number (if appropriate). Items purchased with grant money often require an annual inspection and confirmation that the item is still with the organization.

Levels of inventory of expendable items must be established at both the fire department and the fire station level. The fire officer will monitor supply levels at the work location and assure the proper levels are maintained.

As centers of refuge and organizing points for disaster response, the fire station may maintain a cache of items that would be used during a disaster or catastrophic event. These caches need to be periodically inspected.

Procurement of supplies for the fire station can be accomplished in multiple ways. Some departments have established an automated supply process for consumables. The fire officer may need to complete a requisition form that would go to the logistics section or directly to a vendor. There may be an organizational credit card to cover some purchases. Every transaction needs to be documented.

Inspection and Maintenance

Various equipment at the fire station requires a daily, weekly, or monthly documented inspection as part of a manufacturer, risk management, or regulatory requirement. These inspections create an up-to-date history of the vehicle or device and will be reviewed as part of the preventive maintenance process or during a fire fighter death or injury investigation. The fire officer oversees the proper filling out of these forms and assures that they are properly filed and retained as the jurisdiction requires. The type of documentation includes the following:

- Daily and after-use inspection of SCBA
- Calibration and testing of electronic monitoring and air-quality devices
- Daily inspection and test of biomedical equipment, cardiac defibrillator, and electrocardiogram (ECG) monitor
- Regularly scheduled inspection of personal protective equipment
- Annual fire hose inspection and pressure testing
- Daily and weekly apparatus check

Preventive maintenance is performed to reduce the wear and tear on devices, vehicles, and facilities, usually on a predetermined schedule. Documentation of the preventive maintenance performed is included in that item's document file. For example, analysis after a preventive maintenance may lead to a corrective maintenance if metal was found in the drained engine oil.

Corrective maintenance is repairs done to keep the item, vehicle, or fixed facility operational. Many departments have recent experience trying to extend the service life of an item or vehicle while the jurisdiction was dealing with the low revenues during the Great Recession. Olathe, Kansas, Chief David Dock provides an example, using a 12-year-old pumper with a blown engine. If the cost to replace the engine exceeds the value of the pumper, it may be time to replace the pumper. Other factors include parts availability and advances in technology over the past 10 to 12 years (Hart, 2017).

Training and Credential Records

Fire fighters require accurate and complete records of their training, experience, and certifications to meet the requirements of accreditation programs such as the Center for Public Safety Excellence (CPSE) Fire Officer and the National Fire Academy Managing Fire Officer. The National Board on Fire Service Professional Qualifications (Pro Board), the International Fire Service Accreditation Congress (IFSAC), and the American Council on Education (ACE) have specific documentation requirements:

- Topic of training session
- Time spent in training
- Date of training
- Location of training
- Participants
- Assessment or outcome of training event

State agencies and the Insurance Service Office (ISO) review the training records to confirm that the fire department is meeting its obligation and that individuals are appropriately trained. These records are also reviewed when examining a fire fighter injury, fire fighter death, or liability-incurring event. For example, after a fire apparatus accident, the investigation will include a review of the driver's training and credentials.

Individual Exposure History and Medical Records

There may be a long period before health effects of occupational illnesses are manifested (Fent, 2015). Awareness of this fact requires jurisdictions to maintain documentation of fire fighter exposures and medical records. Depending on the jurisdiction, the medical records may include the following:

- Pre-employment medical examination
- Periodic medical evaluations
- Return-to-duty medical evaluation—after being on light or limited duty
- Exposure reports
- Retirement medical assessment

Many electronic exposure reporting systems are available, and many allow the fire fighter to enter the information on their smartphone. The company

FIRE OFFICER TIP

Recycling Retired Pumpers as Blocker Rigs

After a tractor-trailer struck and totaled an Irving (Texas) Fire Department ladder truck that was operating at an accident on the interstate, Chief Victor Conley explored using older fire apparatus to operate as traffic blockers during interstate accidents. The need was significant, as the region experienced nine apparatus struck on the interstate in 5 years, two of which were totaled.

Using 20-year-old apparatus scheduled for retirement, the department added arrow boards and graphics to redirect interstate traffic. The units went into service in 2017, replacing use of the first-line fire apparatus to block the accident scene. In 2019, while protecting several police officers and fire fighters, a "blocker" pumper was totaled when hit by a high-speed vehicle that ran through an accident scene (Vince, 2019).

officer should encourage every fire fighter to diligently maintain a personal exposure log. Depending on the AHJ, a state-wide or regional reporting system may be available.

The federal Health Insurance Portability and Accountability Act (HIPAA) was established in 1996. Protected health information (PHI) is considered to be individually identifiable information relating to the past, present, or future health status of an individual. PHI can be defined as any medical information that can identify an individual by patient name, identification number, or other means of identification. Fire departments that provide medical care as an emergency medical responder (EMR) or as an EMS agency must follow HIPAA regulations in retaining, managing, and releasing patient information and records.

PHI may only be shared for treatment, payment, or operational needs. Any other use requires written consent and authorization by the patient. This information can be shared by voice, paper, electronic, or telecommunication means. Jurisdiction-specific information is available through the HIPAA compliance officer.

The 2014 Ebola crisis resulted in clarifications of HIPAA privacy in infectious disease or pandemic emergency situations (OCR, 2014). A covered entity, which for fire officers means health care providers who electronically transmit any health information in connection with transactions covered by U.S. Department of Health and Human Services (HHS) standards, is permitted to disclose needed PHI to the following:

- Persons at risk of contracting or spreading a disease or condition if other law, such as state law, authorizes the covered entity to notify such persons as necessary to prevent or control the spread of the disease or otherwise to carry out public health interventions or investigations. (See 45 CFR 164.512(b)(l)(iv).)
- Family, friends, and others involved in an individual's care, and for notification. A covered entity may share protected health information with a patient's family members, relatives, friends, or other persons identified by the patient as involved in the patient's care. A covered entity also may share information about a patient as necessary to identify, locate, and notify family members, guardians, or anyone else responsible for the patient's care, of the patient's location, general condition, or death. This may include, where necessary to notify family members and others, the police, the press, or the public at large. (See 45 CFR 164.510(b).)

Reporting

Some reports are prepared on a regular schedule, such as daily, weekly, monthly, or yearly. Other types of reports are prepared only in response to specific occurrences or when requested. To create a useful report, the fire officer must understand the specific information that is needed and provide it in a manner that is easily interpreted.

Types of Reports

Some reports are presented orally, whereas others are prepared electronically and entered into computer systems. Some reports are formal, whereas others are informal. This chapter covers some of the widely used elements of reporting and provides examples of the most common types of reports that a fire officer might have to prepare. More experienced fire officers can also provide assistance in mastering the reporting requirements of a new position.

Written Reports

Routine reports provide information that is related to fire department personnel, programs, equipment, and facilities. Most fire departments also require company officers to maintain a **company journal** or log book (**FIGURE 5-6**). A fixture in fire stations since the 19th century, the company journal provides an extemporaneous record of all emergency, routine, and special activities that occur at the fire station. Some fire departments have advanced from keeping the journal in a handwritten, hard-covered book to maintaining it as a database on the station computer.

The company journal serves as a permanent reference that can be consulted to determine what happened at that fire station at any particular time and date, as well as who was involved. The general orders usually list all of the different types of information that must be entered into the company journal. The company journal is also the place to enter a record of any fire fighter injury, liability-creating event, and special visitors to the station.

Morning Report to the Battalion Chief. Most career fire officers are required to provide some type of morning report to their supervisor. This report is made by telephone, electronically as part of a staffing software program, or on a simple form that is transmitted by fax or e-mail.

One purpose of the morning report is to identify any personnel or resource shortage as soon as possible. For example, if the driver/operator calls in sick, the driver/operator from the off-going shift might have to remain at work to keep the engine company in service

148 Fire Officer: Principles and Practice, Fourth Edition

Fire Station 100 Company Journal

out	in	inc#	address / comments			sign
	0700		Thursday, February 12 - A shift on duty. Captain Davis OIC			
			Engine	Quint	Medic	
			Cpt. Davis	Lt. Williams	Lt/PM Turner	
			AO Anders	AO Rollo	FF/PM Olliges	
			FF/PM Thompson	FF Grynski		
			FF Sorce (hold)	FF/HM Schultz		
			Status:			
			FF Kinders on Sick Leave			
			FF Source on involuntary hold over			
			FF Wirth on Exchange of Shift until 1900 with FF/HM Schultz			
			FF/PM O'Brien on Annual Leave			
			Vehicle #130654 at Maintenance			
			Vehicle #030112 running as Engine 100			
			Vehicle #110065 running as Medic 44			
			Portable Radio	Anders - 4	Rollo - 4	Olliges - 2
			Keys	Davis	Williams	Turner
			Controlled drugs	Thompson	xxxxxx	Turner
	0730		Paramedic Intern Bartlow (Community College) on Medic 100			Turner
0813	1011	0345	7111 Hogarth St. - Injury		Medic	Olliges
	0820		FF Cegar detailed in from Fire Station 47			Ceg
	0830		FF Sorce off duty			Sorce
0855	0914	0456	9103 Cross Chase Rd - Trouble Breathing		Engine	Anders
1122	1255	0499	4110 Green Springs Dr. #101 - sick		Medic	Olliges
1141	1217	0512	8902 Harrivan Ln - Kitchen fire		Quint	Rollo
1230	1645		Quint to Academy			Williams
	1300		12-lead monitor #1314 damaged on incident #0499			Turner
1314	1655		Facilites working on furnace			Davis
	1345		EMS3 delivers loaner 12-lead #1133 and picks up #1314			Turner
1355	1506	0701	10614 Hampton Rd - STEMI		Medic	Olliges
1355	1422	0710	10614 Hampton Rd -STEMI		Engine	Davis
1430	1450	0715	I-95 North at Franconia exit - MVC		Engine	Davis
1455	1511	0733	6600 Springfield Mall - Injury		Engine	Davis
	1815		Assistant Chief Arrow and City Manager in quarters			
1824	1930	1135	10400 Richmond Hwy - ALS		Medic	Olliges
	1830		FF Wirth on duty			Wirth
1844	0020	1148	7401 Eastmoreland Ave - 2nd Alarm		Quint	Williams
1909	2253	1148	7401 Eastmoreland Ave - 3rd Alarm		Engine	Davis
	1922		Engine 44 filling Fire Station 100			Jones
			Friday, February 13			
0112	0223	0078	9788 Whispering Meadow Ln - stroke		Medic	Olliges
0249	0311	0101	6600 Springfield Mall - Alarm Bells		Quint	Williams

FIGURE 5-6 Example of a company journal.
Courtesy of Mike Ward.

until a replacement arrives. The battalion chief needs to know that the company is short a driver/operator and has a person who is working involuntary overtime or holdover, or that a replacement needs to be found.

Monthly Activity and Training Report. The monthly activity and training report documents the company's activity during the preceding month. This report typically includes the number of emergency responses, training activities, inspections, public education events, and station visits that were conducted during the previous month. Some monthly reports include details such as the number of feet of hose used and the number of ladders deployed during the month.

In many cases, the officer delegates the preparation of routine reports that do not involve personnel actions or supervisory responsibilities to subordinates. In such a case, the officer is always responsible for checking and signing the report before it is submitted.

Two other routine reports are the annual fire fighter performance appraisals and the fire safety inspection. These reports are covered in detail in Chapter 6, *Preincident Planning and Code Inspection*, and in Chapter 10, *Application of Leadership*.

Some fire companies or municipal agencies post a version of a monthly report on their public website. The reports often include digital pictures or other information about the incidents and the personnel involved.

Special events or unusual situations that occurred during the month might also be included in the monthly report. For example, this report might note that the company provided standby coverage for a presidential visit or participated in a local parade.

Incident Reports. An incident report is required for every emergency response. The nature and the complexity of the report depend on the situation: Minor incidents generally require simple reports, whereas major incidents require extensive reports.

The **National Fire Incident Reporting System (NFIRS)** is a nationwide database managed by the U.S. Fire Administration that collects data related to fires and other incidents. Generally, the first-arriving officer (or the incident commander) completes an NFIRS report for each response, including a narrative description of the situation and the action that was taken. The full incident report includes a supplementary report from the officer in charge of each additional unit that responded.

Many departments use a narrative format when reporting on incidents (**FIGURE 5-7**). Some incidents require an **expanded incident report narrative**, in which all company members submit a narrative description of their observations and activities during an incident. The fire officer should anticipate the need to provide expanded incident reports for the following types of incidents:

- The fire company was one of the first-arriving units at a fire with a civilian fatality or injury.
- The fire company was one of the first-arriving units at an incident that has become a crime scene or an arson investigation.
- The fire company participated in an occupant rescue or other emergency scene activity that would qualify for official recognition, an award, or a bravery citation.
- The fire company was involved in an unusual, difficult, or high-profile activity that requires review by the fire chief or designated authority.
- A fire company activity occurred that may have contributed to a death or serious injury.
- A fire company activity occurred that may have created a liability.

Civilians Rescued From Burning Home

Twenty-eight fire fighters responded to an early morning house fire in the Grandview district. The first 9-1-1 call was at 10:47 pm on Monday December 10, 2014, reporting smoke coming from a three-story townhome.

Metro County fire fighters arrived at 11:07 pm, encountered smoke and flames coming from the first floor windows at 1928 Braniff Boulevard. Two elderly females were found on the third floor. They were taken out of the building through a window via an aerial tower, treated by paramedic/fire fighters, and transported to University Hospital.

It took 17 minutes of aggressive fire suppression before Battalion Chief Devon Jones declared the fire "under control." The first floor of the townhouse was extensively damaged, with heat and smoke damage to the adjacent townhouses.

The cause of the fire remains under investigation. Preliminary results show that the fire started in the kitchen. Smoke detectors were in the home, but batteries were removed.

The monetary loss has not been calculated.

Submitted by T. L. Gaines, Metro County Fire Department

MCFDPIO@metrocounty.gov (111) 555-3473

FIGURE 5-7 Sample incident report.
© Jones & Bartlett Learning.

- A fire company activity occurred that has initiated an internal investigation.

The author of a narrative should describe any observations that were made en route to the incident or at the scene, and should fully document his or her actions related to the incident. The narrative should provide a clear mental image of the situation and the actions that were taken, and may include pertinent negatives—elements noted not to be present or actions not taken. For example, if no gas-powered tools were used on the fire floor, this fact may influence the effectiveness of an arson investigation and be cited in future court proceedings.

Mary Sovick, a retired fire fighter who is an advocate on improving fire service writing, points out that a poorly written report hurts your credibility by making you appear less competent and professional. This can impact the outcome of a court case, where the incident report is the official description of the incident or situation. It could also influence the determination of a worker compensation claim. Sovick identifies the characteristics of a good report (Sovick, 2013):

- Accurate and specific
- Factual
- Objective
- Clear
- Complete
- Concise
- Well-organized
- Grammatically correct
- Light on abbreviations

Infrequent Reports. Infrequent reports usually require a fire officer's personal attention to ensure that the information in the report is complete and concise. These special reports include the following documents:

- Fire fighter injury report
- Citizen complaint
- Property damage or liability-event report
- Vehicle accident report
- New equipment or procedure evaluation
- Suggestions to improve fire department operation
- Response to a grievance or complaint
- Fire fighter work improvement plan
- Request for other agency services

Some situations require two or more different report forms to cover the same event. For example, if a fire engine collides with a private vehicle at an intersection while responding to an emergency call, the fire officer will spend significant time completing a half-dozen forms, even if no injuries occur. A supervisor's report must be filed, and the apparatus driver must complete an accident report. Another form is required to identify the damage to the fire apparatus and any repairs that are needed. The AHJ's risk management division may also require a report. The insurance company that covers the fire apparatus usually requires the completion of another form.

A **supervisor's report** is required by state worker's compensation agencies whenever an employee is injured. This report serves as the control document that starts the state file relating to an injury or a disability claim (**FIGURE 5-8**).

The supervisor's first report of an injury must be submitted within 24 to 72 hours of the incident or in line with company policy, after conducting an investigation into the facts. This report often identifies the injured person, the nature of the injury, the means by which the incident occurred, practices or conditions that contributed to the incident, any potential loss, the typical frequency of occurrence, and actions that will be taken to prevent the same incident from occurring again. A **chronological statement of events** is a detailed account of activities, such as a narrative report of the actions taken at an incident or accident, and this should be included in the supervisor's report. Emotional statements, opinions, and hearsay have no place in a formal report that will remain on permanent record and potentially be subject to subpoena as a public record.

Using Information Technology

Most reports are completed using a computer and software. The role of the fire officer will range from selecting preformatted information in an interactive online form, as with an NFIRS report, to composing

FIRE OFFICER TIP

Use a Template and Write without Emotion or Bias

A fire officer was severely injured during ladder truck operations at a structure fire with a civilian rescue. The incident report submitted by a chief officer was returned by the city manager with a directive to rewrite the report ". . . absent of any personal opinion and/or biases" and to include the following elements:

- Chronology of events
- The report from the state investigation of the incident
- Ladder truck maintenance reports

Three of the four points in the original report cited global issues that resulted in a severe injury of the fire officer during the rescue effort (Buteau, 2019). Always use a standardized format or a template for report writing.

SUPERVISOR'S ACCIDENT REPORT WORKERS' COMPENSATION CLAIMS	Claims Management Inc. PO Box 342 Sacramento, CA 95812-3042 (916) 631-1250 FAX (916) 635-6288	DATE & TIME REPORTED:

OSHA CASE NO:

COMPANY *Scotts Valley Fire Protection District*	LOCATION *7 Erba Lane, Scotts Valley, CA, 95066*	LOCATION CODE NO: *1100*

A. EMPLOYEE	NAME		JOB TITLE	
	DEPARTMENT *Scotts Valley Fire Protection District 1100*		☒☒ LOST TIME ☒☒ NO LOST TIME	☒☒ FIRST AID

B. TIME AND PLACE OF ACCIDENT	DATE	HOUR	DEPARTMENT *1100*	IMMEDIATE SUPERVISOR

IDENTIFY EXACT LOCATION WHERE ACCIDENT OCCURRED *(Be specific)*

JOB OR ACTIVITY AT TIME OF ACCIDENT *(Be specific)*

C. WITNESSES - List of Names and Addresses

Name	Address

D. DESCRIBE THE ACCIDENT/ACCIDENT CAUSE - *Please be specific*

E. UNSAFE ACT/CORRECTIVE ACTION TAKEN - *Include both employee and supervisor corrective actions to prevent future occurrences.*

EMERGENCY - WENT TO THE DOCTOR ☒☒ YES ☒☒ NO	If yes, please fill out the following information: Name of Doctor: _____ Address of Doctor: _____

☒☒ NON-EMERGENCY, BUT PLAN ON SEEING A PHYSICIAN
   Physician's Name: _____

FIGURE 5-8 Supervisor's accident report for worker's compensation.
Courtesy of Claims Management, Inc.

a narrative and following a guide sheet to make a suggestion to improve fire department operations. The physical resources available within a fire station for reporting purposes typically include a computer, printer, and network connection.

Two types of software are found on the computer: the operating system and applications. The operating system handles the input, manages files, and manages all of the activity. It is like the drivetrain of a quint. Commonly used operating systems include Microsoft Windows and Apple OS. Application programs are designed to provide a specific task, much like a pump panel controls the delivery of water. Fire officers commonly use word processing, spreadsheet, and presentation applications.

Word processors are used to produce memos, reports, and letters. Many programs provide templates that serve as premade layouts for routine reports. Most provide spell-checking and grammar guides. Popular word processing programs include Microsoft Word, OpenOffice Writer, Google Docs, WordPerfect, and Pages.

Spreadsheets are used to tabulate numbers and information in columns and rows, often as a means to analyze data and develop charts. The display resembles a ledger sheet, with columns represented by letters and rows identified by numbers. The intersection of a column and row, called a cell, can contain numbers, letters, or a formula. Formulas are used to make calculations. For example, in cell A20 you might see "=A19*D03," which means the software multiplies the number in cell A19 with the number in cell D03 and places the result in cell A20. Spreadsheets have immense calculation capabilities. Widely used spreadsheet applications include Microsoft Excel, Open Office Calc, Google Spreadsheet, Apple Numbers, and Libre Office Calc.

Presentation software is used to display information in a format that allows the audience to understand visually the message presented. The images can be displayed on a screen or website, and images, videos, and sounds can be integrated as desired. Commonly used presentation software includes Microsoft PowerPoint, OpenOffice Impress, and Apple Keynote.

You Are the Fire Officer Conclusion

The lieutenant conducts a private meeting with the fire fighter. The lieutenant reviews the work hours SOP, pointing out that the no-call/no-show incident will generate a proposed 12-hour suspension without pay. In addition, the lieutenant notes the excessive use of annual/sick leave and work exchanges. There appears to be an increasing pattern of work performance issues, any of which could lead to termination. What can the department do to help the fire fighter?

The fire fighter explains that he recently separated from his wife and is now a single parent. This morning the babysitter did not show up. The lieutenant acknowledges that the situation is difficult and reviews the discussion from the January mid-year informal review, in which the fire fighter rejected reaching out to the employee assistance program or requesting a transfer to day work. The lieutenant concludes the session by urging the fire fighter to use the EAP and to consider alternative ways of resolving the outside issues that make him late, use up his leave time, and distract him during his emergency driving.

After-Action REVIEW

IN SUMMARY

- The fire officer is in the best position to lead change within the fire department.
- The process for implementing new policies varies, and the fire officer should follow departmental procedures.
- Every fire department has a budget that defines the funds that are available to operate the organization for one year.
- The budget document describes where the revenue comes from (input) and where it goes (output) in terms of personnel, operating, and capital expenditures.

- Most municipal governments use a base budget in the financial planning process. The base budget is the level of funding that would be required to maintain all services at the currently authorized levels, including adjustments for inflation and salary increases.
- The AHJ will require that a capital budget request form be completed, along with appropriate documentation.
- It is a challenge to get new capital equipment using existing funding. The justification will have to vividly describe the value of the purchase in terms of quality of service or improved capacity of service.
- The fire officer should meet with the purchasing officials to understand the process and the terminology used in budget requests.
- The revenue stream depends on the type of organization that operates the fire department and the formal relationship between the organization and the local community.
- Planned expenditures must be balanced against anticipated revenues a year or more in advance.
- Sometimes the results of budget reductions drastically affect fire department operations. Fire departments have to make difficult choices when faced with declining revenues.
- The proper administration of the human resources function is one of the most important and sensitive areas of jurisdictional management.
- The human resources department issues rules and regulations. It also provides a range of services including hiring; coordinating discipline; and managing Family and Medical Leave Act (FMLA), disability, payroll, and retirement services.
- Regular evaluations are performed to ensure that each fire fighter knows which type of performance is expected while on the job and where he or she stands in relation to those expectations.
- The fire officer must determine each new fire fighter's skills, knowledge, aptitudes, strengths, and weaknesses and then set specific expectations for each new fire fighter.
- Regular feedback from the fire officer should keep fire fighters aware of how they are doing, particularly after incidents or activities that present a special challenge.
- Discipline is divided into positive and negative, or corrective, sides. Positive discipline is based on encouraging and reinforcing appropriate behavior and desirable performance. Corrective discipline is based on punishing inappropriate behavior or unacceptable performance.
- Municipal personnel rules usually require all of the official records of an employee's work history to be in a secured central repository.
- The fire officer is expected to maintain reports, logs, and files within the records management system in the manner outlined in the jurisdiction's policies and procedures.
- Every capital item, from the fire station to the industrial dryer, needs to be documented.
- Documented inspection records are reviewed as part of the preventive maintenance process or during a fire fighter death or injury investigation.
- State agencies and the Insurance Service Office (ISO) review the training records to confirm that the fire department is meeting its obligation and that individuals are appropriately trained.
- Jurisdictions maintain documentation of fire fighter exposures and medical records.
- Fire departments that provide medical care as an EMR or as an EMS agency must follow HIPAA regulations in retaining, managing, and releasing patient information and records.
- To create a useful report, the fire officer must understand the specific information that is needed and provide it in a factual manner that is easily interpreted.
- The most common form of reporting is verbal communication from one individual to another, either face-to-face or via a telephone or radio.

KEY TERMS

Base budget The level of funding required to maintain all services at the currently authorized levels, including adjustments for inflation, salary increases, and other predictable cost changes.

Budget An itemized summary of estimated or intended expenditures for a given period, along with proposals for financing them.

Capital budget An allocation of money for the acquisition or maintenance of fixed assets such as land, buildings, and durable equipment.

Chronological statement of events A detailed account of the fire company activities related to an incident or accident.

Company journal A log book at the fire station that creates an up-to-date record of the emergency, routine activities, and special activities that occurred at the fire station. The company journal also records any fire fighter injuries, liability-creating events, and special visitors to the fire station.

Expanded incident report narrative A report in which all company members submit a narrative on what they observed and which activities they performed during an incident.

Expenditure Money spent for goods or services.

Fire tax district A special service district created to finance the fire protection of a designated district.

Fiscal year A 12-month period during which an organization plans to use its funds. Local governments' fiscal years generally run from July 1 to June 30.

National Fire Incident Reporting System (NFIRS) A nationwide database at the National Fire Data Center under the U.S. Fire Administration that collects fire-related data in an effort to provide information on the national fire problem.

Operating budget An estimate of the income and expenditures of the jurisdiction.

Oral reprimand, warning, or admonishment The first level of corrective discipline. Considered informal, this discipline action remains with the fire officer and is not part of the fire fighter's official record.

Revenues The income of a government from all sources.

Supervisor's report A form that is required by most state worker's compensation agencies and that is completed by the immediate supervisor after an injury or property damage accident.

Supplemental budget Proposed increases in spending to provide additional services.

REFERENCES

Anderson, Joseph M. 1998. *Justifying the Purchase of New Fire Apparatus*. Emmitsburg, MD: U.S. Fire Administration, Executive Fire Officer Program.

Belker, Loren B., Jim McCormick, and Gary S. Topchik. 2018. *The First-Time Manager, Seventh edition*. New York: Harper Collins Leadership.

Bureau of Labor Statistics (BLS). 2012. "The Recession of 2007–2009." *BLS Spotlight on Statistics*. Washington, DC: Bureau of Labor Statistics. Accessed June 25, 2019. https://www.bls.gov/spotlight/2012/recession/pdf/recession_bls_spotlight.pdf.

Buteau, Walt. 2019, June 20. Asst. "Chief Suspended for Not Following Order to Change Report on Firefighter's Amputation Injury." WPRI [Providence, RI]. Accessed July 11, 2019. https://www.wpri.com/target-12/asst-chief-suspended-after-not-following-order-to-change-report-about-firefighter-amputation-injury/.

Carter, Harry R. 2015. *Running a Volunteer Fire Department*. Howell Township, NJ: Dr. Harry R. Carter.

Daly, Peter H., Michael D. Watkins, and Cate Reavis. 2006. *The First 90 Days in Government: Critical Success Strategies for New Public Managers at All Levels*. Boston, MA: Harvard Business School.

Farmer, Liz. 2019, May 30. "The Baltimore Cyberattack Highlights Hackers' New Tactics." *Governing*. Accessed July 11, 2019. https://www.governing.com/topics/public-justice-safety/gov-cyber-attack-security-ransomware-baltimore-bitcoin.html.

Fent, Kenny. 2015. "Firefighter's Exposure to Potential Carcinogens." [Presentation] John P. Redman Symposium. National Harbor, MD: International Association of Fire Fighters. Accessed July 11, 2019. https://services.prod.iaff.org/ContentFile/Get/10166.

Fischer, Ronald C. 2016. *State and Local Public Finance*. New York: Routledge.

Gordon, Tracey. 2012. "State and Local Budgets and the Great Recession." Stanford, CA: The Russell Sage Foundation and the Stanford Center on Poverty and Inequality. Accessed July 11, 2019. https://web.stanford.edu/group/recessiontrends-dev/cgi-bin/web/sites/all/themes/barron/pdf/StateBudgets_fact_sheet.pdf.

Hart, Kerri. 2017, December 18. "Q&A: The Art of Fire Department Fleet Management. The Chair of the IAFC's Emergency Vehicle Management Section Shares His Tips for Funding, Replacing the Fire Department Fleet." FireRescue1.com. Accessed July 11, 2019. https://www.firerescue1.com

/fire-products/fire-apparatus/articles/370902018-Q-A-The-art-of-fire-department-fleet-management/.

Kemp, Roger L. 1998. *Managing America's Cities: A Handbook for Local Government Productivity*. Jefferson, NC: McFarland and Company.

Manning, William. 2007 "Change Fire Service Culture—Attitudes and Behavior" Novato, CA: 2nd National Firefighter Life Safety Summit. https://www.everyonegoeshome.com/wp-content/uploads/sites/2/2014/08/novato.pdf

Menifield, Charles E. 2017. *The Basics of Public Budgeting and Financial Management: A Handbook for Academics and Practitioners, Third edition*. Lanham, MD: Hamilton Books.

National Association of State Procurement Officials (NASPO). 2019. *State and Local Government Procurement: A Practical Guide, Third edition*. Lexington, KY: National Association of State Procurement Officials.

Office for Civil Rights (OCR). 2014, November. *Bulletin: HIPAA Privacy in Emergency Situations*. Washington, DC: U.S. Department of Health and Human Services.

Reese, C. Richard. 2005. *Records Management Best Practices Guide*. Boston, MA: Iron Mountain.

Scott, Erik. 2018. "SAFER Grant Provides LAFD with Additional Resources. Los Angeles, CA: Los Angeles Fire Department." Accessed July 11, 2019. https://www.lafd.org/news/safer-grant-provides-lafd-additional-resources.

Smeby, L. Charles. 2013. *Fire and Emergency Services Administration: Management and Leadership Practices*. Burlington, MA: Jones and Bartlett Learning.

Sovick, Mary. 2013. *Writing Effective Incident Reports: Tips for Writing Reports That Are Accurate and Professional*. Campbell, CA: Firebelle Productions.

Truchil, Barry E. 2018. *The Politics of Local Government: Governing in Small Towns and Suburbia*. Lanham, MD: Lexington Books.

Vince, Joe. 2019, May 17. Aging TX Apparatus Find Success as "Blockers." *Firehouse.com News*. Accessed July 11, 2019. https://www.firehouse.com/apparatus/news/21081084/irving-dallas-tx-fire-department-blocker-apparatus-success-firefighters.

Yukl, Gary A. 2013. *Leadership in Organizations*. Boston, MA: Pearson.

Fire Officer in Action

A lieutenant, frustrated at the continuing problem of responding to an excessive number of defective fire alarm system activations in 24 hours, is recommending a new policy for handling the issue.

1. NFPA 1021, *Standard for Fire Officer Professional Qualifications*, describes multiple administrative activities. Which of the following activities is *not* found in that description?
   A. Develop and manage the budget process.
   B. Identify and monitor revenue sources.
   C. Develop a political power base.
   D. Function as a documenter.

2. Recommending a new policy requires the fire officer to do all of the following tasks, *except*:
   A. engage in informal consultation with the fire chief before preparing the recommendation.
   B. gather statistical evidence.
   C. determine how widespread the perceived problem is.
   D. identify any potential negative effects with the new policy.

3. A draft of the policy should be shared with whom before it is submitted?
   A. Fire chief
   B. All personnel
   C. Operations (firefighting) personnel
   D. Administrative services personnel

4. After completing the internal fire department process, a new "Faulty Fire Alarm Policy" is issued by the fire department. Elements of this policy are different from the lieutenant's original proposal. What should the lieutenant do?
   A. Utilize the elements of the original proposal, since they better resolve the original problem.
   B. Review the new policy with the fire fighters and follow the policy.
   C. Review the new policy with the fire fighters and add "local amendments" based on the original proposal.
   D. Ignore the new policy because it does not solve the problem.

Fire Lieutenant Activity

NFPA Fire Officer I Job Performance Requirement 4.4.3

Prepare a budget request, given a unit level need, so that the request is in the proper format and is supported with data.

Application of 4.4.3

1. In a municipality you are familiar with, what information is needed to request a new vehicle?
2. Identify the productivity or cost savings that a smaller response vehicle would have on fire department operations.
3. What emergency equipment would be required on the smaller response vehicle?
4. What would be the service life of a smaller rig responding to 9 to 15 responses per day?

Access Navigate for flashcards to test your key term knowledge.

CHAPTER 6

Fire Officer I

Preincident Planning and Code Enforcement

KNOWLEDGE OBJECTIVES

After studying this chapter, you will be able to:

- Explain the importance of fire inspections and preplans as part of a comprehensive community risk reduction program. (p. 159)
- Compare and contrast the purpose and authority for preplan site visit and fire code inspections. (**NFPA 1021: 4.5.1, 4.5.2**) (pp. 160–166)
- Identify benefits of preplanning for the department and the community. (**NFPA 1021: 4.5.1**) (p. 159)
- Describe the standardized method for completing a preincident plan. (**NFPA 1021: 4.5.1**) (pp. 160–164)
- Explain aspects of buildings and grounds considered during preplanning. (**NFPA 1021: 4.5.2**) (p. 161)
- Explain aspects of building occupancy considered during preplanning. (**NFPA 1021: 4.5.2**) (pp. 161–162)
- Explain aspects of a water supply and fire protection systems considered during preplanning. (**NFPA 1021: 4.5.2**) (p. 162)
- Explain special hazard considerations and conditions during preplanning. (**NFPA 1021: 4.5.2**) (pp. 162–165)
- Determine the resources and actions that may be required for an incident at a specific building or event based on the information gathered during the preplan process. (**NFPA 1021: 4.5.2**) (p. 164)
- Explain documentation requirements for a preplan. (**NFPA 1021: 4.5.2**) (p. 166)
- Identify types of fire codes and their use. (pp. 166–168)
- Differentiate building codes from fire codes. (p. 167)
- Identify types of built-in fire protection systems. (**NFPA 1021: 4.5.1, 4.5.2**) (pp. 168–171)
- Explain the purpose of a fire code compliance inspection. (pp. 171–172)
- Explain how to determine building construction type. (**NFPA 1021: 4.5.1, 4.5.2**) (pp. 172–174)
- Explain how to confirm building or space use matches an occupancy classification. (**NFPA 1021: 4.5.1, 4.5.2**) (pp. 172–177)
- Describe steps the fire officer may take to prepare a company for inspection duties. (**NFPA 1021: 4.5.1, 4.5.2**) (pp. 177–179)
- Describe steps to prepare for an inspection. (**NFPA 1021: 4.5.1**) (pp. 177–179)
- Describe steps taken to conduct an inspection. (**NFPA 1021: 4.5.1, 4.5.2**) (pp. 179–180)
- Explain the necessity of written inspection/correction reports. (**NFPA 1021: 4.5.1, 4.5.2**) (p. 181)
- Identify general inspection items to be checked during an inspection. (**NFPA 1021: 4.5.1**) (pp. 181–182)

- Identify access and egress inspection items to be checked during an inspection. (**NFPA 1021: 4.5.1**) (p. 181)
- Identify exit sign and emergency lighting inspection items to be checked during an inspection. (**NFPA 1021: 4.5.1**) (pp. 181–182)
- Identify portable and built-in fire protection system inspection items to be checked during an inspection. (**NFPA 1021: 4.5.1**) (p. 182)
- Explain special hazard inspection considerations. (**NFPA 1021: 4.5.1**) (p. 182)
- Identify inspection concerns based on occupancy/use groups. (**NFPA 1021: 4.5.1, 4.5.2**) (pp. 182–185)
- Explain search and seizure laws related to fire cause investigation. (**NFPA 1021: 4.5.3**) (p. 185)
- Identify reasons for securing a fire scene. (**NFPA 1021: 4.5.3**) (pp. 185–186)
- Identify methods of securing a scene. (**NFPA 1021: 4.5.3**) (pp. 185–186)
- Explain chain of custody. (pp. 185–186)
- Explain the importance of preventing unnecessary disturbance or destruction at a fire scene. (p. 186)
- Identify methods of protecting evidence at fire scenes. (pp. 186–187)
- Determine when to request a fire investigator. (pp. 187–188)

SKILLS OBJECTIVES

After studying this chapter, you will be able to:

- Create a complete preplan for a building or event. (**NFPA 1021: 4.5.2**) (pp. 160–165)
- Document a preplan. (**NFPA 1021: 4.5.2**) (p. 166)
- Prepare a company to conduct a fire inspection. (pp. 177–179)
- Conduct a fire inspection. (**NFPA 1021: 4.5.1, 4.5.2**) (pp. 179–180)
- Write an inspection/correction report. (**NFPA 1021: 4.5.1, 4.5.2**) (p. 181)
- Determine a probable area of origin of a fire. (**NFPA 1021: 4.5.3**) (p. 185)
- Determine preliminary cause of a fire. (**NFPA 1021: 4.5.3**) (p. 185)
- Secure a scene to prevent unauthorized persons from entering. (**NFPA 1021: 4.5.3**) (pp. 185–186)
- Document area of origin and probable cause for small fires. (pp. 186–187)

ADDITIONAL NFPA STANDARDS

- **NFPA 1**, *Fire Code*
- **NFPA 101**, *Life Safety Code®*
- **NFPA 1620**, *Standard for Pre-Incident Planning*

You Are the Fire Officer

A long-empty big-box retail building is undergoing extensive reconstruction to become a mixed-use residential and retail occupancy, including restaurants.

The local fire company is visiting the completed site to develop a preincident plan.

1. What is the best way to perform a preincident plan for a complex facility?
2. How can you prepare fire fighters for fire inspection duties?
3. How does the fire preincident plan interact with the emergency management and business continuity plan?

Access Navigate for more practice activities.

Introduction

The fire officer considers the building environment from several viewpoints. If it is burning, damaged, or expelling hazardous materials, the fire officer is expected to command the incident, rescue those in harm's way, mitigate the situation, and render the scene safe.

To accomplish those tasks, the fire officer looks at the building from two different perspectives. First, the fire department prepares to handle an emergency in the building by developing a preincident plan. Second, members of the fire department perform a fire and life-safety inspection to ensure that the building meets the appropriate fire prevention code requirements.

Although they are separate activities, preincident planning and code enforcement require similar skill sets, including an understanding of fire growth and development, building construction, and built-in fire protection systems.

The Fire Officer's Role in Community Fire Safety

Fire department community risk reduction activities may include fire prevention, preincident planning, and fire and life-safety education. Fire officers play multiple roles in relation to properties within their communities, including handling the following critical tasks:

1. Identifying and correcting fire safety hazards through safety checks or code enforcement
2. Developing and maintaining preincident plans
3. Promoting fire safety through public education

In most areas, fire inspectors and fire officers must understand fire inspections and code enforcement duties (**FIGURE 6-1**). Fire suppression companies are usually involved in preincident planning. In some areas, the local fire suppression company also conducts code enforcement inspections. Fire and life-safety education activities are often performed by a combination of staff personnel and fire companies (Diamantes, 2015).

Even where the role of fire companies does not include code enforcement, fire officers and fire fighters should conduct regular visits to properties to develop preincident plans. During preincident planning, they should be on the lookout for fire and life-safety hazards. A fire officer should always take proactive steps to reduce the impact of any potential emergency that could occur; this includes identifying and correcting conditions that could contribute to fire spread, or restrict or prevent occupants from leaving the building. Limited egress could ultimately increase the risk to citizens and fire fighters in the event of an emergency (Solomon, 2012).

Built-in Fire Protection

The goal of fire protection is to contain the fire within a building or area. The two primary methods used in big cities that were being built in the late 1880s were compartmentation and automatic fire sprinklers. Compartmentation is a passive fire protection method using fire-rated floors, protection of vertical openings such as stairs and elevator shafts, compartmentation in the form of fire walls and fire partitions, and fire-resistive construction (e.g., rated columns and beams to prevent collapse and spread of fire) (Corbett and Brannigan, 2013). An automatic fire sprinkler system is an active fire protection method, as there has to be some action or motion in order to work effectively in the event of a fire. Fire alarm systems are another example of an active fire protection method.

The status of the built-in fire protection features is essential to occupant and fire fighter safety. Look for fire protection elements that have been disabled or tampered with when doing a preincident plan, when conducting a walkthrough, or when doing a 360-degree size-up at a fire incident.

Common Built-in Fire Protection Code Violations

The five most common issues discovered during a preincident plan or code enforcement visit are:

1. Blocked access to fire department system connections
2. Improper storage in the fire pump and riser rooms
3. Items hanging from fire sprinklers and piping
4. Not enough clearance to allow a fire sprinkler to fully operate
5. Incorrect fire sprinkler coverage if the occupancy or contents have changed (ORFS, 2017)

Items 1 through 4 can be immediately corrected at the request of the fire officer. Follow your jurisdiction's reporting and documentation procedures when identifying a problem with the automatic sprinkler system.

FIGURE 6-1 Fire officers often conduct preincident planning and fire inspection activities during the course of their workday.
© Jones & Bartlett Learning. Photographed by Glen E. Ellman.

Preincident Planning

A **preincident plan** is described by NFPA 1620, *Standard for Pre-incident Planning*, as a document developed by gathering data used by responding personnel in effectively managing emergencies for the protection of occupants, participants, responding personnel, property, and the environment (**FIGURE 6-2**).

The original purpose of a preincident plan was to provide information that would be useful in the event of a fire at a high-value or high-risk location. A **high-value property** contains equipment, materials, or items that have a high replacement value. Examples include properties containing agricultural equipment, electronic data processing equipment, or scientific equipment; fine arts centers; and storage or manufacturing sites.

A **high-risk property** has the potential to produce a catastrophic property or life loss in the event of a fire. Examples include nuclear power plants, bulk fuel storage facilities, hospitals, and jails. Preincident plans include information that could apply to a variety of situations that could potentially occur at the location covered in the plan.

Tactical Priorities

Address: 1500, 1510, 1520		
Occupancy Name:		
Preplan #: 02-N-01	Number Drawings: 1	Revised Date: 12/2002
District: E275A	Subzone: 60208	By: ACEVEDO
Rescue Considerations: Yes ( ) No (X)		
Occupancy Load Day:		Occupancy Load Night:
Building Size:		Best Access:
Knox Box: NONE	Knox Switch: NONE	Opticom: NONE
Roof Type: X	Attic Space: Yes ( ) No ( )	Attic Height: X
Ventilation Horizontal:		Ventilation Vertical:
Sprinklers: Yes (X) No ( )	Full (X)	Partial ( )
Standpipes: Yes (X) No ( )	Wet ( )	Dry ( )
Gas: Yes ( ) No ( )	Lpg ( )	
Hazardous Materials: Yes (X) No ( ) DIESEL GENERATORS 1,000 GALLON TANKS BATTERY ROOM		
Firefighter Safety Considerations: ELEVATOR PIT		
Property Conservation And Special Considerations: VENTILATION: AUTOMATIC SMOKE REMOVAL SYSTEM 3 OFFICE BLDGS; 2 PARKING STRUCTURES 6 FLRS - 1230 W. WASHINGTON ST. 4 FLRS - 1500 N. PRIEST DR. 4 FLRS - PARKING GARAGE		

FIGURE 6-2 An example of a preincident plan.
© Jones & Bartlett Learning.

Facilities that store or handle hazardous materials are required to submit information about those materials and the threats they pose to the fire department and the Local Emergency Planning Council (LEPC); this information is then incorporated into preincident plans for those facilities. Some plans include responses to natural or human-made catastrophic incidents.

Today, preincident plans are used for all types of buildings and occupancies within a fire department's response area. A preincident plan is meant to identify in advance the strategies, tactics, and actions that should be considered if a predictable situation occurs, and to make the fire fighters familiar with the building. Preincident plans are useful at the company level for practicing initial operations for buildings in the company's district. They provide first-due companies vital information that is needed during the initial response.

A Systematic Approach

Data collection requires a systematic approach. The fire officer should use a standardized method for completing each preincident plan. NFPA 1620 outlines six considerations to take into account when completing a preincident plan:

1. Identify physical elements and site considerations.
2. Identify occupancy considerations.
3. Identify water supply and fire protection systems.
4. Identify special considerations.
5. Identify emergency operation considerations.
6. Document findings.

1: Identify Physical Elements and Site Considerations

The first step is to evaluate the physical elements and site considerations. A **plot plan** provides a representation of the exterior of a structure, identifying site access, doors, utilities access, and any special considerations or hazards. **Floor plans** are interior views of a building. Rooms, hallways, cabinets, and other features are drawn in the correct relationship to each other in such a plan.

The building's size and dimensions, including its overall height, number of stories, length, width, and square footage, should be determined and included in the plan. Connections between buildings and distances to exposures should be noted. Access routes and points of entry should be clearly indicated. In addition, the preincident plan should identify concealed spaces and windows that could be used for ventilation or rescue.

The preincident plan should also include detailed information about the construction of the building's roof, floor, and walls. Structural integrity should also be assessed and documented. Focus on factors that could lead to collapse of building components, fire spread, or the release of toxic gases. Any factor that might affect the ability of responding fire fighters to enter and effectively perform interior operations should be highlighted. Conditions that affect the safety of fire operations on the roof should also be noted.

This part of the preincident plan should also document the location of utilities, including gas, electrical, and domestic water entry locations and shut-offs, and heating, ventilating, and air-conditioning (HVAC) controls. Note the presence of flammable liquids, compressed or liquefied gases, chemicals, and steam lines. Information about elevators should include their location in the building, floors served, type of elevator, and type of recall or override service.

Note any conditions that might potentially delay response time and/or access, including weight-restricted bridges, low-overhead clearances, and roads subject to natural or human-made blockages.

Security information, such as the presence of fences, 24-hour security forces, and guard dogs, should be part of the plan. The fire officer must also determine the environmental impact of any contaminants that could be released from the facility. Any release could impact approach and will be influenced by wind direction.

During a test of the preincident plan, the fire officer should test two-way radios and document any interference or poor coverage of the two-way radio system. For example, when conducting a preincident planning survey of a high-rise, be sure to test your portable radios to ensure that you can contact dispatch or the outside unit when in the inner core as well as in the basement. For many radio systems, these areas prove to be dead spots. Such a problem could be mitigated by having a fixed antenna system or repeaters installed in the building.

2: Identify Occupancy Considerations

During a fire, the fire department is expected to provide a rapid and safe evacuation or plan for protecting in place. Protecting in place would require the officer to determine which areas within the structure are resistant to the potential of fire growth and how the ventilation systems can be controlled.

If the preincident planners determine that the occupants should be removed, the plan should identify how that will be accomplished. It should describe both the process and the fire department and building resources that will be required for occupant protection, including occupant escape routes.

The fire officer must know how to access information about the number of occupants and their ages, their physical or mental conditions, and any need for assisting them. Building-specific information should include hours of operation, occupant load, and location of occupants. These points have bearing on how many companies would be needed and what they can expect in terms of evacuation and rescue.

The preincident plan should note the locations of exits and any special locking devices, such as a delayed release or stairwell unlocking system. The fire officer should also ensure that the preincident plan provides contact information for facility staff. Expectations should be communicated. Case studies have proven the need to clearly define the role of the facility and the role of the fire department. This might require a meeting between the fire department and the facility staff to predetermine the role of the facility staff and the emergency responders when handling a fire or disaster. The preincident plan should identify the locations of stair chairs, stretchers, and lifts for those occupants that require assistance.

When large numbers of people are relocated, they need basic services, such as food, water, and sanitation facilities. In some communities, the public school system can assist the fire department by providing areas of shelter and transportation. Joint agreements and defined roles and responsibilities are part of the pre-planning process.

As part of the incident command system, a tracking system for the occupants of a building should be established, particularly for locations where the number of occupants varies with the facility's operations and time of day. If evacuees will be relocated, their locations must be tracked. For larger evacuations during weather-related events, the American Red Cross has had considerable experience with tracking evacuees.

Some facilities, such as high-rise, healthcare, or detention facilities, cannot be completely evacuated in a short time. Consequently, the response plan could be to relocate occupants within other areas of the building. These types of facilities are usually designed to relocate occupants to certain areas until they can be evacuated.

During all of these operations, the products of combustion must be segregated from the occupants. Move those in the greatest peril to areas away from the fire. Close fire and smoke doors, and seal the individual rooms with wet towels to prevent smoke from filtering into the room. The HVAC system is often designed to shut down automatically upon activation of a fire alarm. Sophisticated HVAC systems, such as those found in hospitals and high-rises, are often divided into zones. With such systems, exhaust fans in unaffected zones can be redirected to eject smoke from the zone that is in alarm.

3: Identify Water Supply and Fire Protection Systems

The required water flow is determined by evaluating the size of the building or buildings, their contents, construction type, occupancy, exposures, fire protection systems, and any other features that could affect the amount of water needed to control the fire. The adequacy of available water for sprinkler systems, hose streams, and any other special requirements should also be considered.

Identify the available water supply. Document the locations of fire hydrants, their flow rates, and the distribution design. The ideal hydrant would be fed from a large main that is part of a grid that allows water to flow from several directions. Water supply test data should be obtained and tests should be conducted in accordance with NFPA 291, *Recommended Practice for Fire Flow Testing and Marking of Hydrants*. When the demand exceeds the available supply, the preincident plan should identify an appropriate response to mitigate the deficiency. This might include a water shuttle operation or the initiation of relay operations. Some sites may have a private water system, including their own water tower and pump house.

The preincident plan should identify the location and details of every fire department connection, fire pump, standpipe system, and automatic sprinkler system (**FIGURE 6-3**). Smoke management and special hazard protection systems should also be detailed on the plan. Finally, the preincident plan should include data on the protective signaling system (fire alarm).

4: Special Considerations—Special Hazards

Document special hazards in the facility and develop a plan to send the proper resources during an emergency. Special hazards might include flammable or combustible liquids, explosives, toxic or biological agents, radioactive materials, and reactive chemicals or materials. Request data on the maximum inventory of hazardous materials and highly combustible products found in the building. In addition, the preincident plan should note contact information for the facility

CHAPTER 6 Preincident Planning and Code Enforcement   **163**

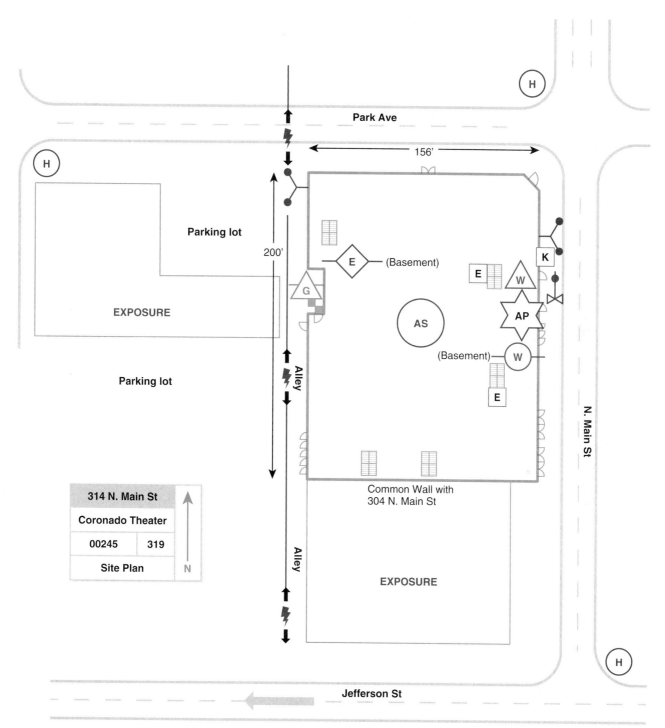

FIGURE 6-3 Example of a building preplan.
Courtesy of the Rockford, Illinois, Fire Department. Retrieved from https://rockfordil.gov/city-departments/fire/fire-prevention/commercial-pre-plans/.

hazardous materials coordinator and the location of material safety data sheets.

Some buildings contain specialized operations, processes, and hazards that can pose unique challenges during an emergency (**FIGURE 6-4**). The preincident plan should identify any area of the occupancy that contains gases or vapors that could present a hazard to emergency responders. This includes confined spaces, inert atmospheres, ripening facilities, and special equipment–treating atmospheres. Document emergency operating procedures and identify personnel who can provide technical assistance during emergency incident mitigation.

The preincident plan should include instructions for de-energizing electrical systems and isolating and securing mechanical systems. These systems should

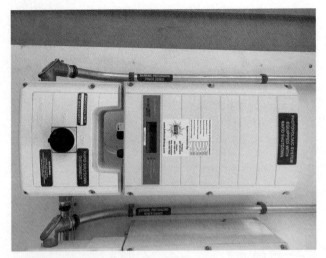

FIGURE 6-4 Photovoltaic power system.
Courtesy of Bill Larkin.

> **FIRE OFFICER TIP**
>
> **Worcester, Massachusetts, Vacant Building Marking System**
> - An X in the box on a building denotes that only exterior operations should be performed, and only enter the dangerous building if there is a known life hazard.
> - A single line in the box on a building denotes that any interior operations should be done using extreme caution.
> - An empty box on a building denotes that there was normal stability at the time of marking (**FIGURE 6-5**).

be shut down or monitored to protect emergency responders from electrocution, mechanical entrapment, or other perils. Some systems may be remotely monitored and controlled by an off-site service. Digital photos should be considered to document the system on the preincident plan—and to use for training.

Special Considerations—Vacant and Abandoned Structures. The U.S. Fire Administration recommends four actions that can reduce arson in vacant and abandoned buildings and improve fire fighter safety:

1. Monitor. Monitor all vacant properties. Properties that are secure and well-maintained, even though they are unoccupied, are not the problem. Those that have no viable owner and are accessible to unauthorized entry require immediate attention to prevent fires and other criminal activity.
2. Secure. Prevent unauthorized access to vacant and abandoned buildings by providing security or high-visibility surveillance. Unsecured vacant or abandoned buildings are intrinsically more dangerous than occupied structures. Securing them is vital. Of all fires in these types of buildings, 72 percent are of incendiary or suspicious origin. Keeping unauthorized occupants out of vacant and abandoned buildings is the key to preventing fires. Boarding a building up is one of the most effective ways to do this.
3. Inspect. Inspect and evaluate vacant or abandoned buildings to identify potential safety issues that first responders would face if they responded to a fire. The authority to inspect comes from building codes and ordinances adopted by the jurisdiction. It is important to check department policy and know what the inspector has to do to legally enter a property.
4. Mark. Mark buildings after they have been secured and inspected. This provides a visual cue to fire fighters responding to a fire, indicating that the property is vacant, has been evaluated, and was found to contain hazards. (FEMA, 2018)

Special Considerations—Mass-Casualty Events. Mass gatherings may be scheduled or spontaneous, both have the potential of requiring significant fire department support. This text uses the National Association of EMS Physicians' definition of a mass gathering event: when there are 1000 or more participants (Jaslow, Yancy, and Milsten, 2000). Mass gathering events pose four problems for fire departments:

1. Produce a workload strain on existing emergency services
2. Present a "soft" target for a terror incident
3. Provide potential for high numbers of casualties due to a natural or man-made situation
4. Difficult to quickly access all areas. (FEMA, 2010)

Planning for a scheduled mass gathering event includes considering security-related incidents. The study of past mass gathering events provides an idea of the frequency and probability of a range of events that need to be considered (Bellavita, 2007). In order of probability, here are the incidents that a fire department preplan should consider:

- Hoaxes and threats—such as a bomb threat
- Minor medical injuries

- Vehicle and pedestrian movement problems
- Fire code violations
- Weather-related problems
- Public health concerns
- Demonstrators, some potentially violent
- Natural disasters
- "Lone wolf" bombings
- Attacks on vital infrastructure
- Terrorist attack

When there is advanced notice, the planning process starts with determining the potential size and nature of the crowd activity. Is this a rally, concert, or march? This is followed by a hazard analysis that considers the type of incidents that may occur and develops the response structure for each type of incident. Finally, physical issues with the venue are identified, such as access, egress, and special hazards.

This planning is a shared effort among local law enforcement, healthcare officials, emergency medical services, emergency management, state representatives, and the Department of Homeland Security. All agencies work under the National Incident Management System (NIMS). An incident action plan (IAP) will be developed every 12 hours.

5: Identify Emergency Operation Considerations

The IAP should be based on the priorities of life safety, incident stabilization, and property conservation, in that order. The preincident plan should address the appropriate and adequate departmental response to a working fire or emergency incident. The planned location of the incident command post and emergency operations center, if provided, should be noted.

The number of required fire companies is affected by fuel loading. **Fuel load** refers to the total quantity of all combustible products found within a room or space. The fuel load determines how much heat and smoke will be produced by a fire, assuming that all of the combustible fuel in that space is consumed. The size and the shape of interior objects and the types of materials used to create them have a tremendous impact on the objects' ability to burn and on their rate of combustion. The same factors influence the rate of fire spread to other objects.

Fires develop in four stages: incipient stage, growth stage, fully developed stage, and decay stage. The incipient stage of a fire is the starting point of a fire. A fire in this stage usually involves only the object of origin. A preincident plan may identify what would be required to handle a fire at the incipient stage. For example, a loaded stream-type portable fire extinguisher may be required for the combustibles present in the building. A fire in the growth stage involves other objects in the fire. The preincident plan might indicate that a growth-stage fire would require a minimum of a 1¾-in. (44-mm) handline.

Once a fire has entered the fully developed stage, depending on occupancy and construction factors, the preincident plan might recommend that a fire in this stage be fought defensively on arrival, or it might indicate that an attack with large handlines be initiated. When the fire has consumed all of the oxygen but has retained its heat and still has fuel available to it, it has entered the decay (smoldering) stage. Plans

FIGURE 6-5 Worcester, Massachusetts, uses a marking system to identify the safety of vacant buildings.

should indicate how the building would be ventilated when this situation is encountered.

The preincident plan should identify anticipated areas of fire spread. Open pipe chases, elevator shafts, and balloon construction, for example, all contribute to vertical fire spread. Large open areas may conceal the magnitude of a fire that has only light smoke showing from the outside yet has the potential for significant horizontal fire spread. False ceilings and cocklofts may conceal horizontal fire spread.

The fire department's preincident plan should be fully coordinated with the internal evacuation or emergency operations plan. The facility should provide the fire department with an on-site liaison as soon as command is established.

6: Document Findings

The goal of preincident planning is to develop a written plan that would be valuable to both the owner of the building and the fire department if an incident occurs at that location. This plan should provide critical information that could be advantageous for responding personnel, in a format appropriate for emergency conditions. This plan identifies the address, type of occupancy, construction type, size of the occupancy, and required water flow for firefighting operations all in one section. In another section, the plan identifies the resources available, including responding units and water sources. Another section includes any special hazards or considerations, such as a fire department connection, utility controls, hazardous material storage, and rapid-entry key box locations. It is important that written plans follow a standardized format and include standardized information to allow for quick access by the officer, whether the plan is in an electronic form or a hard copy.

Many fire departments maintain preincident plans in electronic form instead of printing out hard copies. Data storage systems allow the needed information to be automatically retrieved when the dispatch system processes an alarm for a particular location. This information can be printed at the fire station, or on-board mobile data equipment can provide access to the information from a command post or while en route to the incident. Additional detailed information, such as building plans and fire alarm drawings, can be kept in a lock box or other secured area at the site.

The preincident plan includes a plot plan. The plot plan should show the relationship of the building to other buildings, streets, hydrants, utility controls, and other features. The plot plan visually represents these objects, allowing the officer to identify relevant facts quickly.

Some departments may also maintain drawings of the interior of the buildings. Like the plot plan, a floor plan allows the officer to identify considerations for a fire attack quickly. A floor plan is a drawing of the interior of the structure and is similar to an architect's blueprints. It notes stairwell locations, elevators, standpipe connections, hazardous material storage areas, fire alarm panel locations, and points of entry.

Understanding Fire Codes

Fire officers often perform inspections to enforce a fire code (or fire prevention code). A fire code establishes legally enforceable regulations that relate specifically to fire and life safety, including related subjects such as regulation of hazardous materials and process protection and operating features.

A variety of codes may be adopted by different jurisdictions. A state fire code applies everywhere in the state, whereas a locally adopted code can be enforced only within that particular jurisdiction. In many cases, a state or provincial fire code sets a minimum standard, and local jurisdictions have the option of adopting more stringent requirements. The local jurisdiction may not be able to exceed the state minimum (called a mini/max code).

Fire code requirements are often adopted or amended in reaction to fire disasters, an approach known as the **catastrophic theory of reform** (Corbett and Brannigan, 2013). Many code requirements can be traced back to disasters, such as the following:

- 1903 Iroquois Theatre (Chicago, Illinois: 602 dead)
- 1911 Triangle Shirtwaist Factory (New York, New York: 146 dead)
- 1930 Ohio State Penitentiary (Columbus, Ohio: 320 dead)
- 1942 Cocoanut Grove Nightclub (Boston, Massachusetts: 492 dead)
- 1944 Ringling Brothers–Barnum and Bailey Circus (Hartford, Connecticut: 168 dead)
- 1958 Our Lady of Angeles (Chicago, Illinois: 95 dead)
- 1977 Beverly Hills Supper Club (Southgate, Kentucky: 165 dead)
- 1990 Happy Land Social Club (New York, New York: 87 dead)
- 2003 Station Nightclub (West Warwick, Rhode Island: 100 dead)

Authority having jurisdiction (AHJ) is a term used in NFPA documents to refer to "an organization, office, or individual responsible for enforcing the requirements of a code or standard, or for approving equipment, materials, an installation, or a procedure." The AHJ for a state fire code is usually the state fire marshal. In the case of a provincial fire code, the AHJ would be the provincial fire marshal or fire commissioner. The local fire chief, fire marshal, or code enforcement official would be the AHJ for a local fire code.

The regulations contained in a fire code are enforced through code compliance inspections. These inspections could be conducted by the state fire marshal's office, by the local fire department, or by code enforcement officials who might or might not be a part of the fire department. The AHJ delegates the power to enforce the code to the fire officers, fire inspectors, and other individuals who actually conduct inspections (Solomon, 2012).

Building Code versus Fire Code

Both building codes and fire codes establish legally enforceable minimum safety standards within a state, province, or local jurisdiction. A building code contains regulations that apply to the construction of a new building or to an extension or major renovation of an existing building, whereas a fire code applies to existing buildings and to situations that involve a potential fire risk or hazard. For example, the building code might require the installation of automatic sprinklers, a fire alarm system, and a minimum number of exits in a new building; the fire code would require the building owner to maintain the sprinkler and alarm systems properly and to keep the exits unlocked and unobstructed at all times when the building is occupied. Sometimes, a fire code includes certain requirements that apply to new buildings that are beyond the scope of the building code, such as a regulation requiring the installation of fire lanes and hydrants (Corbett and Brannigan, 2013).

State Fire Codes

Most U.S. states and Canadian provinces have adopted a set of safety regulations that apply to all properties, without regard to local codes and ordinances. Where a state or provincial fire code has been established, it is generally the minimum legal standard in all jurisdictions within that state or province. The state or provincial fire marshal usually delegates enforcement authority to local fire officials.

Most states allow local authorities the option of adopting additional regulations or a more restrictive code. A few states have adopted **mini/max codes**, which mean that local jurisdictions do not have the option of adopting more restrictive regulations (Writ, 1989).

The NFPA *Fire Protection Handbook* identifies seven different organizational patterns for state fire marshal organizations in the United States. The state fire marshal may work in any of these organizations (Farr and Sawyer, 2008):

- The department of insurance
- The department of public safety
- A separate government department
- A regulatory agency
- The state police
- A cabinet-level office
- The state fire commission

Local Fire Codes

At the local level, fire and safety codes are enacted by adopting an **ordinance**, which is a law enacted by an authorized subdivision of a state, such as a city, county, or town. The local jurisdiction adopts an ordinance that establishes the fire code as a set of legally enforceable regulations and empowers the fire chief to conduct inspections and take enforcement actions. This authority can then be delegated to fire officers, fire inspectors, and other individuals.

FIRE OFFICER TIP

Life Safety Code

NFPA 101, *Life Safety Code®*, is a model code document that contains requirements specifically related to protecting the lives of building occupants, covering detailed information on means of egress. When NFPA 101 is adopted by a jurisdiction, it can be applied to both new and existing buildings.

Model Codes

Model codes are documents developed by a standards-developing organization, such as the National Fire Protection Association (NFPA), and made available for adoption by AHJs. In 1905, for example, the National Board of Fire Underwriters published the *National Building Code*. A model code is developed through a consensus process using a network of technical committees. Most jurisdictions use model codes developed by the NFPA and the International Code Council.

States and local jurisdictions may adopt a nationally recognized model code either with or without amendments, additions, and exclusions. A complete set of model codes includes a building code, electrical code, plumbing code, mechanical code, and fire code. The primary advantages of a model code are that the same regulations apply in many jurisdictions, and all of the requirements are coordinated to work together without conflicts.

The process in which a model code is adopted by a local jurisdiction may follow one of two paths. **Adoption by reference** occurs when the jurisdiction passes an ordinance that adopts a specific edition of the model code. For example, a local jurisdiction might adopt NFPA 1, *Fire Code*, by reference. The requirements specified in NFPA 1 then become local requirements that can be enforced by designated local officials. A fire officer would need to obtain a copy of NFPA 1 to read the specific requirements.

Adoption by transcription occurs when the jurisdiction adopts the entire text of the model code and publishes it as part of the adopting ordinance. For example, a city might copy the language of the code and include it in its entirety within the ordinance. A fire officer would then need to read only the city's ordinance to identify the specific requirements rather than having to find the appropriate code book (Cote and Grant, 2008).

The fire officer must know which code and which annual edition are used by the local jurisdiction. Although the model code process updates the code every 3 to 5 years, the AHJ must specifically adopt the new edition of a model code before it becomes legally enforceable. Different codes or different editions of the same code might apply to different occupancies. Some communities adopt only selected portions of one or more model codes, but defer to a state code or locally written ordinances to cover other issues. Some jurisdictions choose to maintain their own independent codes.

Retroactive Code Requirements

Regulations that applied to a particular building at the time it was built remain in effect as long as it is occupied for the same purpose. The fire officer may have to determine the specific code document, title, and year used when the building was built to determine whether a building is still in compliance with the applicable code requirements. Most codes include provisions that can be applied to buildings that were constructed before a code was adopted.

If a building is remodeled or extensively renovated, or if its occupancy use changes, most codes specify that all of the current requirements of the code must be met. New code requirements, meaning those adopted after a certificate of occupancy has been issued, do not apply unless specific language is included in the adopting ordinance. For example, Las Vegas, Nevada, established a mandatory requirement for sprinkler systems in all new residential construction in 2018 (City of Las Vegas, 2018). The ordinance did not require retroactive installation of sprinkler systems in residential structures built before that date.

On occasion, a state or local AHJ passes a code revision that is specifically identified as applying retroactively to all affected occupancies. After the 1980 Las Vegas MGM Grand Fire, which killed 85 people and injured 650, the state of Nevada required fire suppression systems to be installed in all existing casinos and hotels. Rhode Island and Massachusetts adopted retroactive sprinkler requirements for nightclubs in response to the 2003 fire in The Station nightclub, which killed 100 people and injured more than 200 patrons.

Understanding Built-in Fire Protection Systems

After clarifying means of access and egress, the status of the built-in fire protection features is the second reason for a fire company to perform inspections. Built-in fire protection systems are designed as tools to assist fire fighters in combating a fire. The officer should understand how systems work and how the codes are altered when fire protection systems are in place.

A jurisdiction's codes often allow more flexibility in the design of a building when built-in fire protection systems are included. A building with a sprinkler system can be larger and taller, the travel distance to exits can be longer, and the access for fire apparatus could be restricted. All of these trade-offs depend on a properly functioning sprinkler system.

If a fire occurs in such a building, fire fighters are depending on the built-in fire protection systems to assist them. The NFPA report, *U.S. Experience with Sprinklers,* found that sprinklers operated effectively in 92 percent of all reported structure fires that were large enough to activate sprinklers, excluding buildings under construction and buildings without sprinklers in the fire area. When sprinklers operated, they were effective 96 percent of the time, resulting in a combined performance of operating effectively in 88 percent of all reported fires where sprinklers were present in the fire area and fire was large enough to activate them. The dependability of these systems is of prime importance to fire fighters' safety. Inspection is the best method of ensuring these systems will work as intended. Los Angeles, Phoenix, and Fairfax County, Virginia, started testing existing fire protection systems in the late 1980s. In the first year, they encountered significant numbers of failures in built-in fire protection systems. To address this finding, a periodic retesting program was put in place to improve the performance of built-in fire protection services (Jones, 2013).

Water-Based Fire Protection Systems

Automatic sprinkler systems, standpipe systems, and fire pumps are the three primary components of water-based fire protection systems. An **automatic sprinkler system** consists of a series of pipes with small discharge nozzles (sprinkler heads) located throughout a building. When a fire occurs, heat rising from the fire causes one or more sprinkler heads to open and release water onto the fire (**FIGURE 6-6**). When the water starts flowing, a water flow alarm is activated. This alarm may be monitored by a central station alarm service or onsite safety/security service. Some systems are "local alarms" that sound a gong, horn/strobe device, or bell only at the outside of the building (**FIGURE 6-7**).

Depending on the usage and climate, automatic sprinkler systems may be wet pipe, dry pipe, deluge, or preaction. In a wet-pipe system, water is present in all of the pipes throughout the system. When a sprinkler opens, water is discharged immediately. In general, wet-pipe systems require less maintenance than dry-pipe systems. Because of the faster reaction, fewer sprinklers are activated to control most fires.

Dry-pipe systems are used in locations where a wet-pipe system would be likely to freeze, such as unheated storage facilities and parking garages. Instead of containing water, these pipes are filled with compressed air or nitrogen until a sprinkler opens. When the air pressure drops, the dry-pipe valve opens and water is released into the system. Dry-pipe systems require higher maintenance because activation of the sprinkler system requires the entire sprinkler system to be drained. Sometimes, instead of installing a dry-pipe system, an antifreeze solution is added to the water in a wet-pipe system to protect an unheated area (e.g., a freezer or a loading dock). This type of system is being phased out due to environmental concerns and costs.

Deluge systems are a special version of a wet- or dry-pipe system for locations in which large quantities of water are needed to control a fast-developing fire quickly. Deluge systems are most often found in

FIGURE 6-7 Some systems are "local alarms" that sound a gong, horn/strobe device, or bell only at the outside of the building.
© Jones & Bartlett Learning. Photographed by Kimberly Potvin.

ordnance plants, aircraft hangers, and occupancies with flammable liquid hazards. All of the sprinklers are open and ready to discharge water as soon as the control valve opens. A combination of smoke, flame, and/or fire detectors are used to sense a fire and trigger the system.

Preaction sprinkler systems are similar to dry-pipe systems, but include a separate detection system that triggers the dry-pipe valve and fills the sprinkler pipes with water. At this point, it becomes equivalent to a wet-pipe system. A preaction system is designed to reduce the risk of water damage due to accidental sprinkler discharge or a broken pipe.

Some sprinkler systems are designed to discharge protein or aqueous film-forming foam as the extinguishing agent. They are used in high-hazard areas

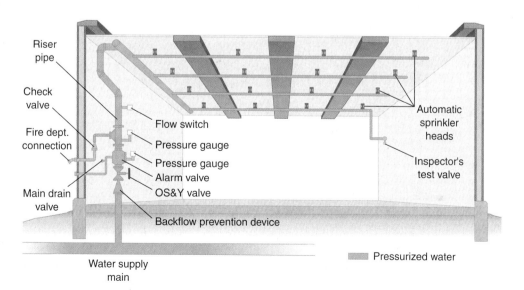

FIGURE 6-6 The basic components of an automatic sprinkler system.
© Jones & Bartlett Learning.

where the contents to be protected are flammable liquids, such as fuel storage and chemical process facilities. These complex systems require more frequent inspections and are custom designed for the storage area, processing plant, or fueling facility that is protected. Often there is an onsite or nearby technician who has been trained in the maintenance of these systems (Isman, 2008).

A **standpipe system** provides the ability to connect fire hoses within a building. Standpipes are an arrangement of piping, valves, hose connections, and allied equipment that allow water to be discharged through hoses and nozzles to reach all parts of the building. As with automatic sprinkler systems, dry standpipes may be installed in unheated areas. Standpipes are subdivided into three classes based on their expected use:

- Class I provides 2½-inch (65-mm) male coupling, intended for use by fire department or fire brigade members trained in the use of large hose streams.
- Class II provides a 1½-inch (38-mm) hose coupling with a preconnected hose and nozzle in a hose station cabinet. The hose is designed for occupant use.
- Class III provides both 1½-inch (38-mm) and 2½-inch (65-mm) connections. The 1½-inch connection may have a preconnected hose line that can be used by the occupants until the fire department arrives.

During building surveys and inspections, you should pay particular attention to the condition of the standpipe system; it is one of the few built-in fire protection devices specifically designed to help fire fighters (Hague, 2008). Fire officers should identify fire hose access and ensure proper calibration of pressure-regulating devices. A pressure-regulating device limits the discharge pressures from standpipe hose outlets. During a 1991 high-rise office fire at One Meridian Plaza in Philadelphia, fire fighters discovered that the pressure-regulating devices were improperly set, and only a weak flow could be obtained from each outlet. Usually found on the lower floors of high-rise standpipes, the devices were installed on floors 26 through 30 of One Meridian Plaza and their poor performance impeded firefighting operations. Three fire fighters died while the fire consumed the 22nd through 30th floors in 19 hours (FEMA, 1991).

Fire Pumps

Fire pumps increase the water pressure in standpipe and automatic sprinkler systems (**FIGURE 6-8**). They are designed to start automatically when the water

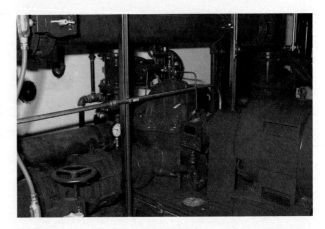

FIGURE 6-8 A fire pump may be needed to maintain or increase water pressure in standpipe and automatic sprinkler systems.
© Jones & Bartlett Learning. Courtesy of MIEMSS.

pressure drops in a system or a fire suppression system is activated. A field inspection of such a device generally is confined to a visual inspection to confirm that the pump appears to be in good condition and is free of physical damage. The fire officer should confirm that the fire pump has passed the annual performance test.

Special Extinguishing Systems

Four types of special extinguishing systems may be used in various structures: carbon dioxide, dry or wet chemical, Halon/clean agent, or foam. Make a note of these specialized systems in the preincident plan. In general, the ongoing code compliance inspection consists of visual inspection, looking for physical damage, and confirming that the safety seals are intact, as well as verification of the documentation for required inspection, testing, and maintenance. A safety seal is placed on any handle that would activate the system. If this safety seal is broken, the system could have been discharged.

Carbon Dioxide

Carbon dioxide systems are fixed systems that discharge carbon dioxide from either low- or high-pressure tanks, through a system of piping and nozzles, either to protect a specific device or process (e.g., a printing press) or to flood an enclosed space. Carbon dioxide extinguishes fire by displacing oxygen. This gas is heavier than air, so it settles in low spaces. Fixed systems are generally required to comply with NFPA 12, *Standard on Carbon Dioxide Extinguishing Systems*. Your preincident plan should require the use of self-contained breathing apparatus (SCBA) and atmospheric monitoring to evaluate oxygen levels if a fixed system has become activated.

Dry or Wet Chemical

Fixed chemical extinguishing systems discharge a chemical extinguishing agent through a system of piping and nozzles. These systems can be found protecting commercial cooking devices and industrial processes where flammable or combustible liquids are used. Observe the condition of the nozzles. Many contain a protective cap that is intended to protect the nozzle from becoming obstructed with cooking grease.

Wet chemical systems are preferred for protecting cooking equipment. The wet chemical agent reacts with hot grease to form a foam blanket, reducing the release of combustible vapors. The foam blanket cools the grill and reduces the possibility of a rekindle. Dry chemical systems leave a residue that is difficult to clean up.

Both types of systems may be activated either of two ways: (1) a fusible link that melts on flame contact or (2) a manual pull station. Activation of the system also turns off the cooking device by closing the cooking fuel valve or turning off the electricity.

Halon

Halon 1301 was the extinguishing agent of choice for fire protection in computer rooms and to protect electronic equipment from the 1950s to the 1990s. Based on the weight of the agent, Halon 1301 is about 250 percent more efficient than carbon dioxide for extinguishing fires (Ford, 1975). Unfortunately, Halon also depletes the ozone layer in the atmosphere. Since 1994, Halon has not been allowed to be manufactured or imported into the United States, however legacy systems may still be recharged (DiNenno and Taylor, 2008).

NFPA 2001, *Standard on Clean Agent Fire Extinguishing Systems,* covers the use of alternative agents that have replaced Halon systems in protecting electrical and electronic telecommunications systems. Most of these systems are designed to flood a room or an enclosure and, like Halon, are toxic. The room or enclosure must be sealed, and the occupants must leave before the agent is discharged. The system can be automatically or manually fired; both methods include a pre-alert warning for the occupants to leave the room before the agent discharges. The preincident plan should identify the chemical used, the duration of the discharge (10 seconds to 1 minute), the enclosure protected, the automatic activation sequence, and the location of the manual activation station.

Foam Systems

A low-expansion foam system is used to protect hazards involving flammable or combustible liquids, such as gasoline storage tanks. These systems discharge foam bubbles over a liquid surface to create a smothering blanket that extinguishes the fire and suppresses vapor production. High-expansion foam is used in areas where the goal is to fill a large space with foam, thereby excluding air from the area and smothering the fire (Scheffey, 2008).

Fire Alarm and Detection Systems

A fire alarm system consists of devices that monitor for a fire and notify the appropriate personnel. Manual fire alarm boxes, smoke detectors, or heat detectors may activate the fire alarm system. The alarm also may be activated by the water flow or pressure switch in a sprinkler system. The activated system notifies the appropriate personnel, including the building occupants, with audible and visual signals. These signals may be transmitted outside the building and to the fire department or a central station monitoring firm.

Understanding Fire Code Compliance Inspections

The objective of a **fire code compliance inspection** is to determine whether an existing property is in compliance with all of the applicable fire code requirements. Some codes call this a maintenance inspection. The goal for the inspector is to observe the housekeeping to ensure that no fire hazards exist and to confirm that all of the built-in fire protection features, such as fire exit doors and sprinkler systems, are in proper working order.

The fire department's authority and responsibilities for conducting code compliance inspections are usually included in the ordinance that adopts the local fire code. The responsibility for code enforcement is usually assigned to the fire chief or fire marshal. The fire chief can delegate the actual code enforcement activities to different individuals or units within the department.

In many fire departments, fire officers and fire inspectors who are assigned to a fire prevention bureau or code enforcement division conduct inspections of specific types of properties. In departments where fire companies conduct code compliance inspections, an individual or a group is usually assigned to coordinate these activities and provide technical assistance.

Fire Company Inspections

The purpose of conducting a fire inspection is to identify hazards and to ensure that the violations are corrected. This process helps the fire department be proactive in preventing fires. One method of maximizing the effectiveness and efficiency of inspections is for fire companies to conduct them.

Since the inception of *NFPA Quarterly* (published for over 100 years), NFPA has advocated for company-level inspections. Through inspections, local fire companies have the opportunity to become familiar with their response areas and take actions to prevent the start of fires and, if a fire starts, to minimize the fire spread and risk of destruction of the neighborhood.

Before conducting an inspection, it is important to understand the scope of code enforcement authority that is delegated to a fire officer in your jurisdiction. There is wide diversity in the scope and authority of inspections conducted by fire companies. In some areas, fire suppression companies are authorized to enforce all fire code **regulations**; one metropolitan city authorizes fire officers to issue corrective orders that would require the installation of automatic sprinklers in a business under renovation. In other fire departments, fire suppression companies are assigned to inspect only certain types of occupancies, while the fire marshal's office or fire prevention division inspects other types of occupancies. Some fire departments can provide only safety recommendations to citizens and property owners, whereas others can issue citations and compliance orders. The fire officer needs to determine the source and the scope of his or her code enforcement authority before conducting any fire safety inspection, as well as which particular codes are used by the jurisdiction (Diamantes, 2015).

Except in case of fire emergencies, a fire officer generally cannot enter private property without the permission of the owner or occupant. Most fire codes include a section that authorizes code enforcement officials to enter private properties at any reasonable time to conduct fire and life-safety inspections; however, the scope of this authority usually depends on the type of property involved. In most cases, the permission of the owner or occupant is required for entrance into a dwelling unit, whereas access to public areas is less restricted. The fire code often contains a section that, if necessary, allows for the issuance of a court order requiring the owner or occupant to allow the fire department agent to enter the occupancy to conduct an inspection.

Classifying by Building or Occupancy

Many of the code requirements that apply to a particular building or occupancy are based on its classification. The codes may classify a building by construction type, occupancy type, and use group. In turn, fire officers need to know how to classify a building before code requirements can be applied. In addition to these three classifications, local zoning ordinances often regulate where certain types of occupancies or buildings are permitted.

Construction Type

The building itself is classified by **construction type**, which refers to the design and the materials used in construction. NFPA 220, *Standard on Types of Building Construction,* addresses building construction. The type of building construction has a significant influence on firefighting strategy. Construction type is a fundamental size-up consideration when firefighting strategy for a burning structure is being determined.

The most commonly used model codes classify construction into five basic types:

- Type I Construction—Fire resistive: The construction elements are noncombustible and are protected from the effects of fire by encasement, using concrete, gypsum, or spray-on coatings (**FIGURE 6-9**). Depending on the model code used, Type I construction is divided into subtypes, based on the level of fire protection provided. The level of protection is described by the number of hours a building element can resist the effect of fire. Features of Type I construction include the following:
  - Type I is the most durable and lasting structure. Extensive fire suppression operations may be carried out in such a building before it will collapse.
  - This type of construction often uses compartmentation instead of fire sprinklers to control fire spread, which in turn may create unsurvivable interior temperatures

FIGURE 6-9 A Type I building.
© John Foxx/Alamy Stock Photo.

FIGURE 6-10 A Type II building.
© Dennis Tokarzewski/Shutterstock.

FIGURE 6-11 A Type III building.
© Ken Hammond/USDA.

when fire fighters are trapped on the fire floor. Metal structural elements can fail due to age and rust.

- Type II Construction—Noncombustible: The structural elements can be made from either noncombustible or limited-combustible materials (**FIGURE 6-10**). Like Type I, Type II construction is classified into subdivisions, based on the level of fire resistance. Although the buildings are assembled from noncombustible components, the structural elements have limited or no fire resistance. A Type IIA structural frame is expected to resist fire for 1 hour, whereas the structural frame in a Type IIB building is not expected to resist the effects of fire. A strip shopping center with cinderblock walls, unprotected steel columns, and steel bar joists supporting a steel roof deck is an example of a Type IIB building. Characteristics of Type II construction include the following:
  - Type II is a common 20th-century construction method and the type of building that Francis Brannigan referenced when describing the 20-minute interior firefighting rule in 1971.
  - Such a structure is durable, but is not a legacy building—it will require replacement in 30 to 40 years. This type of building is frequently updated with Type V structural elements.

- Type III Construction—Ordinary: The exterior load-bearing walls of the building are noncombustible masonry (**FIGURE 6-11**). A **masonry wall** may consist of brick, stone, concrete block, terra cotta, tile, adobe, or concrete. The interior structural elements may be combustible or a combination of combustible and noncombustible. As with Types I and II, different levels of fire protection are possible in Type III buildings. The structural frame of a Type IIIA building is protected, which means it is encased in concrete, gypsum, or spray-on coatings and is expected to have a fire-resistive rating of 1 or 2 hours. A Type IIIB structural frame is unprotected and has no fire resistance rating. Features of Type III construction include the following:
  - Type III was used to build commercial, multiple-family, and mercantile types of buildings through the 1980s. Brannigan classifies these buildings as "Main Street USA."
  - This kind of building is usually no higher than four stories; it was designed to preserve the load-bearing walls if fire consumed the building. The connection between the floor and the load-bearing wall is designed to pull out without damaging the wall.

- Type IV Construction—Heavy timber: The exterior walls are noncombustible (masonry), and the interior structural elements are unprotected wood beams and columns with large cross-sectional dimensions (**FIGURE 6-12**). Mill construction, which was used in many New England textile buildings built in the 1800s, is an example of heavy timber construction. Mill construction features massive wood columns and wood floors. Characteristics of Type IV buildings include the following:
  - As durable as Type I structures, most surviving Type IV buildings have been converted to residential, mercantile, or mixed-use spaces.
  - A well-seated fire in a nonsprinklered Type IV building may exceed the capability of the municipal water supply.

FIGURE 6-12 A Type IV building.
Courtesy of APA - The Engineered Wood Association.

FIGURE 6-13 A Type V building.
© Jones & Bartlett Learning. Courtesy of MIEMSS.

- Type V—Wood frame: The entire structure may be constructed of wood or any other approved material (**FIGURE 6-13**). Sometimes, a masonry veneer is applied to the exterior, but the structural elements consist of wood frame. Characteristics of Type V buildings include the following:
  - Type V buildings are the most common structures and include single-family and multifamily residential, and mercantile and low-rise commercial buildings.
  - Such a structure can injure or kill the first-arriving fire fighters if they enter the structure and fall through a fire-weakened floor. Underwriters Laboratory (UL) research in 2011 showed Type V residences achieving flashover in 3:30 to 4:45 minutes, compared to flashover times of 29:30 to 34:15 minutes for legacy residences (Kerber, 2012).
  - Type V construction techniques and approach will be the foundation for future innovative building methods and materials.

It is often difficult to determine the type of construction from the exterior of a building. Many low-rise offices, apartment buildings, and residences appear to be ordinary construction, but are actually wood-frame buildings with a brick, stone, or masonry veneer applied to the exterior walls. This type of wall covering does not enhance the strength or fire resistance of the building (Corbett and Brannigan, 2013).

Occupancy and Use Group

The **occupancy type** refers to the purpose for which a building or portion of a building is used or is intended to be used. The code requirements are determined by the structure's **use group**. Occupancies are classified into use groups based on the characteristics of the occupants, the activities that are conducted, and the risk factors associated with the contents. Within each occupancy type are dozens of use groups.

Assembly

An assembly occupancy is used for the gathering of people for deliberation, worship, entertainment, eating, drinking, amusement, or awaiting transportation. This classification may be further divided into more specific types of assemblies. Examples of assembly-type occupancies include:

- Churches
- Taverns or bars
- Nightclubs
- Basketball arenas
- Restaurants
- Theaters

Business

A business occupancy is used for account and record keeping or transaction of business other than mercantile (**FIGURE 6-14**). Examples of business occupancies include:

- Dental offices
- Banks
- Architects' offices
- Hair salons
- Colleges and universities
- Doctors' offices
- Investment offices
- Insurance offices
- Radio and television stations

FIGURE 6-14 Example of a business occupancy.
© Konstantin L/Shutterstock.

FIGURE 6-15 Example of an industrial occupancy.
© John A. Rizzo/Photodisc/Getty Images.

Educational

An educational occupancy is used for educational purposes through the 12th grade. Typically, this designation refers to schools. Educational occupancy may also cover some daycare centers for children older than 2½ years.

Industrial

In an industrial occupancy, either products are manufactured or processing, assembling, mixing, packaging, finishing, decorating, or repair operations are conducted (**FIGURE 6-15**). This includes occupancies such as the following:

- Automobile assembly plants
- Clothing manufacturers
- Food processing plants
- Cement plants
- Furniture production facilities

Institutional

An institutional occupancy is used for purposes of medical or other treatment or for care of four or more persons, where such occupants are mostly incapable of self-preservation due to age, physical or mental disability, or security measures not under the occupants' control. Such buildings would include hospitals, nursing homes, and correctional facilities.

A detention and correctional occupancy is used to house four or more persons under varied degrees of restraint or security, where such occupants are mostly incapable of self-preservation because of security measures not under the occupants' control. This category could include prisons, jails, and detention facilities.

Mercantile

A mercantile occupancy is used for the display and sale of merchandise (**FIGURE 6-16**). This group includes the following:

- Retail stores
- Convenience stores
- Department stores
- Drug stores
- Shops

Residential

A residential occupancy provides sleeping accommodations for purposes other than health care, detention, or corrections. This type of occupancy includes five subcategories:

- One- and two-family dwelling units: Buildings that contain no more than two dwelling units with independent cooking and bathroom facilities
- Lodging or rooming houses: Buildings that do not qualify as one- or two-family dwellings but that provide sleeping accommodations for a total of 16 or fewer people on a transient or permanent basis, without personal care services, with or without meals, but without separate cooking facilities for individual occupants
- Hotels: Buildings under the same management in which there are sleeping accommodations for more than 16 persons and that are primarily used by transients for lodging with or without meals
- Dormitories: Buildings in which group sleeping accommodations are provided for more than 16 persons who are not members of the same family in one room, or a series of closely

FIGURE 6-16 Examples of mercantile occupancies.
Top: © Jones & Bartlett Learning. Photographed by Kimberly Potvin; Bottom: © Pavel L Photo and Video/Shutterstock.

associated rooms, under joint occupancy and single management, with or without meals, but without individual cooking facilities
- Apartment buildings: Buildings containing three or more dwelling units with independent cooking and bathroom facilities

Storage

A storage occupancy is used primarily for storing or sheltering goods, merchandise, products, vehicles, or animals. Examples include:
- Cold storage plants
- Granaries
- Lumber yards
- Warehouses

Mixed Occupancies

A mixed-use property has multiple types of occupancies within a single structure. An example would be an old commercial building that has been renovated to include multifamily residential loft apartments on the second and third floors and a bakery on the first floor.

Special Properties

Some structures do not fit neatly into the regular occupancy categories. These occupancies can be placed in a miscellaneous category that represents unusual structures, such as towers, water tanks, and barns.

Once the use group classification is known, the allowable height, floor area, and construction type can be found in the building code. The building code also includes the requirements for the number of exits, the maximum travel distance to an exit, and the minimum width of each exit. Requirements for the installation of fixed fire protection systems in new buildings appear in the building code as well. Additional requirements for fire protection systems are sometimes included in the fire code, particularly for existing buildings.

Code enforcement starts by confirming that the occupancy is used for the original, approved purpose. For example, if a tenant space in a shopping center has been converted from a clothing store to a bar, the exit requirements and the built-in fire protection system requirements could be different.

NFPA 704 Marking System

Buildings with significant quantities of hazardous materials may be required to use a marking system. The most widely recognized standard is NFPA 704, *Standard System for the Identification of the Hazards of Materials for Emergency Response*. This marking system consists of a color-coded array of numbers or letters arranged in a diamond shape (**FIGURE 6-17**).

Each color represents a specific type of hazard. Blue represents health hazards, red represents flammability hazards, and yellow represents the material's instability hazard. Within each color diamond, there is a number from 0 to 4 that represents the relative hazard of each. The number 0 means that the material poses essentially no hazard, whereas a 4 indicates extreme danger. For example, on the exterior of a building, there might be a marking with a 2 inside the blue area, a 4 inside the red area, and a 0 inside the yellow area. This means that inside the building, there are materials that pose a moderate health hazard, a severe flammability hazard, and no real instability hazard. The last quadrant of the diamond is white and is used to indicate special hazards. This area could include letters or numbers. For example, it might include a placard that indicates to the responder that water-reactive material is present.

FIGURE 6-17 Document any special hazard in the facility.
Reproduced from NFPA 704-2017, *Standard System for the Identification of the Hazards of Materials for Emergency Response*, Copyright © 2017, National Fire Protection Association.

This marking system also requires labels to be affixed to containers inside the structure to indicate the hazards of the substance. Such labels might identify materials as "corrosive," "flammable," or "poison." In general, the system requires a NFPA 704 marker at each entrance to the building, on doorways to chemical storage areas, and on fixed storage tanks.

FIRE OFFICER TIP

Neatness Counts

Make sure that your uniform is neat and clean. Some areas require wearing a safety helmet to complete a life-safety inspection or preincident plan. Many members utilize their fire helmet to meet this safety requirement.

Preparing for an Inspection

Preparation is required before conducting a code enforcement inspection. The assumption in this section is that the fire officer has the responsibility, authorization, and training necessary to conduct this activity. The focus here is how to prepare to conduct such an inspection.

The fire officer should regard the owner or occupant as a professional partner. The interaction begins with the fire officer preparing for the inspection by reviewing the applicable code provisions and any information on previous inspections that can be found in the occupancy file. For the business owner or occupant, a visit by the local fire company is a major and disruptive event. Some jurisdictions send a preinspection or self-survey form to the business or property a few weeks before an official fire department inspection will occur.

Reviewing the Fire Code

Before conducting an inspection, you should review the sections of the fire code that apply to the specific property. In some cases, you may discover unanticipated situations during an inspection that require you to research an additional section of the code. The NFPA's *Fire and Life Safety Inspection Manual* provides general information about a large variety of processes, equipment, and systems.

Review Prior Inspection Reports, Fire History, and Preincident Plans

Review the inspection records from previous inspections. Doing so might detect a trend or a chronic problem. If a certain facility always requires two follow-up inspections to correct the violations that are noted during inspections, then that fact might indicate that the fire officer needs to turn up the salesmanship. For example, when calling to schedule the inspection, the fire officer might take the opportunity to review the violations noted during the previous inspection and encourage the owner to check those items.

Look at the fire history for the occupancy. If a barbeque restaurant has a history of a flue or hood fire every 6 to 9 months, you should plan to concentrate on the grill and hood system during your inspection. In this type of case, you should think about which measures you might recommend to reduce the frequency of accidental fires.

Bring a copy of the preincident plan. A code enforcement inspection is an excellent time to update contact information, such as names, e-mail addresses, and phone numbers. Use the preincident plan to identify any building modifications and additions that have been made to the occupancy.

Coordinate Activity with the Fire Prevention Division

Coordination is needed in departments that implement both a fire prevention division and a company-level inspection program. Some jurisdictions assign a list of occupancies to be inspected by each fire suppression company on a monthly or quarterly basis. In buildings where a process, storage, or occupancy is required to have an annual fire prevention permit, the local fire company's **ongoing compliance inspection** could be

Voice of Experience

One Friday I was informed by a fire suppression contractor that the fire suppression system for a large warehouse in our district would be out of service for repairs due to a leak near one of its onsite private fire hydrants. Periodically throughout the day, from the afternoon to the early evening, I received status reports on the progress of the repairs.

Late in the evening, I stopped by to see the actual repair work, as I had not received any confirmation that the project was completed and that the system had been placed back in service. Upon my arrival, the fire suppression contractor said the water line feeding the fire hydrant had been repaired and was ready for the water line to be charged with pressure. There was a large, excavated hole exposing only some of the plastic pipe and hardware on the line. The workers at the site stated that they were unsure/confused about this piping grid because another pipe was exposed and was actually connected to this water line in this same hole.

During the charging of the underground fire service water line, it ruptured, shooting a large, violent stream of water up in the air, flooding the hole immediately, and launching debris everywhere. Even the most mundane, routine event can turn into an incident of injury.

The previous work was now negated, and the entire fire suppression service for the warehouse was out of service for the next 12 hours while further repairs were undertaken. The confusion over the pipe was resolved through a set of plans produced by a supervisor who was called to the site. The pipe was part of the entire grid that provided additional needed water supply to the fire pump on site. The actual cause of the rupture was that the water line's valve connection used only locking lugs (e.g., Mega Lugs) and had not been secured with any additional thrust blocking and/or tie rods.

This incident illustrates why all relevant drawings should be reviewed prior to work being started on a repair. Third-party inspectors such as a private contractor, a fire official, a building official, and/or a utility official should be available for inspection of all repairs.

If the workers and I had not moved away from the hole during the charging of this water line, we would have certainly suffered significant injuries. It is crucial to understand and respect the power of water, especially under pressure, which may catch fire fighters and others off guard. Charging handlines, opening hydrants, and performing hose testing are just some of the scenarios that can result (and have resulted) in unexpected failures and injuries.

Situational awareness is an attribute that is an absolute must in the fire service, whether you are a line fire fighter, a fire officer, or "just the fire prevention guy." Always prepare for the worst—maintain safe practices and wear your safety gear and apparel.

Dave Belcher
President, Ohio Society of Fire Service Instructors
Lieutenant, Violet Township Fire Department, Fire Prevention Bureau
Pickerington, Ohio

scheduled for six months after the fire prevention division issues the permit.

The AHJ usually determines the frequency of inspections for each type of occupancy. For example, daycare facilities might be required to have an annual inspection from the fire marshal's office, whereas business occupancies might be inspected by a local fire company once every two or three years. The fire officer commanding the fire suppression company is expected to follow the inspection schedule and coordinate his or her actions with the fire marshal's office.

Arrange a Visit

It is good practice to contact the owner or business representative to schedule a day and a time for the fire safety inspection. Many businesses are cyclical, and some days or months are more difficult to accommodate a fire department inspection than others. For example, early April may not be the best time to inspect an accounting firm's office because it is the peak of accountants' annual work cycle.

In some cases, a time that is inconvenient for a business is an important time in terms of fire safety. Between October and December, retail stores are packed with extra stock for the holiday buying season. Some retail businesses earn 30 percent or more of their annual revenue during this period. A fire safety inspection of a store in an enclosed mall on December 1st may reveal boxes stored from floor to ceiling, obstructing the sprinklers. Empty cardboard boxes may be pushed up tight against the electrical panel, instead of maintaining a 36-in. (91-cm) clearance. The electrical panel may be hotter than usual because of the extra power needed to run the holiday displays. Trash may have piled up in the storage room and blocked the rear emergency exit.

Assemble Tools and References

Once you have reviewed the information on the occupancy to be inspected and the applicable fire code, you are ready to perform the inspection. In addition to your knowledge, you will need to bring equipment to assist you during the visit. The following tools may be helpful:

- Computer or tablet
- Pen or pencil
- Inspection form
- Graph paper
- Clipboard
- Hand light
- Camera
- Coveralls
- Measuring device, tape measure, or walking meter
- Fire department business cards
- Reference code books or files

Plan to wear your safety shoes and bring your fire helmet, eye protection, and protective gloves if you will be inspecting an area that is under renovation or an occupancy that requires protective equipment.

FIRE OFFICER TIP

Conducting Inspections and Surveys

A four-step system that meets the organization's need for fire company inspections and preincident surveys will help the fire company focus on what is important.

- Step 1: Schedule the inspections based on use group or occupancy. For example, inspections of all of the service stations and auto repair facilities in the district might be scheduled for November.
- Step 2: Hold an in-station drill to review the fire code sections applicable to the types of items that might be found during a typical code compliance inspection in that type of occupancy. Continuing the preceding example, in October, the fire officer can schedule an in-station drill to review the fire code requirements for a service station or automotive repair shop and also delegate a senior fire fighter to schedule the inspections.
- Step 3: Break the fire company into two-person inspection teams. One team member will focus on the code enforcement, while the other will update the preincident survey form.
- Step 4: The company reviews the findings. This review is an ideal time to discuss any changes that are required on the incident action plans for the specific facilities, based on the inspections. For example, if an auto repair shop has added a vehicle spray booth, the plan for that occupancy should be revised.

Conducting the Inspection

Some departments require that the entire fire company perform the fire inspection together. Business owners, in turn, may complain about the disruption that occurs when four to six fire fighters come into a business. Other departments deploy the fire company in two- or three-person teams to conduct inspections in adjacent occupancies. Make sure you know which procedures and practices your organization follows.

The conduction of a fire inspection should be approached in a systematic fashion. The fire officer should begin every inspection in the same manner. First, circle the area as you park the apparatus to get a general overview of the property. Second, meet the property owner or manager to let him or her know that you have arrived and will begin the inspection. Third, begin the inspection at the exterior of the building and work systematically throughout the inside of the building, beginning at the lowest level and working up. Fourth, conduct an exit interview with the contact person. Finally, write a formal report on the inspection.

When you arrive, circle the property and observe all four sides of the building. During this review, look for any obvious access or storage problems and any new construction since the last visit. Confirm the locations of hydrants, sprinkler/standpipe hook-ups, and other outside features with the preincident survey sheet.

Park the fire apparatus in a location that does not disrupt the business and allows the fire company to respond if they are dispatched to an emergency. Some departments require that one fire fighter remain with the apparatus both to listen to the radio and to protect the rig from vandals or thieves. The apparatus should not be parked in a fire lane. It is hard to convince the business owner to comply with the fire code when the fire engine is violating the code by sitting in a fire lane when there is no emergency.

Meet with the Representative

Enter the business through the main door and make contact with the appropriate representative. If you have called ahead to schedule this visit, you already have a name and office location. Introduce your crew and briefly explain the goal of this visit. This is a great time to review and update all of the contact names, phone numbers, and information found in your preincident survey sheet. Ask to have a representative with the appropriate access cards and keys accompany the inspection team.

Inspecting from the Outside In, Bottom to Top

A fire company–level ongoing compliance inspection confirms the built-in fire protection systems are fully operational and makes sure the area is free of fire ignition sources. This type of assessment begins with a walk around the exterior of the premises. Confirm that the address is present and properly identified on both the front and the rear of the building. If the building has a fire department connection, is it free from foreign objects inside the piping, and are the threads in workable condition? Verify that all means of access and egress are clear and in proper operating order—there is no more important issue in this type of inspection. Exit problems require immediate correction.

The fire officer should ensure that the location of dumpsters does not present a fire hazard. Outside storage buildings should be checked for compliance, particularly with the storage of hazardous materials. Are required markings, like those specified in NFPA 704, present?

After the outside walk-around, go to the basement, where you will likely find utility rooms, fire pumps, fire protection system control valves, backup generators, and laundry rooms. These areas are susceptible to improper storage of combustibles near electrical panels, open junction boxes, improperly stored hazardous material, and blocked exits.

The fire officer should systematically work through the building, checking for fire safety issues. Like the exterior and the basement, the ability of occupants to exit a building quickly is a primary concern. Problems may include inventory stock that is blocking exits, locked doors, and exit and emergency lights that are not working.

Look for conditions that are prone to starting fires: open electrical wiring, use of extension cords, and storage of combustibles too close to heat sources. Watch for storage of flammable liquids, as well as the use of candles, portable heating units, and fireplaces or wood stoves.

Last, consider items that would assist in extinguishing a fire once it has started. These considerations would include the clearance between sprinklers and objects, the condition of smoke and heat detectors, and the operability of fire extinguishers. The fire officer should confirm that fire doors are not propped open and are in operable condition. If the structure is equipped with a commercial kitchen, the hood and duct system and fire suppression system should be checked for compliance.

Exit Interview

It is important to wrap up your ongoing compliance inspection/preincident survey by meeting with the owner or designated representative to review what was found and issue any required correction orders. Remember that one of your roles is to be an ambassador for the fire department and an advocate for the business. A written report needs to be completed, with one copy going to the occupant.

Writing the Inspection/Correction Report

Several different types of inspection/correction reports are possible. Many use a check-off system, in which the officer puts a check next to the corresponding deficiency. This allows the citation to have the appropriate code without the officer having to look up each violation. If a reinspection reveals that violations have not been corrected, a notice of hazard is described in a narrative rather than noted in a check-off form.

The report needs to clearly describe any needed corrections and to quote the appropriate sections of the code or ordinance. Some communities have developed report forms that list the most commonly occurring violations.

Once the report is complete, you should review it with the owner or representative. If violations are found during the inspection, you must explain to the responsible individual what needs to occur to correct each problem. You must retain the original report for follow-up purposes. The owner or representative should sign the form and keep a copy of the report. The report should also be forwarded to the fire prevention division.

Life-threatening hazards, such as locked exits, must be corrected immediately. Less critical issues can be corrected within a reasonable time period, generally 30 to 90 days. You should arrange for any needed follow-up inspections to verify that the corrections have been made. Here is the procedure used by one fire department for non–life-threatening code enforcement issues:

- Fails first fire company–level inspection: Schedule follow-up inspection in 30 days.
- Fails second fire company–level inspection: Follow-up inspection in 15 to 30 days.
- Fails third fire company–level inspection: Issue sent to the fire prevention division for resolution. Depending on the nature of the issue and the history of this occupancy, the fire prevention division contacts the owner and makes an inspection within 2 to 30 days.
- Fails first fire prevention division inspection: The fire marshal issues a notice of violation or correction order. The time given to comply with the notice or order is determined by the situation and by local regulations. The owner can file for appeal or equivalency within 10 days of the first fire prevention division inspection.
- Fails second fire prevention division inspection: Second notice issued by the fire marshal. Generally, the time to comply is shorter. The owner is warned of more severe consequences if the issues remain unresolved.
- Fails third inspection: Follow-up inspection in days; the building representative may be subject to a misdemeanor charge punishable by fines or jail, or both. The occupancy may be required to cease operations until the matter is resolved.
- Fails fourth inspection: Fines and legal action are initiated.

General Inspection Requirements

The fire code includes general fire and life-safety requirements that apply to every type of occupancy. These items should be checked during every inspection. The general requirements include properly operating exit doors and unobstructed paths to egress travel. The general requirements also require built-in fire protection systems, such as automatic sprinklers and fire detection systems, as well as portable fire extinguishers, to be regularly inspected and properly maintained in operational condition. In addition, the fire code includes general precautions that are required with the use or storage of hazardous, flammable, or combustible goods. Some fire codes include topics related to emergency planning and conducting of fire drills, as well as features such as rapid-entry key boxes for fire department use.

Access and Egress

In every inspection, the fire department must confirm that there is sufficient means of egress for the occupants. These provisions of the fire code are often violated because improper storage causes the exits to be obstructed.

Proper storage practices mean that the goods for sale or the items used by the business or industrial process are safely stored in accordance with the fire code. In some cases, a fire company may discover a hazardous condition due to excessive, illegal, or dangerous storage practices. This situation requires immediate corrective action and, if available, assistance from the fire prevention division or the fire marshal's office.

Exit Signs and Emergency Lighting

Exit signs and emergency lighting are equally important considerations in occupant evacuation. Exit signs indicate the direction of exit to occupants during a fire. Emergency lights light the path to the exit. These features are particularly important when the occupants may be unfamiliar with the location of exits other than the main entrance, such as at theaters and nightclubs. Often, officers find exit lights that are burned out or for which the backup battery no longer works. The same is true for emergency lights. The occupant is required to maintain documentation of monthly checks performed on these systems.

Portable Fire Extinguishers

Almost every building has portable fire extinguishers, which are provided for the occupants to control incipient fires. The fire code describes requirements for the size, type, and locations of extinguishers required by NFPA 10, *Standard for Portable Fire Extinguishers*, for the details. Verify extinguishers are of the appropriate size and type. In addition to visually inspecting each extinguisher for physical damage, confirm that the instructions are visible, the safety or tamper seal is present, there is current inspection and testing documentation, and the pressure gauge is in the normal range.

Built-in Fire Protection Systems

The officer should verify that all fire department connection caps are in place, and that they are unobstructed and accessible. For sprinkler systems, extra sprinklers and a changing wrench should be readily available. The officer should ensure that all control valves are in the open or correct position (some valves are normally closed) and are locked or have a supervisory alarm to protect the system from being accidentally (or maliciously) shut down.

Only properly trained personnel can test fire protection systems. Los Angeles, Phoenix, and Fairfax County (Virginia) encountered embarrassing situations when they started their retesting programs. Using a pumper to charge a standpipe system, one standpipe was blown right out of the building wall. Another pressure test filled a mercantile basement with hundreds of gallons of water. Issues with liability and repair costs may arise when the local fire company is operating built-in fire protection systems in a non-fire-emergency situation. In some retesting programs, the fire prevention bureau requires a licensed fire protection contractor to perform the retest, and this contractor must submit a certified report of the findings (Carson and Klinker, 2012).

Special Hazards

The nature of an industrial process by itself may be hazardous. For most processes with high hazards, the occupancy is required to have a **fire prevention division or hazardous use permit**. Such a local government permit is renewed annually after the fire prevention division performs a code compliance inspection. The local fire company may be the first to discover a high-risk occupancy. For example, one Los Angeles fire company discovered a fireworks warehouse during a routine inspection. Such surprises are especially likely if your district includes "on spec" or "spec built" industrial parks. In these developments, owners construct buildings that are generic spaces and rent out space to a variety of tenants. These spaces have high occupant turnover, with each occupant being associated with different fire risk factors. A space that was used for storing building materials last year, may be used for assembling computer hardware equipment this year and might be an auto repair shop next year.

Selected Use Group–Specific Concerns

Each use group classification involves special concerns and considerations in addition to the general concerns presented earlier in this chapter. Some of the more significant considerations are described in the following sections.

Public Assembly

The goal for a code compliance inspection in a public assembly occupancy is to ensure that all of the access and egress pathways are clear and in good order (**FIGURE 6-18**). The number and the size of exits should have been approved before the notice of occupancy was given to the business.

FIGURE 6-18 All access and egress pathways must be clear.
© Ulrich Zillmann/Getty Images.

A major problem noted in many assembly inspections is overcrowding. Assembly occupancies should have their occupant load posted, and during the inspection you should check for crowds that exceed this capacity.

Confirm that the exits are not blocked or locked and are in good working order, and that the layout has not changed since the last inspection. The fire code will likely specify when panic hardware must be installed and the direction of swing of the doors. Like the number of exits, these features are not usually an issue unless remodeling has occurred. Storage of combustibles under exit stairs is not permitted.

Confirm use of flame-resistant material in the curtains, draperies, and other decorative materials. Watch for unacceptable storage of hazardous or flammable materials.

Commercial exhaust hoods and ducts also need to be inspected. The fire officer should check for unvented, fuel-fired heating equipment. In addition, these occupancies are prone to occupants propping open fire doors and smoke barriers.

Business

Make sure that access and egress pathways are clear and in good order; this includes exits that are blocked with office furnishings. There is often a lack of maintenance of the fire protection items. Portable fire extinguishers may not have been serviced, exit sign bulbs may be burned out, and the batteries are often dead in the emergency lights. Frequently there is inappropriate use of electrical extension cords and fire doors are chocked open. Older buildings may use too many portable heaters in the winter, overloading the electrical system.

Educational

A single school fire produced the swiftest example of the catastrophic theory of reform. The Our Lady of Angels School burned on December 1, 1958. Despite heroic efforts by the Chicago Fire Department, 95 people died, most of them children. NFPA's Percy Bugbee reported that within days of the disaster, state and local officials ordered the immediate inspection of schools (Bugbee, 1971). The NFPA partnered with Los Angeles Fire Marshal Raymond Hill to conduct 172 fire tests to develop new fire safety regulations for schools (NFPA, 1959). Operation School Burning provided smoke and heat data that guided the technical committee in drafting new standards. It was reported that major improvements had been made in 16,500 schools throughout the United States within 1 year of the new standards' publication (NFPA, 1961).

As in assembly-type occupancies, exit paths are essential for occupant safety in educational buildings. The fire officer should ensure that doors are not blocked or locked, thereby preventing a safe exit. In addition, the fire officer must verify that built-in fire protection systems are in good working order and are properly maintained. Pay attention to cooking facilities, science labs, and shop areas. Confirm that the evacuation plan is up-to-date and practiced. All staff should be trained in fire emergency procedures.

Factory Industrial

Factory industrial occupancies include many of the same hazards as business occupancies. In addition, they may utilize processes that are particular to that type of factory; these should be assessed for specific fire hazards. All too often, factories have improperly stored combustibles. They are also prone to having fire protection systems that have been shut off or are improperly maintained and fire doors that are propped open.

Institutional

Institutional occupancies include hospitals, nursing homes, correctional facilities, and similar occupancies where the occupants are likely to require special assistance to evacuate. For example, a working sprinkler system is essential in correctional facilities because of the inability of the occupants to escape from fires. As a result of the security measures implemented by a correctional facility, the fire department is often delayed in accessing the seat of the fire.

Many of these facilities have standpipe systems because of the travel distance to the exterior. Often, they have "house lines," which are preconnected 1½-inch (38-mm) hose lines attached to the standpipe connection. These lines are used by workers on scene and may be located in a locked metal cabinet.

Verify that the standpipe is accessible and in working order. Confirm that the standpipe threads match those of the hose used by the department. Like the standpipe connections, fire extinguishers may be locked in a metal cabinet; however, they should be properly marked. Typically, these facilities do not have a great deal of combustibles, and the housekeeping is excellent (Carson and Klinker, 2012).

Built-in fixed fire protection systems, including automatic sprinklers, smoke detection systems, and sophisticated alarm systems, are required to provide additional evacuation or relocation time. During an inspection of such an institutional occupancy, fire detection systems should be checked for proper maintenance and operation. These facilities should have a fire evacuation plan that should be up-to-date and practiced.

Mercantile

Mercantile occupancies include retail shops and stores selling stocks of retail goods. Stand-alone 24-hour convenience markets, department stores, and big-box home improvement and discount stores are all mercantile occupancies. Mercantile fires are responsible for a higher-than-average number of fire fighter line-of-duty deaths. As in the assembly classification, the people who are present in mercantile occupancies may not be aware of the locations of exits. Consequently, ensuring that exits are properly marked and illuminated is a prime concern during a fire inspection.

Housekeeping is another concern in these occupancies. Frequently, exits and aisles are blocked with merchandise, sometimes to the point of covering sprinklers, standpipe connections, and fire extinguishers. Inspections should also identify the storage of flammable and combustible material.

Residential

When inspecting residential buildings, only common areas can be inspected unless otherwise requested by the occupant. Hallways, utility areas, entryways, common laundry rooms, and parking areas are all open for inspection. Residential concerns vary by specific occupancy type; however, in almost every situation, exits are a major concern. Fire doors are routinely propped open and may have items stored in their path. Exit and emergency lighting are often not maintained properly.

In residential occupancies, fire protection systems need to be checked for adequate maintenance and operation. Vandals often remove caps and place objects inside standpipes. Valves may be closed, preventing the system from operating. Smoke alarms may have been removed. Hose cabinets may have damaged threads.

Preventing fires from igniting is a prime concern in these buildings. Poor housekeeping, such as bags of trash in hallways, may provide inviting targets for fire setters. Managers of garden apartments often store mowers and gasoline under stairways both inside and outside the structure. Structures in high-crime areas may have bars on windows and doors that prevent occupant egress.

Special Properties

The "special properties" class includes structures that hold a wide variety of hazards. Such occupancies might include the following facilities:

- Covered shopping malls
- Atriums
- Motor vehicle-related occupancies
- Aircraft-related occupancies
- Motion picture projection rooms
- Stages and platforms
- Special amusement buildings
- Incidental use areas
- Hazardous materials

Before inspecting a special-use occupancy, the officer needs to review applicable code requirements that are specific to the occupancy type. For example, special consideration should be given to preventing fires from starting when buildings house hazardous materials. Codes will specify which items and quantities can be stored in a facility, as well as which markings are required for hazardous substances. Signs should be posted that prohibit open flames and smoking. Fire doors and smoke barriers should be in proper working order. Fire protection systems should be regularly inspected and maintained. Fire emergency plans should be in place, along with evacuation plans, which should be practiced.

Manufacturing

Manufacturing facilities cover a wide spectrum of purposes. Some employ many people working in a single building and others are full of machines with few employees. General purpose manufacturing facilities are considered low to moderate hazard occupancies. Some, due to the nature of the materials used or the manufacturing process, will be identified as high hazard facilities. High hazard facilities will fall into hazardous materials occupancy.

In these occupancies, the preinspection preparation requires a review of the specific processes that occur within the building. Some of these facilities will require a fire prevention permit that requires a more extensive inspection by credentialed fire inspectors. The fire company–level inspection should focus on the condition of the automatic fire sprinkler system, the monitored central station fire alarm system, and the preincident plan. These facilities may be operating 24 hours per day or have an onsite security and building engineer.

Storage

Storage facilities often have sprinkler and standpipe systems, which should be checked for proper inspection and maintenance records. Because of the constant movement of stock, be alert for the condition and presence of portable fire extinguishers and standpipe stations. Sometimes stock will block exit pathways or be stacked so high that it blocks the ability of the automatic sprinkler system to properly operate.

These facilities may also store hazardous materials, albeit sometimes for brief periods. During the inspection, the fire officer should determine both the normal amount of onsite hazardous materials and the maximum amount that could be present.

Mixed

Each mixed-use building is considered individually in terms of its requirements, but the building as a whole must meet the most stringent requirement that applies to any of the occupancies inside. Consider a building that has a fireworks storage area in one unit and a barbershop in another unit. Although each occupancy in this building must meet the codes that apply to that specific occupancy, the building in general must meet the most stringent standard—in this case, the requirements that apply to fireworks storage.

Legal Considerations During Investigations

The Fire Officer I level of NFPA 1021, *Standard for Fire Officer Professional Qualifications*, includes performing a fire investigation to determine area of origin and preliminary cause. The goal of this activity is to determine if a fire investigator is needed. If the cause of the fire is evident and accidental, the incident commander would be responsible for gathering the information and filling out the necessary reports.

Although the law states that there is a public interest in determining the cause and origin of a fire, investigators must also respect the competing interests of a citizen's rights to privacy and due process. The fire officer who investigated a fire is often called to testify in court and may be challenged on issues of proper procedure.

Searches

When a fire has occurred and the fire department has been called, the fire department has the right to determine the cause and origin of the fire. This process must be accomplished in accordance with the law. In *Michigan v. Tyler* (1978), the U.S. Supreme Court held: "Fire officials are charged not only with extinguishing fires, but with finding their cause. Prompt determination of the fire's origin may be necessary to prevent its recurrence, as through the detection of continuing dangers such as faulty wiring or a defective furnace. Immediate investigation may also be necessary to preserve evidence from intentional or accidental destruction."

The fire officer must take care to avoid an unlawful search and seizure, which is prohibited by the Fourth Amendment of the U.S. Constitution. Typically, no search warrant is needed to enter a fire scene and collect evidence when the fire department remains on scene for a reasonable length of time to determine the cause of the fire and as long as the evidence is in plain view of the investigator. This principle was reaffirmed by the U.S. Supreme Court in *Michigan v. Clifford* (1984).

The aftermath of a fire often presents exigencies that will not tolerate the delay necessary to obtain a warrant or to secure the owner's consent to inspect fire-damaged premises. Because determining the cause and origin of a fire serves a compelling public interest, the warrant requirement does not apply to such cases.

The plain view doctrine allows for potential evidence to be seized during the processing of a fire scene, if the fire investigator had a legal right to be there and the evidence is in plain view. Under *Michigan v. Clifford*, the court held that as fire fighters remove rubble or search other areas where the cause of a fire is likely to be found, an object that comes into view during such process may be preserved. In the same decision, the court found that once the investigator has determined where the fire started, the scope of the search authority is limited to that area. After the cause and the origin have both been determined, a search warrant or consent is required for any further search.

If reentry is needed after the fire department leaves the scene or to conduct a search for evidence of a crime after the cause and origin of the fire have been determined, the investigator must obtain a search warrant or receive permission from the occupant to conduct such a search. The general provisions related to this process were clarified in *Michigan v. Tyler*:

- No search warrant is needed when fighting a fire or remaining on scene for a reasonable period of time to determine the cause of a fire, and any evidence is admissible under the plain view doctrine.
- Administrative search warrants are needed for reentry that is not a continuation of a valid search when the purpose is to determine the cause of the fire.
- A criminal search warrant is needed when reentry is not a continuation of a valid search and the purpose is to gain evidence for prosecution.

Securing the Scene

A fire officer who conducts a preliminary fire cause investigation and suspects that a crime has occurred should immediately request the response of a fire investigator. In these circumstances, the scene must be secured to protect any evidence that exists. If the fire department leaves the scene unsecured, the validity of any evidence that is collected after that point could be called into question. The fire officer must ensure that fire department personnel maintain custody of the scene until the investigator arrives.

Protecting the scene includes preventing unauthorized personnel from entering the scene. To create a security perimeter, the fire officer can use fire line or police crime scene tape secured to objects. Natural barriers can also be used to aid in securing the area. Objects such as fences or hedges can be used along with barrier tape to completely surround an area.

All access to and from the area must be controlled. To be certain that no unauthorized personnel enter the area, a fire fighter or law enforcement officer may need to be posted to limit access. Posting a guard (or multiple guards) preserves the chain of custody over the scene and any evidence that is present until the fire investigator arrives. Failure to take these steps may require the fire department to get a warrant to return to the fire scene.

Barrier tape may also be placed across a doorway to prevent unauthorized entry into a room or building. To secure and protect smaller areas of evidence, the fire officer may decide to cover them with a plastic sheet or tarp.

In any case, the number of fire personnel who are allowed into the secured area should be limited. The area should be treated as a crime scene, and only activities that are essential to control the emergency or protect the scene should be conducted. Fire fighters should not collect artifacts as souvenirs of their firefighting adventure, particularly when a scene is being secured.

Evidence

Evidence includes material objects as well as documentary or oral statements that are admissible as testimony in a court of law. Evidence proves or disproves a fact or issue. The fire officer must consider three types of evidence:

- **Demonstrative evidence:** Tangible items that can be identified by witnesses, such as incendiary devices and fire scene debris
- **Documentary evidence:** Evidence in written form, such as reports, records, photographs, sketches, and witness statements
- **Testimonial evidence:** Witnesses speaking under oath

If the fire officer has determined that the fire requires a formal investigation, then every effort should be made to protect and preserve the fire scene evidence. The structure, contents, fixtures, and furnishings should remain in their prefire locations, as intact and undisturbed as possible.

Evidence plays a vital role in the successful prosecution of arson cases. To prove that the crime of arson occurred, the fire investigator must rule out all potential accidental and natural causes of the fire. The investigator must consider all possible circumstances, conditions, or agencies that could have brought together a fuel, an ignition source, and an oxidizer, resulting in a fire or combustion explosion.

Artifacts, in the context of fire evidence, could include the remains of the material first ignited, the ignition source, or other items or components that are in some way related to the fire ignition, development, or spread. An artifact could also be an item on which fire patterns are present, in which case preservation of the artifact is not focused on the item itself, but rather the fire pattern that appears on the item (IAFC, IAAI, and NFPA, 2019).

Protecting Evidence

One part of the fire investigator's job is to dig out the fire scene. Like an archaeologist, the fire investigator removes each layer of fire debris. The investigator's ultimate goal is to identify the area of origin and the cause of the fire. Because fire follows the rules of science, an analysis of fire spread assists in the determination of where the fire originated and whether the cause was accidental or intentional.

To determine the area of origin, the fire officer must understand fire behavior, growth, and development. Fire fighters learn the basic concepts of fire behavior in their Fire Fighter I and II courses. The fire officer should also understand the three methods of heat transfer: conduction, convection, and radiation. These concepts must be applied to understand fire growth and interpret the spread of a fire.

Consider a sofa fire. The products of combustion will rise with heat until they reach the ceiling. Then, they spread out laterally until they reach the walls. Next, the heat and smoke bank down lower and lower into the room until the room is filled (or until encountered by an open doorway or window). At that point, the heat and smoke tend to flow outward through the opening into the adjoining room or space, filling that area from the ceiling down. This process continues until the entire structure is filled with heat and smoke or until an opening to the outside is found, allowing the smoke and heat to escape from the structure.

While this process is occurring, the fire is continuing to develop and grow in the area of origin. The fire follows the same growth pattern as the smoke and heat, rising toward the ceiling, spreading out, and banking down. When a room fire reaches the point that flames are spreading across the ceiling, tremendous heat energy is radiated down onto the combustible contents within the room, such as furniture. As these items are heated, they begin to release combustible fuels into the atmosphere and eventually reach their ignition temperatures. As each item ignites, the heat release rate increases dramatically and the temperature increases even faster. Within a few seconds, the entire room flashes over, or becomes fully involved in fire. The fire then quickly spreads to adjoining rooms. This process often causes windows to break or holes to burn through to the outside of the structure, releasing heat and smoke and making the fire visible on the exterior of the structure.

If the fire does not have a fresh supply of oxygen, the fire slowly dies down to the decay stage. This may occur either before or after flashover has occurred. At this point, the fire sometimes completely dies out and self-extinguishes. In other cases, a fresh supply of oxygen is introduced, and the fire resumes the fully developed stage (Gann and Friedman, 2015).

Fire scene reconstruction is the process of re-creating the physical scene before the fire occurred, either physically or theoretically. As debris is removed, the contents and structural elements are replaced in their prefire positions, as much as possible.

If the damage and destruction are too extensive to physically restore the scene, other types of evidence and information are used to fill in the blanks. The investigator interprets the fire scene and documents the fire development by examining the damage to objects, devices, and surfaces. Throughout this process, the investigator must concentrate on locating, examining, and preserving evidence.

The fire officer must determine when to stop fire suppression or overhaul operations as part of the effort to preserve evidence for the investigator. From the fire investigator's viewpoint, the less the fire fighters disturb, the more intact the scene remains.

The worst case occurs when the fire investigator arrives at an apartment fire and discovers that the fire company has removed all of the fire debris, including the fire-damaged ceilings, walls, and doors. These actions prevent the investigator from being able to evaluate the evidence in the context of the fire. The situation would be akin to the police trying to reconstruct a motor vehicle collision from which both cars have been removed to the junkyard and no witnesses remain on the scene: It is possible, but very difficult.

The fire officer is responsible for protecting the fire scene evidence both from the public and from

FIGURE 6-19 The fire officer must protect the fire scene evidence from the public and from excessive salvage and overhaul.
© Jones & Bartlett Learning. Photographed by Glen E. Ellman.

excessive overhaul and salvage (**FIGURE 6-19**). In addition, the fire officer is the first step in the chain of custody that is vital to successful prosecution of arson cases. Once the fire department arrives, it is responsible for preventing evidence contamination—a duty that requires fire fighters to stay until the fire investigator arrives. The chain of custody requires that evidence remain secured and documented, from the fire scene to the courtroom. Most investigators document all physical evidence before collecting it by taking high-resolution photographs.

Requesting an Investigator

The fire officer should be able to determine the area of origin and a cause, or probable cause, for most fires. On small or routine incidents, this is the only investigation that is conducted, and in those cases, the fire officer must carefully document the findings.

A qualified fire investigator has specialized training in determining the cause and the origin of fires and, in most cases, is certified in accordance with NFPA 1033, *Standard for Professional Qualifications for Fire Investigator*. The investigator could be a fire department member, an employee of a county or state fire marshal's office, or a law enforcement officer. In some cases, a fire investigator is responsible for determining the cause and the origin of fires, and a law enforcement agency is responsible for any subsequent criminal investigation. State and federal law enforcement agencies become involved in some investigations, particularly where arson is suspected.

The agency or organization that is responsible for conducting fire investigations will have established a set of guidelines on when to request an investigator. A fire investigator has extensive expertise in determining and documenting fire origin, gathering evidence, and interviewing. The investigator's primary responsibility is to develop a properly documented case and, if needed, to forward it to the prosecutor. The investigator must have credibility and experience in courtroom procedures to ensure that the facts are presented accurately.

A fire investigator should also be called when a death or serious burn injury occurs. If the cause of the fire was deliberate, a crime has occurred and must be fully investigated. If the cause was accidental, the information gathered in the investigation is important as a means to identify and evaluate fire prevention strategies. When any fatality occurs as a result of fire, the coroner or medical examiner's office also must be contacted.

Fire investigators often respond to large-loss fires, even if the cause is known. A fire investigator might also be called in situations that could have caused great harm, such as a fire in a hospital or college dormitory, even if the fire was quickly controlled.

When considering how best to conduct the investigation, the fire officer must evaluate the circumstances of the situation in relation to local guidelines and policies. It is a good practice to request an investigator whenever the facts do not seem to make sense or there is a compelling reason to know the exact cause of the fire.

In many cases, the fire officer is able to determine the exact cause of a fire. Sometimes the cause cannot be established with certainty, but a probable cause can be identified. Nevertheless, the causes of many fires are never determined (or at least are not reported).

You Are the Fire Officer Conclusion

The best way to perform a preincident plan for a complex facility is to start from the outside walls and roof and work inward. For the big-box facility, focus on the new construction and occupancies, rather than what the former occupancy was.

Fire fighters can be prepared for fire inspection duties with three activities:

1. Review the pertinent code enforcement regulations.
2. Discuss the fire history of this type of occupancy.
3. Review the inspection history of this occupancy.

The preincident plan focuses on fire suppression activity in the event of a fire or other emergency. The preincident plan is a component of the emergency management and business continuity plan, and it provides a starting point to preserve the viability of the business by focusing on the essential resources and features within the occupancy that are vital to the business.

After-Action REVIEW

IN SUMMARY

- To determine the area of origin, the fire officer must understand fire growth and development and the three methods of heat transfer: conduction, convection, and radiation.
- Be alert for conditions or situations that might delay the fire department's ability to get to the fire. Malfunctioning keys and key cards, vandalized doors, and materials blocking access are conditions to note.
- A preincident plan is a document developed by gathering general and detailed data used by responding personnel to determine the resources and actions necessary to mitigate anticipated emergencies at a specific facility.
- NFPA 1620 outlines six considerations when developing a preincident plan:

    1. Identify physical and site considerations.
    2. Identify occupant considerations.
    3. Identify water supplies and fire protection systems.
    4. Identify special considerations.

5. Identify emergency operation considerations.
6. Document findings.

- Fire code requirements are often adopted or amended in reaction to fire disasters, an approach known as the catastrophic theory of reform.
- A building code contains regulations that apply to the construction of a new building or to an extension or major renovation of an existing building, whereas a fire code applies to existing buildings and to situations that involve a potential fire risk or hazard.
- Most U.S. states and Canadian provinces have adopted a set of safety regulations that apply to all properties, without regard to local codes and ordinances.
- At the local level, fire and safety codes are enacted by adopting an ordinance, which is a law enacted by an authorized subdivision of a state, such as a city, county, or town.
- States and local jurisdictions may adopt a nationally recognized model code, with or without amendments, additions, and exclusions. A complete set of model codes includes a building code, electrical code, plumbing code, mechanical code, and fire code.
- Regulations that applied to a particular building at the time it was built remain in effect as long as it is occupied for the same purpose.
- Built-in fire protection systems are designed as tools to assist fire fighters in fighting a fire. The fire officer should understand how these systems work and recognize how the codes are altered when fire protection systems are in place.
- Automatic sprinkler systems, standpipe systems, and fire pumps are the three primary components of water-based fire protection systems.
- Several types of special extinguishing systems exist: carbon dioxide, dry or wet chemical, Halon, clean agent, or foam.
- A fire alarm system consists of devices that monitor for a fire and notify the building occupants and appropriate personnel when fire is detected.
- The objective of a fire code compliance inspection is to determine whether an existing property is in compliance with all of the applicable fire code requirements.
- The purpose of conducting fire inspections is to identify hazards and to ensure that any violations are corrected.
- Many of the code requirements that apply to a particular building or occupancy are based on its classification. The codes classify a building by construction type, occupancy type, and use group.
- Occupancies are classified into use groups based on the characteristics of the occupants, the activities that are conducted, and the risk factors associated with the contents.
- Before conducting an inspection, the fire officer should review the sections of the code that apply to the specific property and review the inspection results from past activities.
- Conducting a fire inspection should be approached in a systematic manner.
- The inspection/correction report should describe clearly any needed corrections and quote the appropriate sections of the code or ordinance.
- General inspection requirements pertain to access and egress, exit signs and emergency lighting, portable fire extinguishers, built-in fire protection systems, electrical systems, special hazards, and hazard identification signs.
- Each use group classification involves special concerns and considerations in addition to the general concerns.
- When a fire has occurred and the fire department has been called, the fire department has the right to determine the cause and origin of the fire. This process must be accomplished in accordance with the law.
- A fire officer who conducts a preliminary fire cause investigation and suspects that a crime has occurred should immediately request the response of a fire investigator and secure the scene to protect any evidence.
- Fire scene reconstruction is the process of re-creating the physical scene before the fire occurred, either physically or theoretically.
- The fire officer should be able to determine the area of origin and a cause, or probable cause, for most fires.

KEY TERMS

Adoption by reference Method of code adoption in which the specific edition of a model code is referred to within the adopting ordinance or regulation.

Adoption by transcription Method of code adoption in which the entire text of the code is published within the adopting ordinance or regulation.

Artifacts In the context of fire evidence, could include the remains of the material first ignited, the ignition source, or other items or components that are in some way related to the fire ignition, development, or spread.

Authority having jurisdiction (AHJ) An organization, office, or individual responsible for enforcing the requirements of a code or standard, or for approving equipment, materials, an installation, or a procedure. (NFPA 1951)

Automatic sprinkler system A sprinkler system of pipes with water under pressure that allows water to be discharged immediately when a sprinkler head operates. (Reproduced from National Fire Protection Association, *Installation of Stationary Fuel Cell Power Systems*: 853.)

Catastrophic theory of reform An approach in which fire prevention codes or firefighting procedures are changed in reaction to a fire disaster.

Construction type The combination of materials used in the construction of a building or structure, based on the varying degrees of fire resistance and combustibility. (Reproduced from National Fire Protection Association, *Building Construction and Safety Code*: 5000.)

Demonstrative evidence Tangible items that can be identified by witnesses, such as incendiary devices and fire scene debris.

Documentary evidence Evidence in written form, such as reports, records, photographs, sketches, and witness statements.

Evidence Includes material objects as well as documentary or oral statements that are admissible as testimony in a court of law.

Fire code compliance inspection Periodic check of an occupancy by the authority having jurisdiction to identify if there are any fire code violations and to initiate a repair or restoration to return to fire code compliance.

Fire prevention division or hazardous use permit A local government permit that is renewed annually after the fire prevention division performs a code compliance inspection. A permit is required if the process, storage, or occupancy activity creates a life-safety hazard. Restaurants with more than 50 seats, flammable liquid storage, and printing shops that use ammonia are examples of occupancies that may require a permit.

Fire scene reconstruction The process of recreating the physical scene during fire scene analysis investigation or through the removal of debris and the placement of contents or structural elements in their prefire positions. (Reproduced from National Fire Protection Association, *Guide for Fire and Explosion Investigations*: 921.)

Floor plans Views of a building's interior. Rooms, hallways, cabinets, and the like are drawn in the correct relationship to each other.

Fuel load The total quantity of combustible contents of a building, space, or fire area, including interior finish and trim, expressed in heat units or the equivalent weight in wood. (Reproduced from National Fire Protection Association, *Guide for Fire and Explosion Investigations*: 921.)

High-risk property Structure that has the potential for a catastrophic property or life loss in the event of a fire.

High-value property Structure that contains equipment, materials, or items that have a high replacement value.

Masonry wall A wall that consists of brick, stone, concrete block, terra cotta, tile, adobe, precast, or cast-in-place concrete.

Mini/max codes Codes developed and adopted at the state level for either mandatory or optional enforcement by local governments; these codes cannot be amended by local governments.

Model codes Codes generally developed through the consensus process with the use of technical committees developed by a code-making organization.

Occupancy type The purpose for which a building or a portion thereof is used or intended to be used.

Ongoing compliance inspection Inspection of an existing occupancy to observe the housekeeping and confirm that the built-in fire protection features, such as fire exit doors and sprinkler systems, are in good working order.

Ordinance A law established by an authorized subdivision of a state, such as a city, county, or town.

Plot plan A representation of the exterior of a structure, identifying doors, utilities' access, and any special considerations or hazards.

Preincident plan A document developed by gathering general and detailed data that is used by responding personnel in effectively managing emergencies for the protection of occupants, participants, responding personnel, property, and the environment. (Reproduced from National Fire Protection Association, *Standard for Fire Safety and Emergency Symbols*: 170.)

Regulations Official rules created by government agencies that detail how something should be done. (Reproduced from National Fire Protection Association, *Recommended Practice for Organizing, Managing, and Sustaining a Hazardous Materials/Weapons of Mass Destruction Response Program:* 475.)

Standpipe system An arrangement of piping, valves, hose connections, and allied equipment installed in a building or structure, with the hose connections located in such a manner that water can be discharged in streams or spray patterns through attached hose and nozzles, for the purpose of extinguishing a fire, thereby protecting a building or structure and its contents in addition to protecting the occupants. This is accomplished by means of connections to water supply systems or by means of pumps, tanks, and other equipment necessary to provide an adequate supply of water to the hose connections. (Reproduced from National Fire Protection Association, *Standard for the Installation of Standpipe and Hose Systems:* 14.)

Testimonial evidence Witnesses speaking under oath.

Use group A category in the building code classification system in which buildings and structures are grouped together by their use and by the characteristics of their occupants.

REFERENCES

Bellavita, Christopher. 2007. "Changing Homeland Security: A Strategic Logic of Special Event Security." *Homeland Security Affairs* 3 (3): 1–23.

Bugbee, Percy. 1971. *Men Against Fire: The Story of the National Fire Protection Association 1896–1971*. Boston, MA: National Fire Protection Association.

Carson, Wayne G., and Richard L. Klinker. 2012. *Fire Protection Systems: Inspection, Test and Maintenance Manual*. Quincy, MA: National Fire Protection Association.

City of Las Vegas. 2018. New Ordinance Will Require Fire Sprinkler Systems in All New Homes. *Medium*. Accessed July 3, 2019. https://medium.com/@CityOfLasVegas/new-ordinance-will-require-fire-sprinkler-systems-in-all-new-homes-b98369710b9b.

Corbett, Gary P., and Francis L. Brannigan. 2013. *Brannigan's Building Construction for the Fire Service*. Burlington, MA: Jones and Bartlett Learning.

Cote, Arthur E., and Casey C. Grant. 2008. "Codes and Standards for the Built Environment." In *Fire Protection Handbook*, edited by Arthur E. Cote. Quincy, MA: National Fire Protection Association.

Diamantes, David. 2015. "Fire Prevention Inspection and Code Enforcement." Berryville, VA: David Diamantes.

DiNenno, Philip J., and Gary M. Taylor. 2008. "Halon and Halon Replacement Agents and Systems." In *Fire Protection Handbook*, edited by Arthur E. Cote. Quincy, MA: National Fire Protection Association.

Farr, Ronald R., and Steven F. Sawyer. 2008. "Fire Prevention and Code Enforcement." *Fire Protection Handbook*, Arthur E. Cote. Quincy, MA: National Fire Protection Association.

Federal Emergency Management Agency (FEMA). 1991. "High-Rise Office Building Fire, One Meridian Plaza, Philadelphia, February 23, 1991." Technical Report Series TR-049. Washington, DC: U.S. Fire Administration.

Federal Emergency Management Agency (FEMA). 2010. *Special Events Contingency Planning: Job Aids Manual*. Washington, DC: U.S. Department of Homeland Security.

Federal Emergency Management Agency (FEMA). 2018. "Securing Vacant and Abandoned Buildings: What Communities Can Do." Washington, DC: U.S. Fire Administration.

Ford, Charles J. 1975. "An Overview of Halon 1301 Systems." In *Halogenated Fire Suppressants*. Washington, DC: American Chemical Society.

Fox, Justin. 2019, February 13. "Why America's New Apartment Buildings All Look the Same: Cheap Stick Framing Has Led to a Proliferation of Blocky, Forgettable Mid-rises—and More Than a Few Construction Fires." *Bloomberg Businessweek*. Accessed July 3, 2019. https://www.bloomberg.com/news/features/2019-02-13/why-america-s-new-apartment-buildings-all-look-the-same.

Gann, Richard G., and Raymond Friedman. 2015. *Principles of Fire Behavior and Combustion*. Burlington, MA: Jones and Bartlett Learning.

Hague, David R. 2008. "Standpipe and Hose Systems." In *Fire Protection Handbook*, edited by Arthur Cote. Quincy, MA: National Fire Protection Association.

International Association of Fire Chiefs (IAFC), International Association of Arson Investigators (IAAI), and National Fire Protection Association (NFPA). 2019. "Legal Considerations." In *Fire Investigator: Principles and Practice NFPA 921 and 1033*. Burlington, MA: Jones and Bartlett Learning.

Isman, Kenneth E. 2008. "Automatic Sprinklers." In *Fire Protection Handbook*, edited by Arthur E. Cote. Quincy, MA: National Fire Protection Association.

Jaslow, David, Arthur Yancy, and Andrew Milsten. 2000. "Mass Gathering Medical Care." *Prehospital Emergency Care* 4 (4): 359–360.

Jones Jr., A. Maurice. 2013. *Fire Protection Systems*. Sudbury, MA: Jones and Bartlett Learning.

Kerber, Stephen. 2012. "Analysis of Changing Residential Fire Dynamics and Its Implications on Firefighter Operational Timeframes." *Fire Technology* 48 (4): 865–891.

Michigan v. Clifford, 464 U.S. 287 (1984).

Michigan v. Tyler, 436 U.S. 499 (1978).

National Fire Protection Association (NFPA). 1959. *Operation School Burning: The Official Report on a Series of School Fire Tests Conducted by the Los Angeles Fire Department*. Boston, MA: National Fire Protection Association.

National Fire Protection Association (NFPA). 1961. *Operation School Burning 2: The Official Report on a Series of School*

Fire Tests in an Open Stairway, Multistory School Conducted June 30, 1960 to July 30, 1960 and February 6, 1961 to February 14, 1961 by the Los Angeles Fire Department. Boston, MA: National Fire Protection Association.

National Fire Protection Association (NFPA). 2017, July. "U.S. Experience with Sprinklers." Accessed July 3, 2019. https://www.nfpa.org/News-and-Research/Data-research-and-tools/Suppression/US-Experience-with-Sprinklers.

Quick Response Fire Supply (QRFS). 2017. "#136 – Common Fire Code Violations from the Authority Having Jurisdiction (AHJ): Five fire Code Infractions Facility Managers Need to Look Out For." Accessed July 3, 2019. https://www.qrfs.com/blog/136-common-fire-code-violations-from-the-authority-having-jurisdiction-ahj/.

Scheffey, Joseph L. 2008. "Foam Extinguishing Agents and Systems." In *Fire Protection Handbook*, edited by Arthur E. Cote. Quincy, MA: National Fire Protection Association.

Solomon, Robert E., ed. 2012. *Fire and Life Safety Inspection Manual.* Sudbury, MA: Jones and Bartlett Learning.

Writ, Clay L. 1989. "Dillon's Rule: The Development of Local Governments in Virginia." *Virginia Town and City* 24 (8): 12–15.

Fire Officer in Action

The fire department prepares to handle an emergency in the building by developing a preincident plan. Preincident planning requires an understanding of fire growth and development, building construction, and built-in fire protection systems.

1. What is a class III standpipe system?
   A. Both 1½-in. (38-mm) and 2½-in. (64-mm) fire hose connections
   B. A sprinkler system and 1½-in. (38-mm) fire hose connections
   C. A 2½-in. (64-mm) connection with a reducer and 150 ft (48 m) of 1½-in. (38-mm) fire hose
   D. A class I system with pressure-reducing valves

2. A vacant building would fall under which type of consideration when completing a preincident plan?
   A. Physical and site
   B. Special
   C. Occupancy
   D. Emergency operations

3. A nightclub falls into which type of occupancy category?
   A. Mixed-use
   B. Business
   C. Entertainment
   D. Assembly

4. Which of these is the leading cause of residential structure fires?
   A. Smoking materials
   B. Cooking equipment
   C. Electrical overload
   D. Heating malfunction

Fire Lieutenant Activity

NFPA Fire Officer I Job Performance Requirement 4.5.1

Describe the procedures of the AHJ for conducting fire inspections, given any of the following occupancies, so that all hazards, including hazardous materials, are identified, approved forms are completed, and approved action is initiated:

(1) Assembly

(2) Educational

(3) Health care

(4) Detention and correctional

(5) Residential

(6) Mercantile

(7) Business

(8) Industrial

(9) Storage

(10) Unusual structures

(11) Mixed occupancies

Application of 4.5.1

1. Describe how a fire company would perform an inspection of a mercantile occupancy.

2. Determine if the Local Emergency Planning Committee or the Office of Emergency Management has developed an emergency response plan for a high-hazard chemical or hazardous materials facility within your community.

3. An assembly occupancy was directed to repair a malfunctioning central station fire alarm system that has been generating false alarms for weeks. Describe the jurisdictional process to resolve this issue.

NFPA Fire Officer I Job Performance Requirement 4.5.2

Identify construction, alarm, detection, and suppression features that contribute to or prevent the spread of fire, heat, and smoke throughout the building or from one building to another, given an occupancy, and the policies and forms of the AHJ so that a pre-incident plan for any of the following occupancies is developed:

(1) Assembly

(2) Educational

(3) Institutional

(4) Residential

(5) Business

(6) Industrial

(7) Manufacturing

(8) Storage

(9) Mercantile

(10) Special properties

(11) Mixed Occupancies

Application of 4.5.2

1. Go to a movie theatre or school and identify all of the construction, alarm, detection, and suppression features. Use that information to develop a preincident plan.

2. Go to a service station and determine what their procedure is if there is a spill of 20 gal (76 L) of gasoline.

3. "Podium"-constructed buildings, where the first floor is concrete or steel commercial occupancies and the upper floors are Type V apartments, are the fastest growing type of multifamily buildings. They were responsible for 6 of the 13 large-loss fires in 2017 (Fox, 2019). All of these fires occurred before construction was completed. Identify construction, alarm, detection, and suppression features that could be utilized to reduce the fire loss in this type of building.

Access Navigate for flashcards to test your key term knowledge.

CHAPTER 7

Fire Officer I

Command of Initial Emergency Operations

KNOWLEDGE OBJECTIVES

After studying this chapter, you will be able to:

- Explain how and why the incident command system was created. (pp. 196–197)
- Explain the relationship of the incident command system (ICS) to the National Incident Management System (NIMS). (pp. 197–199)
- Identify the five components of the NIMS. (p. 198)
- Compare and contrast strategic-level, tactical-level, and task-level incident management responsibilities. (**NFPA 1021: 4.6.1, 4.6.2**) (pp. 198–199)
- Describe the fire officer's role in initial incident command. (**NFPA 1021: 4.6.1, 4.6.2**) (p. 199)
- Describe a fire officer's role in commanding incidents. (**NFPA 1021: 4.6.1, 4.6.2**) (pp. 199–203)
- List the three strategic priorities for an incident commander. (**NFPA 1021: 4.6.1, 4.6.2**) (p. 200)
- Explain considerations for selecting an initial mode of operation for an incident. (**NFPA 1021: 4.6.1**) (pp. 200–201)
- List the nine functions of command. (**NFPA 1021: 4.6.1**) (p. 202)
- Identify accountability requirements for incidents. (**NFPA 1021: 4.6.2**) (pp. 202–203)
- Explain the "two-in/two out rule." (**NFPA 1021: 4.6.1, 4.6.2**) (pp. 202–203)
- Describe accounting for personnel using a personnel accountability report (PAR). (**NFPA 1021: 4.6.2**) (p. 203)
- Explain how the results of the full-scale structure fire experiments conducted by the National Institute of Standards and Technology (NIST) and Underwriters Laboratories (UL) inform firefighting strategy and tactics. (**NFPA 1021: 4.6.1**) (pp. 203–204)
- Explain the use of a risk/benefit analysis to make fire attack decisions. (**NFPA 1021: 4.6.1**) (pp. 206–207)
- Describe the purpose of incident size-up. (**NFPA 1021: 4.6.1**) (p. 208)
- Identify information that is gathered during size-up. (**NFPA 1021: 4.6.1**) (pp. 210–211)
- Explain the incident size-up process. (**NFPA 1021: 4.6.1**) (pp. 208–211)
- Identify the two major components of an incident action plan (IAP). (**NFPA 1021: 4.6.1**) (pp. 211–212)
- Identify the three incident action priorities. (**NFPA 1021: 4.6.1**) (p. 212)
- List tactical priorities for an incident. (**NFPA 1021: 4.6.1**) (pp. 212–214)

- Explain development an IAP. (**NFPA 1021: 4.6.1**) (pp. 211–214)
- Identify methods fire officers can use to enhance company safety on the fire ground. (**NFPA 1021: 4.6.2**) (pp. 214–217)
- Describe the standard procedure for transferring command. (p. 217)
- Select an initial command operational mode appropriate for a situation. (**NFPA 1021: 4.6.1, 4.6.2**) (pp. 200–201)
- Utilize a system to maintain accountability. (**NFPA 1021: 4.6.2**) (pp. 202–203)
- Size up an incident. (pp. 208–211)
- Develop an IAP. (**NFPA 1021: 4.6.1**) (pp. 211–214)
- Implement an IAP. (**NFPA 1021: 4.6.2**) (pp. 212–214)
- Transfer command using a standard procedure. (**NFPA 1021: 4.6.2**) (p. 217)

SKILLS OBJECTIVES

After studying this chapter, you will be able to:

- Provide an initial radio report establishing command. (p. 200)
- Utilize the ICS at an incident. (**NFPA 1021: 4.6.1, 4.6.2**) (p. 199)

ADDITIONAL NFPA STANDARDS

- NFPA 1026, *Standard for Incident Management Personnel Professional Qualifications*
- NFPA 1561, *Standard on Emergency Services Incident Management System and Command Safety*

You Are the Fire Officer

The fire fighters are commenting on YouTube videos that show a burning building before the first fire company arrives. They are stating what they would do if it was their fire company arriving.

1. How important is size-up before committing to action at an emergency scene?
2. Is the incident command system only for working structure fires and mutual aid events?
3. Who initiates the incident action plan?

 Access Navigate for more practice activities.

Introduction

A fire officer is expected to perform the duties of a first-arriving officer at any incident, including assuming initial command of the incident, establishing the basic management structure, and following standard operating procedures (SOPs). A fire officer must also be fully competent at working within the incident command system (ICS) at every incident and function as a unit, group, or division leader.

Structural firefighting is a practice built on experience and experiments. The findings from recent experiments have dramatically changed the understanding of fire dynamics and identified the importance of controlling flow path of high-momentum fire gases. In turn, traditional practices have changed. This chapter introduces model procedures for incident management and fire behavior as it applies to structural firefighting practices today.

The History of the Incident Command System

In the past, incident command was a local activity, using unique terms and practices in each community.

The individualized approach did not work well when units from different districts or mutual aid companies responded to a major incident. The effort to develop a standard system began after several large-scale wildland fires in southern California overwhelmed fire departments. Collectively, these agencies established an organization known as **FIRESCOPE (FIre RESources of California Organized for Potential Emergencies)** to develop solutions for a variety of problems associated with large, complex emergency incidents during wildland fires.

FIRESCOPE and Fire Ground Command

FIRESCOPE was used during multiagency wildland incidents in California involving more than 25 resources or operating units. The program developed a standardized method of setting up an incident management structure, coordinating strategy and tactics, managing resources, and disseminating information.

At the same time, Phoenix, Arizona, chief officers Alan Brunacini, Bruce Varner, and Chuck Kime were developing a **Fire Ground Command (FGC)** system to meet the needs of an all-hazards fire department. The FGC program focused on small and medium-sized urban emergencies, such as structural fires, and mass-casualty and hazardous materials events (Brunacini, 2002).

Developing One National System

In the 1980s, the National Incident Management System Consortium (NIMSC) was established to merge California's FIRESCOPE and Phoenix's FGC system into a set of model procedures that could be utilized by any fire department and are compliant within the federal National Incident Management System (NIMS) concepts (Murgallis and Phelps, 2012).

The resulting system, which blended the best aspects of both FIRESCOPE ICS and the FGC system, is now formally known as ICS—the incident command system. An ICS can be used at any type or size of emergency incident and by any type or size of department or agency. Reflecting this change, NFPA 1561 is now called *Standard on Emergency Services Incident Management System and Command Safety*. NFPA 1026, *Standard for Incident Management Personnel Professional Qualifications*, was developed to define the various job performance requirements for each of the positions classified within the NIMS model. The NIMS model is discussed next.

Incident Command and the National Incident Management System Model

The **National Incident Management System (NIMS)** provides a consistent, nationwide framework for incident management, enabling federal, state, and local governments, private sector and nongovernmental organizations, and all other organizations who assume a role in emergency management to work together effectively and efficiently across all emergency management and incident response organizations. It is a core set of doctrines, concepts, principles, terminology, and organizational processes.

The NIMS model can be used regardless of the cause, size, complexity, or type of incident, including acts of terrorism and natural disasters. Along with the basic concepts of flexibility and standardization, the NIMS principles are now taught in every incident management course. You can learn more about NIMS on the Federal Emergency Management Agency (FEMA) website.

ICS Training

Because the NIMS model is intended to be used at all jurisdictional levels, federal grant funding is often reserved for NIMS-compliant communities. The **incident command system (ICS)** is part of the NIMS Command and Management component. Local emergency response agencies were required to adopt ICS to remain eligible for federal disaster assistance. As a result, you will likely receive specific ICS training within your fire department in order to meet compliance requirements. FEMA provides introductory courses that will enhance your ability to work within the NIMS framework at any type of emergency incident (FEMA, 2017). These courses are free of charge, and each course can be completed in approximately 3 hours, either online or in a classroom setting:

- **IS-100: Introduction to Incident Command System** provides an overview of the ICS, including its history, principles, and organizational structure. It also explains how ICS fits into the larger NIMS model of emergency preparedness, response, and recovery. There are no prerequisites for IS-100.
- **IS-200: ICS for Single Resources and Initial Action Incidents** provides training for personnel to function within the ICS at small incidents involving one or two resources, as well as at those incidents that start small but grow in

complexity and size of operation. Completion of IS-100 is required before taking the IS-200 course.

- **ICS-300: Intermediate ICS for Expanding Incidents** is a 2-day, hands-on course that is designed for those who would function in a command or general staff position during a large, complex incident or be part of an incident management team during a major incident.
- **ICS-400: Advanced ICS Command and General Staff—Complex Incidents** is a 2-day, hands-on course that is designed for those who would function as part of an area command, emergency operations center, or multiagency coordination system during a large, complex incident or be part of an incident management team during a major incident.
- **IS-700: National Incident Management System (NIMS), An Introduction** is not required at the Fire Fighter II level, but it is recommended for a better understanding of how NIMS enhances community preparedness, communications and information sharing, resource management, and incident command during emergency incidents.
- **IS-800: National Response Framework, An Introduction** describes the concepts and principles of the National Response Framework and the ways it is applied in actual response situations.
- Appropriate position-specific training (**FIGURE 7-1**)

Each of these courses is updated periodically and is designated by letters such as IS-200.b that indicate the latest version of the course. Be sure to take the latest available version of each course to ensure current information. You will be required to take these courses only once, but it is a good idea to periodically review the information.

NIMS has five components: preparedness, communications and information management, resource management, command and management, and ongoing management and maintenance.

Levels of Command

The ICS includes three levels of command, with a set of responsibilities being assigned to each level. At small-scale incidents, or in the early stages of a larger incident, one company officer may cover all three levels simultaneously. When the incident expands, the management responsibilities are subdivided. A fire officer may have a role at any level.

The overall direction and goals are set at the **strategic level**. The **incident commander (IC)** is the individual who is responsible for the management of all incident operations and always functions at the strategic level. A strategic goal, in a defensive situation, could be to stop the extension of a fire to any adjacent structure.

Objectives at the **tactical level** define the actions that are necessary to achieve the strategic goals. A tactical-level supervisor would manage a group of resources to accomplish the tactical objective.

			Levels of Training		
			Awareness	Advanced	Practicum
Components of NIMS	Preparedness		IS 800 IS 705		
	Communications & Info Management		IS 704		
	Resource Management	IS-700	IS 703 IS 706 IS 707		
	Command & Management	ICS	ICS 100 ICS 200	ICS 300 ICS 400	Position-specific courses
		MACS	ICS 701		
		Public Info	ICS 702		
	Ongoing Management & Maintenance				

FIGURE 7-1 Core curriculum aligned with NIMS components and by level of training.
Courtesy of NIMS/FEMA

In medium- to large-scale incidents, the tactical-level components would be called divisions, groups, or units, and each of these components could include several companies. A company-level officer would be in charge of each company, and all of the company-level officers would report to the tactical-level supervisor. Tactical assignments are usually defined by a geographic area (e.g., one part of a building) or a functional responsibility (e.g., ventilation), or sometimes by a combination of the two (e.g., ventilation in a particular part of the building).

In large incidents, another level of management may be added to maintain a reasonable span of control. Branches are established to group tactical components. An officer assigned to a branch would oversee some combination of divisions, groups, or units.

Assignments at the **task level** are the actions required to achieve the tactical objectives. This is where the physical work is accomplished. Individual fire companies or teams of fire fighters perform task-level activities, such as searching for victims, operating hose lines, or opening ceilings.

The Fire Officer's Role in Incident Management

Every fire officer is expected to function as an initial IC, as well as a company-level supervisor within the ICS. A supervising fire officer functioning as the IC must supervise the work of a group of fire fighters, report to a managing or administrative fire officer, and work within a structured plan at the scene of an incident. The IC position is the *only* position that is always filled. This officer is responsible for completing all tasks that are not delegated. The officer may also be required to initiate the procedure to assemble a multiagency response.

The first-arriving fire officer has the responsibility to establish command and manage the incident until being relieved by a higher-ranking officer. What a first-arriving fire officer does in the first 5 minutes of the incident dictates how the scene will run for the next hour. This task must be accomplished in addition to supervising the members of the officer's own company. Managing an incident requires the fire officer to develop strategies and tactics, determine required resources, and decide how those resources will be used (Page, 1973).

The ICS can be implemented in an incremental fashion. The command structure for an incident should be only as large as the incident requires. The goal is to use the model ICS structure to assign all of the functions that must be performed at that incident (**FIGURE 7-2**). Most fire department tactical and task activities fall under the operations section within the ICS.

The ICS allows the company officer to maintain a manageable span of control. One officer can provide effective supervision to a limited number of subordinates, whether those subordinates are individual workers or supervisors directing groups of workers. For emergency operations, the recommended span of control is three to five individuals reporting to one supervisor. The span of control is maintained by adding more levels of management as an officer's effective span of control is exceeded (Brunacini, 2004).

Initial Incident Command

The first-arriving fire officer needs to focus on the strategic level when he or she arrives at an emergency. Depending on the size and complexity of the incident, the first-arriving officer may quickly move to the tactical or task level as the administrative fire officer and other departmental resources arrive.

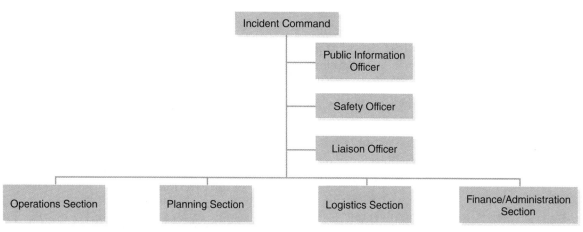

FIGURE 7-2 Incident command system structure.
© Jones & Bartlett Learning.

Responsibilities of Command

The IC is responsible for the completion of three strategic priorities:

- Life safety
- Incident stabilization
- Property conservation

The IC is also responsible for the following aspects of operations:

- Building a command structure that matches the organizational needs of the incident
- Translating the strategic priorities into tactical objectives
- Assigning the resources that are required to perform the tactical assignments

Establishing Command

The first fire officer or fire department member to arrive at the scene is expected to establish command of the incident. The initial IC remains in charge of the incident until command is transferred or the situation is stabilized and terminated. If the incident expands, a higher-ranking officer is likely to arrive and assume command, at which point the initial IC's role changes.

Establish command and use the concepts of ICS at every event. For events that require the services of a single fire company, such as a vehicle fire or medical response, establishing command may simply require notification to the dispatcher that the company is on the scene and is handling the situation. For incidents requiring two or more companies, the first fire department member or company-level officer on the scene must establish command and initiate an incident management structure that is appropriate for the incident.

Activating the command process includes providing an initial radio report and announcing that command has been established. This report should provide an accurate description of the situation for units that are still en route. It should also identify the actions that arriving units will take. In this report, the company officer should include the following information:

- Identification of the company or unit arriving at the scene
- A brief description of the incident situation, such as the building size, height, and occupancy, or the magnitude of a multivehicle collision
- Obvious conditions, such as a working fire, multiple patients, a hazardous materials spill, or a dangerous situation, such as a man with a gun
- A brief description of the action to be taken (e.g., "Engine 2 is advancing an attack line into the first floor")
- Declaration of the strategy to be used (offensive or defensive)
- Any obvious safety concerns
- Assumption, identification, and location of command
- Request for additional resources (or release of resources), if required

Initial Operational Modes

The fire officer taking command has to select one of three operational modes: investigation, fast-attack or command mode. The determination for which mode to use is based what type of situation is encountered on arrival. The mode sets the stage for initial fire-ground activities and will be re-evaluated with the transfer of command or if there is a sudden change in fire-ground conditions.

FIRE OFFICER TIP

"Nothing Showing" Calls Can Be Dangerous

Serious injuries and some fire fighter fatalities have occurred at incidents that started as "nothing showing" events. In many of these cases, after entering the building, the first-in fire company suddenly encountered a rapidly deteriorating situation. Vital firefighting equipment, such as SCBA and tools, were left behind on the apparatus. Fire fighters were unprepared for the situation they encountered.

When incidents, particularly in commercial and industrial buildings, are investigated, consider the following factors:

- Assume that there is a fire in the building and bring and wear the appropriate personal protective equipment (PPE).
- Use thermal imagers.
- Go to the location of a monitored fire alarm activation. Fires in are often detected by the fire alarm system while they are still in an incipient stage. Do not assume that the call is a false alarm if nothing is visible from the exterior. Check the fire alarm panel for the location of the activated device, and be prepared for action when you go there to investigate.
- Use two or three companies to perform the initial investigation in larger buildings.
- Hold the entire first-alarm assignment until the investigation is completed. Many fire departments have the assisting companies reduce their response from emergency to nonemergency when the first-arriving company reports no indication of fire. This policy reduces the risk of accidents during emergency response.

Investigation Mode

When fire fighters arrive, there may be nothing showing or the incident may appear to be a minor situation. In such a case, the first-arriving company will conduct an investigation. The other units assigned to the event will stage and remain uncommitted pending the results of that investigation. The first-arriving company-level officer performs the role of initial IC as well as supervising the company performing the investigation.

Fast-Attack Mode

Some situations require immediate action by the first-arriving fire company to save a life. An example is when a single company arrives at a fire with occupants in imminent danger at upper-floor windows. The company-level officer performs the initial incident command responsibilities through a portable radio while engaged in a fast attack.

The fast-attack mode ends when one of the following occurs:

- The situation is stabilized.
- The situation is not stabilized and the company officer must withdraw to the exterior and establish a command post.
- Command is transferred to another officer.

Command Mode

Some events are so large, complex, or dangerous that they require the immediate establishment of command by the first-arriving company-level officer. In this case, the company-level officer's personal involvement in tactical operations is less important than the command responsibility. The company-level officer should establish a command position in a safe and effective location and initiate a **tactical worksheet** (**FIGURE 7-3**). A tactical worksheet is a form that helps the IC ensure that all tactical issues are addressed and to diagram an incident with the location of resources on the diagram. The role of the initial IC at this type of situation is to direct incoming units to take effective action. While the initial IC remains outside, the rest of the company members should do one of the following:

- Initiate fire suppression or emergency action with one of the members assigned as the acting company officer. The acting company officer must be equipped with a portable radio, and the crew must be capable of performing safely without the initial IC.
- After the initial IC assigns them, the remaining company members work under another company officer.
- Stay with the initial IC to perform staff functions that assist command.

FIRE AND RESCUE DEPARTMENTS OF NORTHERN VIRGINIA
***INITIAL* INCIDENT COMMAND BOARD**

ADDRESS/COMMAND:				
FIRST ALARM		**Floor, Group, Division, Branch**	**Floor, Group, Division, Branch**	**Floor, Group, Division, Branch**
ENGINE				
ENGINE				
ENGINE				
ENGINE				
TRUCK				
TRUCK				
RESCUE				
MEDIC				
AMB				
BFC				
EMS CAPT				
SECOND ALARM		**Floor, Group, Division, Branch**	**Floor, Group, Division, Branch**	**Floor, Group, Division, Branch**
ENGINE				
ENGINE				
ENGINE				
ENGINE				
TRUCK				
TRUCK				
RESCUE				
MEDIC				
AMB				
BFC				
EMS CAPT				

INITIAL RIT ENGINE:	
OPERATIONS CHANNEL:	
TYPE OF OCCUPANCY:	
TASKS	**REQUESTS**
WATER SUPPLY	GAS COMPANY
PRIMARY SEARCH	POWER COMPANY
LADDERS	FIRE INVESTIGATOR
VENTILATION	POLICE
UTILITY CONTROL	LIGHT/AIR UNITS
NOTES	

CHECK ALL OPERATING UNITS AIR SUPPLY. CREWS WITH 2000 *PSI OR LESS* NEED TO BE REPLACED AND SENT TO THE *MEDICAL UNIT.*

RECEO-VS

CHECK EXPOSURES fi

B3	B2	B1	FIRE UNIT	D1	D2	D3

FIGURE 7-3 Tactical worksheet.

Courtesy of the Northern Virginia Regional Commission

Functions of Command

There are nine functions of command:

- Determining strategy
- Selecting incident tactics
- Establishing the IAP
- Developing the ICS organization
- Managing resources
- Coordinating resource activities
- Providing for scene safety
- Releasing information about the incident
- Coordinating with outside agencies

Immediate command functions include determining strategy, selecting incident tactics, and establishing an IAP. These three functions must be completed as part of the first-arriving officer's size-up and initial actions. The initial radio report, as described earlier, will cover these three functions (Brunacini, 2002).

The initial fire-ground assignments and company-level tasks may be predetermined by departmental SOP or practice. The company officer should ensure that the SOP is followed and look for conditions that may require a deviation from the standard initial actions.

Once the initial actions are underway, the IC works on the next four functions: developing the ICS organization, managing resources, coordinating resource activities, and providing for scene safety. Designate an incident safety officer as well as unit, group, or division commanders, as necessary. Request additional resources if required, and manage them through staging and assignments within the incident management system.

Once the incident action plan is fully operating, the IC works on the last two functions: releasing information about the incident and coordinating with outside agencies.

Command Safety

The initial incident commander is responsible for fire fighter accountability and initial rapid intervention resources when operating at an incident with an immediately dangerous to life or health (IDLH) environment.

Fire Fighter Accountability

Occurring concurrently with the evolution of incident management systems was development of systems for accounting for fire fighters and others working at an incident. Legal sanctions accelerated adoption of fire fighter accountability practices.

Additional pressure to improve accountability came from a request by the International Association of Fire Fighters (IAFF) for a clarification of Occupational Safety and Health Administration (OSHA) respiratory protection regulation 29 CFR 1910.134. This section of the Code of Federal Regulations covered industrial employees operating in confined spaces, toxic environments, or oxygen-deficient atmospheres. These work areas are classified as immediately dangerous to life and health (IDLH).

The OSHA standards interpretation was that fire fighters working within a structure fire were operating in an IDLH atmosphere. Consequently, fire departments must comply with 29 CFR 1910.134 while self-contained breathing apparatus (SCBA) is being used. A minimum of two fire fighters must enter the IDLH area together and remain in visual or voice contact with each other at all times (Miles, 1998). In addition, at least two properly equipped and trained fire fighters must be available to assist personnel working in the hazardous area as follows:

- Be positioned outside the IDLH atmosphere.
- Account for the interior teams.
- Remain capable of rescue of the interior team or teams.

FIRE OFFICER TIP

Examples of Initial Radio Reports

- For a single-company incident: "Communications from Engine 2. Engine 2 is on the scene of a fully involved auto fire with no exposures. Engine 2 can handle."
- For an EMS incident: "Radio from Quint 618. Quint 618 is on the scene of a four-vehicle crash with multiple patients and one trapped. Transmit a special alarm for a heavy rescue company, a second suppression company, and an EMS task force. Quint 618 will be Palisade Command."
- For an offensive structure fire: "FireCom from Engine 1034. Engine 1034 has a two-story motel with a working fire in a second-floor room on side Charlie. Engine 1034 has a hydrant and is going in with a handline to start primary search. This will be an offensive fire attack. Engine 1034 will be Cedar Avenue Command."
- For a defensive structural fire: "County from Engine 22. Engine 22 is on the scene of a one-story strip shopping center with an end unit heavily involved. Engine 22 is laying an LDH supply line and will be using a master stream to cover the exposure on side Delta. This is a defensive fire. Strike a second alarm. Engine 22 will be Loisdale Command."

FIRE OFFICER TIP

Rapid Intervention Is Not Rapid

The only large-scale analysis of rapid intervention team performance was conducted by the Phoenix Fire Department as part of their recovery process after Firefighter/paramedic Bret Tarver died at the Southwest Supermarket fire.

Phoenix obtained three vacant commercial buildings: a warehouse, a movie theatre, and a country-western bar. The RIT drill was for the first alarm companies to respond to a report of two fire fighters in trouble. One is disoriented, and the other one is unconscious. The buildings were sealed from outside light and the facemasks were obscured to simulate heavy smoke conditions. The RIT teams were deployed in teams of four or five fire fighters equipped and deployed as if this was a working fire. The department ran through about 200 RIT drills with 1144 fire fighters. Their activities were monitored and timed. An Arizona State University statistician analyzed the data (Kreis, 2003).

The results show that rapid intervention is not rapid:
- Total time for RIC to make contact with downed fire fighter: 5.82 minutes
- Total time inside building for each RIC team: 12.33 minutes
- Total time for rescue: 21 minutes

At the time of the evolutions, the department had few thermal imaging cameras and did not use them for these evolutions. The evolutions also revealed three consistent ratios:
- It takes 12 fire fighters to rescue one.
- One in five RIC members will get into some type of trouble themselves.
- A 3000-psi SCBA bottle has 18.7 minutes of air (plus or minus 30%)

The Asheville Fire Department (North Carolina) conducted 12 RIT evolutions in 2014 looking at time-on-task and air consumption. Their results were similar to those from Phoenix, needing 15 fire fighters to rescue one and taking about 37 minutes to remove a compromised fire fighter (AFD, 2015).

This interpretation became known as the **two-in/two-out rule** and evolved into the rapid intervention team (RIT) or rapid intervention crew (RIC) concept that was incorporated into NFPA 1500. NFPA 1407, *Standard for Training Fire Service Rapid Intervention Crews*, describes basic training procedures and information.

Personnel Accountability Report

A **personnel accountability report (PAR)** is a systematic method of accounting for all personnel at an emergency incident. When the IC requests a PAR, each fire officer physically verifies that all assigned members are present and confirms this information to the IC. A PAR should also be requested at tactical benchmarks, such as a change from an offensive strategy to a defensive strategy. A PAR may be integrated into the incident time clock, requiring a PAR check every 20 minutes while in incident is active.

The fire officer must be in visual or physical contact with all company members to verify their status. The fire officer then communicates with the IC by radio to report a PAR. Anytime a fire fighter cannot be accounted for, he or she is considered to be missing until proven otherwise. A report of a missing fire fighter always becomes the highest priority at the incident scene.

If unusual or unplanned events occur at an incident, a PAR should always be performed. For example, if there is a report of an explosion, a structural collapse, or a fire fighter missing or in need of assistance, a PAR should immediately take place so that it can be determined how many personnel might be missing, the missing individuals can be identified, and their last known location and assignment can be established. This last location would be the starting point for any search and rescue crews.

Fire Research and Risk/Benefit Analysis

Two decades of fire behavior research has expanded our understanding of the combustion process, the value of understanding fire flow, and appreciation of the limitations of some of our tools.

Fire Behavior Graph

In 1908, the American Society for Testing and Materials (ASTM) conducted full-scale fire experiments that led to the development of a **standard time–temperature curve** to guide the testing of building partitions and floors for fire resistance. By 1917, ASTM E119, *Standard Test Methods for Fire Tests of Building Construction and Materials*, provided a time–temperature curve that was used as a source when the fire service described fire behavior as applied to structural firefighting practices (**FIGURE 7-4**).

One hundred years ago, when these standards were developed, a fire within a structure was a **fuel-limited fire**, meaning that, without intervention, fire would consume all of the fuel. Recent full-size fire experiments, however, have shown that a typical modern fire is a **ventilation-limited fire**, resulting in a different time–temperature curve (**FIGURE 7-5**). The National Institute of Standards and Technology (NIST) provided this explanation for the changes (NIST, 2010):

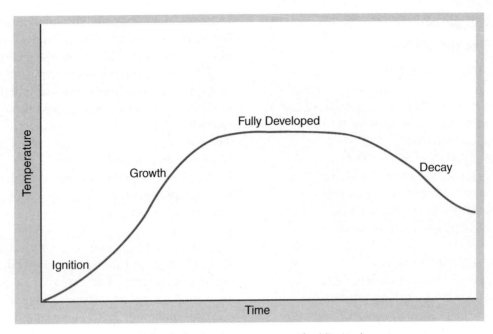

FIGURE 7-4 Traditional fire behavior in a structure: fuel limited.
Courtesy of National Institute of Standards and Technology

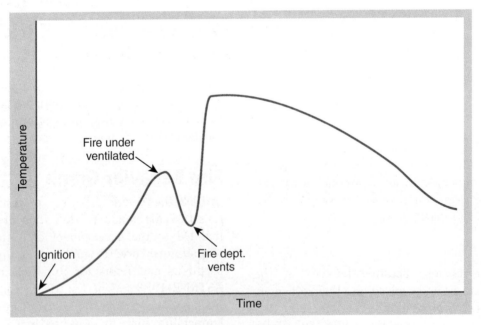

FIGURE 7-5 Fire behavior in a structure: ventilation limited.
Courtesy of National Institute of Standards and Technology

The Fire Behavior in a Structure curve demonstrates the time history of a ventilation limited fire. In this case the fire starts in a structure which has the doors and windows closed. Early in the fire growth stage there is adequate oxygen to mix with the heated gases, which results in flaming combustion. As the oxygen level within the structure is depleted, the fire decays, the heat release from the fire decreases and as a result the temperature decreases. When a vent is opened, such as when the fire department enters a door, oxygen is introduced. The oxygen mixes with the heated gases in the structure and the energy level begins to increase. This change in ventilation can result in a rapid increase in fire growth potentially leading to a flash-over (fully developed compartment fire) condition.

Flow Path

Flow path is the movement of heat and smoke from within the higher-pressure fire area toward lower-pressure areas accessible via doors, window openings, and roof structures. The New York Fire Department conducted a series of full-scale exercises on Governor's Island in 2012 to explore alternative ventilation techniques based on the findings in earlier wind-driven research. Both Underwriters Laboratories (UL) and NIST participated in these experiments (UL FSRI, 2014). Observations at the Governor's Island experiments included the following points when combating ventilation-limited compartment fires, as listed in UL's *Analysis of One and Two-Story Family Home Fire Dynamics and the Impact of Firefighter Horizontal Ventilation* (Kerber, 2013):

- It is essential to control the access door to restrict introduction of air into the fire room and thereby delay flashover.
- The only way to go from a ventilation-limited to a fuel-limited fire is through application of water before vertical ventilation.
- "Softening" the target by applying 30 to 90 seconds of water into the compartment dramatically reduces fire development and improves conditions.
- You cannot make a big enough ventilation hole to localize fire growth or reduce temperatures in ventilation-limited structure fires.

Operations conducted in the flow path place fire fighters at significant risk due to the increased flow of fire, heat, and smoke toward their position (**FIGURE 7-6**). Limiting flow paths until fire suppression water is ready to be applied is an important factor in limiting heat release and temperatures in the house.

Modern versus Legacy Single Family Dwellings

UL has documented changes within the residential fire environment that impact occupant survival and fire fighter safety. In *Analysis of Changing Residential Fire Dynamics and Its Implications on Firefighter Operational Timeframes*, Stephen Kerber identifies four factors that distinguish **modern dwellings** from **legacy dwellings**, meaning structures built before 1980:

- Larger homes
- Open house geometries
- Increased fuel loads
- New construction materials

Modern dwellings are almost twice as large as legacy single-family dwellings. Open geometry in modern homes reduces compartmentalization and allows more air to support rapid fire propagation. The rate of heat released by burning contents is exponentially higher, resulting in the fire within a room rapidly running out of air. Once a new source of air is introduced, such as from the opening of a door or failure of a window, the fire will quickly develop (Kerber, 2012).

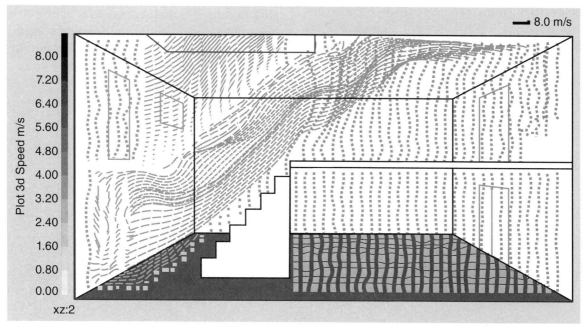

FIGURE 7-6 NIST Fire Dynamics Simulator model of the Cherry Road fire. The fire fighter victims were operating at the top of the stairs, in the middle of the flow path of high-momentum fire gases.

Taking the fire test data and applying a fire department response timeline, UL identified a dramatic reduction in available time when suppressing a modern single-family dwelling fire, as depicted in **FIGURE 7-7**.

Risk/Benefit Analysis

Risk/benefit analysis is an assessment of the risk to rescuers versus the benefits that can be derived from other intended actions. It is a key factor of size-up when selecting the appropriate strategic mode. The strategic mode dictates the degree of risk to fire fighters that is acceptable in a given situation. The degree of risk that is acceptable is determined by the realistic benefits that can be anticipated by taking a particular course of action.

Potential benefits may include the possibility of saving lives or preventing injury to persons who are in danger, preventing property damage, and protecting the environment from harm. The major risk factor is the possibility of death, disability, or injury to fire fighters. Potential occurrences, such as flashover, backdraft, or structural collapse, also have to be evaluated. Each alternative action plan involves a different combination of risks and benefits.

The fire officer can manage these risks. Training, experience, protective clothing and equipment, communications equipment, SOPs, accountability systems, rules of engagement, and the RIT are all factors that allow crews to work with a reasonable level of safety in a hazardous environment.

Fire fighters accept a higher personal risk when it is necessary to save a life; however, there is no justification for risking the lives of fire fighters when no one requires rescue or when interior fire conditions are unsurvivable, such as in fires in vacant or abandoned buildings or situations where significant damage is already done. Building construction features, such as lightweight trusses, can also cause the risks of offensive operations to outweigh the potential benefits. When defensive operations are conducted, the risks to fire fighters are significantly reduced.

The risk/benefit analysis determines the appropriate strategy for an incident: offensive, defensive, or transitional. An **offensive operation**, or an offensive attack, typically consists of an advance into the fire building by fire fighters with hose lines or other extinguishing agents to overpower the fire. Offensive activity drives most fire department training, operations, and organizational structures.

When this mode is selected, the IC believes that the benefits associated with controlling and extinguishing the fire outweigh the risks to the fire fighters. To have any chance of success, such an operation requires sufficient resources to mount an attack that is powerful enough to overwhelm the fire within the time that is available to operate safely inside the burning building. The risks of an offensive attack can be justified only when it may generate realistic benefits. Specifically, there must be lives or valuable property that can be saved to justify taking this risk.

A **defensive operation**, or a defensive attack, is used when the risks outweigh the expected benefits. In this mode, fire fighters are not allowed to enter the structure or to operate from positions that involve avoidable risks. Defensive operations are typically conducted from the exterior, using large streams to contain and, if possible, overwhelm a fire (**FIGURE 7-8**). A defensive approach may be used when there is a risk of a structural collapse or there are inadequate resources to conduct an adequate interior attack to control the fire. A defensive strategy is also the appropriate choice in a situation where the building and contents would be a total loss even if an aggressive interior attack could control the fire.

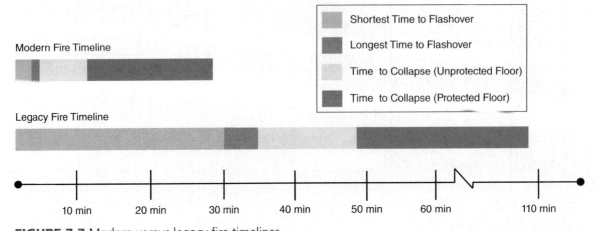

FIGURE 7-7 Modern versus legacy fire timelines.

Reproduced from: Fire Technology 48(4): 865-891, "Analysis of Changing Residential Fire Dynamics and Its Implications on Firefighter Operational TimeFrames", Kerber, S., Copyright © 2012, National Fire Protection Association

FIGURE 7-8 Defensive operations.
Courtesy of Mike Legeros

The primary means of suppressing fires is to apply water. Traditionally, it was thought that attacking the fire from the outside may push the fire through other parts of the building. This has led to the assumption that offensive operations must be made exclusively from the inside and from the unburned side of the structure, effectively "pushing the fire" away from the uninvolved portion of the building and any occupants who might be behind the advancing hose team. This technique requires fire fighters to enter a burning building, often with low visibility and little idea of where the fire is located. NIST and UL research demonstrates that these assumptions are not always accurate.

In 2013, UL released its *Innovating Fire Attack Tactics* report, which outlined the results of a series of tests conducted with NIST and the New York City Fire Department (UL, 2013). During the structure fire experiments, researchers found that applying a limited amount of water through an open window or door into the fire compartment before entering the structure may dramatically improve interior conditions for both occupants and fire fighters (**FIGURE 7-9**).

When the goal is to get inside quickly, and although there is no consensus on a term for this type of operation, the tactics remain the same. A **transitional attack** is an offensive operation initiated by a brief exterior, indirect attack into the fire compartment to initiate cooling and to stop the progress of the fire. This exterior attack quickly transitions to an interior attack to suppress the fire, in coordination with ventilation operations. The transitional attack "softens the target" or cools the fire gases, reduces the risk of flashover, improves visibility, and allows fire fighters to quickly enter the fire compartment. This method can often effectively place water on the seat of the fire sooner, thus making conditions tenable for any trapped occupants.

FIGURE 7-9 Prior to making entry, the nozzle fire fighter directs a straight stream through the top of the door or window and deflects the stream off of the ceiling. Evaluate fire conditions before transitioning to an interior fire attack.
© Jones & Bartlett Learning.

IAFC Rules of Engagement

The mission of the International Association of Fire Chiefs (IAFC) Safety, Health and Survival (SHS) Section is to reduce the number of preventable line-of-duty deaths and injuries in the fire service. After researching line-of-duty deaths and near-misses, the SHS issued "The Incident Commander's Rules of Engagement for Firefighter Safety:"

1. Rapidly conduct, or obtain, a 360-degree situational size-up of incident.
2. Determine the occupant survival profile.
3. Conduct an initial risk assessment and implement a safe action plan.
4. If you do not have the resources to safely support and protect firefighters, seriously consider a defensive strategy.
5. Do not risk firefighter lives for lives or properties that cannot be saved. Seriously consider a defensive strategy.
6. Extend limited risk to protect savable property.
7. Extend vigilant and measured risk to protect and rescue savable lives.
8. Act upon reported unsafe practices and conditions that can harm firefighters. Stop, evaluate and decide.

9. Maintain frequent two-way communications, and keep interior crews informed of changing conditions.
10. Obtain frequent progress reports and revise the action plan.
11. Ensure accurate accountability of every firefighter's location and status.
12. If after completing the primary search, little or no progress toward fire control has been achieved, seriously consider a defensive strategy.
13. Always have a rapid intervention team in place at all working fires.
14. Always have firefighter rehab services in place at all working fires. (IAFC, 2012)

Sizing Up the Incident

Size-up is a systematic process of gathering and processing information to evaluate the situation and then translating that information into a plan to deal with the situation. The art of sizing up an incident requires a diverse knowledge about emergency incidents. The fire officer could be faced with a house fire, a hazardous materials release, and a major automobile collision in one shift. Each situation requires the ability to recognize and analyze the important factors, develop a plan of action, and then implement the plan. Each type of incident also involves specialized knowledge.

Inexperienced fire officers may think that size-up begins when they get their first glimpse of the incident scene and ends when they have decided on a plan of action. In reality, size-up begins long before arrival and continues until the incident is stabilized. Deciding on an initial plan of action after arriving on the scene is only one step in a continuous process. As the plan is executed, new information must be gathered and processed to determine whether the plan is working as anticipated and to make any adjustments that are required.

The initial size-up at the scene of an incident must often be conducted under intense pressure to "do something." This pressure could be coming from the public, from fire companies that are arriving and anxious to get into action, and from within the fire officer who feels the anxiety of being responsible for deciding what to do quickly. The end result of a good size-up is an IAP that considers all the pertinent information, defines strategies and tactics, and assigns resources to complete those tactics.

A comprehensive evaluation of a situation requires information that often is not available to the fire officer. Consequently, the officer must look carefully and assess what can be seen, make reasonable assumptions about what cannot be seen, and anticipate what is likely to happen. An experienced officer will develop an initial plan and then adjust that plan as more information becomes available. Such additional information could, for example, require a major change in the plan. The IC must consciously differentiate among what is known, what is assumed, and what is anticipated in processing size-up information.

There are two models used when conducting a size-up process. Chief Lloyd Layman established the Five-Step Size-Up Process that was published by the NFPA in 1953 and has been the foundation of fireground analysis. In 1999, the National Fire Academy combined size-up practices with the incident management system to establish a three-phase size-up process that is taught in *Managing Company Tactical Operations*. Both models are described in this chapter, with some elements intertwined.

As noted earlier, the first phase of size-up begins long before the incident occurs. Everything the fire officer learns, observes, and experiences goes into his or her memory bank, and those memories are then used to make size-up decisions. Knowledge of the types of building construction used in the area, the available water sources, and the available resources are all factors that go into the size-up. Whenever a preincident plan is updated, including a familiarization visit, the fire officer is performing part of a future incident size-up.

Prearrival Information

The specific size-up for an incident begins with the dispatch. The name, location, and reported nature of the incident all help the fire officer anticipate what might be happening at the scene. Knowledge gained from a preincident plan and awareness of what would normally be found at that day and time are factored into the prearrival data.

On-Scene Observations

The ability to size up a fire situation quickly requires both a systematic approach and a solid foundation of reference information. SOPs list the essential size-up factors, which include building size and arrangement, type of construction, occupancy, fire and smoke conditions, and other factors, such as weather and time of day. The SOPs should guide the officer's systematic thinking to ensure that all of the important factors are considered.

A fire officer must understand and recognize basic fire dynamics. It is easy enough to identify fire coming out of a window and smoke pushing out from the eaves of a house. An understanding of conduction, convection, and radiation, however, is needed to predict where the fire is burning and where it will spread. An officer who has studied and observed fire dynamics is able to quickly process this information, without having to spend an excessive amount of time looking at the building and thinking about what is happening inside.

Visualization is one of the most significant factors in size-up. Every previous situation that the fire officer has experienced or observed, including some that might have been observed only in training sessions, photographs, or videos, is stored in the individual's memory. When a new situation is observed, the mind subconsciously looks for a matching image to create a template for this new observation. Instead of methodically processing the information, the brain can instinctively jump to a similar observation and apply the stored experience to the new set of circumstances. The same process works for other human senses, particularly smell, taste, and touch, fostering rapid recognition and interpretation of many situations (Klein, Calderwood, and Clinton-Circco, 1998).

Experienced fire officers know vigorous, churning dark smoke means a high heat release rate, indicating flashover conditions are present. Darker smoke is generally closer to the seat of the fire than lighter-colored smoke. By evaluating where the smoke is coming from in the building, the fire officer may be able to predict where the fire will be traveling, unless mitigating circumstances occur.

The volume and color of the smoke aid the officer in determining ventilation timing. Dark smoke is also an indication that there are more carbon particles suspended within the smoke and less oxygen. Before introducing additional air into a compartment that is belching dark, billowing smoke, the fire officer needs to "soften the target" with a quick application of water into the compartment and coordinate ventilation efforts with the fire hose crew.

Be cautious when approaching a modern structure with no smoke showing. UL's full-scale experiments showed that once a fire became ventilation limited, the smoke being forced out of the gaps of the structure greatly diminished or completely stopped. No smoke showing during size-up should increase the officer's caution and awareness of conditions inside the structure.

An understanding of fire dynamics is also needed to develop IAPs. Fires in the incipient stage may be extinguished with a portable fire extinguisher. Fires in the growth stage may require a small handline or even multiple large handlines or master streams.

Fuel load is another important factor during size-up. A fire in Class A combustibles normally dictates a direct attack with water. Fires involving Class B combustibles require the use of foams. When using foams, the officer must determine which type and how much foam will be needed to complete the job before beginning the application.

Lloyd Layman's Five-Step Size-Up Process

In 1940, Chief Lloyd Layman authored *Fundamentals of Fire Tactics*, which used the military training model with an emphasis on general principles. Layman presented a five-step process for analyzing emergency situations (Layman, 1953):

1. Facts
2. Probabilities
3. Situation
4. Decision
5. Plan of operation

Facts

Facts are the things that are known about the situation. Chief Layman was an early advocate of preincident planning. This process allows the officer to quickly determine the type of construction, the presence of any fire protection systems, fire flow requirements, water supply sources, special hazards, and other significant factors. Additional facts become known upon dispatch, such as the time of day, location of incident, weather, and specific resources that the fire department is sending. Visible smoke and fire conditions are additional facts. The more facts the fire officer can secure, the more accurately the situation can be sized up.

Probabilities

Probabilities are things that are likely to happen or can be anticipated based on the known facts. For example, the fire officer might know that the structure has a fire sprinkler system but cannot be sure that it will work. The sprinkler system may have been turned off for maintenance or damaged by an explosion. During size-up, the fire officer must consider the most likely possibilities based on the known facts.

The fire officer has to anticipate where the fire is likely to spread. If the fire is burning freely in the attic, how is it likely to affect the stability of the building? This skill requires an in-depth knowledge of factors such as building construction, fire dynamics, hazardous materials, sprinkler system performance, and human behavior in fires.

Another aspect of considering probabilities is the amount of time it takes to accomplish fire-ground tasks. How far is the fire likely to advance before the hose lines are in place to attack? How long will it take to get ground ladders to Charlie side? How fast can a crew with a backup hose line get to your location? How large will the fire be in 10 minutes? Which type of damage is being done to the structural components of the building?

Situation

The situation assessment involves three considerations. The first consideration is whether the resources on scene and en route will be sufficient to handle the incident. The IC will be able to determine whether additional companies will be needed to control the problem. The IC also has to anticipate the time lag between requesting resources and having them on the scene available to use.

Many fire departments have policies that require an IC to request additional resources whenever all of the units will be committed to firefighting operations, to provide an on-scene reserve. Some departments require the first-arriving company to call for a second alarm immediately if any smoke or fire is showing from a high-rise, hospital, or "big-box" mercantile building.

The second consideration is the specific capabilities and limitations of the responding resources in relation to the problem. Consider the amount of water carried, pump capacity, amount and diameter of hose, and length and type of aerial devices. Staffing is an important factor. A company staffed with two fire fighters cannot perform as much work as a company of six fire fighters. The IC might have to call for additional resources or adjust performance expectations downward based on the resources at hand.

The third consideration consists of the capabilities and limitations of the personnel, based on training and experience. Consider your personal experience with this type of incident, as well as the experience and capabilities of your crew members. If two fire fighters out of a four-person crew are rookies, the company requires more direct supervision.

Decision

The fourth step in Layman's size-up procedure is making fire attack decisions. This step requires the fire officer to make specific judgment decisions based on the known facts and probabilities, as well as the situation evaluation. The officer needs to answer four questions:

1. Are there enough resources responding to and on the scene to extinguish the fire or mitigate the situation?
2. Are sufficient resources available, and do conditions allow for an interior attack?
3. What is the most effective assignment of on-scene resources?
4. What is the most effective assignment of responding resources?

The answers to these questions allow the fire officer to develop the overall strategy and the underlying tactics to mitigate the incident.

Plan of Operation

The final step in Layman's decision process is to develop the actual plan that will be used to mitigate the incident. Most other methods of size-up do not include this step as part of the size-up; instead, they refer to the information that is obtained during size-up being used to develop a plan of action.

National Fire Academy Size-Up Process

The National Fire Academy (NFA, 1999) has developed a size-up system that includes three phases:

1. Preincident information
2. Initial size-up
3. Ongoing size-up

Phase One: Preincident Information

Phase one considers what you know before the incident occurs. This consideration closely mirrors the facts step identified by Layman. Preincident plans often provide valuable information for this phase. Information about the building and the occupancy, such as the building layout and construction type, built-in protection systems, type of business, and nature of the contents, is clearly needed to perform an accurate size-up. Preincident information should also identify water supply sources, including their location, accessibility, and capacity.

If preincident plans are not available, this information must be determined through on-scene observation and research. The fire officer's intuition and experience may have to be relied on until this information can be obtained.

A fire officer's memory stores a large amount of useful information that figures into the size-up process. Environmental information includes factors such as heat and cold extremes, humidity, wind, and snow or ice accumulations. The time of day can also be a critical piece of information; responding to fire in a multifamily residential building at 2:00 AM involves different considerations from responding to the same fire at 2:00 PM. A fire officer also needs to be familiar with departmental and mutual aid resources.

Phase Two: Initial Size-Up

The second phase of the size-up begins on receipt of an alarm. Three questions need to be answered:

1. What do I have?
2. Where is it going?
3. How do I control it?

Start with the information that was established in phase one and build upon it. Phase two considers the specific conditions that are present at the incident. The IC's initial actions should include a 360-degree walk-around of a fire building to observe conditions, such as the location and the size of the fire, as well as the volume, color, movement, and location of the smoke. The size-up should also assess the size and construction of the building, identify any exposures that are present, and look for indications of special concerns, such as the possibility that lightweight construction could be involved.

To determine where the fire is going, the fire officer must consider the current location and stage of development of the fire, the direction in which it is likely to spread, and the likely impact of fire suppression efforts. Is this a ventilation-limited fire? Is there an imminent risk of structural collapse? The accuracy of these predictions is heavily dependent on the fire officer's knowledge and experience. This stage is closely associated with the probabilities step noted by Layman.

Answering the third question, "How do I control it?", requires the fire officer to consider different alternatives in relation to the available resources, including apparatus, equipment, and personnel. The development of strategy and tactics must be based on the resources that are available and, if additional resources are needed, the time it will take for them to arrive. This stage is closely associated with the situation step noted by Layman.

Phase Three: Ongoing Size-Up

The third phase addresses the need to continually size up the situation as it evolves. This phase includes an ongoing analysis of the situation and an ongoing evaluation of the effectiveness of the plan being executed. The IC should always be prepared to modify the plan if the situation changes, including switching between offensive and defensive strategies. Additional resources could be required to address unanticipated

FIRE OFFICER TIP

Murrah Federal Building Bombing

The 1995 bombing of the Alfred P. Murrah Federal Building in Oklahoma City, Oklahoma, added a new element to the incident management process. This incident represented the largest response of local, regional, and state agencies to a single incident in the Southwest. It also involved the largest mobilization of the federally funded Urban Search and Rescue Task Forces, with 11 of the 28 teams responding to the incident. The federal after-action report identified the value of establishing a written IAP for every 12-hour work period. Each IAP identified the strategic goals, tactical objectives, and support requirements for the work period. While written IAPs were used in wildland fires, the rescue and recovery operations at Oklahoma City established the practice of using IAPs for all types of emergency incidents.

needs or to replace tired crews. The ongoing evaluation could also indicate that resources can be redeployed from one task to another where they would be more effective (Brunacini, 2004).

The ongoing size-up requires a constant flow of feedback information to the IC. The IC needs to know when:

- An assignment is completed.
- An assignment cannot be completed.
- Additional resources are needed.
- Resources can be released.
- Conditions have changed.
- Additional problems have been identified.
- Emergency conditions exist.

Developing an Incident Action Plan

After size-up, the IC develops an **incident action plan (IAP)** based on the incident priorities. The IAP is a basic component of ICS; all incidents require an action plan. The IAP outlines the strategic objectives and states how emergency operations will be conducted. At most incidents, the IAP is relatively simple and can be expressed by the IC in a few words or phrases.

A written IAP is required for large or complex incidents that have an extended duration. The IAP for a large-scale incident can be a lengthy document that is regularly updated and used for daily briefings of the command staff. There are two major components to the IAP:

1. The determination of the appropriate strategy to mitigate an incident
2. The development of tactics to execute the strategy

Strategies are general, whereas tactics are specific and measurable. In non-fire terms, strategies are equivalent to goals, whereas tactics are the objectives used to meet the goals. The IC identifies the strategic goal and then identifies the tactical objectives that are required to reach this goal. Resources (companies) are assigned to perform individual tasks that allow the tactical objectives to be achieved.

SOPs are used to provide a consistent structure to the process of establishing strategies, tactics, and tasks. Without SOPs, every situation might be managed differently, depending on the individual who was in command and his or her interpretation of the situation, as well as the decisions made by individual company officers. SOPs guide the decision-making process and ensure consistency between officers and events. On arrival at a house fire with smoke showing, the SOP might require the first-arriving officer to assume command, the first-arriving engine company to begin a fire attack, and the second-arriving engine to provide a supply line for the first engine. Thus SOPs assist everyone involved in the process by outlining which actions to take when presented with a given situation. In effect, SOPs are preestablished components of an IAP.

Incident Priorities

There are three basic priorities for an IAP:

1. Life safety
2. Incident stabilization
3. Property conservation

Incident stabilization and property conservation are often addressed simultaneously, although property conservation is ranked lower on the priority list. If the IC does not have sufficient resources to perform both of these operations, property conservation measures might have to be delayed until the incident has been stabilized. In reality, many of the actions that are taken to stabilize an incident also work toward assuring life safety.

The life-safety priority refers to all people who are at risk due to the incident, including the general public as well as fire department personnel. The incident stabilization priority is directed toward keeping the incident from getting any worse. If one structure is fully involved, protecting the exposures is part of incident stabilization. In such a case, the burning structure cannot be saved, but fire fighters can ensure that the fire will not spread beyond that building.

Property conservation is directed toward preventing any additional damage from occurring. Such measures could include minimizing water damage by covering building contents with salvage covers or removing valuable property from harm's way. Property can also be protected by venting smoke and heat from a building and then covering the openings to keep wind and rain from causing additional damage.

Tactical Priorities

At the scene, the IC employs many strategies and tactics to meet the three incident priorities. If an abundance of resources arrived simultaneously, it would be a simple matter to assign them in a manner that would get everything done immediately. Because this rarely happens, fire officers need a method to address the most critical concerns first. Tactical priorities provide a list and an order of priority for dealing with them. The tactical priorities and the information obtained in the size-up are used to develop the IAP.

RECEO-VS

Lloyd Layman developed a general guideline for ICs to systematically address the incident priorities. Layman's method remains one of the most referenced, 70 years after it was first published (Layman, 1953). RECEO-VS is an acronym developed for the IC for accomplishing tactical priorities on the fire ground (**TABLE 7-1**).

- Rescue
- Exposure protection
- Confinement
- Extinguishment
- Overhaul
- Ventilation
- Salvage

S.L.I.C.E.-R.S.

S.L.I.C.E.-R.S. incorporates the proven methodology learned from Lloyd Layman and pairs it with the vitally important data from UL and NIST in an attempt to provide a comprehensive tactical checklist for initially arriving officers operating in a fire attack mode (IAFC, 2013).

- **Size up:** Gather and analyze information to help develop the IAP.
- **Locate:** Determine the location and extent of the fire inside a building.

TABLE 7-1 RECEO-VS

Incident Priorities	RECEO-VS Acronym	Meaning
Life safety	**R**escue	The removal of victims from a life-threatening situation
Stabilize the incident	**E**xposure protection	The protection of surrounding structures from a fire
	Confinement	The containment of a fire to one area or the prevention of further areas from becoming involved in the fire
Property conservation	**E**xtinguishment	The complete extinguishment of the fire
	Overhaul	The process of ensuring that the fire is extinguished completely
	Ventilation	The process of removing smoke, heat, and the products of combustion from a structure
	Salvage	The removal or protection of property that could be damaged during firefighting or overhaul operations

Data from International Association of Fire Chiefs (IAFC). 2013 December 23. *Firefighter Safety Call to Action: New Research Informs Need for Updated Procedures, Policies.* Fairfax, VA: IAFC.

- **I**dentify and control the flow path, if possible.
- **C**ool the space from the safest location: Strategically apply a brief, straight stream of water through an opening to cool the fire before making entry.
- **E**xtinguish the fire: Fully extinguish the fire, including overhaul of void spaces.
- **R**escue: Conduct search and rescue operations if indicated by a risk/benefit analysis.
- **S**alvage: Protect property from further damage.

Size up. A tactical plan for a fire must be developed, communicated and implemented. First-arriving officers/incident commanders are responsible for obtaining a 360-degree view of the structure involved.

Locate the Fire. The location and extent of the fire in the building must be determined and communicated to dispatch and other responding companies. Officers should use all means available to make this determination. Thermal imagers can prove valuable during the initial 360-degree lap of the structure. The location of the fire and current conditions will dictate the best location to attack the fire.

Identify and Control the Flow Path. Identify the presence and/or location of the flow path. Flow path evaluations should include openings acting as intakes and discharges as well as smoke indicators of color, volume, and velocity of the flow path.

Efforts should be taken to control ventilation and the flow path to protect potential building occupants and limit fire growth. If a flow path is visible, consider closing doors and windows to limit air flow. When closing doors and windows, firefighters should be aware of any potential rescues readily accessible via doors/windows. This would be the opportunity for the officers to identify areas available for O-V.E.I.S. (Oriented-Vent-Enter-Isolate-Search) (ISFSI, 2014). O-V.E.I.S. is a high-risk search technique using a team of two to access a room from an outside window, close the door to the hallway and conduct a primary search.

Ventilation is designed to remove the products of combustion from a fire area and allow cool, fresh air to enter. Effective ventilation can improve the survival chances for occupants as well as the conditions that would allow fire fighters to enter and operate inside a building (Traina et al., 2017). When operating at ventilation-limited fires, controlling the door and coordinating ventilation with water application are vital to controlling the flow path early in suppression operations.

Fire and Rescue Departments of Northern Virginia make this observation:

> Perhaps at no other time in our fire service has ventilation played such a critical role in successful operations. Fire department personnel should manage, and control the openings (doors, windows) to the structure to limit fire growth and spread and to control the flow path of inlet air and

fire gases during tactical operations. All ventilation must be coordinated with suppression activities.

Uncontrolled ventilation allows additional oxygen into the structure which may result in a rapid increase in the size and hazards associated with a quickly expanding fire, due to increased heat release rates. Ventilation occurs with many of our actions, even if it is not intended, such as when we perform forcible entry.

This action is necessary to gain access to the seat of the fire but without the proper application of water this form of ventilation can be detrimental to life safety, incident stabilization, and property conservation (NOVAFire, 2014).

Cool the Space from the Safest Location. Given information obtained during the size up and while locating the fire and identifying and controlling the flow path, the officer will determine if high heat/untenable conditions exist inside the structure. When these conditions are present, the officer will determine the safest and most direct way to apply water to the superheated space, or directly onto the fire when possible. The primary goal in this step is to reduce the thermal threat to fire fighters and potential occupants as soon as reasonably possible.

NIST and UL research shows the dramatic impact of initial application of 30 to 60 seconds of water into a compartment: dramatic reduction of heat within the flow path, improved occupant survival, and safer interior suppression conditions (Zevotek, Stakes, and Willi, 2018).

This does not mean every fire must be hit from the exterior prior to entry. It means that the officer must make an educated and informed decision on the quickest and most efficient deployment of water to the seat of the fire with the greatest outcome (NOVAFire, 2014).

Extinguish the Fire. Once the thermal threats have been controlled, the fire should be extinguished in the most direct manner possible. This may mean hitting the fire from the exterior initially with a short blast of water to knock down or "reset" the fire, followed by the rapid deployment of a hose line to the seat of the fire via the interior to fully extinguish the fire.

Rescue. The first-arriving officer must make a rapid and informed choice on the priority and sequence of suppression activities verses occupant removal. The officer should consider the potential for rescues at all times. Fire fighters should be prepared to remove occupants. As life safety is the highest tactical priority, rescue shall always take precedence. The incident commander must determine the best course of action to ensure the best outcome for occupants based on the conditions at that time (NOVAFire, 2014).

Salvage. Fire fighters should use compartmentalization to control fire spread and smoke whenever possible and protect occupant's possessions in the best available means once fire has been extinguished (ISFSI, 2014).

Location Designators

For the purposes of firefighting operations, the exterior sides of a building are designated as sides A (Alpha), B (Bravo), C (Charlie), and D (Delta). The front of the building is designated as side A, with sides B, C, and D following in a clockwise direction around the building. The companies working in front of the building are assigned to division A, and the radio designation for their supervisor is "division A." Similar terminology is used for the sides and rear of the building.

The areas adjacent to a burning building are called exposures. Exposures take the same letter as the adjacent side of the building. A fire fighter facing side A can, therefore, see the adjacent building on the left (exposure B) and the building to the right (exposure D). If the burning building is part of a row of buildings, the buildings to the left are called exposures B, B1, B2, and so on. The buildings to the right are exposures D, D1, D2, and so on.

Within a building, divisions commonly take the number of the floor on which they are working. For example, fire fighters working on the fifth floor would be in division 5, and the radio designation for the chief assigned to that area would be "division 5." Crews performing different tasks on the fifth floor would all be part of this division. Common simple terminology should be used for designations whenever possible, such as "roof division" and "basement division."

Tactical Safety Considerations

Fighting fires is an inherently dangerous activity that exposes fire fighters to a wide variety of risks and hazards. Many advances have been made, including improved protective clothing and equipment, more effective methods of fighting fires, and better procedures for managing fire incidents. The hazards still exist, however, and new threats continue to appear.

Protective clothing and SCBA provide better protection than was available to earlier generations of fire fighters. To benefit from this protection, the full ensemble of personal protective clothing and equipment must be worn whenever fire fighters are exposed to hazardous conditions.

PPE allows fire fighters to enter deeper into burning buildings and stay longer, often without sensing the level of heat in that environment. PPE has saved many lives and prevented disabling injuries. Unfortunately, it also exposes fire fighters to an increased risk of being trapped by a sudden flashover, becoming disoriented inside a burning building, or running out of air in a highly toxic atmosphere.

The weight, bulk, and thermal properties of PPE must be considered during extreme weather conditions. Fire fighters who are wearing PPE need sufficient rehabilitation between work periods.

Scene Safety

The NFPA reports that 64 fire fighters were killed in the line of duty in 2018. These deaths occurred during fire-ground operations; nonemergency incident scenes; and in nonemergency situations, such as in fire stations, during training, and while responding to or returning from emergency situations (**FIGURE 7-10**).

The largest share of deaths occurred from stress, overexertion, and medical issues (**TABLE 7-2**). Cardiac events accounted for 89 percent of these deaths. The second leading cause of death was vehicle crashes. In 2018, 2 fire fighters died in vehicle accidents, a majority of whom died responding to or returning from incidents.

The NFPA estimates that 58,835 fire fighters were injured in the line of duty in 2017, a 5 percent decrease from the previous year. **FIGURE 7-11** shows the breakdown of injuries by type of duty.

Fewer than half of the injuries occurred on the fire ground. The leading cause of fire-ground injuries was overexertion or strain (**TABLE 7-3**).

Many emergency operations are conducted in the street, exposing fire fighters to traffic hazards. Apparatus should be positioned to protect the scene, fire

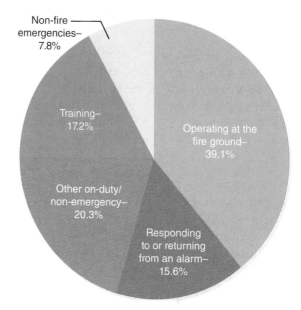

FIGURE 7-10 Fire fighter deaths in the United States by type of duty. Sixty-four fire fighters died in the line of duty in 2018.
Reproduced from National Fire Protection Association, *Fire Fighter Fatalities in the United States*, 2018.

TABLE 7-2 Fire Fighter Deaths by Cause of Injury	
Overexertion, stress, medical	44%
Vehicle accidents	19%
Structural collapse	9%
Rapid fire progress/explosion	8%
Struck by vehicles	5%
Struck by objects	5%
Falls	3%
Exposure	3%
Lost inside	2%
Caught under water	2%
Assault	2%

Reproduced from National Fire Protection Association, *Fire Fighter Fatalities in the United States*, 2018.

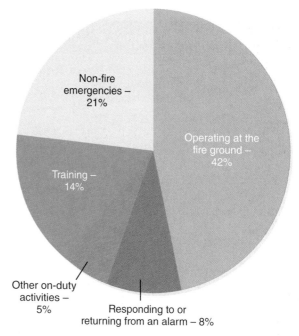

FIGURE 7-11 Fire fighter injuries in the United States by type of duty. In 2017, 58,835 fire fighters were injured in the line of duty.
Reproduced from National Fire Protection Association, *Fire Fighter Fatalities in the United States*, 2018.

fighters need to wear high-visibility safety vests or jackets that comply with American National Standards Institute (ANSI) 207, and the fire officer should request traffic control to reduce the risk of fire fighters being struck by vehicles.

Hazardous areas at the incident scene, such as holes and potential collapse zones, should be clearly identified. Some fire departments use barrier tape to establish exclusion zones around hazards. Many departments utilize hot/warm/cold zones to identify the level of hazard:

- Hot zone: Highest hazard; minimum number of crew allowed in the area. The hot zone could be an area of structural weakness, a hazardous environment, or an area requiring use of full PPE, including SCBA. A hot zone is under direct supervision of a unit, branch, or group leader with an identified entry/exit point.

- Warm zone: Moderate hazard; restricted number of emergency service members allowed in the area. This area remains part of an active incident and requires the use of some protective equipment (helmet, boots, or respirator) or special vigilance.

- Cold zone: Minimum hazard; no restrictions on number of crew or nonemergency service individuals allowed in the area.

The fire officer should encourage fire fighters to stash a spare pair of socks, gloves, a towel, and a knit cap on the fire apparatus. Getting soaked in a winter fire is a common occupational hazard. If they add a T-shirt, shorts, and coveralls to this stash, fire fighters can have dry clothes after a soaking or a technical decontamination.

Exposure to asbestos building components, drug laboratories, and chemical/biological weapons are three situations in which a fire crew may have to disrobe at the incident scene. A technical decontamination may result in the fire fighters losing their uniforms in the warm zone. Without the stash, the only option may be wearing a disposable Tyvek gown.

Fire fighters should change out of their duty uniforms and clean up before they leave the fire station, even after a slow duty day. Infectious diseases, asbestos, industrial chemicals, and products of combustion are hazards that should stay at the fire station and not be brought home to expose family members.

Rapid Intervention Crews

The **rapid intervention crew (RIC)** is "a dedicated crew of firefighters who are assigned for rapid deployment to rescue lost or trapped members." NFPA 1407 describes required capabilities of the RIC. An **initial rapid intervention crew (IRIC)** is two members from the initial attack crew, as defined in NFPA 1710, *Standard for the Organization and Deployment of Fire Suppression Operations, Emergency Medical Operations and Special Operations to the Public by Career Fire Departments*. A RIC, by comparison, consists of four members and is sometimes identified as a rapid intervention team (RIT). This dedicated crew is not to be confused with the IRIC. Most fire departments add an additional fire company to the first alarm assignment for a structure fire, either as part of the initial dispatch or as soon as a working fire is reported, to function as the RIC.

Many departments have developed RIC training and deployment procedures. The standard radio term *mayday* is used to indicate that a fire fighter is lost, missing, or in life-threatening danger. The IC's standard response to a mayday is to activate the RIC. When this occurs, the RIC operation has priority over all other tactical functions.

The RIC is generally positioned outside the building, near an entrance, ready for immediate action. SOPs governing RICs generally specify a list of equipment that will be prepared for use, including forcible entry tools, a thermal imager, a search rope, medical equipment, and a spare SCBA or full replacement cylinder. Every RIC team member should have a portable radio and should be wearing full PPE. While standing by, the RIC members should closely monitor fire-ground radio communications to keep track

TABLE 7-3 Fire-Ground Injuries by Cause

Types of Injury	Percentage of Total Injuries
Overexertion/strain	29%
Fall, jump, slip	210%
Other	16%
Exposure to fire products	11%
Contact with an object	9.7%
Struck by an object	6%
Exposure to chemicals or radiation	4%
Extreme weather	3%

Reproduced from National Fire Protection Association, *Fire Fighter Fatalities in the United States*, 2018.

FIRE OFFICER TIP

On-Duty Wallet, Phone, and Keys

The fire officer should also encourage the crew to carry an on-duty wallet. The on-duty wallet can be lost during decontamination without much disruption to the fire fighter's personal life.

The on-duty wallet includes a duplicate driver's license, copies of any state- or department-mandated certification cards (e.g., paramedic), and, if desired, a credit card. The fire fighter should place enough money in this wallet to get through the day and leave his or her regular wallet locked in the personal vehicle or the fire department locker. The same goes for a smartphone. Consider using a disposable or spare phone while on duty.

Create a set of duty keys. Make duplicates of the keys you need to get into the fire station, your locker, and your car. As with the on-duty wallet, you can lose your duty keys and not have your life disrupted to a great extent. Do not forget to put a durable identification tag on your duty keys so that you can get them back after they are sanitized! Lock up your regular keys, smart/cell phone, wedding ring, and other valuable jewelry before you report for duty. It may be a good idea to use a combination lock to secure your storage space.

of activities and listen for a mayday or any indication of a problem.

Transfer of Command

A chief fire officer assumes command of significant incidents. At a large incident, command could be transferred more than once, depending on the situation and the chain of command, as successively higher-ranking or more qualified officers arrive at the incident scene.

Command should be transferred only to improve the quality of the command organization. It should not be transferred simply because a higher-ranking officer has arrived, and it should not be transferred more times than is necessary to manage the incident effectively.

Transfer of command should follow a standard procedure:

1. The officer assuming command communicates with the initial IC. This can occur over the radio, but a face-to-face meeting is preferred.
2. The initial IC briefs the new IC, including the following information in his or her report:
   - Incident conditions, such as the location and extent of fire, the number of patients, or the status of the hazardous materials spill or leak
   - Tactical worksheet and IAP
   - Progress toward completion of the tactical objectives
   - Safety considerations
   - Deployment and assignment of operating companies and personnel
   - Need for additional resources
3. Command is officially transferred only when the new IC has been briefed.
4. The fact that command has been transferred is communicated to the dispatch center and all units operating on the fire scene.

The initial IC should always review the tactical worksheet with the new IC. This worksheet outlines the status and location of personnel and resources in a standard form. It is especially important on those events where the first-arriving company initially started in the command mode.

After the transfer of command has occurred, the new IC determines the most appropriate assignment for the previous IC. A company-level officer may be assigned as a division or **group supervisor** or may remain with the new IC at the command post. The size and the complexity of the incident determine how much the management structure will need to be expanded. Large or complex incidents requiring a variety of resources may require the use of the National Incident Management System ICS-201 briefing form as more of the ICS roles are staffed (FEMA, 2013).

You Are the Fire Officer Conclusion

It is vital that the first-arriving officer complete a size-up and report the situation to the dispatch center and responding units. The officer is setting the stage and tone for the first hour of this operation. While the federal National Incident Management System provides structure for a greater-than-first-alarm incident, the company officer needs to use the Lloyd Layman or National Fire Academy size-up to establish the initial strategy and tactics to impose order on the chaos that brought the fire department to this incident. An incident action plan provides the structure and documentation for the task-level workers to safely complete their assignments in a coordinated fashion.

After-Action REVIEW

IN SUMMARY

- A fire officer is expected to perform the duties of a first-arriving officer at any incident, including assuming initial command of the incident, establishing the basic management structure, and following standard operating procedures.
- The national incident management system evolved from the southern California FIRESCOPE and the Phoenix Fire Ground Command (FGC) programs.
- The NIMS provides a consistent, nationwide framework for incident management, enabling federal, state, and local governments, private sector and nongovernmental organizations, and all other organizations who assume a role in emergency management to work together effectively and efficiently.
- The NIMS can expand as necessary to handle larger and more complex incidents.
- There are three levels of command in the ICS: strategic, tactical, and task.
- The first-arriving fire officer has the responsibility to establish command and manage the incident until relieved by a higher-ranking officer. This task is accomplished in addition to supervising the members of the officer's own company.
- The first-arriving managing and supervising fire officers are required to focus on the strategic level as they arrive at an emergency.
- The IC is responsible for addressing three strategic priorities: life safety, incident stabilization, and property conservation.
- Command must be established and the ICS must be used at every event.
- There are nine functions of command:
  - Determining strategy
  - Selecting incident tactics
  - Establishing the IAP
  - Developing the ICS organization
  - Managing resources
  - Coordinating resource activities
  - Providing for scene safety
  - Releasing information about the incident
  - Coordinating with outside agencies
- In reality, size-up begins long before arrival and continues until the incident is stabilized.
- The IAP outlines the strategic objectives and states how emergency operations will be conducted.
- There are three basic priorities for an IAP: life safety, incident stabilization, and property conservation.
- RECEO-VS is an acronym developed for the IC for accomplishing tactical priorities on the fire ground.
  - **R**escues
  - **E**xposure protection
  - **C**onfinement
  - **E**xtinguishment
  - **O**verhaul
  - **V**entilation
  - **S**alvage
- S.L.I.C.E.-R.S. is an acronym developed for first-arriving officers operating in a fire attack mode.
  - **S**ize up: Gather and analyze information to help develop the IAP.

- **L**ocate: Determine the location and extent of the fire inside a building.
- **I**dentify and control the flow path, if possible.
- **C**ool the space from the safest location: Strategically apply a brief, straight stream of water through an opening to cool the fire before making entry.
- **E**xtinguish the fire: Fully extinguish the fire, including overhaul of void spaces.
- **R**escue: Conduct search and rescue operations if indicated by a risk/benefit analysis.
- **S**alvage: Protect property from further damage.

- At a large incident, command could be transferred more than once, depending on the situation and the chain of command, as successively higher-ranking or more qualified officers arrive at the incident scene.
- Fighting fires is an inherently dangerous activity that exposes fire fighters to a wide variety of risks and hazards. Many advances have been made, including improved protective clothing and equipment, more effective methods of fighting fires, and better procedures for managing fire incidents.

KEY TERMS

Defensive operation Actions that are intended to control a fire by limiting its spread to a defined area, avoiding the commitment of personnel and equipment to dangerous areas. (Reproduced from National Fire Protection Association, *Standard on Fire Department Occupational Safety, Health, and Wellness Program:* 1500.)

Fire Ground Command (FGC) An incident management system developed in the 1970s for day-to-day fire department incidents (generally handled with fewer than 25 units or companies).

FIRESCOPE Fire Resources of California Organized for Potential Emergencies; an organization of agencies established in the early 1970s to develop a standardized system for managing fire resources at large-scale incidents such as wildland fires.

Flow path The movement of heat and smoke from the higher pressure within the fire area towards the lower pressure areas accessible via doors, window openings, and roof structures. (Reproduced from National Fire Protection Association, *Standard on Training for Emergency Scene Operations:* 1410.)

Fuel-limited fire A fire in which the heat release rate and fire growth are controlled by the characteristics of the fuel because there is adequate oxygen available for combustion. (Reproduced from National Fire Protection Association, *Standard on Training for Emergency Scene Operations:* 1410.)

Group supervisor A person in a supervisory level position responsible for a functional area of operation. (Reproduced from National Fire Protection Association, *Standard on Emergency Services Incident Management System and Command Safety:* 1561.)

Incident action plan (IAP) A verbal or written plan containing incident objectives reflecting the overall strategy and specific control actions where appropriate for managing an incident or planned event. (Reproduced from National Fire Protection Association, *Standard for Incident Management Personnel Professional Qualifications:* 1026.)

Incident commander (IC) The individual responsible for all incident activities, including the development of strategies and tactics and the ordering and release of resources. (Reproduced from National Fire Protection Association, *Standard on Operations and Training for Technical Search and Rescue Incidents:* 1670.)

Incident command system (ICS) The combination of facilities, equipment, personnel, procedures, and communications operating within a common organizational structure that has responsibility for the management of assigned resources to effectively accomplish stated objectives pertaining to an incident or training exercise. (Reproduced from National Fire Protection Association, *Standard on Fire Department Occupational Safety, Health, and Wellness Program:* 1500.)

Initial rapid intervention crew (IRIC) Two members of the initial attack crew who are assigned for rapid deployment to rescue lost or trapped members. (Reproduced from National Fire Protection Association, *Standard for the Organization and Deployment of Fire Suppression Operations, Emergency Medical Operations, and Special Operations to the Public by Career Fire Departments:* 1710.)

Legacy dwellings Single-family dwellings constructed before 1980.

Modern dwellings Single-family dwellings constructed since 1980; they are typically larger structures with an open house geometry, lightweight construction materials, and exponentially increased fuel load.

National Incident Management System (NIMS) A system mandated by Homeland Security Presidential Directive-5 (HSPD-5) that provides a systematic, proactive approach guiding government agencies at all levels, the private sector, and nongovernmental organizations to work seamlessly to prepare for, prevent, respond to, recover from, and mitigate the effects of incidents, regardless of cause, size, location, or complexity, in order to reduce the loss of life or property and harm to the environment. (Reproduced from National Fire Protection Association, *Standard for Incident Management Personnel Professional Qualifications:* 1026.)

Offensive operation Actions generally performed in the interior of involved structures that involve a direct attack on a fire to directly control and extinguish the fire. (Reproduced from National Fire Protection Association, *Standard on Fire Department Occupational Safety, Health, and Wellness Program:* 1500.)

Personnel accountability report (PAR) Periodic reports verifying the status of responders assigned to an incident or planned event. (Reproduced from National Fire Protection Association, *Standard for Incident Management Personnel Professional Qualifications:* 1026.)

Rapid intervention crew (RIC) A minimum of two fully equipped personnel on site, in a ready state, for immediate rescue of disoriented, injured, lost, or trapped rescue personnel. (Reproduced from National Fire Protection Association, *Standard on Fire Department Occupational Safety, Health, and Wellness Program:* 1500.)

Risk/benefit analysis A decision made by a responder based on a hazard identification and situation assessment that weighs the risks likely to be taken against the benefits to be gained for taking those risks. (Reproduced from National Fire Protection Association, *Standard on Operations and Training for Technical Search and Rescue Incidents:* 1670.)

Size-up The process of gathering and analyzing information to help fire officers make decisions regarding the deployment of resources and the implementation of tactics. (Reproduced from National Fire Protection Association, *Standard on Training for Emergency Scene Operations:* 1410.)

S.L.I.C.E.-R.S. An acronym intended to be used by the first arriving company officer to accomplish important strategic goals on the fire ground.

Standard time–temperature curve A recording of fire temperature increase over time.

Strategic level Command level that entails the overall direction and goals of the incident.

Tactical level Command level in which objectives must be achieved to meet the strategic goals. The tactical-level supervisor or officer is responsible for completing assigned objectives.

Tactical worksheet A form that allows the incident commander to ensure all tactical issues are addressed and to diagram an incident with the location of resources on the diagram.

Task level Command level in which specific tasks are assigned to companies; these tasks are geared toward meeting tactical-level requirements.

Transitional attack An offensive fire attack initiated by an exterior, indirect handline operation into the fire compartment to initiate cooling while transitioning into interior direct fire attack in coordination with ventilation operations.

Two-in/two-out rule A guideline created in response to OSHA Respiratory Regulation (29 CFR 1910.134), which requires a two-person team to operate within an environment that is immediately dangerous to life and health (IDLH) and a minimum of a two-person team to be available outside the IDLH atmosphere to remain capable of rapid rescue of the interior team.

Ventilation-limited fire A fire in which the heat release rate and fire growth are regulated by the available oxygen within the space. (Reproduced from National Fire Protection Association, *Standard on Training for Emergency Scene Operations:* 1410.)

REFERENCES

Asheville Fire Department (AFD). 2015. *Report on Rapid Intervention Operations Analysis.* Asheville, NC: Asheville Fire Department.

Brunacini, Alan V. 2002. *Fire Command: The Essentials of IMS.* Phoenix, AZ: Heritage Publishers.

Brunacini, Alan V. 2004. *Command Safety: The IC's Role in Protecting Firefighters.* Stillwater, OK: Fire Protection Publications.

Buchanan, Eddie. (ISFSI) 2014 August 14. *Principles of Modern Fire Attack: SLICE-RS Overview.* 18:49 video. Centreville, VA: International Society of Fire Service Instructors.

Evarts, Ben and Joseph L. Molis. 2018. United States Firefighter Injuries 2017. Quincy, MA: National Fire Protection Association.

Fahy, Rita F., and Joseph L. Molis. 2019. *Firefighter Fatalities in the United States - 2018.* Quincy, MA: National Fire Protection Association.

Federal Emergency Management Agency (FEMA). 2017. *National Incident Management System*. Washington, DC: U.S. Department of Homeland Security.

Federal Emergency Management Agency (FEMA) 2013. *Incident Briefing (ICS 201)*. Washington, DC: U.S. Department of Homeland Security.

International Association of Fire Chiefs (IAFC) Safety, Health and Survival Section. 2012. *Rules You Can Live By*. Fairfax, VA: IAFC.

International Association of Fire Chiefs (IAFC). 2013 December 23. *Firefighter Safety Call to Action: New Research Informs Need for Updated Procedures, Policies*. Fairfax, VA: IAFC.

Kerber, Stephen. 2012. "Analysis of Changing Residential Fire Dynamics and Its Implications on Firefighter Operational Timeframes." *Fire Technology* 48 (4): 865–891.

Kerber, Stephen. 2013. *Analysis of One and Two-Story Family Home Fire Dynamics and the Impact of Firefighter Horizontal Ventilation*. Chicago, IL: Underwriter's Laboratory

Klein, Gary, Roberta Calderwood, and Anne Clinton-Ciricco. 1988. *Rapid Decision Making on the Fire Ground*. Alexandria, VA: U.S. Army Research Institute for the Behavioral and Social Science.

Kries, Steve. 2003, December 1. "Rapid Intervention Isn't Rapid." *Fire Engineering*. Accessed July 2, 2019. https://www.fireengineering.com/articles/print/volume-156/issue-12/features/rapid-intervention-isnt-rapid.html.

Layman, Lloyd. 1953. *Fundamentals of Fire Tactics*. Boston, MA: National Fire Protection Association.

Madrzykowski, Daniel. 2013. "Fire Dynamics: The Science of Firefighting." *International Fire Service Journal of Leadership and Management* 7 (2): 7–15.

Miles, John B. 1998. *Two-In/Two-Out Procedure in Firefighting/IDLH Environments*. Washington, DC: U.S. Department of Labor.

Murgallis, Robert, and Burton Phelps. 2012. *Command and Control: ICS, Strategy Development, and Tactical Selections, Book 1*. Stillwater, OK: Fire Protection Publications.

National Fire Academy (NFA). 1999. *Module 1: The Command Sequence. Managing Company Tactical Operations: Decision making*. Emmitsburg, MD: Federal Emergency Management Agency, U.S. Fire Administration.

National Institute of Standards and Technology (NIST). 2010. *Fire Dynamics*. Washington, DC: U.S. Department of Commerce. Accessed June 19, 2019. https://www.nist.gov/el/fire-research-division-73300/firegov-fire-service/fire-dynamics.

Page, James O. 1973. *Effective Company Command*. Alhambra, CA: Borden Publishing.

Senior Operations Chiefs Committee (NOVAFire). 2014 March 1. *Document to Provide Clarity on the use of RECEO VS and SLICER RS*. Fairfax, VA: Fire and Rescue Departments of Northern Virginia.

Traina, Nicholas, Stephen Kerber, Dimitrios C. Kyritsis, and Gavin P. Horn. 2017. "Occupant Tenability in Single Family Homes: Part II: Impact of Door Control, Vertical Ventilation and Water Application." *Fire Technology* 53 (4): 1611-1640.

UL Firefighter Safety Research Institute (UL FSRI). 2014. *Scientific Research for the Development of More Effective Tactics: Governors Island Experiments July 2012*. [Online training program.] Accessed June 19, 2019. https://ulfirefightersafety.org/research-projects/governors-island-experiments.html.

Zevotek, Robin, Keith Stakes, and Joseph Willi. 2018. *Impact of Fire Attack Utilizing Interior and Exterior Streams on Firefighter Safety and Occupant Survival: Full Scale Experiments*. Columbia, MD: UL Firefighter Safety Research Institute.

Fire Officer in Action

Incident management requires effective size-up and command practices.

1. Which of these objectives define the actions that are necessary to achieve the strategic goals?
   A. Group-level
   B. Company-level
   C. Task-level
   D. Tactical-level

2. The incident commander is responsible for the completion of all of these strategic priorities, *except*:
   a. incident stabilization.
   b. code comqpliance.
   c. property conservation.
   d. life safety.

3. The fire officer should take command and appropriately use elements of the National Incident Management System (NIMS) on events that:
   A. include mutual aid companies.
   B. require a level 1 rapid intervention crew.
   C. go beyond a traditional first-alarm assignment.
   D. involve two or more fire companies.

4. What did the results of the Phoenix and Asheville rapid intervention evolutions show?
   A. It takes 21 to 37 minutes to remove a distressed fire fighter.
   B. Search time is cut by 60 percent when using thermal imaging devices.
   C. Fire fighter removal time is reduced by 50 percent when the search team is increased from three to four fire fighters.
   D. Search time is cut by 40 percent when using two search teams from different entry points.

Fire Lieutenant Activity

NFPA Fire Officer I Job Performance Requirement 4.6.1

Develop an initial action plan, given size-up information for an incident and assigned emergency response resources, so that resources are deployed to control the emergency.

Application of 4.6.1

1. Describe what type of incidents would require the first-arriving fire company to assume the command mode.

2. You are the first fire company to arrive at the scene of an auto-versus-tractor-trailer-truck collision. A saddle tank has ruptured on the tractor, and the fuel has ignited. The car that was struck is under the front axle of the truck. You see an adult lying on the ground taking agonal breaths. Bystanders scream that there are people in the burning car. The next fire company is 15 minutes away. What task will you do first?

3. Your fire company is operating a large handline in a big-box commercial building with burning contents. When you are asked to respond to the personnel accountability report (PAR) check, you cannot see one of your fire fighters. What do you report to command?

4. Using a nursing home or hospital in your community, develop an incident action plan for responding to a fire in the food preparation area.

Access Navigate for flashcards to test your key term knowledge.

CHAPTER 8

Fire Officer I

Safety and Risk Management

KNOWLEDGE OBJECTIVES

After studying this chapter, you will be able to:

- Describe the fire officer's role in risk management. (pp. 224–225)
- Identify and explain the 16 Firefighter Life Safety Initiatives. (pp. 226–227)
- List the most common causes of injury and deaths to fire fighters. (**NFPA 1021: 4.7.1, 4.7.3**) (p. 226)
- Describe methods for reducing fire fighter deaths from cardiac arrest. (**NFPA 1021: 4.7.3**) (pp. 227–228)
- Describe the benefits of medical examinations and physical fitness activities. (**NFPA 1021: 4.7.3**) (pp. 227–228)
- Recognize the prevalence of suicide among fire fighters. (p. 228)
- Explain steps that can be taken by fire fighters to reduce the risk of developing cancer. (**NFPA 1021: 4.7.1**) (pp. 228–229)
- Identify actions that reduce the risk of fire fighter death and injury from motor vehicle collisions. (pp. 229–230)
- Describe methods for reducing the risk of injury and death to fire fighters during suppression operations. (**NFPA 1021: 4.7.1, 4.7.3**) (pp. 230–231)
- Describe factors that contribute to training deaths. (**NFPA 1021: 4.2.3**) (pp. 231–233)
- Explain the role of an incident safety officer within the incident management system. (**NFPA 1021: 4.7.1**) (pp. 233–234)
- Describe the responsibilities of an incident safety officer at an incident. (**NFPA 1021: 4.7.1**) (pp. 233–234)
- Identify qualifications for an incident safety officer. (pp. 234–235)
- Identify ways that officers can ensure adherence to policies and procedures designed to maintain a safe work environment. (**NFPA 1021: 4.7.1**) (p. 236)
- Describe methods preventing injuries to fire fighters at incident scenes. (**NFPA 1021: 4.7.1**) (pp. 236–237)
- Describe safety considerations for the fire station. (**NFPA 1021: 4.7.1**) (pp. 237–238)
- Describe the components of an infectious disease control program. (**NFPA 1021: 4.7.1**) (pp. 238–240)
- Describe procedures for conducting and documenting an accident investigation. (**NFPA 1021: 4.7.2**) (pp. 239, 241)
- Describe the role of the incident safety officer during the postincident analysis phase. (**NFPA 1021: 4.6.3**) (p. 241)

224 Fire Officer: Principles and Practice, Fourth Edition

SKILL OBJECTIVES

After studying this chapter, you will be able to:

- Utilize safety policies and procedures to maintain workplace safety. (**NFPA 1021: 4.2.3, 4.7.1**) (p. 236)
- Conduct and document an accident investigation. (**NFPA 1021: 4.7.2**) (pp. 239, 241)

ADDITIONAL NFPA STANDARDS

- NFPA 1250, *Recommended Practice in Fire and Emergency Service Organizational Risk Management*
- NFPA 1451, *Standard for a Fire and Emergency Service Vehicle Operations Training Program*
- NFPA 1500, *Standard on Fire Department Occupational Safety, Health, and Wellness Program*
- NFPA 1521, *Standard for Fire Department Safety Officer Professional Qualifications*
- NFPA 1581, *Standard on Fire Department Infection Control Program*
- NFPA 1583, *Standard on Health-Related Fitness Programs for Fire Department Members*
- NFPA 1851, *Standard on Selection, Care, and Maintenance of Protective Ensembles for Structural Fire Fighting and Proximity Fire Fighting*
- NFPA 1975, *Standard on Emergency Services Work Apparel*

You Are the Fire Officer

The discussion around the kitchen table at the fire station is somber. A fire officer was killed while operating at a vehicle crash on the interstate. A distracted driver plowed through the cones and collided into the accident scene. It killed the lieutenant and a civilian and hospitalized two fire fighters and a police officer.

1. What are the common causes of fire fighter deaths?
2. What can a fire fighter do to improve survivability?
3. What are the primary responsibilities of an incident safety officer?
4. What are the typical tasks expected of an incident safety officer?

 *Access Navigate for more practice activities.*

Introduction

Reducing fire fighter injuries and deaths requires the dedicated efforts of every fire fighter, of every fire department, and of the entire fire community working together. It also requires a safety program that integrates important components such as regulations, standards, procedures, personnel, training, and equipment.

Fire Officer's Role in Risk Management

Risk management refers to the process of planning, organizing, directing, and controlling the resources and activities of an organization in order to minimize detrimental effects on that organization. Performing risk management includes the following:

- Identifying and analyzing risk exposures
- Formulating risk management solutions
- Selecting risk management solutions
- Implementing risk management solutions
- Monitoring risk management solutions

The fire officer's role encompasses risk management in the community, within the fire department, and during emergency response operations.

Community Fire Risk

The fire service role in controlling community risk has evolved from the risk of conflagrations in the 1880s. The fire department continues to suppress fires, provide occupant rescue from harm, and protect property from loss (USFA, 2018). From the fire chief's perspective, there are three responsibilities:

1. Manage the community's fire risk
2. Provide a set of services that provide risk management measurements
3. Maintain fire department readiness to perform it mission every hour of every day

The methods of managing risks fall into three types of control: administrative, engineering, and personnel protection. These are similar to the methods used in quality improvement programs. Administrative controls include guidelines, policies, and procedures that are established to limit losses. For example, establishing and enforcing a brush control ordinance in the urban-wildland interface areas is a method of limiting the loss by reducing the potential spread of fire to structures by maintaining a wide area free of combustible vegetative materials. Engineering controls mean using built-in systems that remove or limit hazards. Adding fire sprinklers in single family residences both improves occupant life safety and controls the spread of fire, reducing property loss. Personnel protection control means equipment, clothing, and devices to protect the person. A smoke hood to be used by an occupant when escaping a high-rise fire is an example of personal protection.

Five Principal Risk Management Steps

As a working supervisor, the fire officer has the experience and access to provide a detailed evaluation of the risks within the workplace and sufficient authority to establish control (Lindsey, 2014).

Step 1: Identify Risk Exposure

The goal of this step is to identify the elements that create risk for the fire department. The NFPA annual fire fighter death and injury reports provide an industry-level perspective. Your department's Occupational Safety and Health Administration (OSHA)-required reporting provides a more focused perspective. Benchmarking your department's injury and loss experience with similar sized departments provide a comparison of need.

Step 2: Evaluate Risk Exposure

Determine the likelihood that an injury or loss event could occur, and then determine the consequences that would result. For example, using the reserve tiller aerial apparatus at the main fire station requires the tiller operator to lower the windshield and duck down to avoid hitting a ceiling mounted heater when the apparatus leaves or returns to the station. The possible consequence of decapitation of the tiller operator should preclude parking the reserve tiller apparatus in the main fire station.

Less catastrophic risks will benefit from answering the following questions:

- What is our local experience?
- What do we know about the national experience?
- What are the probabilities of different events happening?
- What are the probable consequences if they do occur?

Step 3: Rank and Prioritize Risks

Create a list of risks, along with with the probabilities of occurrence and probable outcomes. Events with the most devastating probable outcomes will be considered first, paying attention to those situations that occur most frequently. For example, failure of the two-way portable radios to receive messages when working in a basement occur every time fire fighters are in a basement. The probable outcome while operating at a fire suppression event in a basement would be catastrophic for any fire victim or fire fighter. Focusing on high-frequency events that would have catastrophic results will create the greatest risk management outcome.

Step 4: Determine and Implement Risk Management Control Actions

When deciding on and taking actions to mitigate risks, consider administrative, engineering, and personal protection controls. Some controls may require budget allocations that will not be approved until the next fiscal year. Fire officers may need to develop an interim first step that can be taken while waiting for funding or new equipment to be procured.

Step 5: Evaluate and Revise Risk Control Actions

Establishing risk control includes monitoring the progress. Do the controls actually reduce the risk? Have the controls produced an unanticipated problem? For example, one department's attempt to better track the location of on-scene fire fighters at a structure fire overwhelmed the tactical radio channel used by the incident commander (IC) and sector chiefs.

Fire Fighter Injury and Death Trends

Fire fighters must be fully prepared to work safely in high-risk situations. NFPA's *United States Firefighter Injuries–2017* report summarizes the injuries that occurred during fire-ground operations (Evarts and Molis, 2018):

- Strains and sprains: 48 percent
- Wounds, cuts, bleeding, and bruises: 15 percent
- Smoke or gas inhalation: 7 percent
- Thermal stress: 5 percent

According to NFPA's *Firefighter Fatalities in the United States–2018* report, there have been approximately 26 fire fighter fire-ground fatalities per year since 2007 (Fahy and Molis, 2019). This is a significant improvement over the average of more than 80 fire-ground deaths per year in the 1970s. The nature of all 2017 line-of-duty fire fighter deaths is as follows:

- Sudden cardiac death: 39 percent
- Internal trauma/crushing: 36 percent
- Asphyxia: 9 percent
- Other: 8 percent
- Burns: 5 percent
- Unspecified medical: 3 percent

Prevention depends on the ability to halt the cascade of events that leads to a serious injury or death. The National Fire Protection Association (NFPA), the U.S. Fire Administration (USFA), the National Institute for Occupational Safety and Health (NIOSH), and the International Association of Fire Fighters (IAFF) all publish reports and statistical analyses that provide information about the causes and circumstances of these events. The NFPA's *United States Firefighter Injuries–2017* report shows that the number of fire-ground injuries per 1000 fires has declined from 26.9 in 2006 to 18.6 in 2017.

Everyone Goes Home

Everyone Goes Home is a program developed by the National Fallen Firefighters Foundation (NFFF) to prevent line-of-duty death and injuries. The NFFF held a Firefighter Life Safety Summit in 2004 that resulted in 16 Firefighter Life Safety Initiatives (**TABLE 8-1**) (NFFF, 2010).

TABLE 8-1 16 Firefighter Life Safety Initiatives

1. **Cultural change**: Define and advocate the need for a cultural change within the fire service relating to safety and incorporating leadership, management, supervision, accountability, and personal responsibility.
2. **Accountability**: Enhance the personal and organizational accountability for health and safety throughout the fire service.
3. **Risk management**: Focus greater attention on the integration of risk management with incident management at all levels, including strategic, tactical, and planning responsibilities.
4. **Empowerment**: All firefighters must be empowered to stop unsafe practices.
5. **Training and certification**: Develop and implement national standards for training, qualifications, and certification (including regular recertification) that are equally applicable to all firefighters based on the duties they are expected to perform.
6. **Medical and physical fitness**: Develop and implement national medical and physical fitness standards that are equally applicable to all firefighters based on the duties they are expected to perform.
7. **Research agenda**: Create a national research agenda and a data collection system that relates to the 16 Firefighter Life Safety Initiatives.
8. **Technology**: Utilize available technology whenever it can produce higher levels of health and safety.
9. **Fatality, near-miss investigation**: Thoroughly investigate all firefighter fatalities, injuries, and near-misses.
10. **Grant support**: Grant programs should support the implementation of safe practices and procedures and/or mandate safe practices as an eligibility requirement.
11. **Response policies**: National standards for emergency response policies and procedures should be developed and championed.
12. **Violent incident response**: National protocols for response to violent incidents should be developed and championed.
13. **Psychological support**: Fire fighters and their families must have access to counselling and psychological support.
14. **Public education**: Public education must receive more resources and be championed as a critical fire and life safety program.
15. **Code enforcement and sprinklers**: Advocacy must be strengthened for the enforcement of codes and the installation of home fire sprinklers.
16. **Apparatus design and safety**: Safety must be a primary consideration in the design of apparatus and equipment.

Reproduced from National Fallen Firefighters Foundation (2010). Understanding and Implementing the 16 Firefighter Life Safety Initiatives. Tulsa OK, Fire Protection Publications.

The goal of the Everyone Goes Home program is to raise awareness of life-safety issues, improve safety practices, and allow everyone to return home at the end of his or her shift. The 16 Firefighter Life Safety Initiatives describe steps that need to be taken to change the current culture of the fire service to help make it a safe work environment.

A follow-up summit was held in 2007 to develop key recommendations for each of these initiatives. A standard approach to safety incorporates best practices that should be part of every operational situation.

Fire fighters must work in teams at emergency incidents, and fire officers must maintain accountability at all times for the location and function of all members working under their supervision. Every company must operate within the parameters of an incident action plan, under the direction of the IC.

Reliable two-way communications must be maintained through the chain of command. Also, adequate backup lines must be in place to ensure that a safe exit path is maintained for crews working inside a fire building and that any sudden flare-ups can be controlled. Rapid intervention crews must be established to provide immediate assistance if any fire fighter is in danger. In addition, air supplies must be monitored to ensure that fire fighters leave the hazardous area before their low-air alarm activates and they run out of air. All fire fighters must watch for, recognize, and communicate any indications of impending building collapse. A result of these efforts has been a trend of less than 70 line-of-duty deaths from 2014 to 2018. This is a significant improvement over the average of 100 deaths from 2002 to 2008 (Fahy and Molis, 2019).

National Firefighter Near Miss Reporting System

The International Association of Fire Chiefs (IAFC) launched a web-based system to report near misses. Developed in 2005, the national Firefighter Near Miss Reporting System provides a method for reporting situations that could have resulted in injuries or deaths. This system provides a means for all fire fighters to identify trends and to share the information with other fire fighters in a confidential and nonpunitive way. This program is based on the Aviation Safety Reporting System, which has been gathering reports of close calls from pilots, air traffic controllers, and flight attendants since 1976.

Reducing Deaths from Sudden Cardiac Arrest

The NFFF Firefighter Life Safety Summit noted that a disproportionate number of fire fighters older than age 49 years die of cardiac arrest while on duty. NFPA's *Firefighter Fatalities in the United States–2018* noted that sudden cardiac death account for the largest share of on-duty fire fighter deaths since NFPA began this study in 1977 (Fahy and Molis, 2019).

Every fire fighter candidate should undergo a medical examination before he or she is allowed to respond to incidents. Regular physical examinations should be scheduled for as long as the fire fighter is engaged in performing emergency duties, with an emphasis on identifying risk factors that could lead to a heart attack under stressful conditions. Fire officers should look for indications that a member is in poor health or is unfit for duty and, if necessary, arrange for a special evaluation by a fire department–approved physician.

Physical fitness activities should be considered an essential component of every fire fighter's training regimen. Changes in lifestyle can often reduce the risk of a fatal heart attack; such changes include stopping smoking, lowering high blood pressure, reducing high blood cholesterol level, maintaining a healthy weight, and managing diabetes. Fire officers must understand that these changes are as important to the body as wearing full protective equipment.

The NFPA, IAFC, NFFF, National Volunteer Fire Council (NVFC), and IAFF have developed resources to help the fire officer encourage healthy living. Fire officers should advocate methods of making positive lifestyle changes to increase fire fighter safety. NFPA 1583, *Standard on Health-Related Fitness Programs for Fire Department Members*, provides a structure and resources to help the fire officer develop a health-related fitness program. The IAFC partnered with the IAFF to develop *The Fire Service Joint Labor Management Wellness-Fitness Initiative* (WFI), which in turn produced the Candidate Physical Ability Test (CPAT) as well as a peer fitness training certificate program with the American Council on Exercise.

FIRE OFFICER TIP

Habits to Improve Safety

A fire officer can develop four simple habits to improve the safety of his or her crew:
- Be physically fit.
- Wear seat belts.
- Practice safety through training and personal example.
- Maintain fire company integrity at emergency incidents.

FIGURE 8-1 The fire officer should strive to be physically fit.
© Jones & Bartlett Learning.

Regardless of any fire department mandates that are handed down, every fire officer should strive to be physically fit (**FIGURE 8-1**). Fitness should be a personal priority and an expression of leadership.

Reducing Deaths from Suicide

The preliminary data from NFPA's *Firefighter Fatalities in the United States–2018* were that 82 fire fighters and 21 emergency medical technicians (EMTs) and paramedics died by suicide in 2018, compared to 64 line-of-duty deaths (Fahy and Molis, 2019). Recent estimates reveal that a fire department is three times more likely to experience a suicide in any given year than a line-of duty death (Luecke, 2017). As discussed in Chapter 3, *Leading a Team*, the rate of fire fighter suicide exceeds the rate of line-of-duty deaths.

An approach to reducing fire fighter suicides that is showing promising results is the idea of peer support programs. There are a variety of programs available to teach departments how to adopt and incorporate peer support systems. These programs include tools to assist officers and fire fighters in understanding the parts of the fire service culture that research indicates are contributing to the problem, including the stigma associated with mental health, coupled with guidance and tools to enable departments to create supportive environments that add protective layers aimed at preventing fire fighters from contemplating suicide.

Behavioral health can be described as behaviors that affect physical or mental health. It is a historical fact that behavioral health issues in fire and emergency medical services (EMS) have seldom been discussed and even more rarely addressed. It is essential for every organization to identify and utilize the resources that are available to assist in maintaining behavioral health. Some of these resources are provided by employers. Other resources and assistance are provided to first responders by professional organizations:

- Share the load is an effort by the NVFC that connects fire fighters, EMTs, and their families with resources and support for mental well-being (NVFC, 2015).
- Medical University of South Carolina has collaborated with the NFFF to develop a "Helping Heroes" training course for counselors who work with fire fighters (MUSC, 2019).
- The IAFF offers a guide on establishing a peer support program (IAFF, 2018a).
- The IAFC issued a yellow ribbon report entitled "Under the Helmet: Performing an Internal Size-Up—A Proactive Approach to Ensuring Mental Wellness" (Anderson-Fletcher et al., 2017).

Chapter 3, *Leading a Team*, covers behavioral and physical health issues in detail, with a focus on individual resiliency.

Reducing Deaths from Cancer

An increase in the use of synthetic products has led to an increase in the toxicity of today's modern fires. Cancer, now considered to be the leading cause of death among fire fighters, can be caused by a wide variety of cancer-causing substances (carcinogens) entering the body (IAFF, 2018b).

Cancer has been identified as a significant risk for fire fighters. The Firefighter Cancer Support Network estimates that fire fighters have a 9 percent higher risk of being diagnosed with cancer than the general U.S. population (FCSN, n.d.). Similarly, a study conducted by NIOSH concluded that the nearly 30,000 participants had a greater number of cancer diagnoses and cancer-related deaths than the general U.S. population.

FIRE OFFICER TIP

Physical Fitness Tips

The NFFF summarizes the importance of physical fitness with the following recommendations that support fire fighter health maintenance:

- Regular medical check-ups: Yes, they can be a pain, but if you do not do it for you, do it for those who need you.
- Regular exercise: Even walking makes a *big* difference! Walk a mile per day and watch the changes.
- Eat healthy: Think about what you are eating, and then picture operating interior at a working fire 30 minutes later. Now, what do you want to eat?

In 2018, the president signed legislation to establish a Firefighter Cancer Registry. This is the next step in identifying and documenting occupational diseases and mortality. The registry is after-the-fact, however, fire officers can initiate day-to-day activities to reduce the exposure to occupational cancer. The NVFC Cancer Subcommittee and the IAFC Volunteer and Combination Officers Section (VCOS), along with the Fire Service Occupational Cancer Alliance, the FCSN, and with support from the California Casualty insurance company, have developed a poster detailing the following best practices for preventing fire fighter cancer:

1. Full personal protective equipment (PPE) must be worn throughout the entire incident, including self-contained breathing apparatus (SCBA) during salvage and overhaul.
2. A second hood should be provided to all entry-certified personnel in the department.
3. Following exit from the area immediately dangerous to life or health (IDLH), and while still on air, you should begin immediate gross decontamination of PPE using soap and water and a brush, if weather conditions allow. PPE should then be placed into a sealed plastic bag and placed in an exterior compartment of the rig, or if responding in personally owned vehicles, placed in a large storage tote, thus keeping the off-gassing PPE away from passengers and self.
4. After completion of gross decontamination procedures as discussed earlier, and while still on scene, the exposed areas of the body (neck, face, arms, and hands) should be wiped off immediately using wipes, which must be carried on all apparatus. Use the wipes to remove as much soot as possible from head, neck, jaw, throat, underarms, and hands immediately.
5. Change your clothes and wash them after exposure to products of combustion or other contaminates. Do this as soon as possible and/or isolate in a trash bag until washing is available.
6. Shower as soon as possible after being exposed to products of combustion or other contaminans—"Shower within the Hour."
7. PPE, especially turnout pants, must be prohibited in areas outside the apparatus floor (i.e., kitchen, sleeping areas, etc.) and never in the household.
8. Wipes, or soap and water, should also be used to decontaminate and clean apparatus seats, SCBA, and the interior crew area regularly, especially after incidents where personnel were exposed to products of combustion.
9. Get an annual physical, as early detection is the key to survival. The NVFC outlines several options at www.nvfc.org. "A Healthcare Provider's Guide to Firefighter Physicals" can be downloaded from http://www.iafc.org/healthRoadmap.
10. Tobacco products of any variety, including dip and e-cigarettes, should never be used at any time, on or off duty.
11. Fully document ALL fire or chemical exposures on incident reports and personal exposure reports. (NVFC, 2018)

NFPA 1851, *Standard on Selection, Care, and Maintenance of Protective Ensembles for Structural Fire Fighting and Proximity Fire Fighting*, focuses on the person who actually uses the protective ensemble. The user should be constantly aware of the protective ensemble's condition and need for cleaning, repair, or more in-depth inspection. Users can perform the simple actions needed to improve the condition of the protective ensemble. The more involved actions of advanced inspection, evaluation, cleaning, decontamination, and repair are handled by the organization's designated staff who are trained and authorized to perform more advanced duties.

Reducing Deaths from Motor Vehicle Collisions

Deaths resulting from responding or returning from alarms continues a 5-year trend of low occurrences. All of the 2018 deaths occurred among volunteer fire fighters, and 8 of the 10 motor vehicle accidents resulted in death. Obeying traffic laws, using seat belts, driving sober, and controlling speed would prevent most of the fire fighter fatalities resulting from road collisions.

Section 6.2.1 of NFPA 1500, *Standard on Fire Department Occupational Safety, Health, and Wellness Program,* states that "Fire apparatus shall be operated only by members who have successfully completed an approved driver training program commensurate with the type of apparatus the member will operate or by trainee drivers who are under the supervision of a qualified driver." Putting an untrained driver behind the wheel of an emergency vehicle places both the occupants of that vehicle and the general public in immediate danger. The driver of an emergency vehicle is legally authorized to ignore certain restrictions that apply to other vehicles, but only when operating the emergency vehicle in a manner that provides for the safety of everyone using the roadways. Specific procedures for safe emergency response must be practiced. The fire officer is responsible for ensuring that the driver consistently follows the rules of the road for emergency response.

Driver minimum qualifications are established in NFPA 1002, *Standard for Fire Apparatus Driver/Operator Professional Qualifications;* NFPA 1451, *Standard for a Fire and Emergency Service Vehicle Operations Training Program;* and NFPA 1500. The fire officer should set high expectations for driver training and performance. In particular, the apparatus operator should be required to have nonemergency driving experience with a specific piece of apparatus before being assigned to emergency response driving duties. There are tremendous differences between the handling characteristics of fire apparatus versus automobiles or light trucks. Stopping distances are directly related to the weight of the vehicle, so fire apparatus take a relatively longer distance (and time) to come to a complete stop. In addition, heavy fire apparatus can turn over easily in a collision. Driving a ladder truck is vastly different from driving an engine, and a mobile water supply apparatus or tender has a different set of handling characteristics from either of these apparatus. As this brief summary of handling characteristics suggests, driving fire apparatus can be quite challenging, and all of the necessary skills must be learned and practiced under controlled conditions before an operator is qualified to drive under emergency response conditions.

Requiring fire fighters to wear seat belts is a simple requirement that could prevent three to six fatalities every year (Fahy, LeBlanc, and Molis, 2007). Many of the fire fighters killed in motor vehicle crashes are thrown from the vehicle. A fire officer who allows company members to respond while unrestrained in the fire apparatus has no excuse if a fatal accident results. The mandatory use of seat belts by fire fighters, like many other safety procedures, may require a change in the culture (**FIGURE 8-2**). A fire officer must be prepared to accept the responsibility and provide the leadership to bring about positive changes.

Reducing Deaths from Fire Suppression Operations

Internal trauma/crushing is the primary noncardiac cause of death during suppression operations, followed by asphyxia and burns (Fahy and Molis, 2019). Even though they may wear protective clothing and SCBA, fire fighters routinely operate in situations where a problem, a procedural error, or an equipment failure could result in a fatality.

Fire officers must fully understand all local policies and procedures that should guide their actions. The established standard operating procedures (SOPs) and safety practices should always be followed in situations that meet the criteria for their use. If a situation

FIGURE 8-2 Always fasten your seat belt and leave it fastened until you reach your destination.
© Jones & Bartlett Learning. Photographed by Glen E. Ellman.

does not fall under an established SOP, the fire officer must determine an appropriate course of action and give specific directives to subordinates to clearly indicate how the situation is to be handled. In these situations, the fire officer must provide an even greater level of supervision to ensure that the crew members understand and follow the plan. While doing so, the fire officer must always be prepared for changing conditions and unanticipated hazards.

Maintaining Crew Integrity

Maintaining the integrity of the fire company while operating at an emergency incident is a consistent challenge. The officer must know the location and function of every crew member at all times. The USFA's *Firefighter Fatality Retrospective Study* indicated that 82 percent of the fatal fire suppression incidents tallied in the report involved the death of a single fire fighter. Many of these fire fighters became lost or disoriented and died before the fire officer or IC was aware that a fire fighter needed help (USFA, 2002). An analysis of fire fighter fatality investigations from 2004 to 2009 suggested four higher-order root causes (Kunadharaju, Smith, and DeJoy, 2011):

- Under-resourcing
- Inadequate preparation for/anticipation of adverse events during operations
- Incomplete adoption of incident command procedures
- Suboptimal personnel readiness

NIOSH investigates fire fighter line-of-duty deaths as part of its Fatality Assessment and Control

Evaluation program. The lack of an effective incident management system that includes a fire fighter accountability component is a common scenario in many of the single fire fighter deaths investigated by NIOSH.

Air Management

A SCBA provides a safe and reliable air supply that allows a fire fighter to operate in an **immediately dangerous to life and health (IDLH)** atmosphere for a finite length of time. The length of time that a SCBA air tank will last depends on the rate at which the individual consumes the available air supply. This varies considerably depending on the user and the nature of the work being performed. The rated time is based on a standard consumption rate for an individual at rest and does not reflect typical experience in firefighting operations. Running out of air in an IDLH atmosphere can result in an unconscious fire fighter who will die of asphyxiation.

Some departments have replaced their 30-minute-rated SCBA air tanks with 45- or 60-minute air tanks to increase the amount of time a fire fighter can operate in IDLH atmospheres. Other departments have established or strengthened incident management procedures to monitor individual fire fighter air use more closely.

Low-pressure warning devices, which provide an indication when the remaining air supply reaches a set point, are effective only if the fire fighter heeds the warning and is able to exit to a safe atmosphere in the time that is provided. The low-pressure alarm might not provide sufficient exit time for a fire fighter who is deep inside a burning building or unable to exit immediately. Many departments monitor the air levels of fire fighters on entry and exit. If a fire fighter fails to exit within the expected time period, an accountability roll call is taken. Fire fighters should know the amount of air in their cylinders at all times and develop an exit strategy before the low-air alarm activates (Gagliano et al., 2008).

Teams and Tools

The minimum size of an interior work team is two fire fighters. Every team member must have full PPE, including SCBA and a personal alert safety system (PASS) device unit. At least one member (if not every member) of every team should have a radio.

If such equipment is available, every interior operating team should also have a thermal imaging device, which is a critical component of a comprehensive safety program. The thermal imaging device allows crews to navigate, locate victims, evaluate fire conditions, locate hazards, and find escape routes in smoke-filled buildings or total darkness (**FIGURE 8-3**).

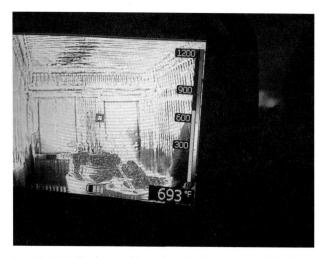

FIGURE 8-3 Thermal imaging devices are a critical component of a scene safety program.
© Andrew Starnes.

NIOSH's investigation of line-of-duty deaths revealed that some departments had thermal imaging devices but did not use them. Either the fire fighters did not know how to use the imager, the device was inoperable, or the device was not used (Starnes, 2018). Thirty years ago, the same observation could have been made regarding breathing apparatus, as SCBA units were often left sitting in the apparatus compartments while fire fighters made interior attacks. The fire officer must provide frequent training to ensure that the company members are fully prepared to integrate the thermal imager into interior firefighting operations.

Directing Members during Training Evolutions

Training ensures that every fire fighter can perform competently and every fire company can operate as a high-performance team. Fire service training must anticipate high-risk situations, urgent time frames, and difficult circumstances. A wide variety of methods and practices are included in the overall category of training.

Initial training leads to basic skill certifications, such as NFPA Fire Fighter I and Emergency Medical Responder. These certifications are often required before a fire fighter is authorized to participate in field operations. Certification usually involves a formal training program conducted by a training academy or equivalent organization that includes both classroom and skills practice (**FIGURE 8-4**). The trainee must pass both skills and knowledge evaluations to become certified.

FIGURE 8-4 A formal training program includes both classroom and skills practice.
© Jones & Bartlett Learning. Photographed by Glen E. Ellman.
Courtesy of Captain David Jackson, Saginaw Township Fire Department.

FIRE OFFICER TIP

Performance in Context

Whole-skill training, or training in context, is an effective tool allowing experienced fire fighters to improve their group performance at standard evolutions. Athletes use visualization and sequential task mastery to develop consistent and high-level performances. Members who prepare for the Firefighter Combat Challenge, Transportation Emergency Rescue Committee (TERC) extrication challenges, or other timed skill events also use training in context to improve performance (Hoffman, Crandell, and Shadbolt, 1998).

Conducting these activities as a group is an excellent method of improving coordination between companies. Multicompany drills should be conducted periodically for the same reason.

Reducing Deaths during Training

Fire fighter deaths during training activities represented 46 percent of the nonemergency annual line-of-duty death total in 2018 (Fahy and Molis, 2018). Deaths and serious injuries that occur during live fire training are caused by the following factors (Fahy, 2012):

- Inadequate training of instructors and assistant instructors
- Inadequate number of instructors for size of class
- Inadequate orientation or training of recruits
- Inadequate planning of the training evolutions by the lead instructor
- No incident management system or no designated incident safety officer
- No provision of rehabilitation or emergency medical services at the drill site

Many factors that result in fire-ground deaths also kill fire fighters on the training ground. The majority of the recent training deaths have been due to sudden cardiac arrest, sometimes secondary to dehydration or heat stroke.

NFPA 1041, *Standard for Fire Service Instructor Professional Qualifications* covers qualifications for instructors. Instructors of live fire training should meet the qualifications of an Instructor I and Live Fire Instructor at a minimum, and the instructor-in-charge should meet the qualifications of an Instructor II and Live Fire Instructor in Charge. Safety and instructional techniques are especially important in live fire training, and the standard outlines the skills and knowledge that an instructor should possess.

Several important components of training occur at the fire station or company level under the supervision of fire officers. This type of training is directed toward practicing basic skills and improving both individual and team performance. Many departments have a standard set of evolutions that each company is expected to be able to perform flawlessly, such as advancing a fire attack line over a ground ladder and into a third-floor window. Additional training at the company level often includes learning how to use new tools and equipment as well as refreshing, reinforcing, or updating knowledge and skills that are related to different aspects of the firefighting craft.

Additional company-level training often includes preincident planning and familiarization visits to different locations in a company's response area. These activities sometimes include a group of companies that would normally respond together to that location.

FIRE OFFICER TIP

Fire Officer Convicted of Criminally Negligent Manslaughter after Live Fire Training

Alan G. Baird III was the instructor-in-charge of a live fire training session in an acquired structure that led to the death of a 19-year-old rookie and severely burned two other fire fighters. The NIOSH report identified eight major mistakes made by Baird in setting up and running the evolution (NIOSH, 2002).

The Oneida County district attorney filed charges of second-degree manslaughter. Part of Baird's defense was that he did not know about NFPA 1403, *Standard on Live Fire Training Evolutions*. He was convicted of criminally negligent manslaughter. While sentencing Baird to 75 days in jail and 5 years of probation, Judge Michael Dwyer said that the incident was not an accident, but rather a series of bad decisions—decisions that never should have been made (ODD staff, 2002).

FIRE OFFICER TIP

Situational Awareness

In *Understanding and Implementing the 16 Firefighter Life Safety Initiatives* (NFFF, 2010), the NFFF summarizes situational awareness by specifying the following items that support the interior firefighting plan:

- Work as a team.
- Stay together.
- Stay oriented.
- Manage your air supply.
- Get off the apparatus with tools and a thermal imager for *every* interior operating team.
- Provide a radio for *every* member.
- Provide regular updates.
- Constantly assess the risk/benefit model.

Live Fire Training: Principles and Practice provides a definitive guide on how to ensure safe and realistic live fire training for both students and instructors.

Incident Safety Officer

An **incident safety officer** is a designated individual at the emergency scene who performs a set of duties and responsibilities that are specified in NFPA 1521, *Standard for Fire Department Safety Officer Professional Qualifications*. The incident safety officer functions as a member of the incident command staff, reporting directly to the IC. The IC is personally responsible for performing the functions of the incident safety officer if this assignment has not been assigned or delegated to another individual.

Many fire departments assign a designated officer to respond to emergency scenes to fill this position. In the absence of a designated incident safety officer, this duty may be assigned to a qualified individual by the IC. NFPA 1500 and NFPA 1521 also specify that the fire department must have an SOP to define the criteria for the response or appointment of an incident safety officer. The same principles should be applied to any situation, whether a designated incident safety officer has been dispatched or an officer has been assigned to fulfill this function by the IC.

The fact that an incident safety officer has been assigned does not relieve any officer or fire fighter of the responsibility to operate safely and responsibly. The incident safety officer is simply an additional resource to ensure that the safety priorities of the situation are being addressed. Every fire officer shares the responsibility to act as an incident safety officer within his or her scope of operations.

Incident Safety Officer and Incident Management

The incident safety officer is a key component of the incident command system (ICS). The ICS is the standard organizational structure that is used to manage assigned resources to accomplish the stated objectives for an incident. Every incident requires one individual, known as the IC, to be in charge at all times to coordinate resources, strategies, and tactics. This requirement comes into force with the initial arriving officer, who functions as the IC until he or she is relieved by a higher-ranking officer. The incident management structure can become larger and more complex, depending on the nature and the magnitude of the incident.

The incident safety officer reports to the IC and is required to monitor the scene, to identify and report any **hazard** (capable of causing harm or posing an unreasonable risk to life, health, property, or environment) to the IC, and, if necessary, to take immediate steps to stop unsafe actions and ensure that the department's safety policies are followed. In most situations, the exchange of information between the IC and the incident safety officer is conducted verbally and quickly at the command post.

Incident safety officers have specific authority and a special set of responsibilities under the ICS. One of the primary responsibilities of an incident safety officer is to identify hazardous situations and dangerous conditions at an emergency incident and recommend appropriate safety measures. In most cases, the incident safety officer acts as an observer, monitoring conditions and actions and evaluating specific situations. When an unsafe condition is observed that does not present an imminent danger, the incident safety officer consults with the IC and with other officers to determine a safe course of action.

If a situation creates an imminent hazard to personnel, the incident safety officer has the authority to immediately suspend or alter activities. When this special authority is exercised, the incident safety officer must immediately inform the IC of the hazardous situation and his or her actions. It is ultimately up to the IC to either approve or alter the action taken by the incident safety officer.

Qualifications to Operate as an Incident Safety Officer

NFPA 1521 outlines the criteria for an incident safety officer. Every fire officer should be trained to perform the basic duties of an incident safety officer and be prepared to act temporarily in this capacity if he or she is assigned to this position by the IC.

According to NFPA 1521, the incident safety officer must be a fire department officer and at a minimum must meet the requirements for Fire Officer I as specified in NFPA 1021, *Standard for Fire Officer Professional Qualifications*. The incident safety officer also must be qualified to function in a sector officer position under the local incident management system.

The general knowledge requirements for an effective incident safety officer are as follows:

- Safety and health hazards involved in emergency operations
- Building construction
- Local fire department personnel accountability system
- Incident scene rehabilitation

Incident safety officers at a special operations incident require additional specialized knowledge and experience. Special operations are emergency incidents to which the fire department responds that require specific and advanced training and specialized tools and equipment, such as water rescue, extrication, confined-space entry, hazardous materials situations, high-angle rescue, and aircraft rescue and firefighting. For example, the incident safety officer at a hazardous materials incident needs to have an advanced understanding of this type of situation, perhaps by training to the Hazardous Materials Technician level of NFPA 470, *Hazardous Materials Standards for Responders*.

In many fire departments, specialized teams have their own designated safety specialists who are trained to work directly with the incident management team.

Typical Incident Safety Officer Tasks

The specific duties that an incident safety officer must perform at an incident depend on the nature of the situation. The following is a partial listing of functions that may need to be addressed at incidents:

- Ensure that safety zones, collapse zones, and other designated hazard areas are established, identified, and communicated to all members present on scene.
- Ensure that hot, warm, decontamination, and other zone designations are clearly marked and communicated to all members.
- Ensure that a rapid intervention crew is available and ready for deployment.
- Ensure that the personnel accountability system is being used.
- Evaluate traffic hazards and apparatus placement at roadway incidents.
- Monitor radio transmissions and stay alert to situations that could result in missed, unclear, or incomplete communication.
- Communicate to the IC the need to appoint assistant incident safety officers because of the need, size, complexity, or duration of the incident.
- Immediately communicate any injury, illness, or exposure of personnel to the IC and ensure that emergency medical care is provided.
- Initiate accident investigation procedures and request assistance from the health and safety officer in the event of a serious injury, fatality, or other potentially harmful occurrence.
- Survey and evaluate the hazards associated with the designation of a landing zone and interact with helicopters.
- Ensure compliance with the department's infection control plan.
- Ensure that incident scene rehabilitation and critical incident stress management are provided as needed.

- Ensure that food, hygiene facilities, and any other special needs are provided for members at long-term operations.
- Attend strategic and tactical planning sessions and provide input on risk assessment and member safety.
- Ensure that an incident safety briefing, including an incident action plan and an incident safety plan, is developed and made available to all members on the scene.

Additional duties that the incident safety officer must perform when fire has involved a building or buildings are as follows:

- Advise the IC of hazards, collapse potential, and any fire extension in such buildings.
- Evaluate visible smoke and fire conditions to advise the IC, tactical-level management unit officers, and other officers on the potential for flashover, backdraft, or any other fire event that could pose a threat to operating teams.
- Monitor the accessibility of entry and egress of structures and the effect this has on the safety of members conducting interior operations.

Assistant Incident Safety Officers at Large or Complex Incidents

Some incidents, based on their size, complexity, or duration, require more than one incident safety officer. Assistant incident safety officers can be assigned to subdivide responsibilities for different areas and functions at events such as high-rise fires, hazardous materials incidents, and special rescue operations. In these cases, the incident safety officer should inform the IC of the need to establish a **safety unit** as a component of the incident management organization. Under the overall direction of the incident safety officer, assistant incident safety officers can be assigned to various functions, such as scene monitoring, action planning and risk management, interior operations, or special operations teams. During extended-duration incident operations, a relief rotation can be established to ensure that safety supervision is maintained at all times.

Incident Scene Rehabilitation

Rehabilitation is the process of providing rest, rehydration, nourishment, and medical evaluation to members who are involved in strenuous or extended-duration

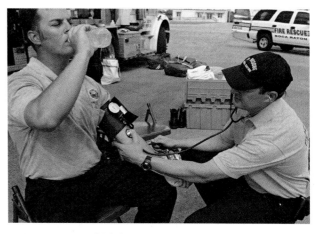

FIGURE 8-5 Rehabilitation is the process of providing rest, rehydration, nourishment, and medical evaluation to members who are involved in extended-duration or extreme incident scene operations.
© Jones & Bartlett Learning. Courtesy of MIEMSS.

incident scene operations (**FIGURE 8-5**). Part of the incident safety officer's role is to ensure that an appropriate rehabilitation process is established. **Incident scene rehabilitation** is the tactical-level management unit that provides for medical evaluation, treatment, monitoring, fluid and food replenishment, mental rest, and relief from climatic conditions of the incident.

Fire fighters are aggressive by nature. They want to do a good job and want to be where the action is occurring. This action orientation makes them susceptible to exceeding the physical limitations of their own bodies. When they overdo it, dehydration may contribute to sudden cardiac arrest on the fire ground. As a fire officer, you must constantly monitor the health and welfare of your crew. The incident safety officer is a third line of defense, after the individual and the fire officer, for ensuring that fire fighters obtain appropriate rehabilitation.

FIRE OFFICER TIP

Fire Fighter Rehabilitation

The NFFF summarizes incident scene rehabilitation with the following items that support the fire fighter rehabilitation guidelines:
- Stop before you drop.
  - Cool down when hot.
  - Warm up when cold.
  - Dry off when wet.
- Stay hydrated.
  - Drink only noncaffeinated drinks.
- Monitor vital signs.

Creating and Maintaining a Safe Work Environment

A safety program based solely on preventing fatalities is far from adequate. Every injury or near miss should be viewed as a potentially fatal or disabling situation, and injury prevention should be an equally important concern of the fire officer. The fire officer must model good behavior to help develop the subordinates' attitudes about injury prevention.

Safety Policies and Procedures

Most fire departments have policies that regulate safety practices at the company level. These policies are designed to address routine circumstances, and many have been developed in reaction to a previous accident or injury. The fire officer is on the front line in ensuring compliance with all safety policies. The fire officer needs to fully understand each policy, follow all safety policies and procedures, and ensure that all subordinates fully understand and follow them.

A number of methods may be used to ensure that fire fighters understand safety policies and procedures. Many departments require all members to sign a document acknowledging that they have read and understand each policy and any new or amended policy. Some departments leave it to the officer to read and explain each policy to the crew members. This can be done at the morning safety briefing; however, there may be little retention of information at this time. A more effective method would be to have the members individually read the policy and then lead a group discussion to ensure that it is understood.

One method to reinforce safety policies is to watch videos of incidents and critique them based solely on safety policies. This exercise often reveals significant differences between the way things are supposed to be done and what actually occurs at an emergency incident. It can be a significant learning experience for everyone involved.

Five other sources that may be accessed to review safety policies are the "Near Miss Report of the Week" from IAFC on their Near Miss website, "The Secret List" compiled by the Firefighter Close Calls website, information posted by the Emergency Responder Safety Institute (ERSI), incident videos collected by STATter911.com, and NIOSH case studies. In addition to reporting on incidents and trends, the NFFF Near-Miss Program, Firefighter Close Calls, and ERSI provide training programs and teaching resources on their websites.

The fire officer must conscientiously ensure that all safety policies are followed in training activities. Training provides the luxury of time to learn and correct errors. On emergency scenes, when there is only one opportunity to do things properly, fire fighters will perform the way they have been trained.

It is impossible to write a policy to cover every conceivable hazard. For this reason, the fire officer should use good judgment to identify hazardous situations and implement mitigating measures. These efforts might include requiring fire fighters to use safety glasses while mowing the lawn or ensuring that food preparation surfaces are properly cleaned to prevent the spread of germs. The fire officer should make sure that company members understand the need for safe work practices and develop an attitude that internalizes safety, rather than relying on the fire officer to be a "safety police officer."

Emergency Incident Injury Prevention

Many of the same techniques that are used to prevent fire fighter deaths will also help prevent fire fighter injuries; however, they are not necessarily the same. The following principles should be implemented to prevent injuries as well as deaths.

Physical Fitness

Fire fighters who are in good physical condition are less prone to injury as well as at reduced risk for heart attack. Through strength and flexibility training, the body becomes more resistant to sprains, strains, and other injuries and is more capable of responding to critical situations. Sprains and strains are the leading type of fire fighter injury. A physical fitness program that includes cardiovascular, strength, and flexibility training is needed to reduce fire fighter deaths and injuries.

Personal Protective Equipment

PPE performance has improved tremendously over the past 40 years. The protective ensemble for structural firefighting includes a protective coat and pants, boots, gloves, hood, helmet, SCBA, and PASS. NFPA standards define the minimum performance standards for each item.

Of course, the best protective equipment is of no use if it is not worn. The fire officer should monitor the proper use of protective clothing and take immediate action to correct any deficiencies. In some departments, fire fighters routinely enter buildings to investigate fire alarm activations and other seemingly minor situations without fully donning their PPE and SCBA, and sometimes without bringing tools, hose packs, or extinguishers. This practice has proved deadly when the situation turned out to be more serious than anticipated. The officer must set a good example and require every crew member to treat every call—no matter how seemingly benign—like an actual fire.

The need for protective clothing and equipment does not end when the fire is extinguished. SCBA must be used until carbon monoxide and particulate matter levels have been reduced. Coats, pants, gloves, goggles, and helmets are needed during overhaul to prevent wounds. The IC and the incident safety officer are responsible for determining whether it is safe to reduce the level of PPE that will be used at any time.

In addition to the protective clothing and equipment, fire fighters operating in hazardous areas should carry several safety-related items. A radio is needed to maintain contact, to receive instructions, or to request assistance in an emergency; at least one member of each team should have a radio, and preferably every fire fighter should carry one. A flashlight is needed to help prevent fire fighters from becoming lost or disoriented and to signal for help. A forcible entry tool should be carried to provide a method of escape in an emergency. Personal wire cutters can also prove valuable if a fire fighter becomes entangled in wires. A personal escape rope should be carried by every fire fighter as well. Each item provides an additional measure of safety, and the fire officer should always set the example by making sure he or she carries this equipment.

Many non-fire incidents also require the use of appropriate protective equipment. Fire fighters must pay attention to the task at hand and use proper eye protection during forcible entry or vehicle extrication.

Although station uniforms are not included in most definitions of PPE, the clothing that is worn under turnout gear can provide an additional level of protection. NFPA 1975, *Standard on Emergency Services Work Apparel*, provides a set of performance requirements for station clothing. The use of clothing that meets this standard under protective gear will further reduce the risk of burn injuries in a severe situation.

With EMS calls accounting for more than 60 percent of responses in many fire departments, the fire officer must also understand the requirements for medical PPE (Evarts, 2018). In many cases, this requirement is as simple as wearing appropriate gloves; healthcare providers must wear gloves for all patient contact situations. When blood or other body fluids are present, eye and face protection and full-body protection also may be needed. An effective education program is essential for fire fighters to understand the importance of wearing the appropriate protection for every situation.

Fire Station Safety

In addition to emergency situations, safety considerations apply to the fire station environment. The fire station and other fire department facilities are a workplace, and the fire department is fully responsible for maintaining a safe work environment. Every fire department should have a comprehensive workplace safety program that applies to all of its facilities. The station setting allows the fire officer to be even more proactive in enforcing safety policies than is possible at an incident scene.

Safety hazards at the fire station can have the same consequences as hazards encountered at emergency incidents. The most important difference is that the fire department has control over the fire station environment at all times and has the ability, as well as the responsibility, to identify and correct any safety problems that are present. The fire department does not have the same type of control over the conditions that are typically encountered at the scene of an emergency incident.

Clothing

Protective clothing can become contaminated at incident scenes and should never be worn in the living quarters of the fire station. Gear that is known to be contaminated with fire residue, chemicals, or body fluids should be cleaned as soon as possible. When washing gear, be sure to follow the manufacturer's recommendations. Cleaning gear requires the use of a commercial extractor-type washer rather than a residential washing machine. Many fire departments provide special washing machines for protective clothing at the fire station or send gear out for professional cleaning and decontamination. Duty uniforms should also be washed at the fire station or sent out to a commercial laundry to prevent cross-contamination of the family laundry.

Protective clothing should be inspected regularly by the fire officer, and any items that are worn or damaged should be repaired or replaced. Properly fitting gear is also important, both for comfort and for protection. Gear inspection should include an exterior evaluation while the clothing is being worn and a close examination of the condition of all interior layers.

Housekeeping

General housekeeping around the fire station is important in injury and accident prevention. Apparatus bay floors that become slippery when wet have caused many fire fighters to slip and fall. A squeegee should be used to remove standing water, and "wet floor" warnings should be used. The same is true for interior station floors that require mopping.

Leaving equipment lying around can present a tripping hazard to fire fighters when they are in a hurry to respond to an alarm. All walking traffic flow areas should always be kept clear.

The same fire hazards that exist in other types of occupancies can be found in many fire stations. Leaving food unattended on the stove and improperly storing flammable liquids can lead to a publicly embarrassing situation if they lead to a fire at the facility. Every fire station should have the appropriate number and types of fire extinguishers, which must be properly maintained. The fire officer should ensure that working smoke alarms are present and checked regularly. Automatic fire suppression sprinklers are highly recommended.

Food preparation activities are responsible for many relatively minor injuries to fire fighters. Inappropriate use of kitchen knives has caused injuries resulting in lost time and worker's compensation claims. Proper decontamination of all food preparation surfaces is often overlooked. Regular hand washing while cooking prevents the transmission of illness.

Diesel exhaust systems installed in apparatus bays reduce the exposure of fire fighters to hazardous gases. If the fire station does not have an exhaust system, the fire officer should ensure that vehicles are not left running inside the building.

Lifting Techniques

Back injuries can occur while lifting and moving patients, dragging fire hoses, setting ladders, or working at the station. No matter what the cause, back injuries are serious and potentially career ending.

Proper lifting techniques should be used to reduce the risk of back injuries. In particular, a fire fighter should never bend at the waist to pick up items or victims. Instead, the fire fighter should bend at the knees and lift by standing straight up. The back should be kept in a natural position rather than locked in a hyperextended manner. When an object is too bulky, in an awkward position, or too heavy for one fire fighter, additional help should be sought to lift it. Fire officers should always reinforce the use of proper techniques and procedures to avoid injuries.

Some injuries have resulted from horseplay around the station. This is an area where the fire officer must exercise supervisory control to reduce the risk of unjustifiable injuries.

Infection Control

Every fire department should have an infection control program that meets NFPA 1581, *Standard on Fire Department Infection Control Program*. This standard identifies six components of a program:

1. A written policy with the goal of identifying and limiting exposures
2. A written risk management plan to identify risks and control measures
3. Annual training and education in infection control
4. A designated infection control officer
5. Access to appropriate immunizations for employees
6. Instructions for handling exposure incidents

Proper decontamination procedures are followed after emergency medical incidents or any situation where equipment could have become contaminated. Disposable items, gloves, and contaminated expendable equipment should be disposed of in a specially marked bag that is designed for that use.

Patient compartments in ambulances should be properly decontaminated after every transport. Likewise, patient cots should be decontaminated after every call. Equipment that is designed to be decontaminated and reused should be cleaned only in an approved decontamination sink, typically in a designated area of the fire station. Contaminated equipment should *never* be taken into the living area of the station or cleaned in a sink where food is prepared.

Always follow written procedures and the manufacturer's guidelines for decontamination of medical equipment. A 1 percent bleach and water solution is typically used for this purpose; however, this solution should *not* be used on turnout gear. Many of the materials commonly used in protective clothing can be seriously damaged or destroyed by this type of cleaning agent.

NFPA 1581 provides specific information on establishing an infection control program, including guidelines on equipment cleaning and storage, facility requirements, and methods of protection. The fire officer should also consult departmental guides and policies for further details.

Infectious Disease Exposure

NFPA 1581 provides a model program for situations where a fire fighter has been exposed to an infectious or contagious disease. Most fire departments have established policies that govern postexposure procedures. The fire officer must be prepared to fulfill the duties of the initial supervisor when a member has experienced an infectious disease exposure.

The most important first step with any exposure is to wash the exposed area immediately and thoroughly with soap and running water. If soap and running water are not available, waterless soap, antiseptic wipes, alcohol, or other skin cleaning agents can be used until soap and running water are obtained.

The fire department infection control officer should be immediately notified after an exposure incident has occurred. The designated infection control officer should arrange for the member who has experienced an exposure to receive medical guidance and treatment as soon as is practical. An exposure can be a stressful event for a fire fighter, so it is important to provide confidential postexposure counseling and testing. Because the fire fighter may not seek out this information on his or her own, the department should proactively inform the individual of this service.

Document all exposures as soon as possible using a standardized reporting form (**FIGURE 8-6**).

At a minimum, the record should include the following information:

- Description of how the exposure occurred
- Mode of transmission
- Entry point
- Use of personal protective equipment
- Medical follow-up and treatment

The record of an exposure incident will become part of the member's confidential health database. A complete record of the member's exposure incidents must be available to the member on request. Data on exposure incidents should also be maintained by the department but without personal identifiers to allow for analysis of exposure trends and to develop strategies to prevent such incidents.

Due to the hazardous nature of some communicable diseases, individuals who have been exposed to these risks are often required to report to the infection control officer. This information must be maintained with the strictest of confidence. It is up to the fire department physician to determine the member's fitness-for-duty status after reviewing documentation of his or her exposure.

Accident Investigation

The fire department **health and safety officer** is charged with ensuring that all injuries, illnesses, exposures, fatalities, or other potentially hazardous conditions and all accidents involving fire department vehicles, fire apparatus, equipment, or fire department facilities are thoroughly investigated. The initial investigation of many situations is often delegated to a local fire officer.

An accident is any unexpected event that interrupts or interferes with the orderly progress of fire department operations. This definition includes personal injuries as well as property damage. An accident investigation should determine the cause and circumstances of the event and identify any corrective actions that are needed to prevent another injury, accident, or incident from occurring.

The fire officer must ensure that all required federal, state, and local documentation is complete and accurate. The result of an accident investigation should always include recommended corrective actions that are presented to the fire chief or the chief's designated representative.

Accident Investigation and Documentation

Most fire departments have established procedures for investigating accidents and injuries. The fire officer is usually responsible for conducting an initial investigation and for fully investigating many minor accidents. Accidents that involve serious injuries, fatalities, or major property damage are usually investigated by the health and safety officer or by other qualified individuals.

An officer could be expected to conduct a full investigation for a simple ankle sprain, a broken pike pole, or a dented rear step on the apparatus. In each case, it is the fire officer's responsibility to protect the physical and human resources of the department. To accomplish this goal, the officer must understand the basic principles of investigation.

An investigation normally consists of three phases:

1. Identification and collection of physical evidence
2. Interviews with witnesses
3. Written documentation (at the end of an investigation)

Each of these steps is essential for the development of a comprehensive report that can be used to initiate safer work practices and conditions. The investigation

Sample Blood and Body Fluid Exposure Report Form

Exposure Event Number _____

Facility name: _____

Name of exposed worker: Last _____ First: _____ ID #: _____

Date of exposure: _____/_____/_____   **Time of exposure:** _____:_____ AM   PM (Circle)

Job title/occupation: _____ **Department/work unit:** _____

Location where exposure occurred: _____

Name of person completing form: _____

Section I. Type of Exposure (Check all that apply.)

☐ **Percutaneous** (Needle or sharp object that was in contact with blood or body fluids)
  (Complete Sections II, III, IV, and V.)

☐ **Mucocutaneous** (Check below and complete Sections III, IV, and VI.)
  ___ Mucous Membrane   ___ Skin

☐ **Bite** (Complete Sections III, IV, and VI.)

Section II. Needle/Sharp Device Information
(If exposure was percutaneous, provide the following information about the device involved.)

Name of device: _____   ☐ Unknown/Unable to determine

Brand/manufacturer: _____   ☐ Unknown/Unable to determine

Did the device have a sharps injury prevention feature, i.e., a "safety device"?

☐ Yes   ☐ No   ☐ Unknown/Unable to determine

If yes, when did the injury occur?

☐ Before activation of safety feature was appropriate   ☐ Safety feature failed after activation

☐ During activation of the safety feature   ☐ Safety feature not activated

☐ Safety feature improperly activated   ☐ Other: _____

Describe what happened with the safety feature, e.g., why it failed or why it was not activated: _____

Section III. Employee Narrative (Optional)

Describe how the exposure occurred and how it might have been prevented:

_____
_____
_____

NOTE: This is not a CDC or OSHA form. This form was developed by CDC to help healthcare facilities collect detailed exposure information that is specifically useful for the facilities' prevention planning. Information on this page (#1) may meet OSHA sharps injury documentation requirements and can be copied and filed for purposes of maintaining a separate sharps injury log. Procedures for maintaining employee confidentiality must be followed.

A-7 Sample Blood and Body Fluid Exposure Report Form

FIGURE 8-6 Exposure report form.
Courtesy of CDC.

report can also provide evidence that could be used to support or refute future claims. If a department has a standardized procedure for investigating accidents, the fire officer should always use the recognized procedure.

Examination of the physical evidence includes documenting the time, day, date, and conditions that existed at the time of the incident. Such factual information should be verifiable and should not include any interpretation. The weather conditions should be noted. The scene of the accident should be fully documented, with drawings noting the locations of relevant artifacts that could provide information about how or why the accident occurred. These details are needed to help determine prevention methods.

Witnesses to an accident should also be interviewed, including all of the individuals who were directly or indirectly involved, as well as anyone who simply observed what happened. When interviewing a witness, explain the rationale for the interview and note that this is a fact-finding process. Advise the person that it is okay not to know the answer to every question. Begin by asking the individual to give a verbal chronology of whatever occurred, being sure not to interrupt. Watch for nonverbal communication by the witness. Once the opening report has been given, ask questions to clarify the details, but do not ask leading or misleading questions. If the level of PPE being worn at the time of the incident has not been mentioned, determine what was being worn, how it was being worn, and whether it performed as designed. Last, restate previous questions in a different order and from a different perspective to verify previous answers.

The last step in an accident investigation is the documentation of the findings and the conclusions and recommendations. The facts should be presented in a logical sequence with the results of the interview or interviews attached. A determination as to the most likely cause or causes of the accident should be included, along with a recommendation on how this type of accident could be prevented. Some departments require a determination about the likely frequency as well as the likely severity of this type of accident—an assessment intended to optimize prevention efforts.

As a fire officer, you have a duty to perform an initial accident investigation that is fair and unbiased. Both the department and the employee have a vested interest in the final report. An employee who applies for a disability pension based on an accident that occurred several years earlier could be denied the pension if adequate documentation is not maintained. For this reason, most departments require a report to be filed whenever there is a potential exposure to an infectious disease, when injury has occurred to an employee or a citizen, and when property damage has occurred to either private property or departmental property.

Postincident Analysis

The incident safety officer provides a written report for the department that includes pertinent information relating to safety and health issues involved with the incident. This would include information about the use of protective clothing and equipment, personnel accountability system, rapid intervention crews, rehabilitation operations, and other issues that directly affect the safety and welfare of members at the incident scene.

You Are the Fire Officer Conclusion

There has been a noticeable increase in distracted drivers crashing into apparatus parked at an incident scene with their emergency lights on. Sudden cardiac arrest, however, remains the most common cause of fire fighter on-duty deaths: on the fire ground, in the fire station, and on the drill site. But sudden cardiac arrest has been overshadowed by the impact of fire fighter suicide, with more fire fighters dying by suicide than all types of line-of-duty deaths combined.

Fire fighters need to concentrate on individual resiliency (body, mind, family) and aggressively maintain team cohesion during emergency incidents and training.

The incident safety officer is part of the initial incident command team, reporting to the IC, and is required to monitor the scene, to identify and report hazards and, if necessary, to take immediate steps to stop unsafe actions and ensure that the department's safety policies are followed.

The incident safety officer acts as an observer, monitoring conditions and actions and evaluating specific situations, including taking immediate action to mitigate a dangerous situation.

After-Action REVIEW

IN SUMMARY

- Risk management refers to the process of planning, organizing, directing, and controlling the resources and activities of an organization in order to minimize detrimental effects on that organization.
- Fire fighters must be fully prepared to work safely in high-risk situations.
- Everyone Goes Home is a program developed by the National Fallen Firefighters Foundation to prevent line-of-duty death and injuries; it includes the 16 Firefighter Life Safety Initiatives.
- The goal of www.firefighternearmiss.com, launched by the International Association of Fire Chiefs in 2005, is to track incidents that avoided serious injury or death, to identify trends, and to share the information with other fire fighters in a confidential and nonpunitive way.
- Sudden cardiac arrest accounts for the largest share of on-duty fire fighter deaths.
- The rate of fire fighter suicide exceeds the rate of line-of-duty deaths.
- Deaths from responding or returning from alarms continues a 5-year trend of low occurrences.
- An incident safety officer is a member of the incident command staff who monitors the scene, identifies and reports hazards to the IC, and takes immediate steps to stop unsafe actions and ensure that the department's safety policies are followed.
- Every fire officer should be trained to perform the basic duties of an incident safety officer and prepared to act temporarily in this capacity if he or she is assigned to fill this position by the IC.
- Some incidents, based on their size, complexity, or duration, require more than one safety officer. Assistant incident safety officers can be assigned to subdivide responsibilities for different areas and functions at incidents such as high-rise fires, hazardous materials incidents, and special rescue operations.
- Part of the incident safety officer's role is to ensure that an appropriate rehabilitation process is established. Incident scene rehabilitation provides for medical evaluation, treatment, monitoring, fluid and food replenishment, mental rest, and relief from climatic conditions of the incident.
- A safety program that is based solely on preventing fatalities is far from adequate. Every injury or near miss should be viewed as a potentially fatal or disabling situation, and injury prevention should be an equally important concern of the fire officer.
- The fire officer needs to fully understand each policy, follow all safety policies and procedures, and ensure that all subordinates fully understand and follow them.
- Principles that should be implemented to prevent injuries and deaths include physical fitness and consistent use of personal protective equipment.
- Safety hazards at the fire station can have the same consequences as hazards encountered at emergency incidents. The fire department, however, has control over the fire station environment at all times and has the ability, as well as the responsibility, to identify and correct any safety problems.
- Every fire department should establish an infection control program that meets NFPA 1581, *Standard on Fire Department Infection Control Program*.
- The most important first step with any infectious disease exposure is to wash the exposed area immediately and thoroughly with soap and running water. If soap and running water are not available, waterless soap, antiseptic wipes, alcohol, or other skin cleaning agents can be used until soap and running water are obtained.
- The fire department health and safety officer ensures that all injuries, illnesses, exposures, fatalities, or other potentially hazardous conditions and all accidents involving fire department vehicles, fire apparatus, equipment, or fire department facilities are thoroughly investigated.

- The fire officer is usually responsible for conducting an initial investigation and for fully investigating many minor accidents.
- The incident safety officer provides a postincident analysis for the department in the form of a written report that includes pertinent information relating to safety and health issues involved with the incident.

KEY TERMS

Behavioral health Behaviors that affect physical or mental health.

Hazard Capable of causing harm or posing an unreasonable risk to life, health, property, or environment. (NFPA 1072)

Health and safety officer The member of the fire department assigned and authorized by the fire chief as the manager of the safety and health program.

Immediately dangerous to life and health (IDLH) Any condition that would pose an immediate or delayed threat to life, cause irreversible adverse health effects, or interfere with an individual's ability to escape unaided from a hazardous environment. (NFPA 1670)

Incident safety officer A member of the command staff responsible for monitoring and assessing safety hazards and unsafe situations and for developing measures for ensuring personnel safety. (NFPA 1500)

Incident scene rehabilitation The tactical-level management unit that provides for medical evaluation, treatment, monitoring, fluid and food replenishment, mental rest, and relief from climatic conditions of an incident.

Rehabilitation The process of providing rest, rehydration, nourishment, and medical evaluation to members who are involved in extended or extreme incident scene operations.

Risk management The process of planning, organizing, directing, and controlling the resources and activities of an organization in order to minimize detrimental effects on that organization.

Safety unit A member or members assigned to assist the incident safety officer; the tactical-level management unit that can be composed of the incident safety officer alone or with additional assistant safety officers assigned to assist in providing the level of safety supervision appropriate for the magnitude of the incident and the associated hazards.

REFERENCES

Anderson-Fletcher, Elizabeth, John M. Buckman, Jeff Dill, Charles Flynn, Scott Geiselhart, Robin Grant-Hall, Steve Heitman, et al. 2017. *Under The Helmet: Performing an Internal Size-Up—A Proactive Approach to Mental Wellness.* Chantilly, VA: International Association of Fire Chiefs.

Evarts, Ben. 2018, October. *Fire Loss in the United States during 2017.* Quincy, MA: National Fire Protection Association.

Evarts, Ben, and Joseph L. Molis. 2018. *United States Firefighter Injuries–2017.* Quincy, MA: National Fire Protection Association.

Fahy, Rita F. 2012. *U.S. Firefighter Deaths Related to Training 2001–2010.* Quincy, MA: National Fire Protection Association.

Fahy, Rita F., Paul R. LeBlanc, and Joseph L. Molis. 2007. *What's Changed over the Past 30 Years?* Quincy, MA: National Fire Protection Association.

Fahy, Rita F., and Joseph L. Molis. 2019. *Firefighter Fatalities in the United States–2018.* Quincy, MA: National Fire Protection Association.

Firefighter Cancer Support Network. n.d. "Who We Are." Accessed August 21, 2017. https://firefightercancersupport.org/%20Who%20We%20Are/.

Gagliano, Mike, Casey Phillips, Phillip Jose, and Steve Bernocco. 2008. *Air Management in the Fire Service.* Tulsa, OK: Fire Engineering.

Hoffman, Robert R., Beth Crandall, and Nigel Shadbolt. 1998. "Use of the Critical Decision Method to Elicit Expert Knowledge: A Case Study in the Methodology of Cognitive Task Analysis." *Human Factors* 40 (2): 254–276.

International Association of Fire Fighters (IAFF). 2018a. "10 Steps to Build Your Peer Support Program." Washington, DC: International Association of Fire Fighters. Accessed July 12, 2019. https://services.prod.iaff.org/ContentFile/Get/41972.

International Association of Fire Fighters (IAFF). 2018b. "IAFF Cancer Summit, 2018." Accessed July 10, 2019. www.iaff.org.

Kunadharaju, Kumar, Todd D. Smith, David M. DeJoy. 2011. "Line-of-duty Deaths among U.S. Firefighters: An Analysis of Fatality Investigations." *Accident Analysis and Prevention* 43 (3): 1171–1180.

Lindsey, Jeffrey T. 2014. *EMS Safety and Risk Management*. London, UK: Pearson Education.

Luecke, Craig. 2017. "Checking In: A Behavioral Health Size-Up" [podcast]. Crofton, MD: National Fallen Firefighters Foundation.

Medical University of South Carolina (MUSC). 2019. "When Firefighters Need Rescuing Too." Accessed July 12, 2019. https://education.musc.edu/colleges/nursing/about/news-media/college-of-nursing-news/2019/march/when-firefighters-need-rescuing-too.

National Fallen Firefighters Foundation (NFFF). 2010. *Understanding and Implementing the 16 Firefighter Life Safety Initiatives*. Tulsa, OK: Fire Protection Publications.

National Institute for Occupational Safety and Health (NIOSH). 2002. *Volunteer Fire Fighter Dies and Two Others Are Injured During Live-Burn Training—New York F2001-38*. Atlanta, GA: National Institute for Occupational Safety and Health.

National Volunteer Fire Council (NVFC). 2015. "Share the Load: A Support Program for Fire and EMS." Accessed July 12, 2019. https://www.nvfc.org/share-the-load-may-2015/.

National Volunteer Fire Council and International Association of Fire Chiefs' Volunteer and Combination Officers Section. 2018. "Lavender Ribbon Report: Best Practices for Preventing Firefighter Cancer." www.nvfc.org/cancer, www.vcos.org/BeatFFCancer.

ODD staff. 2002, May 23. "Lairdsville Firefighter Guilty of Lesser Charge." *Oneida Daily Dispatch*. Accessed July 16, 2019. https://www.oneidadispatch.com/news/lairdsville-firefighter-guilty-of-lesser-charge/article_11dac638-1dc2-58ea-8b31-505dcfac5c76.html.

Starnes, Andrew. 2018, April 3. "Thermal Imaging Cameras in the Fire Service: Asset or Detriment? You Decide." *Fire Apparatus and Emergency Equipment*. Accessed July 16, 2019. https://www.fireapparatusmagazine.com/articles/print/volume-23/issue-3/features/thermal-imaging-cameras-in-the-fire-service-asset-or-detriment-you-decide.html.

U. S. Fire Administration (USFA). 2002. *Firefighter Fatality Retrospective Study*. Emmitsburg, MD: U. S. Fire Administration.

U. S. Fire Administration (USFA). 2018. *Risk Management Practices in the Fire Service*. Emmitsburg, MD: U. S. Fire Administration.

Fire Officer in Action

Let's review the fire fighter mortality information provided in this chapter.

1. Based on the statistics provided in this chapter, an active duty firefighter is most likely to die from which of these?
   A. Vehicle accident
   B. Occupational cancer
   C. Suicide
   D. Opioid exposure

2. Which of these is most frequent cause of fire fighter death on the fire ground?
   A. Cyanide poisoning
   B. Crush trauma
   C. Sudden cardiac arrest
   D. Carbon monoxide poisoning

3. To reduce the impact of smoke on occupational cancer development, a fire fighter should do all of the following, *except*:
   A. initiate gross decontamination while on SCBA air as soon as you clear the IDLH environment.
   B. use wipes to clean exposed areas of the body after a gross decontamination.
   C. shower within the hour.
   D. remain on SCBA air until all members of your company are wiped down and you have cleared the fire incident scene.

4. Which of these is the second most frequent cause of fire fighter death?
   A. smoke explosion
   B. burns
   C. internal trauma or crushing
   D. closed head injury

Fire Lieutenant Activity

NFPA Fire Officer I Job Performance Requirement 4.7.2

Conduct an initial accident investigation, given an incident and investigation process, so that the incident is documented and reports are processed in accordance with policies and procedures of the AHJ [authority having jurisdiction].

Application of 4.7.2

1. While responding to a working structure fire, a pumper failed to stop at the incident address and collided with a civilian's parked SUV. Properly document this accident, including the required forms and reports of the AHJ.

2. A truck company is investigating a fire alarm sounding in a one-story commercial building. After breaking a display window, the officer notices the Knox-Box above the door. Properly document this incident, including the required forms and reports of the AHJ. Explain how you will secure the business before the fire department leaves.

3. Assisting the paramedics, your fire fighters have placed the patient on the ambulance stretcher. The patient is not secured to the stretcher. While moving from the home to the ambulance, the patient falls off the stretcher, suffering a broken nose and possible closed head injury. Properly document this incident, including the required forms and reports. Include a description of the notifications you will make to the fire department.

4. There was an explosion in the fire station mechanical room, requiring the services of an engine and a truck company to control. The electrical power and the heating, ventilation, and air conditioning (HVAC) services require repair before the fire station can be re-occupied. You were the officer-in-charge at that station as well as the initial IC for the explosion. Properly document this incident, including the required forms and reports. Include a description of the notifications you will make to the fire department.

Access Navigate for flashcards to test your key term knowledge.

Continuous Case Study

Textile City Fire Department Overview

The Textile City Fire Department is a combination fire department providing service to a suburban/rural community of 34 square miles. The department is staffed with 15 full-time fire fighters, five per 24-hour shift. Each shift has a lieutenant and fire fighter assigned to each station and a shift captain who is charged with the day-to-day management of the shift, including station assignments and scheduling of part-time personnel. Additional staffing is provided by part-time fire fighters who sign up for additional weekend and/or evening time slots and fill in for full-time fire fighters on leave. In addition to the full-time fire fighters, who are members of a union, there is a full-time nonunion deputy chief and fire chief.

PART 1: Lieutenant (Fire Officer I)

During a training session, the lieutenant observes a heated argument between a male full-time fire fighter and a female part-time fire fighter. The lieutenant separates the two fire fighters and speaks to them separately. The lieutenant learns that they were involved in a personal relationship, and that relationship was ending.

The female fire fighter claims that the male fire fighter has posted "suggestive" photos on a private social media page with derogatory allegations about her life choices. The female fire fighter provides a screen shot of the pictures, along with comments made by other people who have access to the social media page.

The lieutenant notes that one of the pictures was taken in the fire station and all of them include images of the fire department logo, some with fire department equipment. The lieutenant recognizes that some of the respondents are other fire department members. Some of the comments are vile and demeaning.

Questions for discussion or assignment:

1. Which employee behavior requires an immediate response, the argument between the two fire fighters during the training session or the posting of suggestive pictures with derogatory remarks on a private social media page?
2. If there is no written policy regarding social media, what action can the Textile City Fire Department formally take to discipline the male fire fighter that posted the picture or the other fire fighters who responded to the post?
3. How does social media play a role in creating a hostile work environment?
4. What should the lieutenant do to assure that the workplace does not contribute to the drama of an ending personal relationship?

SECTION 2

Fire Officer II

CHAPTER **9 The Fire Officer II as a Manager**

CHAPTER **10 Applications of Leadership**

CHAPTER **11 Managing Community Risk Reduction Programs**

CHAPTER **12 Administrative Communications**

CHAPTER **13 Fire Cause Determination**

CHAPTER **14 Managing Major Incidents**

CHAPTER 9

Fire Officer II

The Fire Officer II as a Manager

KNOWLEDGE OBJECTIVES

After studying this chapter, you will be able to:

- Identify the requirements for Fire Officer II. (**NFPA 1021: 5.1, 5.1.1**) (p. 252)
- Describe the roles and responsibilities of the Fire Officer II. (**NFPA 1021: 5.1.1**) (pp. 252–253)
- Identify other organizations a fire department officer may work with. (**NFPA 1021: 5.1.2**) (pp. 258–259)
- Describe additional challenges for the fire captain. (**NFPA 1021: 5.1.1, 5.1.2**) (pp. 256–258)
- Describe the legislative framework for collective bargaining. (**NFPA 1021: NFPA 5.1.1**) (pp. 259–262)
- Explain the fire service's relationship with labor unions through history. (**NFPA 1021: 5.1.1**) (p. 262)
- Explain the historic use of strikes in the fire service. (**NFPA 1021: 5.1.1**) (pp. 262–264)
- Explain the emergence of labor organizations as a means of exerting political influence. (**NFPA 1021: 5.1.1**) (pp. 264–265)
- Explain the role of labor–management alliances and list four national examples. (**NFPA 1021: 5.1.1**) (pp. 262–264)

SKILL OBJECTIVES

There are no skill objectives for this chapter.

You Are the Fire Officer

It is 45 minutes before the start of the shift, and the new captain is meeting the crew. The captain will be commanding the engine company as well as supervising the truck company lieutenant and the paramedic ambulance lieutenant.

1. What duties does a captain have that a lieutenant does not?
2. How is the captain's role different under the incident management system?
3. How does the captain's role differ with regard to fire fighter professional development?

Access Navigate for more practice activities.

Introduction

The Fire Officer II classification begins with meeting all of the requirements for Fire Officer I as defined in National Fire Protection Association (NFPA) 1021, *Standard for Fire Officer Professional Qualifications*. (See Chapter 1, *The Fire Officer I as a Company Supervisor*.) The scope of responsibility for the Fire Officer II is wider than the Fire Officer I, with the Fire Officer II identifying issues and developing plans that a Fire Officer I implements. An additional responsibility for the Fire Officer II is developing professional development plans for each Fire Officer I and fire fighter that reports to the Fire Officer II.

Requirements for Fire Officer II

The requirements for Fire Officer II begin with meeting all of the requirements for Fire Officer I as defined in NFPA 1021. As is true for the Fire Officer I, the duties of the Fire Officer II can be divided into administrative, nonemergency, and emergency activities.

Administrative duties include evaluating a subordinate's job performance, correcting unacceptable performance, and completing formal performance appraisals on each member. Other duties include developing a project or divisional budget, including the related activities of purchasing, soliciting, and awarding bids, and preparing news releases and other reports to supervisors.

Nonemergency duties include conducting inspections to identify hazards and address fire code violations; reviewing accident, injury, and exposure reports to identify unsafe work environments or behaviors; and taking approved action to prevent reoccurrence of an accident, injury, or exposure. Other duties could include developing a preincident plan for a large complex or property; developing policies and procedures appropriate for this level of supervision; and analyzing reports and data to identify problems, trends, or conditions that require corrective action and then developing and implementing the required actions.

Emergency duties include supervising a multiunit emergency operation using the **incident command system (ICS)** and developing an operational plan to deploy resources to mitigate the incident safely. The ICS defines the roles and responsibilities to be assumed by personnel and the operating procedures to be used in the management and direction of emergency operations. The Fire Officer II is also expected to determine the area of origin and preliminary cause of a fire and to develop and perform a postincident analysis of a multicompany operation.

The International Association of Fire Chiefs (IAFC) calls the Fire Officer II level a **managing fire officer** within its *Officer Development Handbook*. In this textbook, the Fire Officer II will be referred to as a *captain*.

Additional Roles and Responsibilities for Fire Officer II

The captain has a complex role as a supervisor and manager. As a manager, the captain develops or implements a strategy to accomplish a goal or mission. This occurs by managing lieutenants who are directly supervising work teams performing task-level activities. The fire service may also expect the captain to directly supervise a small group of fire fighters engaged in task-level activities.

An example of how this complex role could play out would be a fire station with an engine company and a truck company. The engine company is supervised by a captain and the truck company is supervised by a lieutenant. Both officers directly supervise a team of fire fighters who are assigned to their company. The captain will have additional managerial and administrative responsibilities both on the emergency incident scene and at the fire station. The captain is a working manager, overseeing multiple crews that are operating in a hazardous, life-threatening, and dynamic environment. During these events, the captain is constantly monitoring the progress of these task-level activities in meeting the strategic goal and is prepared to change tasks or strategy, as needed.

The fire service is a decentralized organization, with fire stations spread throughout the community operating under a hierarchical command system. Fire companies, command staff, and specialized resources respond from all corners of the community and assemble to handle an emergency incident. Once the incident is mitigated, the responders disassemble into their individual teams and leave. Administrative work groups or task forces are created to resolve an organizational problem or handle a temporary need. These activities create unique managerial leadership needs that are similar to military operations with small work group teams. (Ward, 2018)

A Fire Officer II has the same roles and responsibilities as a Fire Officer I, along with the following additional items:

- Supervises and directs the activities of a multiunit station
- Completes employee performance appraisals
- Creates a professional development plan for members of the organization
- Leads water rescue, hazardous materials, or other special teams as assigned
- Ensures the safe and proper use of equipment, clothing, and protective gear, and enforces departmental policies
- Participates in the formulation or evaluation of departmental or agency policies as assigned, implements new or revised policies, and encourages team efforts of fire personnel
- Participates in the formulation of the departmental budget and makes purchases within it
- Develops emergency incident operational plans requiring multiunit operations
- Prepares written reports so major causes of local service demand are identified for various planning areas within the service area of the organization

This text covers only the roles and responsibilities of Fire Officers I and II according to NFPA 1021. Fire Officer III, IAFC's Administrative Fire Officer, and Fire Officer IV, IAFC's Executive Fire Officer, require additional training and responsibilities.

Authorization to Function as a Fire Department

A fire department is a public organization that provides fire prevention, fire suppression, and associated emergency and nonemergency services to a jurisdiction such as a county, municipality, or organized fire district (Evarts and Stein, 2019). The type of fire department organization varies by how it is organized (Varone, 2014):

- City
- County
- Town
- Fire district
- Regional organization
- State or federal fire department
- Volunteer fire department
- Industrial fire department
- Fire brigade

State Police Powers

Each state has the authority to govern matters related to the welfare and safety of the residents, this is known as **state police powers**. This covers all powers not specifically delegated by the Constitution to the U.S. government (Darling, 2012). These powers include the right and responsibility of a state to protect its citizens from fire. State activities include the following:

- Adoption of fire and building codes
- Identification of who will enforce code compliance
- Firefighting services

Although the authority for a fire department to carry out its responsibilities comes from the state, it is sometimes more closely defined or described in local law. States differ on how they delegate responsibility for fire protection. In some states, there are differences based on community resources. For example, the state fire marshal investigates fires in rural and unincorporated communities, whereas the city fire marshal investigates fires in municipalities.

Administrative Law

Most states and local governments have developed a body of statutes that provides a series of rules that govern how a state agency operates and how it promulgates rules (JUSTIA, 2019). The elected officials of a state or local jurisdiction enact basic government policies and then administrative agencies create rules to implement the details of these policies. Under the Model State Administrative Procedure Act (MSPA), the reason for delegating the rulemaking power by the legislature to an administrative agency is to bring scientific and technical details into the rules that are created (USLEGAL, n.d.).

FIRE OFFICER TIP

Example of Delegation: California Fire Protection District Law of 1987

13801. The Legislature finds and declares that the local provision of fire protection services, rescue services, emergency medical services, hazardous material emergency response services, ambulance services, and other services relating to the protection of lives and property is critical to the public peace, health, and safety of the state. Among the ways that local communities have provided for those services has been the creation of fire protection districts. Local control over the types, levels, and availability of these services is a long-standing tradition in California which the Legislature intends to retain. Recognizing that the state's communities have diverse needs and resources, it is the intent of the Legislature in enacting this part to provide a broad statutory authority for local officials. The Legislature encourages local communities and their officials to adapt the powers and procedures in this part to meet their own circumstances and responsibilities. (SGFC, 1987)

Supervising Multiple Companies

The first-arriving officer at a fire incident assumes the role of incident commander (IC). When operating as the IC, the fire officer has an even greater level of responsibility, because the IC is responsible for every company on the scene and for management of the overall operation. The initial IC's responsibilities include conducting a size-up, developing an action plan to mitigate the situation, assigning the resources to execute the plan, developing a command structure to manage the plan, and ensuring that the plan is completed safely. (This is covered in detail in Chapter 7, *Commanding Local Incidents*.) The fire officer who is functioning as the IC is responsible for all of these functions.

The fire officer might also be assigned as a division/group/unit leader, supervising multiple units, or even as a branch director within the ICS. In each of these situations, the officer serves as a relay point within the command structure. Direction comes down through the system, from the IC, through the intermediate levels, to the individual companies or units. At the same time, information is transmitted upward from subordinates to the officer, who either acts on the information directly or relays the information up to the next level. For example, a division supervisor could be assigned to oversee operations directed toward keeping the fire out of an exposure. The division supervisor would be expected to manage that operation and supervise the assigned companies, giving regular progress reports to the IC.

Determining Task Assignments

Tactical priorities are subdivided into tasks and assigned to companies. Tasks are specific assignments that are typically performed by one company or a small number of companies working together.

During the early stages of an incident, there are likely to be more tasks to be performed than there are companies available to do the work. The IC makes assignments based on tactical priorities and available resources. The IC must prioritize assignments and distribute them to companies as they arrive or become available. Companies that have completed one task assignment may have to be reassigned to another. The IC should use the tactical priorities to determine the relative importance of each task that needs to be performed, in the context of the specific situation. Standard operating procedures (SOPs) often guide these decisions, as do the staffing and standard equipment on each piece of apparatus.

The normal function of the company should also be considered, when possible. In most cases, an engine company would be assigned to attack the fire, whereas a paramedic ambulance would provide medical care, and a truck company would be assigned to ventilation duties. Sometimes, however, the IC has to assign companies to perform tasks that might not fit their normal role, simply because of the importance of the task and the limited resources available. For example, if no paramedic ambulance is on scene, the IC might have to assign an engine company to provide medical care. If the truck company will be delayed, an engine company could be assigned to perform ventilation (Brunacini, 2004).

The rescue priority could be assigned to different companies. Such tasks might include performing primary and secondary searches of the structure, raising ladders, removing trapped occupants, providing medical care and transport, and establishing a rapid intervention crew.

- Task assignments for the exposure priority typically include establishing a water supply and setting up master streams for use on the fire building or on the exterior of the exposure. Placing a handline on the unburned side of a firewall or in the cockloft of an exposure building could also be a task assignment. Removing combustible material from the windows of the exposed buildings is another example.
- Task assignments for fire confinement typically include advancing handlines to the room of origin, into stairways, and into the attic or the floor above the fire. Ventilation tasks could be performed to support confinement efforts.
- Task assignments for extinguishing a fire typically include establishing a water supply, advancing a handline to the seat of the fire, and applying water or other extinguishing agents. When making assignments for hose lines, the fire officer should indicate the line size as well as placement, unless these decisions are predetermined by departmental SOPs.
- Task assignments for overhauling a fire typically include pulling ceilings and walls in the burned areas, removing door and floor trim where charred, checking the attic and basement for extension, checking the floors above and below the fire, and removing or wetting all burned material.
- Task assignments for ventilating a fire typically include vertical ventilation, horizontal ventilation, positive-pressure ventilation, negative-pressure ventilation, and natural ventilation. When making assignments for ventilation, the officer should indicate the location of entry points, the potential victims, and the location of the fire to ensure that proper techniques are used.
- Task assignments for salvage typically include throwing salvage covers over large, valuable items; removing lingering smoke; soaking up water from floors; deactivating sprinklers; and removing important documents or memorabilia (Brunacini, 2002).

Assigning Resources

The resources assigned to an incident vary greatly and are influenced by history, tradition, and budgets. One department may respond to a structure fire with a single engine and tanker, whereas another might send four engines, two aerial apparatus, a rescue or squad company, and two command officers.

FIGURE 9-1 Incident commanders should use tactical priorities when assigning tasks.
Courtesy of Robert B. Rodriguez, Chief Fire Marshal, Alexandria Fire Department.

Some situations exceed the capabilities of any organization and require assistance from other agencies or jurisdictions. In most cases, the first level of assistance for fires is mutual aid from surrounding fire departments. Most fire departments also have working relationships with other local agencies, such as law enforcement and public works, to obtain the resources that are likely to be needed in most situations.

When the need for resources exceeds the normal capabilities of the fire department and involves numerous other agencies, a fire officer may have to activate the local emergency plan. This plan defines the responsibilities of each responding agency and outlines the basic steps that must be taken for a particular situation. For example, firefighting is normally the responsibility of the fire department, along with response to technical rescue and hazardous materials incidents. Traffic control, crowd control, perimeter security, and terrorist incidents are usually the responsibility of the police or sheriff's department (**FIGURE 9-1**). Every fire officer should be familiar with the local plan and the role of the fire department within it.

The method most commonly used by a fire officer to activate the local emergency plan is to notify the dispatch center, which should have the information and procedures needed to either activate the local emergency plan or notify the emergency management office. The emergency management office normally is responsible for maintaining and coordinating the plan, as well as facilitating the response to emergency situations (Deal, de Bettencourt, and Deal, 2012).

The local emergency plan usually includes an evacuation component that can be used for any situation where the residents or occupants of an area must be protected from a dangerous situation. A hazardous materials release or major fire event, for example, could require the evacuation of a local area. In many cases, the evacuation plan can be activated on a limited scale, without activating the overall local emergency plan.

The police department is the agency that has the primary responsibility for notifying people within an evacuation area regarding what to do. The use of the public address feature on vehicle sirens, the emergency broadcast system, or a reverse 911 telephone system can aid in this process. For large-scale or long-term evacuations, the Red Cross may need to be contacted to establish emergency shelters for the residents. Depending on the situation, fire crews could be available to assist evacuating residents, or they might be occupied with fighting a fire or other mitigation actions. Mutual aid companies or other agencies could be needed to assist with evacuation.

Consider the nature of the event when establishing an evacuation area. A natural gas leak in a street generally requires short-term evacuation of residents in the immediate area. An apartment building fire could result in several residents being displaced for a long period of time, but it affects only those who live in the building or complex. A major hazardous materials release could result in the evacuation of a large area and thousands of people. The fire officer should consult the *Emergency Response Guidebook*, the local hazardous materials team, or CHEMTREC for guidance in determining evacuation distances, based on the product, quantity, and environmental conditions (PHMSA, 2016).

Challenges for the Captain

The role of the managing fire officer has become more engaged over time when it comes to working with other organizations and groups. Demographic, economic, environmental, and community changes have created unprecedented challenges that will require changes in the structure, task, and mission of the fire department.

Fire Fighter Emotional Health

The increased emergency service workload is taking a toll on fire fighters (Jahnke, Poston, Haddock, and Murphy, 2016). In one study, between 7 and 37 percent of fire fighters met the criteria for posttraumatic stress disorder (Tull, 2019). Similarly, the number of fire fighter suicides has been rising (Heitman, 2016).

Supervision and Motivation

The paramilitary structure of the fire department was established in the 1860s, when cities were replacing independent volunteer companies with municipal fire departments based on the Civil War military deployment model (Hensler, 2011). A rigid command and control process remains essential when operating at emergency scenes. Away from emergencies, however, departments are using the concepts of employee empowerment, decentralized **decision making**, and delegation to fully engage fire fighters in the required tasks to prepare and maintain readiness for a wide range of community needs. This "all-hazards" approach is especially important when considering nonfire incidents, a crumbling built environment, homeland security, cultural diversity, and ethics.

Increase in Non-fire Incidents

Battling a structure fire is a common perception of a "typical day at work" for fire fighters, and both the Fire Fighter I and II levels focus on this task. The NFPA, however, notes that firefighting is actually one of the activities least frequently performed by fire companies. It accounts for only 4 percent of the response workload based on data submitted to the National Fire Incident Reporting System (NFIRS). Of the runs requiring fire suppression tasks, almost half are for structure fires.

Emergency medical services (EMS) calls are now the most frequent activity undertaken by the fire service, accounting for 64 percent of fire company responses. In the last decade, the increase in EMS workload has exceeded growth on a demographic basis, with some cities noticing the number of EMS calls increasing even while the population was declining (Evarts, 2018). In some fire departments, EMS calls represent more than 80 percent of a fire company's response workload (Cannuscio et al., 2016; Jao, 2017)

Activated fire protection system alarms are the second most common reason for fire service response. The majority of these activations are due to faulty alarm systems, good intentions, or false calls. The company officer must work to remain vigilant in events that result in no service in most of the responses, with an occasional incipient fire or an inferno occurring in fewer than 1 out of 100 activated fire alarm responses (Aherns, 2016).

Investigating an odor, a hazardous condition, or other service call is the third most common reason for fire department response. Fire fighters encounter many opportunities to be creative problem solvers and deliver outstanding customer service during these events.

Voice of Experience

Fire attack . . . seems so simple: put the wet stuff on the red stuff. If it was just that simple, as a fire officer arriving on scene, it may be up to you to make that happen. But as you have learned in your training, it isn't quite that simple. Development of an incident action plan (IAP) is required to take a systematic approach to a fire, so the type of fire attack must also be systematic and dynamic to reach our objective: putting out the fire.

Whether you're making a positive-pressure attack, offensive or defensive, remember that the fire attack has to be carefully considered and then maintained. Stay flexible in the face of the possible changes that can occur during fire operations. I teach that the single most important component of fire attack is to maintain situational awareness. Paying attention to your surroundings at all times is imperative to being safe—that is, noticing changing situations ranging from structure stability to fire behavior. Changes in these and other situations can mean the difference between safely mitigating the situation and having a disaster to deal with. I'm reminded of one of my first calls. I was on interior attack. I can still see the outlines of the 2 by 4s burning bright red (the building was under some remodeling), and, of course, the heat and smoke were rather intimidating as well. But my partner and I made our attack, and with the help of another attack team we brought the fire under control. It wasn't until overhaul that I realized I had not been paying attention to everything. We went in, and moving my flashlight around and retracing our steps inside, I looked up—and my eyes must have grown to the size of baseballs. What I saw changed my thought process for fire attack forever: right above our heads was a balcony-type section that was leaning severely enough that even under postfire conditions it made me back up. Right above exactly where we were standing on that leaning part of the structure was a small workout area, with a weight machine and a couple of hundred pounds of weights. If that section had gone . . . well, I stopped thinking about what could have happened and made it a point to make sure that I was never in that situation again. All aspects of the fire attack plan are important: ICS, IAP, safety, ventilation, watching for abnormal fire behavior—and the list goes on and on. But knowing where you are during fire attack—maintaining situational awareness—is so important. As an officer, you might conduct training or run an academy. Remember that the basic fundamentals of fire attack must include every aspect of that attack and how better to control our scenes and personnel to make the best outcome happen every time: everyone goes home.

Jim Lovell
Captain
Fire Training Division
Santa Fe County Fire Department
Santa Fe, New Mexico

Deterioration of the Built Environment

Although the number of structure fires continues to decline, the vast majority of fire fighter injuries still occur while fighting structural fires (Campbell, 2016). Flashover and structural collapse are the primary causes of noncardiac death within a burning structure (Fahy and Molis, 2019). Decades of deferred maintenance and repair all too often make structures unstable and dangerous in which to operate.

Century-old buildings, even those that appear sturdy, may have many worn-out or rusted-out building components. Fire escapes may be pulling out of the walls (Biddle, 2018) and structural components may be crumbling. Many modern-day renovations involve the substitution of lightweight wood structural elements, some without the benefit of a fire code inspection after the changes are made to the occupancy or use of the building (Hoefferle, 2006).

Catastrophic Weather Events

Extreme weather events have increased in frequency over the past decades. Annual civilian deaths from flooding have increased from 86 people thirty years ago to more than 100 since 2015 (Lam, 2018) doubling since 2004. While the number of wildfires has remained steady since 1985, the size and intensity of each fire has dramatically increased (Koerth-Baker, 2018).

This affects fire departments, because they are the first responders to every type of weather catastrophe. A review of the Federal Emergency Management Agency (FEMA) disaster declarations over the past 40 years shows that four times as many counties were hit by disaster-scale hurricanes, storms, and floods between 1997 and 2016 than between 1976 and 1996 (Symons, 2017). From 2014 to 2018, the United States has averaged 13 disasters per year requiring a billion dollars or more in FEMA disaster relief (Stein and Van Dam, 2019).

Diversity

Fifty years ago, the fire service was virtually all made up of white males. This composition began to change in the 1960s. The initial integration of women and minorities focused on assimilation—that is, the fire service desired to make those who were different fit into the mold of the traditional fire service. When the fire service began including a few women and minorities, the new employees either assimilated to the existing culture or they left.

Inclusion—being valued and having a sense of belonging—remains a challenge for non-white and non-male fire fighters. Professor Corinne Bendersky, PhD, worked with LA Woman in the Fire Service along with the Los Angeles county and city fire departments to target qualified candidates from underrepresented groups to advance through the selection process. Her research finds that reframing the professional prototype of what it means to be a fire fighter to emphasize the importance of legitimate, stereotypically feminine traits, like compassion, has promising effects on creating a more inclusive environment for women (Bendersky, 2018).

Integration of diverse employees is far from complete, and many departments continue to struggle with bridging these relationships. To be successful, the fire officer must look beyond the physical attributes of individuals and match each individual's strengths with the needs of the organization, the individual, and the officer.

Working with Other Organizations

Fire departments are part of the structure of the community. To fulfill its mission, a fire department must often interact with other organizations. A motor vehicle crash provides a good example of this need. In an area with a centralized 911 call center, the fire department, a separate emergency medical services provider, law enforcement officials, and tow truck operators might all be dispatched to the same incident. At the scene, all of these personnel must work together to solve the problem. Fire officers frequently have to request assistance from and then interact with other agencies.

Conversely, other public safety agencies are entering areas that were once the sole domain of the fire department. Hundreds of police departments have created hazardous materials response teams, training police officers to the hazardous materials technician level and procuring equipment that may exceed the fire department resources (FBI, n.d.).

To manage this kind of change effectively, a fire officer must understand the roles played by other agencies and recognize how they interact with the fire department. Federal, state, and local response plans identify which organizations are responsible for each area of the incident. For a local incident, the local fire service is often given primary responsibility for search, rescue, and fire extinguishment. Law enforcement is given responsibility for criminal investigation and scene security. The jurisdiction's Office of Emergency Management is responsible for evacuation notification. The American Red Cross may be responsible for establishing evacuation shelters (FEMA, 2017).

As incidents grow, the Federal Bureau of Investigation (FBI) may take a lead role in an investigation. FEMA may become the primary agency responsible for the coordination of the incident, which could include the use of urban search and rescue (USAR) teams. Fire officers must understand their role within the local, state, and federal response plans.

Fire officers must also be aware that some individuals and organizations wish to create chaos and harm emergency service workers. They research fire department activities to exploit weaknesses and identify opportunities. The fire officer must be vigilant for threats to fire fighters and the department.

FIRE OFFICER TIP

Tent Cities

Homeless encampments experienced a 1000 percent increase from 2007 to 2017, and they are occurring throughout the United States. Half of the encampments have 11 to 50 residents; 17 percent of the encampments have more than 100 occupants (NLCHP, 2017). Fire departments are responding to calls for violence, fires, and infectious diseases like typhus and tuberculosis (Gorman, 2018)

Most communities utilize the National Incident Management System (NIMS) to coordinate, track and assign public, nonprofit, and commercial resources for working with the encampments. Fire departments generally send two crews to any EMS event, to have one captain function as the scene safety officer while the others work at the task-level (Goldfeder, Long, and Schaeffer, 2019).

Unique response units that service the encampments have included the following:

- Community care response unit. Consisting of a physician's assistant and a captain/paramedic
- Behavioral response unit. Consisting of a paramedic with a licensed mental health counselor
- Alternative response unit. Staffed with EMT/fire fighters to perform initial assessment and triage patients to nonurgent transportation options or a paramedic ambulance

Legislative Framework for Collective Bargaining

Collective bargaining is the process in which individuals (through their unions) negotiate contracts with their employers in order to determine terms of employment. Terms include such items as salary, benefits, leave, hours, job health and safety policies, and more. The process is regulated by a complex system of federal and state legislation. Federal legislation establishes a basic framework that applies to all workers, and the states have discretionary powers to adopt labor laws and regulations that do not violate the federal requirements. The application of this basic model to governmental employees is much more complex. Some of the federal labor laws do not apply to federal government employees, state government employees, or employees of local government agencies within the states.

Four major pieces of federal legislation have established the groundwork for the rules and regulations of the present collective bargaining system. Before the adoption of these federal laws, each labor case was decided by a judge in a local court, who applied the broad concepts used in common-law decisions. The federal legislation provides a set of guidelines for how each state or commonwealth can regulate collective bargaining. Fire fighters employed by local government agencies are subject to state law that can require, permit, or prohibit collective bargaining for local public employees. The four federal laws that regulate the collective bargaining system are the Norris-LaGuardia Act of 1932, the Wagner-Connery Act of 1935, the Taft-Hartley Labor Act of 1947, and the Landrum-Griffin Act of 1959. These federal laws, together with the Railway Labor Act of 1926 and some antitrust legislation, created the legal foundation for collective bargaining in the United States. Like most federal legislation, each act was designed to address a specific aspect of collective bargaining or to correct a problem. Each subsequent act built on the earlier legislation. Like a pendulum, each subsequent act may change the direction of the federal government in relation to collective bargaining issues (Dilts, Dietsch, and Rassuli, 1992).

Norris-LaGuardia Act of 1932

The Norris-LaGuardia Act of 1932 specified that an employee could not be forced into a contract by an employer as part of obtaining and keeping a job. Prior to this act, some employers required workers to sign a pledge that they would not join a union as long as the company employed them; these pledges were called **yellow dog contracts**.

The local courts generally sided with management in any labor dispute where a yellow dog contract was in effect. The judge would order an injunction that either prohibited striking or prohibited picketing during a strike. The police would enforce these injunctions.

In reaction to this practice, the Norris-LaGuardia Act of 1932 said that yellow dog contracts were not enforceable in any court of the United States. This act made it almost impossible for an employer to obtain an injunction to prevent a strike.

President Franklin Roosevelt took many steps to bolster economic growth during the Great Depression.

One of his initiatives was the National Industrial Recovery Act (NIRA) of 1933. Section 7a of the NIRA guaranteed unions the right to collective bargaining as part of an effort to keep wages at a level that would maintain the purchasing power of the worker. After the NIRA passed, workers flocked to join both the American Federation of Labor (AFL) and the new Congress of Industrial Organizations (CIO). The Supreme Court struck down the NIRA as unconstitutional in 1935.

Wagner-Connery Act of 1935

When the NIRA was overturned, employers were free to use **unfair labor practices** in the management of their employees. The ensuing abuses led to the Great Strike Wave of 1935–1936. During this period, labor organized against management and conducted city-wide strikes and factory takeovers in numerous industrial sectors. In response to the ongoing labor unrest, Senator Robert Wagner of New York introduced the Wagner-Connery Act, which was quickly passed by Congress in 1935, to mitigate the revolutionary labor climate and avert further economic disruption. A 1936 strike in the automobile industry quickly brought the Wagner-Connery Act before the U.S. Supreme Court, where it was upheld as constitutional. The Wagner-Connery Act established the procedures that are commonly called collective bargaining.

The Wagner-Connery Act forms the basis of formal labor relations in the United States. It grants workers the right to decide, by majority vote, which organization will represent them at the labor–management bargaining table. The act also requires management to bargain with duly elected union representatives and outlaws yellow dog contracts.

Provisions of the Wagner-Connery Act also established the National Labor Relations Board (NLRB), which has the power to hold hearings, investigate labor practices, and issue orders and decisions concerning unfair labor practices. The act defined five types of unfair labor practices and declared them illegal:

1. Interfering with employees in a union
2. Stopping a union from forming and collecting money
3. Not hiring union members
4. Firing union members
5. Refusing to bargain with the union

The Wagner-Connery Act and favorable court decisions resulted in some unions becoming extremely powerful. Union membership swelled from 4 million members in 1935 to 16 million members in 1948. The next two acts described here were adopted to balance the power between unions and management.

Taft-Hartley Labor Act of 1947

The Taft-Hartley Labor Act of 1947, which was passed by Congress over the veto of President Harry Truman, was formulated during the period that followed World War II. At that time, industry-wide strikes threatened to undermine a smooth return to civilian production. The Taft-Hartley Labor Act was designed to modify the Wagner-Connery Act and swing the pendulum back toward the middle by reducing the power of unions. It also spelled out specific penalties, including fines and imprisonment, for violation of the act.

Taft-Hartley gave workers the right to refrain from joining a union and applied the unfair labor practice provisions to unions as well as to employers. It specifically prohibited a union from forcing management to fire antiunion or nonunion workers. Unions were required to engage in **good faith bargaining**, and a 60-day "cooling off" period was created, which comes into effect when a labor agreement ends without a new contract. The act also regulated many of the unions' internal activities.

The most significant provision of the Taft-Hartley Labor Act that affects fire fighters is referred to as "strikes during a national emergency." In the event that an imminent strike could affect a major part of an industry and imperil the health and safety of the nation, the president is granted certain powers to help settle the dispute. The president can order employees back to work; compel **arbitration**; or provide economic, judicial, or political pressure to achieve a resolution of the dispute.

Landrum-Griffin Act of 1959

After the AFL and CIO merged in 1955 to become the AFL-CIO, Senator John McClellan conducted hearings that revealed evidence of crime and corruption in some of the older local unions. The Labor-Management Reporting and Disclosure Act, otherwise known as the Landrum-Griffin Act, passed in 1959, at the height of the furor.

Landrum-Griffin established a bill of rights for members of labor organizations. It required that unions file an annual report with the government listing the assets of the organization as well as the names and assets of every officer and employee. Minimum election requirements were mandated, as were the duties and responsibilities of union officials and officers. This act also amended portions of the Taft-Hartley Labor Act.

Collective Bargaining for Federal Employees

Collective bargaining rights for public employees have traditionally lagged behind those in the private sector. Federal legislation passed in 1912 prohibits federal employees from striking. There were fewer than 1 million unionized government employees in the United States in 1956. Federal legislation during the 1950s and 1960s, however, allowed unions to grow in the public sector. By 1970, the number of unionized government employees had grown to 4.5 million. Relations between management and unions had matured.

President John F. Kennedy issued Executive Order #10988 in 1963, which granted federal employees the right to bargain collectively under restricted rules. This step represented a significant milestone for public-sector unions. President Richard Nixon further expanded the rights of employee unions within the federal government and established the Federal Labor Relations Council, which is similar to the National Labor Relations Board for private-sector unions.

State Labor Laws

In addition to the federal laws, each state exercises legislative control over several aspects of collective bargaining. Each state determines whether it will engage in collective bargaining with state government employees and whether local government jurisdictions within the state may engage in collective bargaining with their employees. Some states require municipal governments to bargain collectively with fire fighters' unions, some permit collective bargaining, and others limit or prohibit collective bargaining (Piskulich, 1992).

Strikes by state employees are illegal in all but 10 states, and many states also prohibit local government employees from striking. Forty percent of local governments have forbidden employee strikes; even so, numerous municipal strikes occurred between 1967 and 1980. After this period, several states and municipalities adopted legislation to prohibit strikes or other adverse labor actions.

Fire fighters have the right to bargain collectively in half of the states. Those states authorize municipalities to recognize a local fire fighter labor organization as the bargaining unit and enter into a binding contract that covers fire fighter pay, benefits, working conditions, conflict resolution, promotions, and department practices. The remaining states place restrictions on bargaining with fire fighters.

Right to Work

The collective bargaining rights of public employees in different states are related to the right-to-work issue. Under the Taft-Hartley Labor Act, workers have the right to refrain from joining a union. In 2019, **right-to-work** laws were in effect in 29 states (**TABLE 9-1**).

TABLE 9-1 Right-to-Work States

Alabama	Nebraska
Arizona	Nevada
Arkansas	North Carolina
Florida	North Dakota
Georgia	Oklahoma
Idaho	South Carolina
Indiana	South Dakota
Iowa	Tennessee
Kansas	Texas
Kentucky	Utah
Louisiana	Virginia
Michigan	West Virginia
Mississippi	Wisconsin
	Wyoming

Data from The National Right to Work Committee.

In those states, a worker cannot be compelled, as a condition of employment, to join or pay dues to a labor union. The same states tend to limit or restrict collective bargaining for local government employees.

The purpose of the original right-to-work legislation in 1947 was to prohibit the practice of closed shops, in which a worker must be a member of a particular union to work for the company. Open shops provide the worker with the option of remaining outside the union. Proponents of the open shop statute believe that the practice eliminates union collusion and exclusionary practices, which are now deemed illegal, and protects an individual's right to refrain from joining an organization.

Opponents of open shops express the opinion that right-to-work statutes reduce a union's bargaining power and place an unfair burden on the union members. They state that, in essence, the union has to bargain for the entire workforce, even for nonmembers. A worker who chooses not to join the union is afforded the same benefits as union members, but without paying dues to support union activities, pay for attorney's fees, or conduct research studies. Under the federal regulations, however, an individual who chooses not to join a union can still be compelled to pay a share of the cost of the union's representation in collective bargaining.

Right-to-work laws remain a source of controversy among labor organizations. Although unions support overturning existing right-to-work statutes, advocacy groups defend the existing statutes and promote their adoption in additional states. Workers entering the labor market should be aware of the prevailing labor laws that pertain to them.

Organizing Fire Fighters into Labor Unions

Regardless of the historical period, statutory environment, or work culture, labor–management relationships have been present in every work environment that includes an employer and employees. The first fully paid fire department in the United States was established in 1853 in Cincinnati, Ohio.

The career fire fighter's work environment at the start of the 20th century was grim. New York City fire fighters worked "continuous duty," 151 hours per week, with just 3 hours off each day to go home for meals. San Francisco fire fighters received one day off after five consecutive 24-hour duty periods. The federal courts prohibited union representation of multiple labor groups. Consequently, individual local unions could not organize to form regional or national labor organizations. This prohibition lasted until the Clayton Act was passed in 1914.

Federal Labor–Management Conflicts

Two notable strikes involving federal government employees marked the beginning and the end of a significant era in the evolution of labor–management relations: the postal service workers strike of 1970 and the air traffic controllers strike of 1981.

Postal Workers' Strike: 1970

In 1970, U.S. postal workers initiated an illegal strike. At that time, the postmaster general was legislatively restricted from negotiating with the striking workers; even so, negotiations were conducted and a settlement was reached. The striking postal workers were reinstated without penalty, and the negotiated wage increases were adopted. Subsequently, Congress recognized the postal workers' union for the purposes of collective bargaining. This strike set the tone for future civil service strikes, including several that involved fire fighters.

PATCO Air Traffic Controllers' Strike: 1981

One reason why public employees rarely strike today could be the result of the 1981 strike involving U.S. air traffic controllers. In that year, the Professional Air Traffic Controllers Organization (PATCO) went on strike after reaching an impasse in contract negotiations with the Federal Aviation Administration.

President Ronald Reagan cited the example of the 1919 Boston, Massachusetts, police strike when he described the actions he would take to end the walkout. When the PATCO workers failed to report to work as ordered, Reagan fired every striking member and decertified the union. Military air traffic controllers were brought in as temporary replacements. Air traffic volume was reduced for 18 months while the Federal Aviation Administration hired and trained replacement workers.

Unlike in the 1970 postal workers' strike, none of the PATCO strikers was hired back, and many of the union officials were subjected to years of aggressive litigation by the federal government. This action set the tone for employer–employee relations during the remainder of the Reagan administration. The frequency of public-sector strikes in the United States decreased significantly over the following two decades.

Labor Actions in the Fire Service

The fire service has experienced many adverse labor actions in response to poor labor–management relations. A strike—the act of withholding labor for the purposes of effecting a change in wages, hours, or working conditions—is one of the most drastic labor actions. Although a strike is precipitated by a conflict between labor and management, its impact is often felt beyond the parties that are directly involved in the relationship. In the private sector, the consumer is often the ultimate victim of a strike, through a combination of inconvenience and economic impact. In the public sector, the safety and the welfare of the general public may be directly affected by major labor actions.

The potential impact of a strike on public safety is so severe that many states prohibit fire fighters from walking out. In many of the cases in which fire service strikes have occurred, the public and media have questioned labor's right to strike as a matter of ethics. The International Association of Fire Fighters (IAFF) started with the premise that fire fighter strikes are inadvisable, and from 1930 to 1968 the IAFF charter included a no-strike clause. Nevertheless, in three periods during the 20th century, municipal fire fighters did go on strike in various U.S. cities.

As paid fire departments were established, fraternal or benevolent fire fighter support organizations gradually became political advocates for the workers and began to define the labor–management relationship. The genesis of these organizations occurred entirely at the local level. Antitrust protection, legislated by the 1890 Sherman Act, was interpreted.

Striking for Better Working Conditions: 1918–1921

The 1918–1921 strikes occurred during a period of economic instability following World War I. Most of the fire fighter strikes were efforts to establish a two-platoon system, obtain more pay, or simply gain the right to form a labor organization that would be recognized by the municipality (**FIGURE 9-2**). Organized labor reached a turning point in 1919. First, there was an industry-wide strike in the steel industry. Then, the country witnessed the first "general strike" that shut down Seattle, Washington, on February 6. A similar strike occurred in Winnipeg, Manitoba, Canada, on May 15, 1919.

FIGURE 9-2 A fire officer is generally the first point of contact between the workers and the organization.
© Jones & Bartlett Learning. Photographed by Glen E. Ellman.

Fire fighters in Memphis, Tennessee, threatened to strike in 1918 and succeeded in obtaining a two-platoon work schedule, allowing fire fighters more time at home. The old system in Memphis had allowed just 1 to 3 days off each month; the new system allowed fire fighters to spend every other day away from the fire station.

Many IAFF local unions were able to gain a two-platoon schedule and/or a pay raise by threatening to strike. Fire fighters on the two-platoon system could finally have a home away from the fire station, although they were still required to work 84 hours per week. In turn, they could marry and raise a family. Today, most municipal fire fighters work between 42 and 56 hours per week. The workweek for most federal and military fire fighters still exceeds 60 hours.

Striking to Preserve Wages and Staffing: 1931–1933

The Great Depression began with the stock market crash of 1929. For the next several years, local governments reduced wages, initiated furloughs (unpaid leave), and reduced workforce size. In Seattle, fire fighters had to take two 25-day furloughs in 1933. In addition to the furloughs, the city closed seven fire stations and eliminated eight engine companies, two hose companies, two squad companies, two fireboats, and one truck company to reduce expenditures.

Several fire fighter strikes between 1931 and 1933 occurred in reaction to wage reductions and elimination of fire fighter jobs. One local union went on strike because the fire fighters had not been paid by the city for 2 months.

Organized labor and fire fighter strikes during this time seem to have accomplished the IAFF's mission of improving working conditions. Davis Ziskind's 1940 book, *One Thousand Strikes of Government Employees*, examined 39 fire fighter strikes and lockouts from 1900 to 1937 (Ziskind, 1940). Approximately half of the strikes resulted in partial or complete victory by labor. In contrast to the chaos associated with the Boston police riot, Ziskind could find only one large-loss fire that occurred during a labor action—a large industrial fire in Pittsburgh, Pennsylvania, during the first IAFF organized strike in 1918.

Staffing, Wages, and Contracts: 1973–1980

The most recent series of fire fighter strikes occurred between 1973 and 1980. The IAFF voted to eliminate the 50-year-old no-strike clause in its constitution at its 1968 convention, at a time when the United States was mired in political turmoil.

Most of the ensuing strikes occurred after labor and management reached an impasse while negotiating a new contract. An **impasse** occurs when the parties have reached a deadlock in negotiations. None of the strikes resulted in a net gain for organized labor.

In the 1970s, local governments were facing another recession that hit hard at their finances. Several large cities were facing bankruptcy. In the depths of the recession, New York City laid off more than 40,000 city employees on July 2, 1975, including 1600 fire fighters. Although 700 of the fire fighters were hired back within 30 days, 900 others lost their permanent fire fighter jobs. It would take 2 years before the city could rehire all of the laid-off fire fighters.

Negative Impacts of Strikes

There have been fewer fire fighter strikes in the United States and Canada since the 1980s. The negative public reaction to such labor actions, the political response of changing legislation, and the lasting legacy of lost trust have caused fire fighter strikes to be viewed as counterproductive (Mateja, 2002). Measures such as picketing and alternative pressure tactics that maintain emergency services have been used to influence public opinion in several fire departments (**FIGURE 9-3**). Fire service strikes, or job actions, have occurred more recently among other industrialized nations throughout the world, particularly in Europe, where military personnel have been called upon to fill in for fire fighters (Ward, 1992).

The Wisconsin Challenge

Wisconsin was the first state to allow public employees bargain with their employers in 1959. In 2011, the newly elected Governor Scott Walker repealed most labor laws in the state. He fought off subsequent laws that were passed by the state legislature. The governor survived a 2012 recall election (Stein and Marley, 2013). Wisconsin became a right-to-work state, one of five states added since 2012.

The Growth of IAFF as a Political Influence

The economic turmoil and wage reductions that occurred in the 1990s were as severe as those in the 1930s; however, this time organized labor took a different path. Instead of striking, labor worked to elect political candidates who were supportive of labor to local and national political offices. The attempted

> **FIRE OFFICER TIP**
>
> **When the City, Then the Citizens, Voted Down Seattle Medic One**
> The Seattle Medic One paramedic ambulance service began as a university research project on March 7, 1970. The Seattle City Council declined to fund continuation of the Medic One project within the fire department budget in 1972. Members of IAFF Local 27 scrambled to fund the life-saving project through a special countywide tax levy. Two years later, in 1974, the CBS news magazine *60 Minutes* profiled Seattle as "the best place to have a heart attack." Organized labor had to scramble again in 1997 when voters defeated a plan to continue the levy, requiring a special referendum in February 1998 to keep 22 paramedic ambulances running in Seattle and King County. Both the jurisdiction and labor were aggressive in promoting the levy, which passed in 2007.

recall of elected officials in Memphis in 1979 may have been the beginning of this new era of labor influence.

Fire fighters are particularly respected in political circles for the efforts they can put behind a political candidate who supports the objectives of the IAFF or a particular local union. In addition to assisting with political campaigns, organized labor found that "money

FIGURE 9-3 Picketing is one of the most visible labor actions.
© Clive Chilvers/Shutterstock

talks." The creation in 1978 of FIREPAC, the IAFF's political action committee, was a natural progression of the labor organization's activity as a major influence in improving fire fighters' work environment. A **political action committee (PAC)** is a special-interest group that can solicit funding and lobby local and national elected officials on behalf of its cause. Funded by donations from individual IAFF members, FIREPAC promotes the legislative and political interests of the IAFF. The money is used to educate members of Congress about issues important to fire fighters and EMS personnel and to elect candidates to office who support those issues. In the 2018 election cycle, FIREPAC raised more than $6.49 million in voluntary donations. Its support was not completely one-sided in regard to political parties; 21 percent of the funds donated to candidates for federal elections went to Republicans.

Using money and off-duty fire fighters to assist political candidates or causes may appear unseemly. Hal Bruno, former director of political reporting for ABC television and radio networks, repeatedly pointed out in his *Firehouse* magazine columns that the fire service must be fully engaged in the political process.

You Are the Fire Officer Conclusion

The biggest distinction between a captain and a lieutenant is that a captain is a managing supervisor. As a manager, the captain develops or implements a strategy to accomplish a goal or mission. This occurs by managing lieutenants who are directly supervising work teams performing task-level activities. In addition, the captain assumes a larger role in the budget development and management process, including procurement.

The captain will be given a larger assignment within the IMS, developing strategy and supervising multiple companies as a division or group leader. Some departments require captains to be credentialed in ICS 400: Advanced ICS Command and General Staff—Complex Incidents.

The captain establishes and administers professional development programs for lieutenants and fire fighters—a combination of skills, experiences, formal education, and assignments to help everyone progress with their professional growth.

After-Action REVIEW

IN SUMMARY

- The administrative duties of the Fire Officer II include evaluating a subordinate's job performance, correcting unacceptable performance, and completing formal performance appraisals on each member.
- Other duties of a Fire Officer II include developing a project or divisional budget, including the related activities of purchasing, soliciting, and rewarding bids, and preparing news releases and other reports to supervisors.
- The fire department is part of the structure of its community. To fulfill its mission, the fire department must often interact with other organizations. The managing fire officer is more engaged in working with other organizations and groups.
- A rigid command and control process remains essential when operating at emergency scenes. Away from emergencies, however, departments are using the concepts of employee empowerment, decentralized decision making, and delegation.
- Firefighting is one of the activities least frequently performed by fire companies; it accounts for only 5 percent of the response workload.
- Flashover and structural collapse are the primary causes of noncardiac death for fire fighters within a burning structure.

- As community expectations change, fire departments are providing a wider variety of services, often as a part of a multiagency effort.
- The fire officer of the future must look beyond the physical attributes of individuals and match each individual's strengths with the organization's needs for the organization, the individual, and the officer to be successful.
- Four major pieces of federal legislation have established the groundwork for the rules and regulations of the present collective bargaining system:
  - The Norris-LaGuardia Act of 1932 made yellow dog contracts unenforceable.
  - The Wagner-Connery Act of 1935 established the National Labor Relations Board and the procedures commonly called collective bargaining.
  - The Taft-Hartley Labor Act of 1947 established good faith bargaining and a 60-day "cooling off" period.
  - The Landrum-Griffin Act established a bill of rights for members of labor organizations.
- Federal legislation passed in 1912 prohibits federal employees from striking.
- Each state determines whether it will engage in collective bargaining with state government employees and whether local government jurisdictions within the state may engage in collective bargaining with their employees.
- Regardless of time, statutory environment, or work culture, labor–management relationships have been present in every work environment that includes an employer and employees.
- Although a strike is precipitated by a conflict between labor and management, the impact of a strike is often felt beyond the parties that are directly involved in the relationship.
- From 1918 to 1921, fire fighter strikes focused on better working conditions.
- Between 1931 and 1933, fire fighter strikes focused on preserving wages and staffing.
- Fire fighter strikes between 1973 and 1980 focused on staffing, wages, and contracts.
- The negative public reaction to strikes, the political response of changing legislation, and the lasting legacy of lost trust have caused fire fighter strikes to be viewed as counterproductive.
- Fire fighters are particularly respected in political circles for the efforts they can put behind a political candidate who supports the objectives of the IAFF or a particular local union.
- New fire research challenges many strategy and tactic traditions with the fire service.

KEY TERMS

Arbitration Resolution of a dispute by a mediator or a group rather than a court of law. Any civil matter may be settled in this way; some labor–management agreements include a binding arbitration clause.

Collective bargaining The process in which working people, through their unions, negotiate contracts with their employers to determine their terms of employment.

Decision making The process of identifying problems and opportunities and resolving them.

Good faith bargaining Legal requirement of both the union and the employer arising out of Section 8(d) of the National Labor Relations Act. Enforced by the National Labor Relations Board, the parties are required to meet regularly to bargain collectively for wages, hours, and other conditions of employment.

Impasse A situation in which the parties in a dispute have reached a deadlock in negotiations; also described as the demarcation line between bargaining and negotiation. A declaration of a deadlock in labor–management negotiations brings in a state or federal negotiator who will start a fact-finding process that will lead to a binding arbitration resolution.

Incident command system (ICS) A system that defines the roles and responsibilities to be assumed by personnel and the operating procedures to be used in the management and direction of emergency operations; also referred to as an incident management system (IMS).

Managing fire officer The description from the IAFC *Officer Development Handbook* for the tasks and expectations for a Fire Officer II. In this role, the company officer is encouraged to acquire the appropriate levels of training, experience, self-development, and education to prepare for the Chief Fire Officer designation.

Political action committee (PAC) An organization formed by corporations, unions, and other interest groups that solicits campaign contributions from private individuals and distributes these funds to political candidates.

Right-to-work A worker cannot be compelled, as a condition of employment, to join or not to join or to pay dues to a labor union.

State police powers Authority of each state to govern matters related to the welfare and safety of the residents.

Unfair labor practices Employer or union practices forbidden by the National Labor Relations Board or state/local laws, subject to court appeal. It often involves the employer's efforts to avoid bargaining in good faith.

Yellow dog contracts Pledges that employers required workers to sign indicating that they would not join a union as long as the company employed them. Such contracts were declared unenforceable by the Norris-LaGuardia Act of 1932.

REFERENCES

Aherns, Marty. 2016, May 2. "The Unwanted Conundrum: In an Education Session Panel Preview, a Host of Experts Takes on the Problem of Unwanted Alarms." *NFPA Journal*. Quincy, MA: National Fire Protection Association.

Bendersky, Corinne. 2018, December 7. "Making U.S. Fire Departments More Diverse and Inclusive." *The Harvard Business Review*. Cambridge, MA: Harvard Business School Press.

Biddle, Pippa. 2018, February 25. "Fire Escapes Are Evocative, But Mostly Useless." *The Atlantic*. Accessed July 25, 2019. https://www.theatlantic.com/technology/archive/2018/02/how-the-fire-escape-became-an-ornament/554174/.

Brunacini, Alan V. 2002. *Fire Command: The Essentials of IMS*. Phoenix, AZ: Heritage Publishers.

Brunacini, Alan V. 2004. *Command Safety: The IC's Role in Protecting Firefighters*. Stillwater, OK: Fire Protection Publications.

Campbell, Richard. 2016. "Patterns of Firefighter Fireground Injuries." *National Fire Protection Association*. Accessed July 25, 2019. https://www.nfpa.org/News-and-Research/Data-research-and-tools/Emergency-Responders/Patterns-of-firefighter-fireground-injuries.

Cannuscio, Carolyn C., Andrea L. Davis, Amelia D. Kermis, Yasin Khan, Roxanne Dupuis, and Jennifer A. Taylor. 2016. "A Strained 9-1-1 System and Threats to Public Health." *Journal of Community Health* 41 (3): 658–666.

Darling, Brian. 2012. *Firefighters, Teachers and Police: Not a Federal Responsibility*. Washington, DC: The Heritage Foundation.

Deal, Tim, Michael de Bettencourt, and Vickie Deal. 2012. *Beyond Initial Response: Using the National Incident Management System Incident Command System, Second edition*. Bloomington, IN: AuthorHouse.

Dilts, David A., Clarence R. Dietsch, and Ali Rassuli. 1992. *Labor Relations Law in State and Local Government*. Santa Barbara, CA: Praeger.

Evarts, Ben. 2018. *Fire Loss in the United States During 2017*. Quincy, MA: National Fire Protection Association.

Evarts, Ben, and Gary Stein. 2019. *U.S. Fire Department Profile 2017*. Quincy, MA: National Fire Protection Association.

Fahy, Rita F., and Joseph L. Molis. 2019. *Firefighter Fatalities in the US—2018*. Quincy, MA: National Fire Protection Association.

Federal Bureau of Investigation (FBI). n.d. "Laboratory Services: Technical Hazards Response." Accessed July 25, 2019. https://www.fbi.gov/services/laboratory/forensic-response/technical-hazards-response-.

Federal Emergency Management Agency (FEMA). 2017. *National Incident Management System*. Washington, DC: US Department of Homeland Security.

Goldfeder, Billy, Alan Long, and Brian Schaeffer. 2019, July 9. "Crisis in the Streets: Fire Department Response to Homeless Individuals." [webinar] *Lexipol*. Accessed July 25, 2019. https://info.lexipol.com/webinar-fire-department-response-to-homeless.

Gorman, Anna, and Kaiser Health News. 2019, March 8. "Medieval Diseases Are Infecting California's Homeless: Typhus, Tuberculosis, and Other Illnesses Are Spreading Quickly Through Camps and Shelters." *The Atlantic*. Accessed July 25, 2019. https://www.theatlantic.com/health/archive/2019/03/typhus-tuberculosis-medieval-diseases-spreading-homeless/584380/.

Heitman, Steven C. 2016. *Suicide in the Fire Service: Saving the Lives of Firefighters*. Naval Postgraduate School, Monterey, CA. Accessed July 25, 2019. https://calhoun.nps.edu/handle/10945/48534.

Hensler, Bruce. 2011. *Crucible of Fire: Nineteenth-Century Urban Fires and the Making of the Modern Fire Service*. Washington, DC: Potomac Books.

Hoefferle Jr, Joseph, ed. 2006. *Fire and Building Safety Code Compliance for Historic Buildings: A Field Guide, Second edition*. Burlington, VT: University of Vermont Historical Preservation Program.

International Association of Fire Chiefs (IAFC). 2010. *Officer Development Handbook*. Fairfax, VA: International Association of Fire Chiefs.

Jahnke, Sara A., Walker S. Carlos Poston, Christopher K. Haddock, and Beth Marie Murphy. 2016. "Firefighting and Mental Health: Experiences of Repeated Exposure to Trauma." *Work* 53 (4): 737–744.

Jao, Carren. 2017, March 9. "Inside One of the Country's Busiest Fire Stations That Serves the City of Los Angeles." *KCET/Social Connected*. Accessed July 25, 2019. https://www.kcet.org/shows/socal-connected/inside-one-of-the-countrys-busiest-fire-stations-that-serves-the-city-of-los.

JUSTIA. 2019. "State-Level Administrative Law." Accessed July 15, 2019. https://www.justia.com/administrative-law/state-level-administrative-law/.

Koerth-Baker, Maggie. 2018, July 17. "Wildfires in the U.S. Are Getting Bigger." *FiveThirtyEight*. Accessed July 25, 2019. https://fivethirtyeight.com/features/wildfires-in-the-u-s-are-getting-bigger/.

Lam, Linda. 2018, November 8. "A Concerning Trend: Flooding Deaths Have Increased in the U.S. the Last Few Years." Atlanta, GA: The Weather Channel. Accessed August 20, 2019. https://weather.com/safety/floods/news/2018-11-08-flood-related-deaths-increasing-in-united-states

Mateja, Michael G. 2002. *Fiery Struggle: Illinois Firefighters Build a Union, 1901–1983*. Chicago, IL: Illinois Labor History Society.

National Law Center on Homelessness and Poverty (NLCHP). 2017. *Tent City, USA: The Growth of America's Homeless Encampments and How Communities Are Responding*. Washington, DC: National Law Center on Homelessness and Poverty.

Pipeline and Hazardous Materials Safety Administration (PHMSA). 2016. *Emergency Response Guidebook*. Washington, DC: U.S. Department of Transportation.

Piskulich, John P. 1992. *Collective Bargaining in State and Local Government*. Santa Barbara, CA: Praeger.

Senate Governance and Finance Committee (SGDC). 1987 "The Fire Protection District Law of 1987." Sacramento, CA: State of California.

Stein, Jason, and Patrick Marley. 2013. *More Than They Bargained For: Scott Walker, Unions and the Fight for Wisconsin*. Madison, WI: The University of Wisconsin Press.

Stein, Jeff, and Andrew Van Dam. 2019, April 22. "Taxpayer Spending on U.S. Disaster Fund Explodes Amid Climate Change, Population Trends." *The Washington Post*. Accessed July 25, 2019. https://www.washingtonpost.com/us-policy/2019/04/22/taxpayer-spending-us-disaster-fund-explodes-amid-climate-change-population-trends/?utm_term=.966b5021a7dd.

Symons, Jeremy. 2017, September 13. "New Data Shows Changing Disaster Trends – and Why Congress Should Take Note." [EDF Voices blog.] *Environmental Defense Fund*. Accessed July 25, 2019. https://www.edf.org/blog/2017/09/13/new-data-shows-changing-disaster-trends-and-why-congress-should-take-note.

Tull, Matthew. 2019, June 10. "Development of PTSD in Firefighters." *Verywell Mind*. Accessed July 15, 2019. https://www.verywellmind.com/rates-of-ptsd-in-firefighters-2797428.

USLEGAL. n.d. "Rulemaking Defined Under the Model State APA." Flowood, MS: airSlate Legal Forms, Inc.

Varone, J. Curtis. 2014. *Legal Considerations for Fire and Emergency Services, Third edition*. Tulsa, OK: PennWell Company.

Ward, Michael J. 1992. *Labor Strikes in the American Fire Service: A Special Report for the Japanese Local Government Center*. Fairfax, VA: International Association of Fire Chiefs.

Ward, Michael J. 2018, October 22. "Fire Officer as a Transformational Leader." *Company Commander*. Accessed July 25, 2019. https://companycommander.com/2018/10/22/fire-officer-as-a-transformational-leader/.

Ziskind, David. 1940. *One Thousand Strikes of Government Employees*. New York: Columbia University.

Fire Officer in Action

1. On the evening shift there is an engine company captain, a truck company lieutenant, and a paramedic ambulance lieutenant. Who completes the annual evaluation report for the engine company driver?
   A. Battalion chief
   B. Captain
   C. Truck company lieutenant
   D. Paramedic ambulance lieutenant

2. What is the most frequent cause of death for fire fighters?
   A. Sudden cardiac arrest
   B. Crush trauma
   C. Suicide
   D. Carbon monoxide poisoning

3. According to annual data from the National Fire Protection Association, fire suppression represents _____ of a fire company workload.
   A. 1 percent
   B. 4 percent
   C. 9 percent
   D. 12 percent

4. The number of "right-to-work" states between 2012 and 2019 has:
   A. remained the same.
   B. increased by 5 to 29 states.
   C. decreased by 2 to 20 states.
   D. increased by 4 to 22 states.

Fire Captain Activity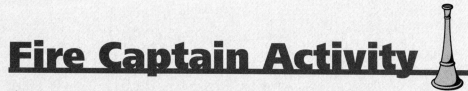

This chapter describes the process of conducting an initial accident investigation.

NFPA Fire Officer II Job Performance Requirement 5.1.2

Intergovernmental and interagency cooperation.

Application of 5.1.2

1. A sinkhole has developed on the street immediately outside the fire station. It is difficult to get fire apparatus out of the fire station to respond to emergencies. Using information from a jurisdiction you are familar with, describe the agencies needed to (1) resolve the immediate issue of fire company response readiness, and (2) mitigate the sinkhole hazard.

2. A senior living facility in your town is calling 911 for every resident that needs medical transportation to make a scheduled appointment at a medical clinic or physician office. Apparently, the callers always describe some type of medical situation that is dispatched as a life-threatening emergency. These calls come in two or three times a week. As the captain you have visited the facility, provided information, and requested that they stop calling 911 for scheduled medical transportation. There has been no change in frequency of the 911 calls; the facility does not want to wait the 3 to 4 hours for a commercial ambulance to arrive. The 911 call center is run by the sheriff's office, the senior living facility is owned by a national firm, and the town fire department has one paramedic ambulance and one quint. Describe how you will engage with your community partners to resolve this situation.

3. A defective valve allowed natural gas to be sent into a community at 1000 psi. There are 127 people displaced from their homes. What is the process in your jurisdiction to provide emergency housing for 88 adults and 39 children? (1) Describe how your authority having jurisdiction can provide emergency shelter, which agencies would be involved, and how long will the process would take. (2) Create an incident action plan for the restoration of natural gas service to the four-story building, with 19 apartments, and the 27 single-family dwellings. Use the organization within your jurisdiction.

4. A hemorrhagic fever has placed 40 percent of your fire fighters on sick leave; either they are sick or their family member is sick. There is an overwhelming number of requests for EMS services. (1) Describe which agencies you would collaborate with to develop a plan to mitigate this situation. (2) Develop an incident action plan for the next 12 hours.

 *Access Navigate for flashcards to test your key term knowledge.*

CHAPTER 10

Fire Officer II

Applications of Leadership

KNOWLEDGE OBJECTIVES

After studying this chapter, you will be able to:

- Explain the officer's role in daily training and performance management. (**NFPA 1021: 5.2.1**) (pp. 274–275)
- Describe reasons officers must understand compensation and benefits. (**NFPA 1021: 5.2.1**) (pp. 283–284)
- Identify reasons for conducting annual evaluations. (**NFPA 1021: 5.2.1**) (pp. 282–283)
- Explain the four steps of the annual evaluation process. (**NFPA 1021: 5.2.1, 5.2.2**) (p. 283)
- Describe the use of progressive negative discipline to correct unacceptable behavior. (**NFPA 1021: 5.2.1**) (pp. 277–280)
- Explain methods of tracking an employee's performance and progress throughout the year. (**NFPA 1021: 5.2.2**) (pp. 273–274)
- Discuss the role of professional development in the fire service. (**NFPA 1021: 5.2.3**) (p. 287)
- List and describe the components of a professional development plan. (**NFPA 1021: 5.2.3**) (pp. 290–291)
- Analyze fire fighter death and injury data and make recommendations. (**NFPA 1021: 5.7.1**) (pp. 284–287)
- Describe the impact of sudden cardiac arrest on the fire service. (**NFPA 1021: 5.7.1**) (p. 284)
- Describe near-miss report analysis using the Human Factors Analysis and Classification System (HFACS) tool. (**NFPA 1021: 5.7.1**) (pp. 284–285)
- Explain the role of data analysis and risk management in fire departments. (**NFPA 1021: 5.7.1**) (pp. 285–286)
- Describe how to use postincident analysis to mitigate hazards. (**NFPA 1021: 5.7.1**) (pp. 286–287)
- Describe the role of data analysis and risk management in fire departments. (**NFPA 1021: 5.7.1**) (pp. 285–286)

SKILLS OBJECTIVES

After studying this chapter, you will be able to:

- Take action to address a member or unit performance issue. (**NFPA 1021: 5.2.1**) (pp. 277–283)
- Conduct and document formal employee evaluations. (**NFPA 1021: 5.2.1, 5.2.2**) (pp. 282–283)
- Create and document a professional development plan. (**NFPA 1021: 5.2.3**) (pp. 287, 290)

You Are the Fire Officer

The sound of the crash was startling. Following the noise, the captain ran out to the front of the station and saw that the garage door had started closing before the spare pumper was clear of the door frame. The front ramp was littered with broken emergency lights and garage door pieces.

1. What is the best way to respond to this incident?
2. What process will be most effective to avoid having this happen again?

 Access Navigate for more practice activities.

Introduction

A Fire Officer II has a much wider sphere of influence than a Fire Officer I. They are expected to develop Fire Officer Is, as well as fire fighters, in an arc of activities that exceed the boundaries of a single fire company. In addition, Fire Officer IIs are tasked with improving individual and company level performance through training, coaching, evaluation, and feedback. This chapter looks at the application of select leadership and management theories covered in Fire Officer I and introduces two helpful concepts: functioning as a transformational rather than a transactional leader and applying the concepts of **just culture** (a change in focus from errors and outcomes to the behavioral choices made by subordinates).

Fire Officer II as Transformational Leader

Transformational leadership, as described by theorist Peter G. Northouse, is a process in which a person engages with others and creates a connection that raises the level of motivation and morality in both the leader and the follower (Northouse, 2018). Transformational leadership changes and transforms people because it aids leaders in understanding individual emotions, values, and ethics; establishing standards; and achieving long-term goals.

Transactional Relationship Versus Transformational Leadership

A **transactional relationship** is a relationship where all parties are in it for themselves and they do things for each other with the expectation of reciprocation. Think back to your school days, or even your experience as a fire recruit. If you performed a task to your teacher's satisfaction, you received a "reward" (a good or passing grade). If you did not perform a task to your instructor's satisfaction, you received a negative response (a poor or failing grade). This linear and analog style of transactional relationship has its roots in the treatment of factory workers at the start of the Industrial Revolution. The transactional relationship remains important when we are engaged in high hazard emergency scene activity, where there is one incident commander and a direct chain of command.

Achieving transformational leadership, on the other hand, includes assessing the follower's motives, satisfying their needs, and treating them as full human beings, not simply workers. When incorporating charismatic and visionary leadership, transformational leadership involves an exceptional form of influence that moves followers to accomplish more than what is usually expected of them. Each of the components of transformational leadership helps build follower commitment in four different ways:

- Idealized influence: Leaders who behave consistently with the values they espouse can more easily build commitment to a group or organization's values, goals, or standards of behavior. A charismatic relationship cultivates trust, respect, admiration, and commitment to the leader and the future of the organization.
- Inspirational motivation: The major characteristics of inspirational motivation include optimistically and enthusiastically articulating a clear and appealing view of the future while expressing confidence, encouraging teamwork, and inspiring high standards of performance. Leaders use inspirational motivation to build emotional commitment to a mission or goal.
- Intellectual stimulation: In order to raise followers' awareness of problems as well as develop their capability to solve such problems, transformational leaders foster a climate that favors critical examination of commonly held notions, beliefs, and the status quo. Leaders

use intellectual stimulation to get followers to reexamine assumptions, seek different perspectives, and encourage nontraditional thinking.

- Individualized consideration: A transformational leader applies individual considerations by listening to each follower's needs and concerns, expressing words of thanks or praise as a means of motivation, making public recognition of achievements and initiatives, and undertaking individualized career counseling and mentoring.

Transformational leaders treat others as individuals with different needs, abilities, and aspirations and compel them to appreciate the benefits of diversity. They also help others develop their knowledge, skills, and abilities. As a result, followers begin to value personal learning and development and may appreciate the breadth of knowledge, skills, and abilities associated when collaborating with a diverse team.

Evaluating Job Performance of Assigned Members

While the jurisdiction or fire department will provide a set of behavioral or task expectations for each job classification, the fire officer has a powerful tool to encourage additional performance. **Goal setting** is the process of establishing of specific goals over a designated period of time. The goals should be specific and challenging, but attainable. Clear, specific, and measurable goals are essential for motivation. Vague goals, such as "get to the apparatus as quickly as you can," do not give the fire fighter clear enough expectations to determine success. Instead, a goal might be to "reduce your turnout time to less than 50 seconds." The most significant goal-setting application is to have fire fighters carefully consider which goals improve the organization and the fire fighter.

Establishing Annual Fire Fighter Goals

The fire officer should require all fire fighters who have completed probation to identify three work-related goals they want to achieve during the next evaluation period. These goals should be consistent with the fire department mission, goals, and objectives. Establishing annual goals provides a focus beyond the minimum day-to-day activities and helps fire fighters prepare for future promotional examinations or assignments. Examples of individual fire fighter goals might include the following:

- Enrolling in and completing a building construction course at the local community college
- Becoming qualified as an aerial operator
- Learning how to use the computer-aided design software to develop tactical preincident plans

The goals should be described in the S.M.A.R.T. format. That means the goal is *specific*, describing the who-what-where-how-when and conditions. The goal is *measurable*, for example, to reduce to 15 percent body fat. The goal is *attainable* during this evaluation period. The goal is *relevant* to the current or future fire department job responsibilities. And finally, the goal has a *timeline* and deadline.

Keeping Track of Every Fire Fighter's Activity

In an **activity log**, the fire officer maintains a list of the fire fighter's activities by date, along with a brief description of performance observations (**FIGURE 10-1**).

The **T-account** is a slightly more sophisticated documentation system, similar to an accounting balance sheet listing credits and debits. A single-sheet form is used to list the assets on the left side and the liabilities on the right side, with the result resembling the letter "T" (**FIGURE 10-2**).

March 11, Incident 3467. Trouble positioning apparatus to hook up soft suction and hydrant. Charged wrong attack line. First structural fire as engine driver.

FIGURE 10-1 Sample notation in an activity log.
© Jones & Bartlett Learning.

+	−
March 11, Incident 3467. Rescue of elderly female from hallway outside burning bedroom in a one-story, single-family dwelling.	March 11, Incident 3467. Had to return to Engine 4 to retrieve forcible entry tools and firefighting gloves during initial actions.

FIGURE 10-2 Sample notation in a T-account.
© Jones & Bartlett Learning.

> **FIRE OFFICER TIP**
>
> **Document "Hot Wash" Reviews**
>
> A **"hot wash"** is a debriefing that occurs immediately after an incident to identify issues needing prompt attention as well as procedural or systemic problems needing later follow-up (Canton, 2016). The hot wash process is a quick process. Questions to ask yourself include the following:
> - What went well?
> - What needs improvement?
> - Remarks or comments about the incident?
>
> Documenting the after-incident company-level hot washes into the activity log or T-account will provide a detailed history for each member under your supervision.

Using either method, the fire officer compiles an extemporaneous "when, what, and how" record of each fire fighter's work history throughout the evaluation period. This record can be a powerful evaluation and motivational tool. The activity log can track fire fighter progress to this year's goals.

Informal Work Performance Reviews

Most civil service and personnel regulations require an annual evaluation, but many departments also encourage fire officers to conduct informal performance reviews with each fire fighter throughout the year. During these informal sessions, the fire officer can review the T-account or activity log with the fire fighter and see what the officer can do to assist the fire fighter in meeting the established goals. Together, they can identify situations or work conditions that impede progress. For example, a fire fighter who is spending 70 percent of the time detailed to cover positions at other stations will have difficulty qualifying as a backup aerial operator. The fire fighter and officer could discuss changing the goal or adjusting assignments so that the fire fighter would have more time at the station to work on meeting the goal.

Correcting Unacceptable Behavior

All members are expected to meet performance standards and behave appropriately. The fire officer uses feedback, coaching, and performance appraisals to improve unacceptable performance or behavior. High performance organizations that have significant dangers utilize a just culture approach to address behavior and performance issues. Just culture changes the focus from errors and outcomes to systems design and management of behavioral choices of all members (Boyson, 2013).

Using Just Culture to Identify Source of Behavior

If there are no consequences, repercussions, or focused practice, humans drift away from complete and perfect performance. Daily training at the fire station and having company operations regularly assessed by the chief are ways that organizations can stay close to complete and perfect performance. Just culture focuses on the behavioral choices that individuals make that lead to unacceptable behaviors (Dekker, 2014). Unfortunately, this type of evaluation may only occur after a high-profile incident or a horrific outcome event.

A behavioral choice is benchmarked against the organization's mission, values, and goals. For example, if the department has a value of effective structural fire suppression, what is causing an increase in rekindle fires? Evaluation is made about the system and procedures that may have led to a specific behavioral decision. The just culture process classifies the spectrum of human intention as follows (Marx, 2001):

- Human error
- At-risk behavior
- Reckless behavior
- Knowingly causing harm
- Purposely causing harm

Most of the fire officer's focus will be on human error, at-risk behavior, and reckless behavior. Some people who *knowingly* cause harm may have a justification. People who *purposely* cause harm are immediately introduced to the criminal justice system. Both of these behaviors would require initial removal from the workplace by a command chief–level officer for evaluation.

Human Error

Human error is an inadvertent action such as a slip, lapse, or mistake—for example, the first instance of oversleeping and arriving late for work. The fire officer has the member identify the factors that created the late arrival. Outside factors may include situations that involve the following:

- Process
- Procedures
- Training
- Design
- Environment

In working with the member, the fire officer emphasizes that the member has the responsibility to make a behavioral choice to avoid being late to work again. This action falls into the informal work performance review.

At-Risk Behavior

At-risk behavior is a behavioral choice that increases risk when risk is not recognized or is mistakenly believed to be justified. Take the example of a lieutenant making the call not to wait for an ambulance, and instead deciding to transport a child in respiratory distress to the hospital via a ladder company. This action would be considered an at-risk behavior. Correcting at-risk behavior is the most significant impact a fire officer can have on a subordinate in terms of recognizing an at-risk situation and providing an informed opportunity to change a behavior.

The first goal in correcting at-risk behavior is to understand the thought process that lead up to the action. What was the situation that the member was encountering? What options were considered? In the previous example, for the ladder company lieutenant handling a child in respiratory distress, there was a lack of options. Dispatch did not respond to multiple radio requests asking how far away the ambulance was. The hospital was 3.2 miles away. The lieutenant remembered a long-ago story where a fire company pulled a child out of a pool and did CPR while transporting the child on the pumper (Remmers, 2016). The second goal is to make certain the member recognizes the risk. Emergency services are regulated through federal and state administrative law, sometimes with significant penalties if an unallowed activity occurs. In addition, the jurisdiction has rules, regulations, and standard operating procedures (SOPs) that document acceptable activities. In the real situation this example is based on, the fire department's state permit to operate as an EMS agency was jeopardized.

Based on the information gathered, the fire officer should consult with a supervisor to determine the most appropriate action. Options include the following:

- Individual training or coaching
- Revision of policy, procedure, or SOP/standard operating guideline (SOG)
- Modification of the organizational structure that led to this at-risk behavior

Reckless Behavior

Reckless behavior is a behavioral choice to consciously disregard a substantial and unjustified risk. In other words, the member recognized the risk and decided to continue the behavior. If your department practices vertical ventilation, and requires fire fighters to wear self-contained breathing apparatus (SCBA) while cutting open a roof during a structure fire, a fire fighter operating a chainsaw on the roof without SCBA is demonstrating reckless behavior. The fire officer needs to understand the fire fighter's thought process leading to this behavior and to determine if the member understood the risk. The determination whether this was an at-risk or reckless behavior is clarified using the "reasonable person" standard. The **reasonable person standard** describes the actions that a similar person with the same background would take given the same situation (Dekker, 2014). Would a reasonable person don SCBA before going to the roof to operate a chainsaw?

Conscious disregard of a substantial and unjustified risk requires negative reinforcement to discourage future reckless behaviors. Consulting with the fire officer's supervisor and considering the factors in each specific incidence, the responses would include the following:

- Remedial training
- Formal discipline
- Punitive action

Advance Notice of a Substandard Employee Evaluation

A fire fighter who receives coaching or training and continues to miss work-task expectations should know there is a problem long before the annual evaluation is undertaken. The subordinate should be given adequate time to change the behavior or improve the skill, particularly if it is jeopardizing a scheduled pay raise or continuing employment (Bender, 2013). Jurisdictions typically require a supervisor to provide a formal notification to an employee who is likely to be evaluated below the level necessary to obtain a pay increase or a "satisfactory" rating. The notice must identify the aspect of the job performance that is substandard and indicate what the employee must do to avoid receiving a substandard annual evaluation. A common practice is to require such a notification as soon as an issue arises, or at least 10 weeks before the annual evaluation is due.

Work Improvement Plan

If the employee receives a substandard annual evaluation, then the municipality might require a **work improvement plan** (**FIGURE 10-3**). This plan should cover a specific span of time, such as 120 calendar

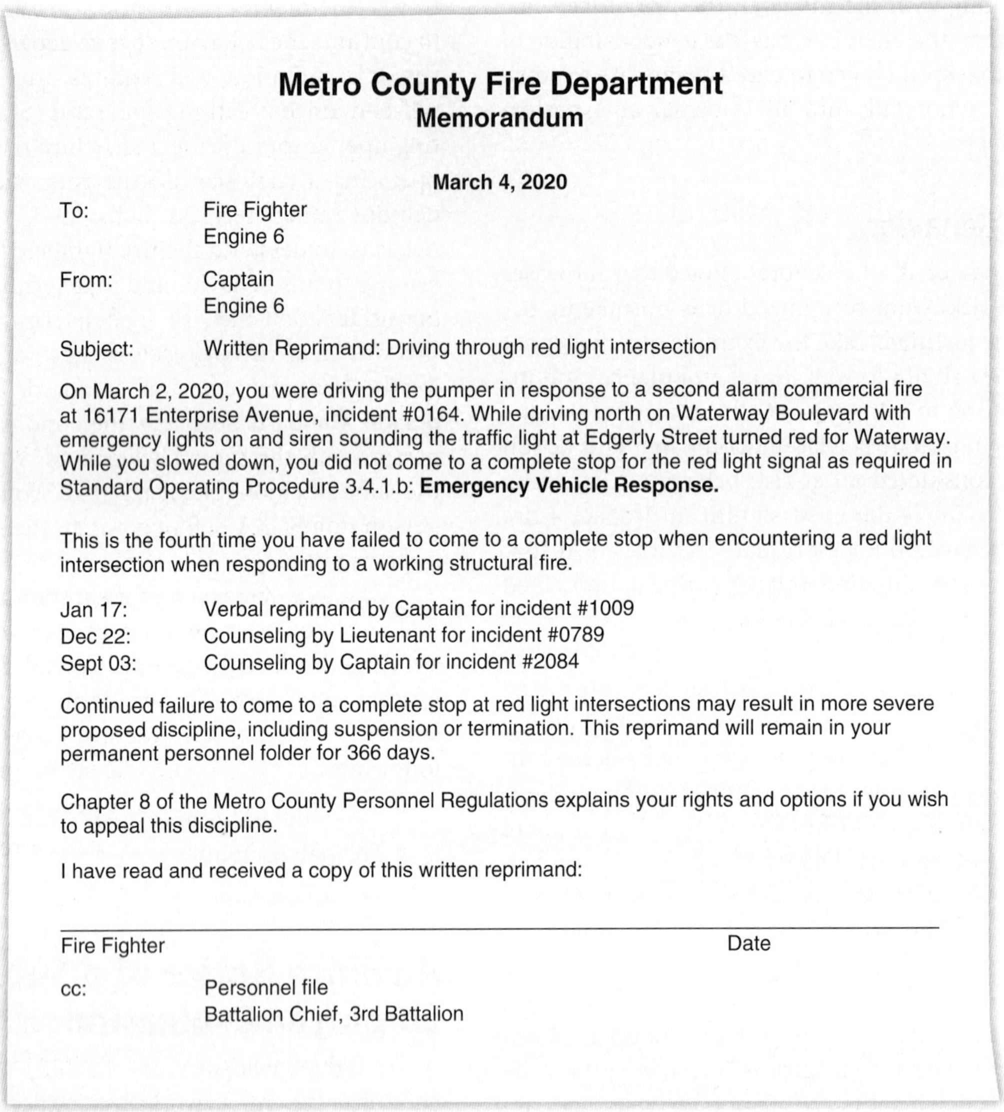

> **Metro County Fire Department**
> **Memorandum**
>
> March 4, 2020
>
> To: Fire Fighter
> Engine 6
>
> From: Captain
> Engine 6
>
> Subject: Written Reprimand: Driving through red light intersection
>
> On March 2, 2020, you were driving the pumper in response to a second alarm commercial fire at 16171 Enterprise Avenue, incident #0164. While driving north on Waterway Boulevard with emergency lights on and siren sounding the traffic light at Edgerly Street turned red for Waterway. While you slowed down, you did not come to a complete stop for the red light signal as required in Standard Operating Procedure 3.4.1.b: **Emergency Vehicle Response.**
>
> This is the fourth time you have failed to come to a complete stop when encountering a red light intersection when responding to a working structural fire.
>
> Jan 17: Verbal reprimand by Captain for incident #1009
> Dec 22: Counseling by Lieutenant for incident #0789
> Sept 03: Counseling by Captain for incident #2084
>
> Continued failure to come to a complete stop at red light intersections may result in more severe proposed discipline, including suspension or termination. This reprimand will remain in your permanent personnel folder for 366 days.
>
> Chapter 8 of the Metro County Personnel Regulations explains your rights and options if you wish to appeal this discipline.
>
> I have read and received a copy of this written reprimand:
>
> _____
> Fire Fighter   Date
>
> cc: Personnel file
> Battalion Chief, 3rd Battalion

FIGURE 10-3 Example of a work-improvement plan.
© Jones & Bartlett Learning.

days, that would be designated as a **special evaluation period**. The employee is expected to participate fully in a work improvement plan during this special evaluation period. He or she is given an opportunity to demonstrate the desired workplace behavior or performance no later than the end of the special evaluation period.

The following is an example of the human resources division requirements that could be applied to a work improvement plan:

- The work improvement plan shall be in writing, stating the performance deficiencies and listing the improvements in performance or changes in behavior required to obtain a "satisfactory" evaluation.
- The work improvement plan has been reviewed and approved by a senior management officer.
- During the special evaluation period, the employee shall receive regular progress reports.
- If, at the end of the special evaluation period, the employee's performance rating is "satisfactory" or better, the time-in-rank pay increase will start at the first pay period after the work improvement period.
- If, at the end of the special evaluation period, the employee rating remains unsatisfactory, then no time-in-grade pay increase will be issued. In addition, the supervisor will determine whether additional corrective action is appropriate.

Regardless of the outcome, a special evaluation period performance evaluation is filled out and submitted as a permanent record in the employee's official file (Belker, McCormick, and Topchik, 2018).

The employee's performance is officially evaluated during the annual review. Conducting informal reviews enables a fire officer to identify any underperforming fire fighter early in the annual evaluation period. The advance-notice procedure should provide the fire fighter with enough time to change the behavior or improve the necessary skills before the formal evaluation is scheduled to occur. If the employee has still not improved, a substandard evaluation would not come as a surprise to either the fire fighter or the fire officer's supervisor.

Negative Discipline: Correcting Unacceptable Behavior

Sometimes, an effort that begins with positive discipline (e.g., motivation, training, and coaching) evolves into negative discipline because of the continuing inability or unwillingness of a fire fighter to meet the required performance or behavioral expectations. If an individual does not respond to positive efforts to correct a problem, the next logical step is to punish continuing unsatisfactory performance. **Progressive negative discipline** moves from mild to more severe punishments if the problem is not corrected. The ultimate goal remains fire fighter performance improvement. In an extreme case in which progressive discipline fails to solve the problem, it may become necessary for the organization to terminate an employee who is ineffective and unwilling to improve.

The Progression of Negative Discipline

The disciplinary process is designed to be consistent and well documented. Typical steps in a progressive negative discipline system include the following:

- Counsel the fire fighter about poor performance and ensure that he or she understands the requirements.
- Ascertain whether there are any issues contributing to the poor performance that are not immediately obvious to the supervisor. Resolve these issues, if possible.
- Verbally reprimand the fire fighter for poor performance.
- Issue a written reprimand and place a copy in the fire fighter's file.
- Suspend the fire fighter from work for an escalating number of days.
- Terminate the employment of a fire fighter who refuses to improve (Antonellis, 2012).

As the penalty increases, negative discipline requires the participation of higher levels of supervision. A fire officer is often required to consult with his or her supervisor before issuing formal negative discipline. Many jurisdictions provide detailed descriptions of the "due process" and just culture principles that guide the administration of progressive negative discipline under a labor–management contract (Edwards, 2010). The following is an example of how the negative discipline process might proceed:

- Informal oral or written reprimand: Reprimand is issued by a supervising or managing officer. Reprimand remains at the fire station level and expires after a time period of no longer than 1 year.
- Formal written reprimand: Document initiated by a fire officer (usually after consulting with his or her supervisor). Some organizations require a battalion chief or other command officer to issue a written reprimand. A copy of the letter goes into the individual's official personnel file. It expires after a set time period, usually 1 year.
- Suspension: May be initiated or recommended by a fire officer. The **suspension** notice is usually issued by a battalion chief or a higher-level command officer after consultation with the fire chief or designee. The record of a suspension remains permanently in the employee's official file. Occasionally, the record is removed after a grievance, arbitration, or civil service hearing, or in the wake of a lawsuit.
- Termination: Although a **termination** is usually recommended by a lower-level command officer, the fire chief issues the formal termination notice after consulting with the personnel office.

Some employee behaviors require the fire officer to implement negative discipline immediately. This is a common policy when there is willful misconduct, as opposed to inadequate performance. Personnel regulations usually provide a list of behaviors that will lead to immediate negative discipline, such as the following acts:

- Knowingly providing false information affecting an employee's pay or benefits or in the course of an administrative investigation
- Willfully violating an established policy or procedure
- Being convicted of a criminal offense that affects the ability of the employee to perform his or her job

- Displaying insubordination
- Behaving in a careless or negligent way that leads to personnel injury, property damage, or liability to the municipality
- Reporting to work when under the influence of alcohol or a controlled substance
- Misappropriating fire department property or funds

As in all cases of discipline, the fire officer will need to consult with a superior and, often, a human resource official before issuing any discipline.

Formal Written Reprimand

A **formal written reprimand** represents an official negative supervisory action at the lowest level of the progressive discipline process. Even when the department requires a command-level officer to issue a formal written reprimand, this document is often prepared by a fire officer. The fire officer is the closest person to the issue and will be involved in the remediation effort. He or she should consult the personnel regulations and departmental guidelines when preparing a written reprimand. In departments in which a fire officer can issue a written reprimand, the fire officer's supervisor should review and approve the document before it is issued to the fire fighter.

The written reprimand should contain the following information:

- Statement of charges in sufficient detail to enable the fire fighter to understand the violation, infraction, conduct, or offense that generated the reprimand
- Statement that this is an official letter of reprimand and will be placed in the employee's official personnel folder
- List of previous offenses in cases in which the letter is considered a continuation of progressive discipline
- Statement that similar occurrences could result in more severe disciplinary action, up to and including termination

A written reprimand starts the formal paper trail of a progressive disciplinary process. If the behavior is not repeated, some jurisdictions allow written reprimands to be removed from the employee's file after a designated time period, usually 1 year after the reprimand is issued.

The written reprimand may potentially be seen by a wide audience. If it is appealed or becomes part of a larger disciplinary action, the fire chief, labor representatives, civil service commissioners, and attorneys will read the reprimand. Some departments require that an impartial third-party panel review formal discipline recommendations. Fire officers must focus on the work-related behaviors and clearly explain the behavior or action that generated the reprimand.

Predisciplinary Conference

In many jurisdictions, a predisciplinary conference or hearing must be conducted before a suspension, demotion, or involuntary termination can be invoked. The degree of investigative effort and the opportunities for fire fighter response before these punishments are issued are set higher than for less severe levels of discipline. Most fire departments require a formal disciplinary hearing before a suspension is issued to provide an opportunity for the fire fighter to formally respond to the charges.

Known as a **Loudermill hearing**, this process resulted from the 1985 U.S. Supreme Court decision in *Cleveland Board of Education v. Loudermill*. The Cleveland Board of Education hired James Loudermill as a security guard in 1979. Loudermill received a letter from the Board on November 3, 1980, dismissing him because of dishonesty in filling out his employment application. He had not listed a 1968 felony conviction for grand larceny on his application. The U.S. Supreme Court combined the Loudermill case with a similar appeal filed by Richard Donnelly, a Parma Board of Education bus mechanic fired in August 1997 after failing an eye examination.

The court ruling indicated that a **pretermination hearing**, including a written or oral notice, in which the employee has an opportunity to present his or her side of the case, and an explanation of adverse evidence were essential to protect the worker's due process rights. Such a hearing serves as a check against a possible mistaken decision, as it seeks to determine whether there are reasonable grounds to believe that the charges against the employee are true and whether they support the proposed action.

Not all disciplinary actions require a Loudermill hearing before suspension; the rules vary from state to state. In 1997, in *Gilbert v. Homar*, the U.S. Supreme Court ruled that East Stroudsburg University did not violate the constitutional due process rights of a university police officer when it immediately suspended him without pay after learning he had been charged with a drug felony.

A predisciplinary hearing can be conducted by a disciplinary board, by the fire chief or another ranking officer, or by a hearing officer. The designated individual or board reviews the case and makes a recommendation to the fire chief. Before the hearing, the fire fighter receives a letter that outlines the offense and the results of the investigation (Corbin and May, 2012). The proposed duration of the suspension without pay could also be stated in the letter. The following is a typical example of the process:

1. The fire officer investigates the alleged employee offenses promptly and obtains all pertinent facts in the case (time, place, events, and circumstances) including, but not limited to, making contact with persons involved or having knowledge of the incident.
2. The fire officer prepares a detailed report outlining the offense, the circumstances, the individual's related prior disciplinary history, and recommended disciplinary action.
3. The report is submitted to a higher-ranking officer in the chain of command.
4. The fire department representative consults with the human resources director or his or her designee, if necessary, when suspensions are contemplated.
5. The disciplinary board hearing is scheduled, and an advance notice letter is prepared.
6. The disciplinary board considers the charges and hears the employee's response.
7. The disciplinary board makes a recommendation to the fire chief, who issues the final decision.

At the hearing, the accused individual has an opportunity to refute the charge or present additional information about mitigating circumstances. In most career fire departments, a union representative is present at the hearing to advise the individual. An alternative or lower level of discipline might be proposed, based on past department practices. The disciplinary board makes a final recommendation to the fire chief.

The fire fighter might or might not have the ability to appeal the disciplinary action through a grievance procedure, depending on the personnel rules of the organization. The final resolution of a disciplinary action usually resides with the civil service commission or the city personnel director. Some municipalities use arbitration to resolve these issues.

Alternative Disciplinary Actions

Depending on the nature and the severity of the offense, a variety of alternative penalties could be imposed such as the following:

- Extension of a probationary period: If the fire fighter is in a probationary period, as occurs with a recruit or after a promotion, the probationary period can be extended until the work performance issue is resolved.
- Establish a special evaluation period: An incumbent fire fighter might be given a special evaluation period to resolve a work performance/behavioral issue. For example, a fire fighter who fails a required recertification examination could be placed in a special evaluation period until he or she is recertified.
- Involuntary transfer or detail: In an **involuntary transfer or detail**, a fire fighter is transferred or detailed to a different or a less desirable work location or assignment.
- Make financial restitution: For example, the fire fighter might be required to pay the insurance deductible after a property damage incident. One department has a $2500 deductible insurance policy on fire and EMS vehicles and equipment. If the fire fighter is judged to be responsible for a department vehicle crash, he or she is required to repay the deductible amount. The restitution payment period can take up to a year and the payment made through payroll deductions.
- Loss of leave: A fire fighter might lose annual or compensatory leave. This is equivalent to the practice of paying a cash fine.
- Demotion: A **demotion** occurs when an individual is reduced in rank, with a corresponding reduction in pay. Demotions are more common in the supervisory ranks.

Suspension

Suspensions are the next step of a progressive negative discipline path but have many different forms. A suspension is a negative disciplinary action that removes a fire fighter from the work location and prohibits him or her from performing any fire department duties. A disciplinary suspension usually results from a willful violation of a policy or procedure or another specific act of misconduct. For career fire fighters, suspension is a personnel action that places an employee on a leave-without-pay status for a specified period. For volunteer fire fighters, a suspension means they are not allowed to respond to emergencies and, in some cases, are prohibited from entering the fire station or participating in other fire department activities. Suspensions usually run from 1 to 30 days.

A fire officer usually recommends a suspension to a higher-level officer, providing the required documentation. In some organizations, a fire officer has the authority to suspend a fire fighter immediately for the balance of a work period, pending a formal disciplinary action. Depending on the nature and the severity of the offense, career fire fighters can also be suspended with pay or placed on **restrictive duty** while an administrative investigation is being conducted.

In a few situations, the municipality can suspend a fire fighter before an investigation is completed. Such a suspension might occur when the employee is being investigated by the fire department or a law enforcement agency for an offense that is reasonably related to fire department employment or when an employee is waiting to be tried for an offense that is job related or a felony.

Termination

Termination means that the organization has determined that the employee is unsuitable for continued employment. In general, only the top municipal official, such as the mayor, county executive, city manager, or civil service commission, can terminate an employee. Many senior municipal officials delegate this task to the agency or department head, such as the fire chief. Terminations are high-stress events, with labor representatives, the personnel office, and the fire chief's office all involved in the process.

FIRE OFFICER TIP

Example of a Predetermined Disciplinary Policy

A predetermined policy for specific offenses may already have been developed for common serial behavioral issues. Here is how one department deals with fire fighters who are late reporting to work, without mitigating circumstances:

- First offense: Oral admonition and leave without pay (LWOP) covering the period of time the employee was late. Admonition remains active for 366 days with the immediate supervisor.
- Second offense (within 366 days of first offense): Written reprimand and LWOP for the period of time the employee was late.
- Third offense (within 366 days of second offense): Suspension for one workday (8 hours for day-work staff or 12 hours for shift workers).
- Fourth offense (within 366 days of third offense): Suspension for four workdays (32 hours for day-work staff, 48 hours for shift workers).
- Fifth offense (within 366 days of fourth offense): Suspension for 10 workdays (80 hours for day-work staff, 120 hours for shift workers).
- Sixth offense (within 366 days of fifth offense): Proposed termination.

Documenting Fire Fighter Performance

Setting annual goals, as well as documenting experiences and hot wash results using the activity log or T-account, provides the fire officer with a rich work history. The informal work performance reviews keep the fire officer and the fire fighter in tune with the level of accomplishment and performance that is necessary during this evaluation period. Two additional time milestones provide clarification for both the evaluator and the member.

Mid-Year Review

The mid-year review is another informal performance review. The officer should have each fire fighter assess the progress on his or her annual goals. This task helps the fire fighter focus on the job description and the personal goals that were set at the beginning of the year. The officer and the fire fighter should discuss how well expectations are being met.

If necessary, the officer can identify resources or assistance that would help the fire fighter meet the mutually established goals. For example, the officer might suggest meeting with another fire fighter who is experienced at working with the preincident planning software or point out that only one semester remains to enroll in the building construction course at the college during the evaluation period.

At this time, the personal goals can also be adjusted based on changes in the work environment. For example, it would be unrealistic to maintain the goal of attending a building construction class at the community college if the department has selected the fire fighter to attend paramedic school.

Six Weeks Before the End of the Annual Evaluation Period

The fire fighter should conduct another self-evaluation approximately 6 weeks before the official annual evaluation is due. The goal of this self-evaluation is to identify how well the fire fighter has met the organizational expectations and personal goals for the year. The fire officer should meet with the fire fighter and provide feedback. Together, the fire officer and fire fighter should develop the final, formal evaluation report for this rating period, which includes developing three new work-related personal goals for the next evaluation period.

Completing annual evaluation reports is an important fire officer responsibility. It offers an opportunity to identify and evaluate the work performed by subordinates. The fire officer should recognize outstanding accomplishments, encourage improvement,

Voice of Experience

Fire service leaders must set clear expectations of their crew members from the start. Many company officers have a list of rules and expectations they present to crew members. My first crew was a group of folks with many years on the job, and all were older than me. I came into that crew with what I call a *conversation of expectations*. I sat down with the crew and had a brief discussion about my expectations. I assumed that this would be a great starting point for them to understand my expectations. Over the next year and a half I learned that many people need tangible written guidelines. This helps you, as a company officer, reinforce your expectations when the time comes—and it will.

One example of when this occurred involved a certain individual who had a small company on the side. From the beginning, I explained to him that our organization had a mission and that he was not to make money elsewhere while making money at the fire house. On the surface, this seemed to me a pretty cut-and-dried explanation of my expectations. The client phone calls stopped, and I thought all was good. However, one day I found him preparing invoices on a fire department computer. His justification was that it was after hours; he stated, "What's the problem? I could just be sleeping or watching TV." I again reinforced my expectation that he would not run his business while on duty.

We had a few other corrective moments, and in his mind most of his actions had a good justification. He would bring up folks who were surfing the Internet for hunting gear or vacation information. He would imply that I was singling him out. In another incident, he was found doing payroll and lining out his employees while the rest of the crew was doing morning checks. His justification once again was vague: "Last weekend Fire Fighter A had his kids at the station for an hour while we were doing chores. What is the big deal?" After this encounter I sat him down and documented what we call a PIP, or performance improvement plan. I had attempted to correct actions verbally over the course of a year, *assuming* my employees understood my expectations and shared the same morals and values. This was a big mistake. My approach made it harder to manage my other crew members, and it reduced my effectiveness as a leader. I once heard that good soldiers know who the bad soldiers are; pay attention.

Since this learning experience, I have made my expectations clear and clearly followed our organization's disciplinary and corrective action policy. This practice is critical to ensuring that the good folks we have—who are the majority—are confident in the organization's ability. Company officers are on the front lines of ensuring this credibility. In the end, our human resources will test us much more than a large commercial fire ever will.

Ed Mezulis
Battalion Chief, Sedona Fire District
Yavapai College
Clarkdale, Arizona

and, in some cases, identify those personnel who should be considering another career option.

The fire officer who maintains a activity log or T-account, conducts bimonthly informal reviews, conducts a mid-year review, and has the fire fighter complete a self-evaluation 6 weeks before the annual evaluation will have a wealth of information with which to present an accurate, well-documented, and comprehensive annual evaluation.

Evaluation Errors

Evaluation is a largely subjective process that is vulnerable to unintentional biases and errors. Having a detailed record of activities, experience, and accomplishments over the evaluation period helps create an accurate assessment. When completing the evaluation, be aware of these seven evaluation errors (Edwards, 2010).

Leniency or Severity

Some fire officers tend to rate all of their fire fighters either higher or lower than their actual work performance. This practice is called leniency or severity, respectively. Leniency reduces conflict, because a positive evaluation is likely to make the evaluation a more pleasant experience and avoids confrontation. Leniency is common when the fire officer is required to conduct a face-to-face meeting with the fire fighter to review the evaluation.

Some fire officers lean in the opposite direction and rate all fire fighters with "needs improvement" or "unsatisfactory." Some officers think that low ratings will cause fire fighters to be motivated to work harder. Newer fire officers sometimes make this mistake when they have not received adequate training in preparing performance evaluations.

Personal Bias

Personal bias is an evaluation error that occurs when the evaluator's perspective skews the evaluation such that the classified job knowledge, skills, and abilities are not appropriately evaluated. Fire officers must not allow an evaluation to be slanted by factors such as race, religion, gender, disability, or age. This is the area most likely to be referenced when an employee files hostile workplace or discrimination charges.

Recency

Recency is an evaluation error in which the fire fighter is evaluated only on incidents that occurred in the last few weeks, rather than on all of the events that occurred throughout the evaluation period. Fire fighters who are aware of this tendency are on their best behavior in the weeks leading up to the fire officer's evaluation.

Central Tendency

Most evaluation systems involve some type of rated scale, ranging from unsatisfactory to outstanding. A fire officer demonstrates a **central tendency** when a fire fighter is rated in the middle of the range for all dimensions of work performance. A central tendency evaluation holds little value for the fire fighter or the evaluation process. Being rated as "OK in all areas" is not very informative or helpful.

Frame of Reference

In a **frame of reference** evaluation error, the fire fighter is evaluated on the basis of the fire officer's personal ideals instead of the classified job standards. For example, a fire officer who spent hours every day fixing, improving, and polishing the engine when he was an apparatus driver/operator might issue "unsatisfactory" or "needs improvement" ratings to apparatus operators who are meeting all of the departmental standards but are not meeting his personal ideals.

Halo and Horn Effect

Like life, fire fighters' performances are not just black and white. Sometimes, however, the fire officer may concentrate on only one aspect of the fire fighter's performance, which is either exceptionally good or bad, and apply that perception across the board to all aspects of the individual's work performance. This type of evaluation error is called the **halo and horn effect**.

Contrast Effect

Contrast effect is an evaluation error that can occur when the fire officer compares the performance of one subordinate with the performance of another subordinate instead of against the classified job standards.

Conducting the Annual Evaluation

The personnel regulations in most career fire departments require that every employee who has completed the initial probationary period receive an annual written evaluation from his or her immediate supervisor. These annual evaluations become a formal part of the employee's work history. The fire officer may receive an annual evaluation form from the human resources or personnel division a month or two before the employee's annual evaluation is due to be submitted. Some organizations expect the officer to keep track of the dates and submit a completed form when the evaluation for each employee is due.

The annual evaluation process consists of four steps:

1. The fire officer fills out a standardized evaluation form. This form asks the officer to evaluate the subordinate on a number of knowledge areas, skills, and abilities appropriate for the subordinate's rank and classified job description. The officer can add comments in a narrative section. The officer's supervisor is required to review the evaluation before it is issued to the subordinate.
2. The subordinate is allowed to review and comment on the officer's evaluation. The subordinate has an opportunity to provide feedback or additional information that would affect the performance rating.
3. A face-to-face feedback interview between the fire officer and the subordinate to discuss the evaluation. This meeting is an opportunity to clarify points and review the performance over the last rating period. A goal of the feedback interview is that both the supervisor and the subordinate understand the results of the evaluation.
4. Establish goals for the subordinate to accomplish during the next evaluation period. These are usually specific, measurable goals that allow the subordinate to improve on the knowledge, skills, and abilities of the current job or prepare for the next higher job level.

The completion of the annual evaluation forms should be viewed as a formality. The fire officer should be providing continual evaluation and feedback to the fire fighter throughout the year, so the information that goes onto the form should not come as a surprise. Unless there is a problem with the employee's performance, the scheduled evaluation should be used primarily as an opportunity to discuss future goals and objectives.

The Four Borders of Human Resources

The fire officer must be familiar with the locations and topical areas that are covered in the fire department's human resources policies and procedures. Specifically, the organization may be subject to federal and state laws, a labor contract, jurisdiction regulations, and fire department policies that must all be followed. These four borders define the fire officer's human resources arena.

Federal and State Laws

When presented with an issue, the fire officer must first determine which laws apply. The laws most commonly involving fire fighters are as follows:

- Fair Labor Standards Act (FLSA)
- Title VII of the Civil Rights Act of 1964
- Age Discrimination in Employment Act (ADEA)
- Americans with Disabilities Act (ADA)
- Family and Medical Leave Act (FMLA)
- Uniformed Services Employment and Reemployment Rights Act (USERRA)
- Freedom of Information Act (FOIA)
- Health Insurance Portability and Accountability Act (HIPAA)

The U.S. Department of Labor provides a great deal of information about the various laws that affect fire fighters. The fire officer may also want to discuss issues that are affected by these laws with the human resources department and legal counsel.

Labor Contract

Once it is determined which laws affect a decision, the fire officer should review the fire department's policies to ensure compliance. If a labor contract exists, the fire officer must determine whether the activity is addressed in the contract. If the fire organization has a human resources department, it may prove a valuable asset in determining the proper action by the fire officer.

Jurisdiction Regulations

The jurisdiction's human resources department manages **compensation and benefits**. Although the fire officer may not determine the compensation policy, the officer must understand how the system is designed and to which benefits the employees are entitled. Most fire departments operate on a step-and-grade pay system. In such an approach, the position of fire fighter is established on a particular pay grade level that comprises a number of steps. If the fire fighter demonstrates satisfactory performance, he or she will progress from one pay step up to the next until he or she eventually reaches the top step of the pay grade. The fire fighter will remain at this pay step and grade until he or she is promoted. He or she will then move to a new pay grade and progress up through the steps until he or she again reaches the top step.

The other most commonly used compensation systems include merit-based pay and skill-based pay. In merit-based pay systems, the fire fighter is typically paid a base amount and then receives additional compensation for good performance. For example, a fire fighter who receives an outstanding evaluation might receive a bonus of five percent of his or her annual pay. In contrast, skill-based pay systems typically pay a base amount and then give additional compensation for any skills the fire fighter can demonstrate. For example, a fire fighter who is also a hazardous materials technician would be paid an additional amount because of those special skills.

The 2009 recession drew attention to and placed pressure on public safety benefits programs, because many fire departments provide more generous packages than private-sector employers. Fewer and fewer systems offer a defined benefit retirement program, that is, one in which the employee contributes a set percentage of his or her pay and receives a guaranteed benefit amount. Today, more systems use a defined contribution retirement plan, in which the employee pays in a percentage of his or her pay and receives a benefit amount that is dependent on the performance of the account. Typically, these defined contribution accounts take the form of a 401(k) plan (Investopedia, 2019). Other benefits include paid holidays; vacation leave; sick leave; and health, dental, vision, and life insurance.

Fire Department Policies

Performance management is the process of setting performance standards and evaluating performance against those standards. The fire service is unlike most other private and public organizations in its ability to spend vast amounts of time solely devoted to developing its employees while at work. Ensuring daily training is one of the most basic responsibilities of the fire company officer. Generally, the standards are set at the fire organizational or fire departmental level; however, the evaluation of the fire fighter's performance is a primary function of the fire company officer.

The fire officer must be able to accomplish the department's work with and through other people. In doing so, he or she must ensure that duties are carried out in accordance with departmental policies. Normally, this is done through direct supervision. **Direct supervision** requires that the fire officer directly observe the actions of the crew. If the crew makes an interior attack, the fire officer is present. When a hose is loaded, the fire officer oversees the reloading. Direct supervision allows the fire company officer to ensure that departmental safety policies are used. The more dangerous the activity, the more direct supervision is required. For nonemergency activities, less direct supervision is needed.

Using Just Culture Concepts to Improve Health and Safety

Health, safety, and security include all activities to provide and promote a safe environment for fire fighters. The fire officer is responsible for many of these activities, such as ensuring that seat belts are utilized, proper procedures are followed, and personal protective equipment is used correctly. The fire department or organization may provide other health activities such as an employee assistance program. The fire department may also offer tobacco cessation classes and incentives or diet and exercise information.

Due to the efforts of the National Fallen Firefighters Foundation, there has been a trend of fewer than 70 fire fighter line-of-duty deaths from 2014 to 2018—a significant improvement over the average of 100 deaths from 2002 to 2008 (Fahy and Molis, 2019). Specifically looking at deaths on the fire ground, an average of 26 fire fighter fire-ground fatalities per year has remained the average since 2007. This is a noteworthy improvement over the average of more than 80 fire-ground deaths per year in the 1970s.

Unfortunately, not all trends are positive ones. Sudden cardiac arrest remains the number one cause of line-of-duty deaths, a statistic that has not budged since NFPA began tracking trends in 1977. In 2018, sudden cardiac arrest represented 39 percent of all line-of-duty deaths. The 2017 NFPA report shows that the number of fire-ground injuries per 1000 fires has declined from 26.9 in 2006 to 18.6 in 2017 (Evarts and Molis, 2018).

Analyzing Near-Miss Reports

Every August, a working group of fire fighters and officers assemble to analyze near-miss reports using a tool modified from the U.S. Navy's Human Factors Analysis and Classification System (HFACS). The results of the analysis are provided in an annual report. This report is divided into four levels.

HFACS Level 1: Unsafe Acts

Level 1 in the HFACS includes two categories: errors and violations. Errors are considered unintentional and can be based on decisions, skills, or perceptions. Decision-based errors include flaws in communicating information to decision makers. Skill-based errors include attention failure (lack of situational awareness), memory failure (forgotten or missed step in a procedure), or technique failure (lack of training). Perception-based errors may include visual illusions.

Violations are considered intentional and are classified as either routine or exceptional. Routine violations include failure to use safety equipment, failure to follow a recommended tactile best practice (sounding the floor before entering), or failure to follow recommended cerebral best practices (conducting a risk/benefit analysis). Exceptional violations include not being qualified to perform an action.

HFACS Level 2: Preconditions to Unsafe Acts

Level 2 analyzes substandard conditions and practices of the individuals involved in the incident. Substandard conditions include factors contributing to adverse mental states, psychological states, and physical limitations. Substandard practices include failure to use elements of crew resource management and personal readiness.

HFACS Level 3: Unsafe Supervision

Unsafe supervision is broken down into four categories: inadequate supervision, allowing inappropriate operations, failure to correct known problems, and supervisory violations **TABLE 10-1**. Level 3 is intended to examine the role (if any) of supervision in a near-miss event.

HFACS Level 4: Organizational Influences

The most difficult HFACS level to analyze in a near-miss report is Level 4. The operating culture of the fire department is often as significant to the near miss as the individual's action. This level examines both resource management (staffing, training, budget resources, and equipment/facility resources) and organizational climate (chain of command, delegation of authority, risk management programs, and safety programs).

Data Analysis

The fire department is required to maintain records of all accidents, occupational deaths, injuries, illnesses, and exposures in accordance with federal, state, and local regulations. In addition to meeting regulations, these reports can assist the fire officer in identifying trends and safety-related issues.

Risk management is the identification and analysis of exposure to hazards, selection of appropriate risk management techniques to handle exposures, implementation of chosen techniques, and monitoring of results, with respect to the health and safety of members. Accident and injury reduction should be a major concern of every fire officer. The fire officer should review all injury, accident, and health exposure reports to identify unsafe acts and work conditions.

When reviewing an accident report, it is critical to determine the root cause of the accident. Were unsafe acts being committed? If so, what must be done to change the behavior or attitude? If the cause was an unsafe condition, how can this condition be corrected? An evaluation of all unsafe acts and conditions and identification of the means to address each is required to prevent future occurrences. Some corrective actions can be made at the company level, whereas others require action at a higher level.

Once the cause has been identified, a report should be filed with your supervisor that outlines the problem, actions taken to correct the problem, and any additional actions that should be implemented. In some departments these reports are sent directly to the department's health and safety officer.

The fire officer may be required to complete a longitudinal study of accidents, injuries, and exposures for the company. The officer must review the data to determine which areas are causing the greatest number of incidents as well as which problems have the greatest potential for loss. This review should look for trends, such as an increase in the number of slips, trips, and falls on the apparatus bay floor. This pattern might point to the condition of the floor surface being a contributing factor, which could be corrected. Similarly, a series of exposures to bloodborne pathogens indicates a need for additional company-level training.

TABLE 10-1 HFACS Level 3: Unsafe Supervision Factors

Inadequate Supervision	Allowing Inappropriate Operations	Failure to Correct Known Problems	Supervisory Violations
■ Guidance not provided ■ Training not provided ■ Qualifications not tracked	■ Personnel not adequately briefed ■ Understaffed ■ Unnecessary hazards permitted	■ Unidentified and unqualified ■ Failure to provide training to unqualified personnel ■ Failure to correct inappropriate or unsafe behaviors	■ Authorization of unnecessary hazards ■ Failure to enforce department rules ■ Authorization of unqualified personnel to perform tasks

Reprinted with permission of Elsevier Public Safety, copyright 2008.

Risk management principles suggest that fire officers should focus attention on the high-risk activities that are performed infrequently. An example would be training on locating and removing lost or trapped fire fighters.

Designing Better Systems

The just culture approach acknowledges that humans make errors and drift in performance. When evaluating a member's behavioral decision, it looks at the environment to see what factors contributed to this decision. This approach can be seen as an effort to plug the holes that show up when considering a systems approach to human error management. Dr. James Reason stated:

> High technology systems have many defensive layers: some are engineered (alarms, physical barriers, automatic shutdowns, etc.), others rely on people (surgeons, anaesthetists, pilots, control room operators, etc.), and yet others depend on procedures and administrative controls. Their function is to protect potential victims and assets from local hazards. Mostly they do this very effectively, but there are always weaknesses. (Reason, 2000)

Reason points out that each layer of defense is more like a slice of Swiss cheese than a solid barrier. The presence of a hole in one defensive layer does not create a bad outcome event, but when the holes in all of the levels of defense align, a bad or catastrophic outcome becomes more likely (**FIGURE 10-4**).

Controlling Contributing Factors

When determining if a behavior was reckless or at-risk, the fire officer may identify existing situations that would lead to a human error or at-risk behavior. For example, does a relentless pressure on reducing travel time increase the reward for running through a stop sign or red light intersection without stopping?

Adding Barriers

Procedural or physical barriers may be added to stop an activity that leads to harm, loss, or risk. Enforcing the requirement to have a spotter when backing up fire apparatus would be an example. Changing the size and color of a medication container to avoid confusion while working a cardiac arrest incident is another example.

Adding Recovery

Adding recovery means adding a system to "catch" an accident trajectory downstream. It may get through one swiss cheese hole but will be stopped at the next level. Fire fighters performing a personal protective clothing "buddy check" before entering a burning structure is an example.

Adding Redundancy

Establishing parallel elements can assure that the task or activity will not be interrupted. The deployment of

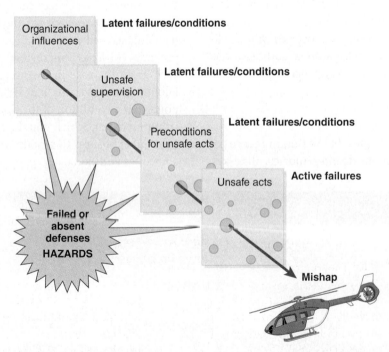

FIGURE 10-4 The Swiss cheese model of how defenses, barriers, and safeguards may be penetrated by an accident trajectory.

Reproduced from Reason J. 1990. *Human Error.* Cambridge, UK.: Cambridge University Press.

backup attack line crews is an example during fire suppression. Using a colorimetric end-tidal CO_2 device to monitor exhaled breaths of an intubated patient is a paramedic example.

The goal of designing a better system is to reduce the impact of human error and at-risk behaviors. When applied to high-risk occupations like health care, this will result in better patient outcomes due to fewer bad outcome events or catastrophic incidents.

Professional Development

Professional development refers to skills and knowledge attained for both personal development and career advancement. Professional development encompasses all types of facilitated learning, ranging from college degrees to certification training, continuing education, skill acquisition, and informal learning opportunities in the field. It has been described as intensive and collaborative, and ideally it will incorporate an evaluative stage (Knipe, 2005).

Each fire officer should develop a personal professional development plan. Obtaining professional designations and developing experience and skill sets makes the officer more valuable to the department and the fire service. More opportunities are available to those personnel with a vibrant professional development plan.

Training Versus Education

The relationship between training and education is confusing. **Education** is the process of imparting knowledge or skill through systematic instruction. Education programs are conducted through academic institutions and are primarily directed toward an individual's comprehension of the subject matter. Training, by comparison, is directed toward the practical application of education to produce an action, which can be an individual or a group activity. Thus, there is an important distinction between these two types of learning.

Training is an essential fire service activity. The first Wingspread Conference on Fire Service Administration, Education and Research was sponsored by the Johnson Foundation and held in Racine, Wisconsin, in 1966. This conference brought together a group of leaders from the fire service to identify needs and priorities. They agreed that a broad knowledge base was needed and that an educational program was necessary to deliver that knowledge base. This agreement became the blueprint for the development of community college fire science and fire administration programs, as well as the bachelor's-level degrees at a distance program.

In 1998, the U.S. Fire Administration (USFA) hosted the first Fire and Emergency Services Higher Education (FESHE) conference. That conference produced a report entitled *The Fire Service and Higher Education: A Blueprint for the 21st Century*, which started a national effort to update the academic needs of the fire service. FESHE participants developed a model fire science curriculum that would go from associate's degree through master's degree. By the 2008 conference, the FESHE model undergraduate fire science curriculum had been adopted by most academic institutions, with each course supported with textbooks available through two or more publishers.

The direction of the USFA changed in 2012 with the emergence of the National Professional Development Matrix (NPDM), which combined the FESHE and Training, Resources and Data Exchange (TRADE) initiatives. The focus of the NPDM is on career enhancement for emergency responders, professional development matrix discussions, succession planning, and national professional development standards recommendations.

Academic Accreditation

Just as the difference between training and education can be confusing, so too can the issue of academic **accreditation**. Consider the following example: A prestigious accomplishment for a fire officer is to attend the 3-week *Senior Executives in State and Local Government* program at the John F. Kennedy School of Government. This on-campus program is one of 34 executive education lifelong learning programs offered at Harvard University. A handful of fire service members are awarded an annual scholarship through the Harvard Fire Executive Fellowship Program sponsored through a partnership among the IAFC, International Fire Service Training Association/Fire Protection Publications (IFSTA/FPP), the NFPA, and the USFA. This outstanding program carries no academic credit but does come with a completion letter.

Three types of academic accreditation are available:

1. Programs that meet a specific profession or vocation. For example, for paramedic programs, the Commission on Accreditation of Allied Health Education Programs (CAAHEP), administered through the Committee on Accreditation for the Emergency Medical Services Professions (CoAEMSP), determines program accreditation.
2. Educational and training organizations that meet federal requirements for tuition reimbursement. This covers truck driver schools, computer training institutions, and colleges and universities.

3. Educational institutions that meet voluntary accrediting requirements to issue degrees and academic transcripts that are acceptable to other educational institutions.

Accreditation in higher education is defined as a collegial process based on self- and peer assessment for public accountability and improvement of academic quality. Peers assess the quality of an institution or academic program and assist the faculty and staff in improvement.

Voluntary accreditation has been in place for more than 100 years. It was linked with federal tuition benefits in the 1952 Korean War GI Bill after issues arose with "diploma mills" following enactment of the original GI Bill in 1944 (Bear and Bear, 2004).

The Council for Higher Education Accreditation (CHEA) was established in 1996 as a national advocate and institutional voice for self-regulation of academic quality through accreditation. CHEA is an association of 3000 degree-granting colleges and universities that recognizes 65 institutional and programmatic accrediting organizations (CHEA, 2019). Accreditation through a regional organization is how most universities and colleges participate in the process. Accreditation through a regional organization means that the degrees awarded and courses completed are recognized for transfer or credit.

Universities and schools may participate in a voluntary regional accrediting process to issue degrees and academic transcripts that are acceptable to other educational institutions.

Under CHEA, six regional organizations provide academic accreditation (CHEA, 2019):

1. Middle States Commission on Higher Education
2. New England Commission of Higher Education
3. Higher Learning Commission
4. Southern Association of Colleges and Schools Commission on Colleges
5. Accrediting Commission for Community and Junior Colleges
6. Western Association of Schools and Colleges (WASC) Senior College and University Commission

A number of universities offer degrees entirely through distance education, delivering academic coursework either from an interactive web-based service or through weekly assignments delivered by mail. Some have achieved regional academic accreditation, including those that have no bricks-and-mortar campus.

For example, American Public University System (APUS) started as American Military University in 1991. APUS has regional and Distance Education and Training Council (DETC) accreditation. Despite its regional accreditation, APUS provides the following guidance to students in its undergraduate catalog: "APUS cannot guarantee that its credit will be accepted as acceptance of credit is always the prerogative of the receiving institution."

Accreditation of Skill Certification

Chief Engineer Ralph J. Scott of the Los Angeles Fire Department (LAFD) is one of the fathers of fire fighter certification training, creating a fire college in 1925. The LAFD training staff researched and documented every task a fire fighter might be required to perform. The list of almost 2000 entries evolved into a document that became known as the *Trade Analysis of Fire Engineering*. While serving as president of the IAFC in 1928, Scott convinced the U.S. Department of Vocational Education to accept this list as an official definition of fire fighter tasks.

The NFPA helped standardize fire fighter training by publishing the inaugural edition of NFPA 1001, *Standard for Fire Fighter Professional Qualifications*, in 1974. For the first time, there was a national consensus on the knowledge and skills a fire fighter should possess. The development of this standard started a trend toward developing national consensus standards on a wide range of fire service occupations. Every state or commonwealth in the United States has some type of fire service professional certification system in place. The particular programs range from local to national, depending on local history, government structure, legislation or regulation, and funding. Most of these systems are based on the NFPA professional qualifications and competency standards.

- NFPA 472, *Standard for Competence of Responders to Hazardous Materials/Weapons of Mass Destruction Incidents*
- NFPA 473, *Standard for Competencies for EMS Personnel Responding to Hazardous Materials/Weapons of Mass Destruction Incidents*
- NFPA 1001, *Standard for Fire Fighter Professional Qualifications*
- NFPA 1002, *Standard for Fire Apparatus Driver/Operator Professional Qualifications*
- NFPA 1003, *Standard for Airport Fire Fighter Professional Qualifications*
- NFPA 1005, *Standard for Professional Qualifications for Marine Fire Fighting for Land-Based Fire Fighters*

TABLE 10-2 Right-to-Work States

Alabama	Nebraska
Arizona	Nevada
Arkansas	North Carolina
Florida	North Dakota
Georgia	Oklahoma
Idaho	South Carolina
Indiana	South Dakota
Iowa	Tennessee
Kansas	Texas
Louisiana	Utah
Michigan	Virginia
Mississippi	Wyoming

© Jones & Bartlett Learning.

- NFPA 1006, *Standard for Technical Rescue Personnel Professional Qualifications*
- NFPA 1021, *Standard for Fire Officer Professional Qualifications*
- NFPA 1026, *Standard for Incident Management Personnel Professional Qualifications*
- NFPA 1031, *Standard for Professional Qualifications for Fire Inspector and Plan Examiner*
- NFPA 1033, *Standard for Professional Qualifications for Fire Investigator*
- NFPA 1035, *Standard for Professional Qualifications for Fire and Life Safety Educator, Public Information Officer, Youth Firesetter Intervention Specialist, and Youth Firesetter Program Manager Professional Qualifications*
- NFPA 1037, *Standard on Fire Marshal for Professional Qualifications*
- NFPA 1041, *Standard for Fire and Emergency Services Instructor Professional Qualifications*
- NFPA 1051, *Standard for Wildland Fire Fighter Personnel Professional Qualifications*
- NFPA 1061, *Standard for Public Safety Telecommunications Personnel Professional Qualifications*
- NFPA 1071, *Standard for Emergency Vehicle Technician Professional Qualifications*
- NFPA 1072, *Standard for Hazardous Materials/Weapons of Mass Destruction Emergency Response Personnel Professional Qualifications*
- NFPA 1081, *Standard for Facility Fire Brigade Member Professional Qualifications*
- NFPA 1521, *Standard for Fire Department Safety Officer Professional Qualifications*

Development of the NFPA professional qualifications standards during the 1970s created a need to validate the many different certification systems that were already in use and to establish criteria for new systems. The NFPA standards define the minimum qualifications that an individual must demonstrate to be certified at a given level, such as Fire Officer I and Fire Officer II. Accreditation establishes the qualifications of the system to award certificates that are based on the standards. A certification that is awarded by an accredited agency or institution is generally recognized by other accredited agencies and organizations.

Accreditation is a system whereby a certification organization determines that a school or program meets the requirements of the fire service. A group of impartial experts is assigned to thoroughly review a given program and determine whether it is worthy of accreditation. Two organizations provide accreditation to fire service professional certification systems: the National Board on Fire Service Professional Qualifications (NBFSPQ) and the International Fire Service Accreditation Congress (IFSAC).

National Board on Fire Service Professional Qualifications

The Joint Council of National Fire Service Organizations created the National Professional Qualifications System (also known as the "Pro Board") in 1972. When the Joint Council dissolved in 1990, the Pro Board evolved into the independent NBFSPQ. The board of directors consists of representatives from national fire service organizations that have an interest in training and certification. The NBFSPQ has accredited the certification programs that are operated by 74 states, provinces, and other agencies, covering 21 different standards.

International Fire Service Accreditation Congress

Established by the National Association of State Directors of Fire Training and Education in 1990

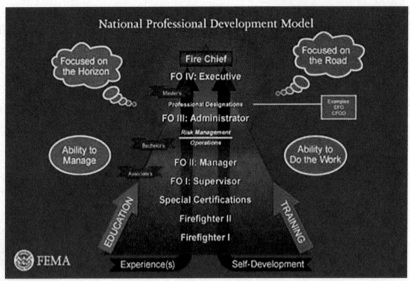

FIGURE 10-5 The National Professional Development Model.
Courtesy of FEMA

(IFSAC, 2007), IFSAC provides accreditation to certificate-issuing entities. IFSAC also accredits fire-related degree programs at the college and university levels. The IFSAC process follows the CHEA format and operates through Oklahoma State University. In 2019, there were 70 IFSAC-accredited fire service training programs and 38 fire science degree programs.

Building a Professional Development Plan

NFPA established the professional qualification standards, adopting NFPA 1021 for fire officers in 1976. The IAFC expanded company officer development, updated in 2010, providing an *Officer Development Handbook* to encourage company officers to acquire the appropriate levels of training, experience, self-development, and education throughout their professional journey to prepare for the Chief Fire Officer (CFO) designation, the pinnacle of professional development. There are two organizations that offer fire credentialing, the Center for Public Safety Excellence (CPSE) and the National Fire Academy (NFA) (**FIGURE 10-5**).

CPSE Fire Officer Credential

The CPSE is a not-for-profit 501(c)(3) corporation that is an international technical organization that works with the most fire and emergency service agencies and most active fire professionals. The CPSE mission is to lead the fire and emergency service to excellence through the continuous quality improvement process of accreditation, credentialing, and education. CPSE provides accreditation to fire agencies and credentialing to individuals. The Commission on Professional Credentialing (CPC) offers five distinct designations covering the various levels and specialties of fire officers. CPC provides (1) an application process that officers use to develop their portfolio, (2) training and support while developing their portfolio, and (3) access to experienced peer reviewers. Fire Officer is one of the designations, open to all junior officers, company officers, or those who have served on an intermittent acting status for 12 months. The CPC model looks at the whole officer:

1. Education
2. Experience
3. Professional development (training and certifications)
4. Professional contributions and recognitions
5. Professional memberships and affiliations
6. Technical competence (depending on the credential, 7 to 20 different competencies)
7. Community involvement

The process includes a self-assessment, professional portfolio, peer review, and interview process. There are over 450 CPC-credentialed Fire Officers (CPSE, 2019).

NFA Managing Officer Program

The NFA is a component of the United States Fire Academy (USFA) that works to enhance the ability of fire and emergency service and allied professionals to deal more effectively with fire and related emergencies. Free training courses and programs are delivered at a campus in Emmitsburg (Maryland), online, and through designated state fire agencies.

The NFA's Managing Officer Program is a multi-year curriculum that introduces emerging emergency service leaders to personal and professional skills in change management, risk reduction, and adaptive leadership. Acceptance into the program is the first step in your professional development as a career, volunteer, part-time, or combination fire/EMS manager, and includes all four elements of professional development: education, training, experience, and continuing education. There are three elements of the program:

1. Five prerequisite courses (online and classroom deliveries in your state)
   a. Q0890, Introduction to Emergency Response to Terrorism
   b. ICS-100, Introduction to ICS for Operational First Responders
   c. ICS-200, Basic NIMS ICS for Operational First Responders
   d. IS-700a, National Incident Management System (NIMS), An Introduction
   e. IS-800c, National Response Framework, An Introduction
2. Four courses at the NFA in Emmitsburg, Maryland
   a. Application of Community Risk Reduction
   b. Applications of Leadership in the Culture of Safety
   c. Analytical Tools for Decision-Making
   d. Training and Professional Development Challenges for Fire and Emergency Service Leaders
3. A community-based capstone project
   a. Lessons learned from one of the core courses required in the Managing Officer Program
   b. Experiences of the Managing Officer as identified in the *IAFC Officer Development Handbook*, Second Edition
   c. An issue or problem identified by your agency or jurisdiction
   d. Lessons learned from a recent administrative issue
   e. Identification and analysis of an emerging issue of importance to the department

A certificate of completion for the Managing Officer Program is awarded after the successful completion of all courses and the capstone project (FEMA, 2019).

IAFC Company Officer Leadership Symposium

Designed for crew leaders, senior station leadership, sergeants, lieutenants, and captains, the Company Officer Leadership Symposium (COLS) is a three-level program that provides what company officers need and what chief officers expect. The content of COLS is based on NFPA 1021, *Standard for Fire Officer Professional Qualifications* (Fire Officer I and II), and the *IAFC Officer Development Handbook*. There are three levels for the Company Officer program, each level consists of 21 contact hours:

Administration and Human Relations	3 hours
Leadership	3 hours
Community Risk Reduction	3 hours
Operations	3 hours
Safety, Health, and Wellness	3 hours
Elective	3 hours
Social Learning and Networking	3 hours

You Are the Fire Officer Conclusion

After making sure that no one is injured, the captain conducts a hot wash with the involved fire fighters and inspects the area. The activity log (or T-account) will memorialize this incident for those fire fighters involved. The evaluation of the behavior revealed that the fire fighter driving the spare pumper out of the station assumed that the garage door was controlled by an electric eye that would interrupt the power to the garage door until the pumper cleared the door frame. Unfortunately, the electric eye for this door was damaged 3 months ago. There remains an open work repair order.

The process failed to account for a nonfunctioning electric eye. In fact, only one member of the crew was aware of the issue. For the driver of the spare pumper, the captain classified the behavior as human error.

The captain has a follow-up meeting with the crew to brainstorm what to do before the electric eye is repaired. Repair may take some time since parts for the 20-year-old device are scarce. The captain made a survey, and found 11 fire department garage doors with broken electric eyes. A report was made to risk management recommending installation of new electric eyes for the 11 broken ones, with a suggestion to replace all of the electric eyes due to age, unreliability, and obsolete technology. An interim plan was made to add signs to all affected doors.

After-Action REVIEW

IN SUMMARY

- Transformational leadership is a process in which a person engages with others and creates a connection that raises the level of motivation and morality in both the leader and the follower.
- Transactional relationship is a relationship where all parties are in it for themselves and they do things for each other with the expectation of reciprocation.
- Transformational leadership builds follower commitment in four ways: idealized influence, inspirational motivation, intellectual stimulation, and individual consideration.
- Goal setting is the process of establishing of specific goals over a designated period of time.
- Fire officers should require all off-probation fire fighters to identify three work-related goals to achieve over the next evaluation period.
- Goals should be S.M.A.R.T.: specific, measurable, attainable, relevant with a timeline.
- Fire officers should keep track of each fire fighter's activity using an activity log or a T-account.
- The just culture approach changes the focus from errors and outcomes to systems design and management of behavioral choices of all members.
- The spectrum of human intention: human error, at-risk behavior, reckless behavior, knowingly causing harm, and purposely causing harm.
- Human error is an inadvertent action such as a slip, lapse, or mistake.
- At-risk behavior is a behavioral choice that increases risk when risk is not recognized or is mistakenly believed to be justified.
- Reckless behavior is a behavioral choice to consciously disregard a substantial and unjustified risk. In other words, the member recognized the risk and decided to continue the behavior.
- Reasonable person standard describes the actions that a similar person with the same background would take given the same situation.
- Response to reckless behavior could include remedial training, formal discipline, or punitive action.
- A common practice is to notify subordinates of a substandard employee evaluation 10 weeks before the annual evaluation is due.
- A work improvement plan is a special evaluation period where the fire fighter is provided an opportunity to demonstrate the desired workplace behavior or performance.
- Progressive negative discipline occurs due to the continuing inability or unwillingness of a fire fighter to meet the required performance or behavioral expectations.
- The steps of progressive negative discipline are informal reprimand, formal written reprimand, suspension, and termination.
- Many jurisdictions require a Loudermill hearing before a suspension, demotion, or involuntary termination can be invoked.
- A pretermination hearing is where an employee has an opportunity to present his or her side of the case, with and an explanation of adverse evidence to protect the worker's due process rights.
- Alternative disciplinary actions include extension of probationary period, establishment of special evaluation period, involuntary transfer or detail, financial restitution, loss of leave, or demotion.
- Suspension is a negative disciplinary action that removes a fire fighter from the work location and prohibits him or her from performing any fire department duties.
- Restrictive duty is usually a work assignment that isolates the fire fighter from the public, often an administrative assignment away from the fire station environment.
- Two informal reviews of firefighter work performance are the mid-year review and 6 weeks before the end of the annual evaluation period.

- Evaluation errors include leniency/severity, personal bias, recency, central tendency, frame of reference, halo and horn effect, and contrast effect.
- The annual evaluation requires review by a higher ranking officer, a subordinate response to the evaluation, a face-to-face discussion with the subordinate, and establishing goals for the next evaluation period.
- The four borders of human resources are federal/state law, labor contract, jurisdiction regulations, and fire department policies.
- Near-miss reports are classified using a modified tool from the U.S. Navy's Human Factors Analysis and Classification System (HFACS) as Unsafe Acts, Preconditions to Unsafe Acts, Unsafe Supervision, or Organizational Influences.
- Risk management is the identification and analysis of exposure to hazards, selection of appropriate risk management techniques to handle exposures, implementation of chosen techniques, and monitoring of results, with respect to the health and safety of members.
- Just culture looks at designing better systems, controlling contributing factors, adding barriers, adding recovery, and adding redundancy as ways to manage risk.
- Professional development refers to skills and knowledge attained for both personal development and career advancement.
- Education is the process of imparting knowledge or skill through systematic instruction.
- Accreditation in higher education is defined as a collegial process based on self- and peer assessment for public accountability and improvement of academic quality.

KEY TERMS

Accreditation Third-party attestation related to a conformity assessment body conveying formal demonstration of its competence to carry out specific conformity assessment tasks. These tasks include sampling and testing, inspection, certification, and registration. (ISO/IEC 17000, 2004)

Activity log An informal record maintained by the fire officer that lists fire fighter activities by date and includes a brief description; it is used to provide documentation for annual evaluations and special recognitions.

At-risk behavior Behavioral choice that increases risk when risk is not recognized, or is mistakenly believed to be justified.

Central tendency An evaluation error that occurs when a fire fighter is rated in the middle of the range for all dimensions of work performance.

Compensation and benefits Human resources system to identify and determine the pay, leave, and fringe benefits for each position in the organization.

Contrast effect An evaluation error in which a fire fighter is rated on the basis of the performance of another fire fighter and not on the classified job standards.

Demotion A reduction in rank, with a corresponding reduction in pay.

Direct supervision A type of supervision in which the fire officer is required to observe the actions of a work crew directly; it is commonly employed during high-hazard activities.

Education The process of imparting knowledge or skill through systematic instruction.

Formal written reprimand An official negative supervisory action at the lowest level of the progressive disciplinary process.

Frame of reference An evaluation error in which the fire fighter is evaluated on the basis of the fire officer's personal standards instead of the classified job description standards.

Goal-setting The process of establishing specific goals over a designated period of time.

Halo and horn effect An evaluation error in which the fire officer takes one aspect of a fire fighter's job task and applies it to all aspects of work performance.

Health, safety, and security Human resources activities intended to provide and promote a safe work environment.

Hot wash Debriefing immediately after an incident to identify issues needing prompt attention as well as procedural or systemic problems needing later follow-up.

Human error An inadvertent action such as a slip, lapse, or mistake.

Involuntary transfer or detail A disciplinary action in which a fire fighter is transferred or assigned to a less desirable or different work location or assignment.

Just culture A change in focus from errors and outcomes to systems design and management of behavioral choices of all employees.

Loudermill hearing A predisciplinary conference that occurs before a suspension, demotion, or involuntary termination is issued. The term "Loudermill" refers to a U.S. Supreme Court decision.

Performance management Oversight of process management to ensure that the ensemble of activities consistently meet organizational goals in an effective and efficient manner.

Personal bias An evaluation error that occurs when the evaluator's perspective skews the evaluation such that the classified job knowledge, skills, and abilities are not appropriately evaluated.

Pretermination hearing An initial check to determine if there are reasonable grounds to believe that charges against an employee are true and support the proposed termination.

Professional development A continuous process of training, education, knowledge, and skills enhancement.

Progressive negative discipline Process for dealing with job-related behavior that does not meet expected and communicated performance standards. The level of discipline increases from mild to more severe punishments if the problem is not corrected.

Reasonable person standard The actions that a similar person with the same background would take given the same situation.

Recency An evaluation error in which the fire fighter is evaluated only on incidents that occurred over the past few weeks rather than on the entire evaluation period.

Reckless behavior A behavioral choice to consciously disregard a substantial and unjustified risk.

Restrictive duty Temporary work assignment during an administrative investigation that isolates the fire fighter from the public; this is usually is an administrative assignment away from the fire station.

Risk management Identification and analysis of exposure to hazards, selection of appropriate techniques to control exposures, implementation of chosen techniques, and monitoring of results to ensure the health and safety of members.

Special evaluation period Designated period of time when an employee is provided additional training to resolve a work performance/behavioral issue. The supervisor issues an evaluation at the end of the special evaluation period.

Suspension Negative disciplinary action that removes a fire fighter from the work location; he or she is generally not allowed to perform any fire department duties.

T-account Documentation system similar to an accounting balance sheet listing credits and debits, in which a single-sheet form is used to list the employee's assets on the left side and liabilities on the right side, so that the result resembles the letter "T."

Termination Situation in which the organization ends an individual's employment against his or her will.

Transactional relationship A relationship where all parties are in it for themselves, and where individuals do things for each other with the expectation of reciprocation.

Transformational leadership A process in which a person engages with others and creates a connection that raises the level of motivation and morality in both the leader and the follower.

Work improvement plan Written document that is part of a special evaluation period. The plan identifies performance deficiencies and lists the improvements in performance or changes in behavior required to obtain a "satisfactory" evaluation.

REFERENCES

Antonellis, Sr., Paul J. 2012. *Labor Relations for the Fire Service*. Tulsa, OK: Fire Engineering Books and Videos.

Bear, John P., and Mariah Bear. 2004. *Bear's Guide to Earning Degrees by Distance Learning, 16th edition*. Berkeley, CA: Ten Speed Press.

Belker, Loren B., Jim McCormick, and Gary S. Topchik. 2018. *The First Time Manager, Seventh edition*. New York, NY: Harper Collins Leadership.

Bender, Lewis. 2013. "Accountability in the Workplace." In *Effective Supervisory Practices: Better Results through Teamwork, Fifth edition*, edited by Michelle P. Flaherty. Washington, DC: International City/County Management Association.

Boysen II, Phillip G. 2013. "Just Culture: A Foundation for Balanced Accountability and Patient Safety." *The Ochsner Journal* 13 (3): 400–6.

Canton, Lucien. 2016, March. "Is Your Hot Wash Just a Gripe Session? 5 Steps for Improving Exercise Evaluations." *LucienCanton.com*. Accessed July 31, 2019. https://lucien-canton.typepad.com/files/is-your-hotwash-just-a-gripe-session.pdf.

Center for Public Safety Excellence (CPSE). 2019. "How to Get Credentialed." Accessed June 21, 2019. https://cpse.org/credentialing/how-to-get-credentialed/.

Corbin, Janice, and Janet May. 2012, March 1. "Taking the Mystery out of Loudermill Meetings." Accessed July 31, 2019. http://mrsc.org/Home/Stay-Informed/MRSC-Insight/Archives/Taking-the-Mystery-out-of-Loudermill-Meetings.aspx.

Council for Higher Education Accreditation (CHEA). 2019. *2017–2018 Annual Report*. Washington, DC: Council for Higher Education Accreditation.

Dekker, Sidney. 2014. *The Field Guide to Understanding 'Human Error,' Third edition*. Burlington, VT: Ashgate Publishing Company.

Edwards, Steven P. 2010. *Fire Service Personnel Management, Third edition*. New York, NY: Brady/Pearson.

Evarts, Ben, and Joseph L. Molis. 2018. *United States Firefighter Injuries 2017*. Quincy, MA: National Fire Protection Association.

Fahy, Rita F., and Joseph L. Molis. 2019. *Firefighter Fatalities in the US–2018*. Quincy, MA: National Fire Protection Association.

Federal Emergency Management Agency (FEMA). 2019. *Managing Officer Program Handbook*. Washington, DC: Department of Homeland Security.

Investopedia. 2019, July 29. "Defined Benefit Pension Plan vs. Defined Contribution Plan." Accessed August 1, 2019. https://www.investopedia.com/ask/answers/032415/how-does-defined-benefit-pension-plan-differ-defined-contribution-plan.asp.

ISO. (2004). *ISO/IEC 17000: Conformity Assessment – Vocabulary and General Principles*. Geneva, Switzerland: International Organization for Standardization.

Knipe, Caroll O., ed. 2005. *Why Can't We Get It Right?: Designing High-Quality Professional Development for Standards-Based Schools, Second edition*. Thousand Oaks, CA: Corwin Press.

Marx, David. 2001. *Patient Safety and the "Just Culture": A Primer for Health Care Executives*. New York, NY: Columbia University.

Northouse, Peter G. 2018. *Leadership: Theory and Practice, 8th edition*. Thousand Oaks, CA: SAGE publications.

Reason, James. 2000, March 18. "Human Error: Models and Management." *British Medical Journal* 320 (7237): 768–70.

Remmers, Vanessa. 2016, March 5. "Stafford Firefighters Suspended After Taking Child to Hospital in Fire Engine." *The Freelance Star*. Accessed July 31, 2019. https://www.fredericksburg.com/news/local/stafford-firefighters-suspended-after-taking-child-to-hospital-in-fire/article_47616207-df11-5492-be95-83b28272b6fe.html.

Fire Officer in Action

The fire company was dispatched for a "haze of smoke in the building." On arrival at a strip shopping center, there is a haze in the building. Before the fire officer can complete a 360-degree survey, the newest firefighter takes out a plate glass window with a flat-head axe. The firefighter did not see the rapid-entry key box that was mounted to the right of the window.

1. A fire fighter who does not recognize a risk when a behavioral decision is made is an example of:
   A. knowingly causing harm.
   B. reckless behavior.
   C. at-risk behavior.
   D. human error.

2. Factors that influence a human intention include all of these, *except*:
   A. environment.
   B. psychological state.
   C. training.
   D. design.

3. An officer's goal when evaluating a subordinate's behavioral decision is to:
   A. make sure the member recognizes the risk.
   B. ensure that the subordinate takes full responsibility of the decision.
   C. facilitate a process of identifying the alternative decisions available to the subordinate.
   D. calculate the cost of the subordinate's behavioral decision.

4. The fire officer has determined that the subordinate demonstrated an at-risk behavior. Supervisory response options include all of these, *except*:
   A. modification of organizational structure.
   B. individual training or coaching.
   C. revision of policy, procedure, or SOP/SOG.
   D. reassignment of subordinate.

Fire Captain Activity

NFPA Fire Officer II Job Performance Requirement 5.2.1

Initiate actions to maximize member performance and/or to correct unacceptable performance, given human resource policies and procedures, so that member and/or unit performance improves or the issue is referred to the next level of supervision.

Application of 5.2.1

1. A fire fighter under your supervision constantly falls asleep during the workday. You have noticed many times that the fire fighter cannot even stay awake during in-station training. In the last month, this fire fighter has failed to get on the fire truck for an emergency response three times. Describe the process you will take to identify the causes of this occupational narcolepsy.

2. A Fire Officer I under your supervision is observed screaming at and berating a fire fighter whose fire company should have been the first to arrive at an incident, yet due to a delay, arrived second. Your initial inquiry has determined that the fire fighter drove down the wrong road and had to turn around, making them late to the incident. Identify the appropriate personnel regulations from a jurisdiction you are familiar with, and describe how you will review this behavior with the Fire Officer I.

NFPA Fire Officer I Job Performance Requirement 5.2.2

Evaluate the job performance of assigned members, given personnel records and evaluation forms, so that each member's performance is evaluated accurately and reported according to human resource policies and procedures.

Application of 5.2.2

1. A senior fire fighter under your supervision has called in sick for 24 hours today. The fire fighter has a sick leave balance of 4 hours and an annual leave balance of 8. Human resources have flagged the sick leave request as the fire fighter is 20 hours short to cover the day. Describe the process required by a jurisdiction familiar to you when an employee is overdrawn on sick leave.

2. For the same fire fighter, leave records show an accelerated use of 208 hours of annual and 312 hours of sick leave in the last 12 months. In the prior 4 years, the fire fighter averaged 24 hours of annual leave and 18 hours of sick leave each year. Describe the process you will use to identify the causes of this change in leave usage.

NFPA Fire Officer I Job Performance Requirement 5.2.3

Create a professional development plan for a member of the organization, given the requirements for promotion, so that the individual acquires the necessary knowledge, skills, and abilities to be eligible for the examination for the position.

Application of 5.2.3

1. Identify the training, certification, credentials, and education required to sit for a first-line fire supervisor position for a jurisdiction you are familiar with. Now consider the following example: Fire Fighter Eager has 2 years with the fire department, a high school diploma, and an emergency medical technician (EMT) card. With this information in hand, lay out a 2-year professional development program for Fire Fighter Eager to prepare to sit for the first-line supervisor promotional exam.

2. Identify the training, certification, credentials, and education required to sit for a battalion or district position for your jurisdiction. Using your actual background and resources available in your community, lay out a professional development program that will qualify you to sit for the battalion chief promotional exam.

NFPA Fire Officer I Job Performance Requirement 5.7.1

Analyze a member's accident, injury, or health exposure history, given a case study, so that a report including action taken and recommendations made is prepared for a supervisor and requirements for reporting and receiving information related to health exposures.

Application of 5.7.1

Situation A: Fire Fighter Adams is a 17-year employee who has developed a recurrent experience with on-the-job injuries.

Date	Nature	Lost Calendar Days
9/15/2023	Right rotator cuff: Lifting large patient onto stretcher	167
10/13/2022	Torn back muscle: Overhaul operations with pike pole and shovel while on SCBA	92
3/3/2022	Torn Achilles tendon: Threw ground ladders and moved smoke ejectors	116
4/1/2020	Fractured femur, fractured radius/ulna, and concussion: Engine 3 roll-over crash	216

1. Compare Fire Fighter Adams's lost calendar days with a regional or national report of fire fighter injuries. What is the difference between calendar days lost in this report when compared to national or regional fire injury experience?
2. Review disability retirement regulations for a jurisdiction or state you are familiar with. Does Fire Fighter Adams qualify for a disability pension? If the answer is yes, write up a recommendation for your chief. If the answer is no, create a rebuttal.

Situation B: Two fire fighters under your supervision provide emergency medical care during a lengthy extrication of a passenger trapped in an SUV after a vehicle crash. During the hot wash, the firefighters comment that the patient was coughing on them and sweaty. The paramedic who transported the patient calls to inform you that the patient has viral meningitis. Using the infection control protocols used by a jurisdiction you are familiar with:

1. Describe the immediate actions you need to take for the exposed fire fighters and the rest of the on-duty team.
2. Describe the notifications you need to make.
3. Identify the reports that are required for this exposure.
4. Describe the process to make the fire apparatus and fire station safe for use.

 *Access Navigate for flashcards to test your key term knowledge.*

CHAPTER 11

Fire Officer II

Managing Community Risk Reduction Programs

KNOWLEDGE OBJECTIVES

After studying this chapter, you will be able to:

- Describe the role of the media in getting fire department information to the community. (pp. 305–309)
- Describe the role of the public information officer (PIO) in working with the media. (**NFPA 1021: 5.3, 5.3.1, 5.4, 5.4.4**) (pp. 305–307)
- Discuss the role of social media in community relations. (pp. 309–310)
- Explain the benefits to the organization of cooperating with allied organizations, given a specific problem or issue in the community. (**NFPA 1021: 5.3.2**) (p. 304)

SKILL OBJECTIVES

After studying this chapter, you will be able to:

- Prepare a press release that conforms to the locally preferred format. (pp. 307–308)
- Conduct a media interview as the fire department representative. (pp. 307–309)

ADDITIONAL NFPA STANDARDS

- **NFPA 1452**, *Guide for Training Fire Service Personnel to Conduct Community Risk Reduction for Residential Occupancies*
- **NFPA 1035**, *Standard on Fire and Life Safety Educator, Public Information Officer, Youth Firesetter Intervention Specialist and Youth Firesetter Program Manager Professional Qualifications*
- **NFPA 1300**, *Standard on Community Risk Assessment and Community Risk Reduction Plan Development*
- **NFPA 1600**, *Standard on Continuity, Emergency, and Crisis Management*

You Are the Fire Officer

The captain and battalion chief are discussing the increased number of dryer fires in apartment buildings. The chief tells the captain to develop a community risk reduction plan to reduce these fires.

1. What research is required?
2. Is this just a fire department issue?
3. What defines success in community risk reduction?

Access Navigate for more practice activities.

Introduction

The explosion of available information and the growth of digital communications has changed both how the fire service views risks and how it communicates with the public. These changes in information and communications technologies have shattered the silo of fire prevention efforts and have made the fire department a partner in a community-wide effort to reduce risks through community risk reduction programs and crisis communications.

Managing Community Risk Reduction Implementation

The process of **community risk reduction (CRR)** begins with identifying and prioritizing local risks (**risk assessment**) and is followed by an integrated and strategic investment of resources to reduce their occurrence and impact. CRR is a community-focused effort that employs all of the available risk reduction techniques. The process integrates both emergency response and prevention into one overall approach.

Gathering Community Risk Assessment Data

NFPA 1300, *Standard on Community Risk Assessment and Community Risk Reduction Plan Development*, describes the process of **community risk assessment (CRA)**. This process involves a comprehensive evaluation that identifies, prioritizes, and defines the risks that pertain to the overall community. The community is profiled using these elements to describe the community:

1. **Demographics** (characteristics of human populations, such as age, race, sex, income, etc.)
2. Geography
3. Building stock
4. Public safety response agencies
5. Community service organizations
6. Hazards
7. Economic factors
8. Past loss/event history
9. Critical infrastructure systems

Demographics

Developing a CRR process includes describing the people who are impacted by the risk or may be part of the problem. The **community demographic profile** reflects the economic and social issues that significantly influence the identified risk. Vision 20/20, a coalition of national organizations that are developing strategies for fire loss prevention, identify these elements when developing a community demographic profile (Stouffer, 2016):

1. Age
2. Gender
3. Income
4. Race and ethnicity
5. Social and cultural information
6. Education
7. Housing type, age, and density (optional)

A significant source for community profile data can be found on the U.S. Census Bureau website and the Explore Census Data website. Many communities may have already developed a community demographic profile through local government, public health departments, school districts, universities, and nongovernmental organizations.

Geography

The physical environment includes collection of available assessments and, if possible, assembling this information into a master geographic overview of the regional factors in the risk analysis. The most common feature is identifying the communities that could be affected by a **100-year flood**. This term is used by the U.S. Geological Survey after performing 10 years of frequency analysis to determine the probability that a given event will be equaled or exceeded in any given year. For example, the probability of a river reaching a flood stage of 20 feet is once in 100 years. Another way of describing it is to say that there is a 1 percent chance of that river reaching a flood stage of 20 feet in any given year (USGS, 2019).

Other geographic factors include identified earthquake faults, landslide-prone areas, volcano locations (ICDO, 2019), industrial/chemical site hazards, and nuclear and radiation hazards (UNISDR, 2018).

Building Stock

Building stock evaluation is an important tool when evaluating a community's risk due to natural disasters such as windstorms, earthquakes, and floods. Evaluation of building stock includes the following:

- Housing density
- Housing age
- High-density structures
- Overcrowding (Shawnee Fire Department, 2017)

The local real estate tax assessment office should have details on housing, including age and density. Information regarding overcrowding may be available through the local social service agencies or public health offices.

Public Safety Response Agencies

What public safety agencies are available, and what are their response capabilities? If your community is on an island that gets fire suppression service from the mainland, via a ferry, your level of coverage will be different from a house in a downtown city. National Fire Protection Association (NFPA) 1710, *Standard for the Organization and Deployment of Fire Suppression Operations, Emergency Medical Operations, and Special Operations to the Public by Career Fire Departments*, and NFPA 1720, *Standard for the Organization and Deployment of Fire Suppression Operations, Emergency Medical Operations, and Special Operations to the*

FIRE OFFICER TIP

Shawnee (Kansas) Fire Department: Graphical Representation of Building Stock Community Assessment

The Shawnee (Kansas) Fire Department created an online dashboard as part of their community risk assessment process. The building stock section included graphical representations of housing density, housing age, high-density structures, and overcrowding.

Used with permission from the Shawnee Fire Department, taken from the "Community Risk Assessment" dashboard. Accessed July 18, 2019. https://dashboards.mysidewalk.com/community-risk-assessment/building-stock.

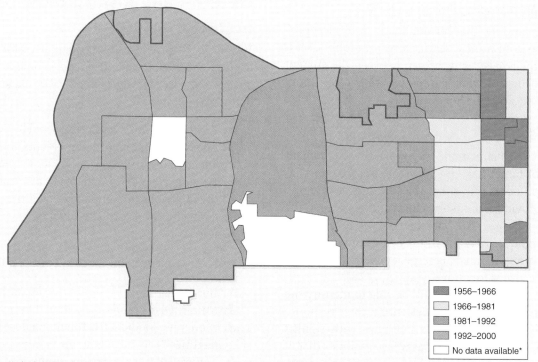

Used with permission from the Shawnee, Kansas Fire Department, taken from the "Community Risk Assessment" dashboard accessed July 18, 2019 https://dashboards.mysidewalk.com/community-risk-assessment/building-stock.

Public by Volunteer Fire Departments, describe coverage in terms of response time and assembly of crews.

A more comprehensive description comes from the 226 agencies that have achieved accreditation through the Center for Public Safety Excellence. These agencies have gone through an all-hazard, quality improvement model based on risk analysis and self-assessment that promotes the establishment of community-adopted performance targets for fire and emergency service agencies (CPSE, 2019).

The management of data is providing additional opportunities to describe how the fire department is meeting the community needs (Dwoskin, 2014). The Fire-Community Assessment/Response Evaluation System (FireCARES) is a collaborative effort between several fire service organizations to develop a new, data-driven tool that is integrated with scientific models to assist with community risk assessment and best firefighting practices (Treibitz, 2017).

Community Service Organizations

Community service organizations already service some of the communities at risk or have identified needs within their area of service. In terms of population, those at risk include the following (Stouffer, 2016):

- Children under six years old
- Adults over 65 years old
- People with disabilities
- People living in poverty
- People that speak little or no English

Understanding the current level of service provided, the community service organization's assessment of their need, and the gap between the current services delivered when compared to the community need provides a snapshot of the risk.

Hazards

This is an expansion of the traditional fire department preplan of target hazards. FEMA defines target hazards as "… facilities in either the public or private sector that provide essential products and services to the general public, are otherwise necessary to preserve the welfare and quality of life in the community, or fulfill important public safety, emergency response, and/or disaster recovery functions" (FEMA, 2018).

The local fire company will be able to identify the critical structures and facilities within their response district. This should include buildings with significant value to the community as well as facilities that, if damaged or destroyed, would have a significant negative impact on the community.

Economic Factors

Losing a major employer due to a fire or disaster would create a significant economic loss to the community. The Phoenix Fire Department determined the economic impact of successful interventions when handling fires in commercial facilities for 1 year. If those 42 facilities had been completely destroyed, there would have been a loss of 6591 private non-farm jobs (Evans, 2014).

Past Loss/Event History

A loss/event history profile describes the community's past experience and trends and how the community's experience compares to local, regional, and national trends. For example, the FireCARES geospatial-based risk assessment for a community requires a decade of local fire loss data (Jay, 2017). Data on deaths, injuries, causation, and dollar loss are important components of a profile. Local, regional, and national statistics can assist in providing data (Twigg et al., 2017).

Critical Infrastructure Systems

Water, waste, communications, energy, and transportation are critical infrastructure systems where a failure would have a significant impact on the community risk profile. Often a failure will trigger a cascading effect of additional failures, creating a high-impact event (Kotzanikolaou, Theoharidou, and Gritzalis, 2013).

Determining Risk Likelihood and Impact Consequences

Vision 20/20 provides three qualitative tools to determine the likelihood and impact of a risk (Stouffer, 2016). Different categories of risk likelihood are defined as follows:

A. Almost certain: Event is expected to occur
B. Likely: Regular recorded incidents and strong anecdotal evidence
C. Possible: Very few incidents in associated organizations or comparable facilities
D. Unlikely: No recorded incidents
E. Rare

Risk consequence of impact is classified as follows:

1. Insignificant
2. Minor: Disruption of less than 24 hours
3. Moderate: Medical treatment needed, but no fatalities
4. Major: Fatalities, extensive number of injuries, significant damage and financial loss

5. Catastrophic: Significant number of fatalities, community unable to function without significant outside support

Analyzing the combination of risk likelihood and risk consequence of impact allows the fire officer to determine the category of risk. There are four levels:

- Extreme risk: Detailed research and risk management planning required by executive officers. Action required to reduce consequences or likelihood.
- High risk: Planning required by chief officer. Some action required to reduce consequences or likelihood.
- Moderate risk: Specific target hazard or critical facility response and monitoring required.
- Low risk: Manage through routine procedures.

More sophisticated risk management probability analysis involves combining local data and national or international experiences (Ostrom and Wilhelmsen 2019).

Developing a Community Risk Reduction Plan

An important concept in developing a CRR plan is identifying the **acceptable level of risk**. In the process of developing a risk reduction plan with community stakeholders, the group establishes a community standard for the acceptable level of risk that citizens and government can tolerate and afford. This community standard is enhanced when the level of risk is well-defined. This analysis involves community engagement in the discussions and decision making by those affected by the risk (FEMA, 2018).

This process involves obtaining comprehensive information about hazards, vulnerabilities, groups affected, frequency, severity, duration, capacity, and more. Community and political leaders, members of the populations affected by risks, agency representatives, and other stakeholders should collaborate to determine the community's acceptable level of risk (Columbus, 2019).

The Five E's of Prevention and Mitigation

The National Fire Academy advocates "The Five E's of Prevention" (USFA, 2013) when developing a community risk reduction plan:

- Education intervention influences behavior by raising awareness and providing information and knowledge to produce the desired behavior. Education is only effective if people do what they are supposed to do, i.e., fire fighters wearing seat belts.
- Enforcement interventions include passing, strengthening, and enforcing laws, as well as issuing and enforcing regulations.
- Engineering involves changes in the physical environment: the design, development, and manufacture of safety products. Changes are the result of advances in technology; some examples are fire sprinklers, smoke alarms, helmets, and air bags.
- Economic incentives are measures used to influence behavior either positively or negatively. Positive economic incentives reward people monetarily for behaving in a certain manner or making certain choices. Negative economic incentives punish people monetarily for behaving in a certain manner or making certain choices. Tickets, fines, and citations are used to discourage people from choosing unsafe behaviors. An example of positive economic incentive is a one-time 50 percent reduction in property taxes for retrofitting a home with a fire sprinkler system in Montgomery County, Maryland.
- Emergency response interventions are used by first responders to mitigate risk. The emergency response capability of the community must be considered during the risk-assessment process. Some risks can only be mitigated by adding new emergency response capability or enhancing current emergency response capability.

Not every risk will require all five interventions, but many will require more than one type of intervention to effectively reduce the risk.

Writing the Community Risk Reduction Plan

The CRR is a document developed by the community, usually written by the agency that is leading the activities. The elements of the CRR plan include the following:

- Description of the community or service area covered by the CRR
- Identification of the risks and ranking of their priority
- The goals and expectations of the program
- Description of the Five E's that will be utilized
- Identification of the resources required and how they will be allocated

- Timeline with identified milestones, by anticipated date
- Organizational chart showing the agencies and individuals involved in the CRR
- Description of the measurement methods used to monitor and evaluate the program

The written plan provides a map that provides the direction for all of the participants' activities and commitments. This helps the fire officer keep all of the team members focused on the tasks and activities of the CRR plan.

FIRE OFFICER TIP

Mt. Lebanon Fire Department: Lessons Learned from a Mostly Volunteer Combination Department

Mt. Lebanon, Pennsylvania, discovered that 36 percent of the homes in the two poorest districts had no working smoke alarms and one-third of the working smoke alarms throughout the community were over 10 years old. Their Five E's plan included annual community outreach training and door-to-door canvassing to install new smoke detectors. They had better results with volunteer fire fighter participation and occupant access when performing an after-the-fire neighborhood canvassing. At the end of 3 years, the percentage of homes with a working smoke detector in the two poorest districts went from 36 to 84 percent. There was also a 50 percent reduction in fire-related civilian injuries, one structure fire was mitigated to object of origin, and there were two carbon monoxide saves (Switala, 2016).

Cooperating with Allied Organizations

The process of gathering risk data places the fire officer in contact with allied organizations that are already addressing community needs, such as social services, health departments, and nongovernmental organizations. The fire department can be a complementary partner or become a force-multiplier in reducing community risks. This type of cooperation works in two ways: the fire department provides a service that complements or reinforces efforts by the allied organization, or the fire department becomes a partner in a combined effort to reduce a community risk.

Partnership may require a memorandum of agreement or a contract for services. If the scope of activity is considerably different from traditional fire fighter tasks, departments with a labor contract will need to meet with the local labor leadership before making a commitment. In some instances, these activities are funded through federal, state, or private grants. Grant-funded activities require a higher level of documentation of the roles, activities, and oversight of the project and often more frequent update reports.

Implementing and Monitoring the CRR

The completed CRR plan identifies the goals, partners, activities, and anticipated outcomes. As the plan begins, the fire officer should make sure that all of the stakeholders and partners involved in the effort understand their responsibilities and have the resources they need to achieve their tasks.

The timeline provides a schedule of goals and objectives as outlined in the plan. On a regular basis, the CRR activities should be checked against the timeline, usually monthly for year-long projects.

Data need to be collected in order to provide analysis on the progress of the plan as well as the outcomes. Are we doing what we said we would do? Are our activities creating the outcome we anticipated? Predetermined communications to partners, stakeholders, and administrators are needed on a schedule. If the CRR program is not meeting the targeted goals, the partners providing the activities need to make adjustments to the plan.

A detailed annual report of the CRR plan that includes description of the implementation and analysis of efficiency and effectiveness shall be provided to the appropriate agencies, partners, and stakeholders for review. Grant-funded programs may require specific reporting elements describing the financial activities of the program. For fire departments, the CRR programs need to be a part of the annual departmental report to the elected officials and the community.

CRR is a community-focused effort that employs the full spectrum of risk reduction tools. It integrates emergency response capabilities and preventive strategies into one overall approach to provide fire protection and risk reduction. Because the tools are the same regardless of risk, it integrates fire and other hazards that any community may face into a multi-hazard approach that provides an overall protection strategy to mitigate risks. CRR is a process that begins with a risk assessment followed by deliberate plans to prevent or mitigate those risks. Because the risks will vary from one community to another, or even one part of a community to another part, there is no set outcome that fits all communities. CRR will always include emergency response, but to improve outcomes (fewer incidents, deaths, injuries, dollars lost) more proactive measures will need to be identified (NVFC, 2018).

Media Relations

Fire departments have frequent interactions with the local news media when reporters want to obtain information about a situation that is ongoing, as well as for newsworthy situations that are related to other fire department programs or activities. In addition, in many situations the fire department has information or an important message to communicate to the public.

There are two types of communications provided to the media. The first type is **risk communications**, and this refers to the exchange of real-time information, advice, and opinions between experts and people facing threats to their health and economic or social well-being (WHO, 2019). Fire prevention messages are an example of risk communications. CRR plans include risk communications. The second type is **crisis communications**; this type of communications covers occurrences that create surprise, present a threat, and require a short response time (Ulmer, Sellnow, and Seeger, 2019). Crisis communications seek to explain the specific event; identify likely consequences and outcomes; and provide specific harm-reducing information to affected communities in an honest, candid, prompt, accurate, and complete manner (Reynolds and Seeger, 2005).

Emergency incidents are high-profile news events that capture public attention. Members of local print, radio, and television organizations can be expected to appear at major emergency incidents or to call the fire department seeking information. Other activities, ranging from inspection and public education programs to promotions, retirements, awards ceremonies, and delivery of a new fire truck, are also likely to generate media interest.

The news media have an important and legitimate mission to obtain and report information to the general public. The fire department is an organization that performs a public function, operates in the public view, and is usually supported by public funds. It is always in the best interest of the fire department to maintain a good relationship with the local media. Providing information to the news media should be a normal occurrence (Garner, 2010).

Social media is changing crisis dynamics by speeding up the development of critical situations and by potentially forming new types of crises. Social media speeds the development of critical situations because it enables rapid sharing of information on an unimaginably huge scale in real time (Austin and Jin, 2018).

Although the news media have a legitimate purpose in seeking information, in many circumstances there are limitations on the information that can be released. The fire officer should know the applicable departmental procedures and guidelines and understand how to respond to a request for information that the fire officer is not authorized to release. The fire officer should recognize that the interaction that occurs between the fire department and the news media is likely to have a direct impact on the department's public image, reputation, and credibility. The fire department depends on public trust and confidence to be able to perform its mission (Coombs, 2019).

FIRE OFFICER TIP

Examples of Risk Communications

Fireplace ashes: Fire departments in cold climates often experience large-loss fires due to the improper disposal of fireplace ashes. In many cases, the ashes are placed in combustible containers and left in a garage or on a deck. The smoldering ashes may then eventually ignite the container and cause a fire. An engineering solution for this problem might be for the fire department to provide free metal ash cans to the community as part of a seasonal fire safety education program.

Dead smoke alarms: Many fire departments encounter single-station smoke alarms with dead or missing batteries in residences. Some departments make it a practice to check the smoke alarm during every fire company response to a residential occupancy in the community. After stabilizing the initial incident, the fire officer may request permission to check the smoke alarms in the home. A simple engineering solution can quickly solve this problem. Every fire company carries a supply of 9-volt batteries that are used to replace any dead or missing batteries. Some fire departments even issue complete smoke alarms to their fire companies with instructions to install one in any home that does not have a functioning smoke alarm. The fact that the fire department takes this problem so seriously conveys an added public education lesson.

Brush abatement: Failure to maintain a clear area around a building is a common cause of structure fire losses in wildland fires. Los Angeles County uses all three factors—education, engineering, and enforcement—in efforts to reduce the impact of brush fires on the built environment. Education starts with seasonal public safety messages conveyed through the media in multiple languages. Engineering includes a description of the size of the required clear area and a program to certify brush clearance contractors. Enforcement includes local government inspection of properties, as well as issuance of fines and work orders if the brush has not been cleared.

Fire Department Public Information Officer

Some fire departments have a full-time or part-time public information officer (PIO) who functions as the media contact person and the source of official fire department information. Where this position exists, most interactions with the news media should go through this individual. Other officers then have limited responsibilities and opportunities to interact with the news media. On some occasions, the PIO may refer a reporter to another fire officer to obtain information about a particular subject or situation.

Many smaller departments rely on a staff officer to function as the PIO and interact with the local newspaper, radio, and television media as the need arises. In these situations, the individual should be guided by the same basic principles as a regularly assigned PIO. NFPA 1035, *Standard on Fire and Life Safety Educator, Public Information Officer, Youth Firesetter Intervention Specialist and Youth Firesetter Program Manager Professional Qualifications,* covers the job requirements of a public information officer.

This section presents basic recommended practices for a PIO who has to establish and maintain an ongoing relationship with the local media. This aspect of community relations is particularly important when the fire department has information it needs to release to the general public. Reaching out to a community is much easier for an organization that has already established solid contacts and relationships with the local media. The NFPA provides a "Get Your Message Out" media primer (NFPA 2010) to help the fire service publicize community outreach campaigns and other events associated with public safety education. The basic concepts are valuable for any fire officer who has to be prepared to interact with media representatives. The NFPA publication recommends three steps when working with the media:

1. Build a strong foundation.
2. Use a proactive outreach.
3. Use measured responsiveness.

Step 1: Build a Strong Foundation

Look at your relationship with the media as a business arrangement: the media have something you need (an audience and the means to communicate with them), and you have something they need (news and information). If you work collaboratively, both parties can achieve their objectives. If you expect too much or if you do not anticipate the media outlet's needs, someone will end up disappointed, which could prove counterproductive.

The most important media asset you will ever have is a good relationship with the media. Members of the local media need to be familiar with you and comfortable working with you; they must have confidence in the information you provide and, most importantly, need to trust you. Whatever else you do, always tell the truth to the media, even if the truth is "I do not know" or "I cannot release that information at this time" or "It is premature to answer that question." Do not guess, speculate, or lie.

Do not wait until you need a reporter to establish media contacts; lay the groundwork in advance to ensure a collaborative working relationship. Build a list of local media contacts and keep it up-to-date (**FIGURE 11-1**). Obtain the names, titles, and contact information for key individuals. Learn who has responsibility for covering fire and life-safety issues at the newspapers, radio stations, and television stations.

Find out the deadlines for different news outlets and know how to reach reporters with late-breaking information. Television reporters need to have the information before their broadcast time, and print reporters need it before the paper is printed.

Be a consumer of your local media. Read the local newspapers, watch television, subscribe to smartphone-based news applications, visit appropriate blogs, and listen to the radio. Pay attention to the kinds of stories these media outlets tend to cover, the angles they use, and the personal style of individual reporters. If a reporter does a good job reporting a fire department story, make a thank-you call. If a story contains inaccurate information, make a call to provide the correct information, keeping the tone positive.

Make sure that the news representatives know how to contact you whenever they need information. If you will not be available, make sure that someone else is accessible. Your objective is to make sure that news media representatives know that your department is interested in working with them and in responding to their inquiries, that you are the person to contact, and that they can have confidence in any information you give them.

FIGURE 11-1 Reaching out to the community is easier when solid contacts and relationships with members of the media have been established.
© Karin Hildebrand Lau/Shutterstock

Step 2: Proactive Outreach

If you have sufficient resources to undertake one, a proactive media communications plan can be much more beneficial than a reactive approach. Simply responding to media inquiries as they come to you is a reactive strategy. You should be as helpful as possible, providing information within the resource capabilities of your department.

Being proactive means that in addition to being responsive, you actively look for opportunities to use the media to communicate your department's objectives and mission. The media need factual, timely stories to report, and the fire department generates broad public interest. You can help them by providing announcements of newsworthy issues, topics, and events.

Once the working relationship has been established, reporters will begin to look to you for story ideas, resources, and quotes. You can call a reporter or editor and suggest a story that could be interesting to him or her. Learn the types of stories that are most interesting to different reporters and call the appropriate person when you have something that fits his or her interests. Reporters will not automatically accept every story idea, but they will listen to what you have to suggest.

When you have information that you want to distribute to several news outlets, use a press release or hold a press conference. A press conference is a staged event to which you invite the news media to come and hear something important that you have to announce. Press conferences should be reserved for topics and situations that definitely have broad interest; on these occasions, you know that the media will want to attend and ask questions.

Step 3: Measured Responsiveness

When dealing with the news media, remember that one of your responsibilities is to present a positive image of the fire department. In certain situations, this is difficult, particularly if something negative has occurred. It is no use denying that something has happened when the facts are clearly evident. Sometimes, the best that you can do is to be honest and factual to maintain the credibility of the organization.

Be wary of situations in which someone could be trying to misrepresent a situation in a manner that reflects badly on the department. The reporter who comes to you may have already been given a highly inaccurate or slanted viewpoint on a story by some other source. Do not allow yourself to be placed in a situation in which you make a statement that is inaccurate, misleading, or damaging to the fire department. Make sure that your response is accurate and that the information is appropriate for release before you react.

Press Releases

A press release is used to make an official announcement to the news media from the fire department PIO (**FIGURE 11-2**). It could highlight a special event, a promotion or appointment, an award ceremony, a fire station opening, a retirement, or any similar occurrence that the department wants to make known to the public. A press release could also provide information about a new or ongoing program, such as the launch of a new juvenile fire-setter counseling program or the release of an annual report showing a 34 percent decrease in residential fires since a major public education program was implemented.

A press release should be dated and typed on department stationery, with the PIO contact information at the top. It should be as brief as possible, ideally one to two pages in length, using an accepted journalistic writing style. Make the lead paragraph powerful to entice the reporter to read on, and answer the basic "who, what, when, where, and why" questions.

Depending on the preference of your local media contacts, you can mail, fax, upload, or e-mail press releases. Reporters and editors are more likely to read your releases if they receive only those that pertain to their particular beat or to topics in which they are interested. Personalize the release to an individual at each destination by name and title. For example, a release announcing the official opening of a new fire station could go to the assignment editor at a television station and to the newspaper reporter who usually covers the fire department. A release on cooking fire safety could be directed to the food editor.

The Fire Officer as Spokesperson

Every fire officer should be prepared to act as a spokesperson for the fire department. Even if the department has a PIO who is the designated official spokesperson, there are likely to be occasions when other fire officers will have to use these skills.

The most common situation in which a fire officer would appear as an official spokesperson is an interview. The NFPA provides the following guidelines for the fire officer conducting an interview with the media:

- Be prepared.
- Stay in control.
- Look and act the part.
- It is not over until it is over.

Be Prepared

Before agreeing to be interviewed, make sure you are authorized to speak on the subject as a departmental spokesperson and be sure you have the appropriate information. Determine the reporter's story angle and identify what he or she already knows about the topic. Ask with whom he or she has already spoken or plans to contact and what he or she hopes to learn from you.

If you are uncomfortable with the situation angle or do not have the information, you do not have to agree to be interviewed. You should be polite but firm if you decline a request for an interview. If possible, refer the reporter to someone else who can provide the information.

Be on time for an agreed-on interview, but do not begin the interview until you are ready. The reporter probably has a deadline that you should try to respect, but not at the expense of your readiness.

When preparing for an interview in which you want to deliver a specific message, identify no more than three key message points in advance and practice saying them in varied ways. Learn to use wording that emphasizes what you are saying is important, such as "The number-one thing to remember is _____" or "More than anything else, people should realize that _____." Practice your talking points with a colleague or a tape recorder. Do not memorize your message points, but be very familiar with them. Check any statistics, trends, or other information you would not know on a casual basis.

Stay in Control

Be cooperative, but stay focused on your key message points. Listen carefully to the reporter's questions and ask for clarification if you do not fully understand a question. If you need a few seconds to think through your answer, take the time necessary to formulate your answer. If you do not know the answer, say so and offer to find the information and get back to the reporter. Avoid jargon and highly technical language that confuses or distracts your audience. When you have answered the question, stop talking.

Remember that the reporter came to you looking for information and expertise. Be confident and authoritative, steering the interview in the direction that you want it to go. Showcase your message points and your department's mission and work.

Look and Act the Part

When doing a television interview, make sure your appearance is clean, tidy, and professional (**FIGURE 11-3**).

FIGURE 11-3 Make sure your appearance is clean, tidy, and professional, and wear your uniform properly.
© Jeff Greenberg/Alamy Stock Photo.

Press Release

Press Release
Public Information Officer
www.fire.jbpub.com
(978) 443-5000

Civilians Rescued from Burning Home
Twenty-eight fire fighters responded to an early-morning house fire in the Grandview District. The 911 call was at 10:47 pm on Monday, December 16, 2023; the caller reported smoke coming from a townhome.
Jones & Bartlett fire fighters arrived at 11:07 pm, encountering smoke and flames coming from the first-floor windows at 10934 Braniff Way. Two elderly females were found by fire fighters on the third floor of the townhome.
An aerial tower was needed to remove the residents from the building. After treatment by paramedic/fire fighters, the women were transported to University Hospital.
It took 22 minutes of aggressive fire suppression before the Battalion Chief declared the fire "under control." There was extensive damage to the first floor of the home.
The cause remains under investigation. No monetary loss has been calculated.

FIGURE 11-2 Example of a press release.
© Jones & Bartlett Learning.

Wear your uniform properly, and show pride in the department that you are representing. Adopt the appropriate demeanor for the subject. If you are on the scene of a fire fatality, a somber yet authoritative tone is appropriate. If you are conducting a television interview from a fire station open house, a more enthusiastic tone is acceptable.

It Is Not Over Until It Is Over

When speaking to a reporter, do not assume that anything you say is "off the record." Instead, assume that everything you say could be quoted, including conversations before and after the actual interview. When you are wearing a microphone, anything you say can be overheard and recorded.

FIRE OFFICER TIP

Grammar Counts

Press releases are public documents that are read by reporters, the public, and local elected officials. The presentation and the content both reflect on the image of the fire department. Invest in a dictionary, thesaurus, and grammar book or style guide to improve your written presentation skills. Check the document's spelling, particularly names, and be sure that ranks, titles, and telephone numbers are correct.

After the interview, double-check to ensure that the reporter has the correct name and spelling for you and your department and accurate information about any program or event that you are promoting. Leave your business card to ensure that the proper name, title, and organization attribution are recorded.

Social Media Outreach

NFPA leadership provided 10 tips on using social media to expand safety message outreach. The goal of social media outreach is to reach people who are on the Internet (Carli, Hazell, and Backstrom, 2012).

1. **Plan a strategy:** Select a digital media and find your customers. Plan on two digital postings per day. NFPA's Director of Internal Communications Mike Hazell suggests a blog as a good way to interact with subscribers and readers.
2. **Commit:** You need to spend time every day getting the message out. NFPA social media manager Lauren Backstrom recommends spending 25 percent of the time listening to customers, 50 percent of the time interacting, and 25 percent of the time creating new material.
3. **Be authentic:** Social media use represents a big departure from earlier message delivery procedures, and creating a human and authentic digital presence is more effective in getting your message across than issuing a cold, stiff news release.
4. **Be current:** Backstrom recommends connecting fire prevention information with current events.
5. **Be social:** Expect responses to digital postings and provide prompt and authentic responses.
6. **Maintain quality:** Provide accurate information that is useful to the reader.
7. **Tailor content to each venue:** Instead of posting the same information on all social media venues, tailor the post to utilize each venue's capability best. Consider establishing focused blogs or websites to meet a particular group's need. NFPA vice president of communications Lorraine Carli noted large growth in the number of age 55 and older residents who are using social media.
8. **Be interesting and entertaining:** Hazell described a contest to get a new voice for Sparky the fire dog. The contest got 26 people to audition for the role.
9. **Consider timing:** Identify the usage patterns for each social media venue and time messages to have the largest impact on the target audience.
10. **Track results:** Determine readership, level of engagement, and any measurable change in reader behavior or impact on the identified community issue.

Social Media Challenges

Social media describes forms of electronic communication through which users create online communities to share information, ideas, personal messages, and other content. Online communities include Facebook, Twitter, and Instagram, as well as specific-interest groups.

The immediacy and far reach of social media can have the same effect on a reputation as a 2½-inch attack line has on a room-and-contents fire. Virtual flash mobs can quickly elevate a minor event into an international incident. Media expert Dave Statter coined the phrase "social media–assisted career suicide syndrome" (SMACSS) to describe the impact of social media videos, pictures, and postings that lead to a sudden (and often unfortunate) change of status for the author of the post (Statter, 2019). Attorney Curt Varone, a retired Deputy Assistant Chief, points out that the limits of free speech for fire fighters are not crystal clear, but two types of social media posts may not be protected by the First Amendment:

- Speech that threatens violence or harm or encourages others to commit violence/harm to any person or group.
- Speech that threatens to or encourages others to withhold public safety services from any person or group (Varone, 2019).

You Are the Fire Officer Conclusion

Community risk reduction integrates emergency response capabilities and preventive strategies into one overall approach to provide fire protection and risk reduction. CRR is a process that begins with a risk assessment and is followed by deliberate plans to prevent or mitigate those risks.

1. The captain assesses the risk by identifying the communities with the most apartment dryer fires and comparing them with other communities of similar size, demographics, and socioeconomic status.
2. The captain needs to reach out to the apartment owners, local government agencies, and community groups that are impacted by these incidents. Develop a coalition of interested stakeholders to participate in developing the community risk reduction plan.
3. The community involved in developing the risk reduction plan defines success by identifying the acceptable level of risk and utilizing the Five E's to develop a plan.

After-Action REVIEW

IN SUMMARY

- Community Risk Reduction (CCR) is the process of a comprehensive evaluation that identifies, prioritizes, and defines the risks that pertain to the overall community.
- Elements used to describe the community include demographics, geography, building stock, public safety response agencies, community service organizations, hazards, economic factors, past loss/event history, and critical infrastructure systems.
- A community demographic profile reflects the economic and social issues that significantly influence an identified community risk.
- Geographic risk factors include 100-year flood areas, identified earthquake faults, landslide-prone areas, volcano locations, industrial/chemical site hazards, and nuclear and radiation hazards.
- Building stock evaluation includes housing density, age, high-density structures, and overcrowding.
- The Center for Public Safety Excellence provides a framework to conduct a comprehensive analysis of public safety response agencies.
- Community service organizations focus on those at risk, including children under six years old, adults over 65 years old, people with disabilities, people living in poverty, and people that speak little or no English.
- Target hazards are facilities in either the public or private sector that provide essential products and services to the general public; are otherwise necessary to preserve the welfare and quality of life in the community; or fulfill important public safety, emergency response, and/or disaster recovery functions.

- Vision 20/20 provides three qualitative tools to determine the likelihood and impact of risk: categories, consequences, and level. This can guide the prioritization of community risk reduction efforts.
- Acceptable level of risk is the community standard for the acceptable level of risk that citizens and government can tolerate and afford.
- The Five E's of Prevention and Mitigation are education, enforcement, engineering, economic incentives, and emergency response.
- A written community risk reduction plan identifies the goals, partners, activities, and anticipated outcomes.
- Implementing a community risk reduction plan includes regular monitoring of progress through a timeline.
- A detailed annual report of the CRR plan that includes description of the implementation and analysis of efficiency and effectiveness shall be provided to the appropriate agencies, partners, and stakeholders for review.
- Risk communications is the exchange of real-time information, advice, and opinions between experts and people facing threats to their health and economic or social well-being.
- Crisis communications cover occurrences that create surprise, present a threat, and require a short response time.
- Crisis communications seek to explain the specific event; identify likely consequences and outcomes; and provide specific harm-reducing information to affected communities in an honest, candid, prompt, accurate, and complete manner.
- NFPA recommends building a strong foundation, using proactive outreach, and using measured responsiveness when working with the media.
- A press release is used to make an official announcement to the news media from the fire department Public Information Officer (PIO).
- During a media interview, the fire officer should be prepared, stay in control of the interview, and look and act the part of a fire department spokesperson.

KEY TERMS

100-year flood A frequency analysis performed by the U.S. Geological Survey to identify the 1 percent probability that a given event will be equaled or exceeded in any given year.

Acceptable level of risk A community standard for the acceptable level of risk that citizens and government can tolerate and afford. This level of risk should be well-defined.

Community demographic profile Statistical description of economic and social issues of a community.

Community risk assessment (CRA) A comprehensive evaluation that identifies, prioritizes, and defines the risks that pertain to the overall community. (NFPA 1300)

Community risk reduction (CRR) A process to identify and prioritize local risks, followed by the integrated and strategic investment of resources to reduce their occurrence and impact. (NFPA 1300)

Crisis communications Communications that seek to explain the specific event; identify likely consequences and outcomes; and provide specific harm-reducing information to affected communities in an honest, candid, prompt, accurate, and complete manner.

Demographics The characteristics of human populations and population segments, especially when used to identify consumer markets; generally includes age, race, sex, income, education, and family status.

Risk assessment Identification of potential and likely risks within a particular community records management system (RMS).

Risk communications The exchange of real-time information, advice, and opinions between experts and people facing threats to their health and economic or social well-being.

REFERENCES

Austin, Lucinda, and Yan Jin, eds. 2018. *Social Media and Crisis Communications*. New York: Routlege.

Carli, Lorraine, Mike Hazell, and Lauren Backstrom. 2012, June 13. "Social Media: What's Your Strategy?" Presentation at NFPA Conference and Exhibition. Las Vegas, NV.

Center for Public Safety Excellence (CPSE). 2019. "Accreditation Overview." Accessed July 19, 2019. https://cpse.org/accreditation/.

Columbus Fire and Emergency Medical Services. 2019, May 30. "Columbus Fire and EMS Standards of Cover/Community Risk Assessment." Columbus, GA: City of Columbus.

Coombs, W. Timothy. 2019. *Ongoing Crisis Communication: Planning, Managing and Responding, Fifth edition*. Thousand Oaks, CA: Sage Publications.

Dwoskin, Elizabeth. 2014, January 24. "How New York's Fire Department Uses Data Mining." *The Wall Street Journal*. Accessed July 26, 2019. https://blogs.wsj.com/digits/2014/01/24/how-new-yorks-fire-department-uses-data-mining/.

Evans, Anthony. 2014. *The Economic Impact of Successful Commercial Fire Interventions: Phoenix Fire Department June 1, 2012 - May 13, 2013*. Phoenix, AZ: Arizona State University.

Federal Emergency Management Agency (FEMA). 2018. *Threat and Hazard Identification and Risk Assessment (THIRA) and Stakeholder Preparedness Review (SPR) Guide: Comprehensive Preparedness Guide (CPG) 201, Third Edition*. Washington, DC: Department of Homeland Security.

Garner, Bruce. 2010, August 6. "PIO 101: The Basics of Media Relations for the Fire Service." *Firehouse.com*. Accessed July 22, 2019. https://www.firehouse.com/prevention-investigation/article/10464906/pio-101-the-basics-of-media-relations-for-the-fire-service.

Gielen, Andrea C., Dina Borzekowiki, Rajiv Rimal, and Anita Kumar. (2010, December) *Evaluating and Creating Fire and Life Safety Materials: A Guide for the Fire Service*. Quincy MA: National Fire Protection Association.

Gorman, Anna, and Kaiser Health News. 2019, March 8. "Medieval Diseases Are Infecting California's Homeless: Typhus, Tuberculosis, and Other Illnesses Are Spreading Quickly Through Camps and Shelters." *The Atlantic*. Accessed July 25, 2019. https://www.theatlantic.com/health/archive/2019/03/typhus-tuberculosis-medieval-diseases-spreading-homeless/584380/.

International Civil Defence Organisation (ICDO). 2019. "Disasters." Accessed July 18, 2019. http://www.icdo.org/en/disasters/.

Jay, Jonathan. 2017, March 6. "Can Algorithms Predict House Fires?" *Data-Smart City Solutions*. Accessed July 19, 2019. https://datasmart.ash.harvard.edu/news/article/can-algorithms-predict-house-fires-990.

Kotzanikolaou, Panayiotis, Marianthi Theoharidou, and Dimitris Gritzalis. 2013. "Cascading Effect of Common-Cause Failures in Critical Infrastructures." In *Critical Infrastructure Protection VII Proceedings*. Amsterdam, Netherlands. Accessed July 26, 2019. https://link.springer.com/content/pdf/10.1007%2F978-3-642-45330-4_12.pdf.

National Volunteer Fire Council (NVFC). 2018. *Volunteer Fire Service Culture: Essential Strategies for Success*. Greenbelt, MD: National Volunteer Fire Council.

Ostrom, Lee T., and Cheryl A. Wilhelmsen. 2019. *Risk Assessment: Tools, Techniques, and Their Applications*. Hoboken, NJ: John Wiley & Sons, Inc.

Reynolds, Barbara, and Matthew Seeger. 2005. "Crisis and Emergency Risk Communication as an Integrative Model." *Journal of Health Communications* (10): 43–55.

Shawnee Fire Department. 2017. "Community Risk Assessment: Building Stock." Accessed July 18, 2019. https://dashboards.mysidewalk.com/community-risk-assessment/building-stock.

Statter, Dave. 2019. "SMACCS." *STATter911*. Accessed July 22, 2019. https://www.statter911.com/tag/smacss/.

Stouffer, John A. 2016. *Community Risk Assessment: A Guide for Conducting a Community Risk Assessment*. Washington, DC: Vision 20/20.

Switala, Christopher K. 2016, March 11. "Smoke Alarm Installations: Windows of Opportunity and Lessons Learned from a Mostly Volunteer Combination Department." Presented at Vision 20/20 Model Performance in Fire Prevention Symposium. Fairfax, VA.

Treibitz, Scott. 2017, May 10. "Big Data Comes to Public Safety: Do You Know Your Community's Safe Grade?" [press release]. *FireCARES.org*. Washington, DC: International Public Safety Data Institute.

Twigg, John, Nicola Christie, James Haworth, Emmanuel Osuteve, and Artemis Skarlatidou. 2017. "Improved Methods for Fire Risk Assessment in Low-Income and Informal Settlements." *International Journal of Environmental Research and Public Health* 14 (2): 139.

Ulmer, Robert R., Timothy L. Sellnow, and Matthew W. Seeger. 2019. *Effective Crisis Communication: Moving from Crisis to Opportunity, Fourth edition*. Thousand Oaks, CA: Sage Publications.

United Nations Office for Disaster Risk Reduction (UNISDR). 2018. "Man-made and Technological Hazards: Practical Considerations for Addressing Man-made and Technological Hazards in Disaster Risk Reduction." Geneva, Switzerland: United Nations Office for Disaster Risk Reduction.

U.S. Fire Administration (USFA). 2013, October 17. "Coffee Break Training: The Five E's of Prevention." Accessed July 26, 2019. https://www.hsdl.org/?abstract&did=745705.

U.S. Geological Survey (USGS). 2019. "The 100-Year Flood." Accessed July 18, 2019. https://www.usgs.gov/special-topic/water-science-school/science/100-year-flood.

Varone, Curt. 2019, July 5. "NC Facebook Post Results in Two Cybercasualties." *FireLawBlog.com*. Accessed July 22, 2019. http://www.firelawblog.com/2019/07/04/nc-facebook-post-results-in-two-cybercasualties/.

World Health Organization (WHO). 2019. "General Information on Risk Communication." Accessed July 22, 2019. https://www.who.int/risk-communication/background/en/.

Fire Officer in Action

1. Community risk reduction is a _____ process to identify and prioritize local risks.
   A. fire department
   B. fire department and authority having jurisdiction
   C. state and local government
   D. community

2. Determining the category of risk when performing a community risk assessment requires:
   A. projecting the frequency of future events based on local emergency incident experience.
   B. estimating of the impact of existing fire prevention and health department efforts.
   C. benchmarking population demographics with national fire experience.
   D. analyzing the combination of risk likelihood and risk consequence.

3. Which of these is a community-developed standard for acceptable hazards?
   A. Acceptable level of risk
   B. Average impact consequence
   C. Worst acceptable outcome
   D. Delphi disaster level

4. Which of these provides specific harm-reducing information to affected communities in an honest, candid, prompt, accurate, and complete manner?
   A. Risk communication
   B. Written community risk reduction
   C. Crisis communications
   D. Community preplan

Fire Captain Activity

This chapter describes the process of developing a community risk reduction program.

NFPA Fire Officer II Job Performance Requirement 5.3.1

Supervise multi-unit implementation of a community risk reduction (CRR) program, given an AHJ CRR plan, policies, and procedures, so that community needs are addressed.

Application of 5.3.1

1. A local CRR program plans to replace missing or old smoke detectors in the fire company's district. The research showed 50% of the homes had missing or old smoke detectors. CRR includes members of a two non-governmental organizations assisting the fire company in door-to-door neighborhood canvassing on Saturday afternoons. The 3-month assessment showed that in 12 Saturdays:
   - 120 homes approached
   - No one home at 75 homes
   - Refused access in 15 homes
   - Replaced smoke detectors in 10 of the 30 homes accessed

   Getting access to 30 homes per month is below the plan's goal to access 100 homes a month. Describe the process to develop an adjustment plan to increase the monthly access to homes.

2. A senior living community, consisting of an apartment totaling 263 residents, is calling 911 for lifting patients from the floor or out of the bathtub. The frequency of calls has gone from once per week 2 years ago to 1.5 times per day now. Describe how you would construct a community risk assessment plan.

NFPA Fire Officer II Job Performance Requirement 5.3.2

Explain the benefits to the organization of cooperating with allied organizations, given a specific problem or issue in the community, so that the purpose for establishing external agency relationships is clearly explained.

Application of 5.3.2

1. A homeless encampment has developed on state land between an interstate and a county high school. There appears to be 75 residents at the site. The fire department has been responding to the site daily for EMS calls and cooking-related fires. Paramedics are encountering residents with typhus (Gorman, 2018). Identify the agencies and the nongovernmental organizations you would contact in developing a CRR to stop the spread of disease at the encampment site.

2. Arson at a place of worship has moved a congregation to use a commercial building. The problem is that the size of the congregation significantly exceeds the maximum occupancy allowed by the fire prevention code for that commercial building. Which nongovernmental organizations and agencies would you consult with to resolve this community risk?

Access Navigate for flashcards to test your key term knowledge.

CHAPTER 12

Fire Officer II

Administrative Communications

KNOWLEDGE OBJECTIVES

After studying this chapter, you will be able to:

- Explain considerations for developing SOPs and SOGs. (**NFPA 1021: 5.4.1, 5.4.6**) (p. 317)
- Explain best practice in leading organizational change. (**NFPA 1021: 5.4.1, 5.4.6**) (pp. 328–329)
- Prepare a report recommending adoption of a new policy or procedure to resolve a problem. (**NFPA 1021: 5.4.1, 5.4.6**) (pp. 328–329)
- Describe the budgetary process. (**NFPA 1021: 5.4.2**) (pp. 319–320)
- Explain capital and operation budgets and their relationship to each other. (**NFPA 1021: 5.4.2**) (pp. 326–328)
- Define *line-item budget*. (**NFPA 1021: 5.4.2**) (p. 326)
- Explain the process of preparing a budget. (**NFPA 1021: 5.4.2**) (pp. 319–322)
- Identify information that can be used to provide budget justification. (**NFPA 1021: 5.4.2**) (pp. 319–322)
- Explain the purpose of equipment purchasing specifications. (**NFPA 1021: 5.4.3**) (pp. 323–324)
- Identify the necessary components of a purchasing specification. (**NFPA 1021: 5.4.3**) (pp. 323–324)
- Identify types of formal communications and explain the appropriate use of each. (**NFPA 1021: 5.4.5**) (pp. 317–318)
- Describe considerations for writing accurate and concise reports. (**NFPA 1021: 5.4.5**) (p. 318)
- Describe how to give an oral presentation of a written report. (**NFPA 1021: 5.3**) (p. 319)

SKILLS OBJECTIVES

After studying this chapter, you will be able to:

- Prepare a budget. (pp. 321–322)
- Create a plan to implement an organizational change. (**NFPA 1021: 5.4.1, 5.4.6**) (pp. 328–329)
- Prepare a proposed budget for a program or division. (**NFPA 1021: 5.4.2**) (pp. 321–322)
- Write a budget justification to support budget needs. (**NFPA 1021: 5.4.2**) (pp. 319–320)
- Prepare accurate reports and documents. (**NFPA 1021: 5.4.5**) (p. 318)
- Present a report or proposal. (**NFPA 1021: 5.4.5**) (p. 319)

ADDITIONAL NFPA STANDARDS

- NFPA 1452, *Guide for Training Fire Service Personnel to Conduct Community Risk Reduction for Residential Occupancies*
- NFPA 1035, *Standard on Fire and Life Safety Educator, Public Information Officer, Youth Firesetter Intervention Specialist and Youth Firesetter Program Manager Professional Qualifications*
- NFPA 1300, *Standard on Community Risk Assessment and Community Risk Reduction Plan Development*
- NFPA 1600, *Standard on Continuity, Emergency, and Crisis Management*

You Are the Fire Officer

Engine 2 is an 11-year-old pumper that may not make it to its planned replacement at 20 years. The unit responds to almost 40 calls every day: approximately 24 emergency medical services (EMS) first responder calls; 10 activated fire, smoke, or carbon monoxide responses; and 5 other responses such as structure fires, outside fires, and hazardous condition investigations. Last budget year, a request was made to replace the pumper at the 15-year mark. The budget director denied the request. The battalion chief has told the captain to request an alternative response unit that will handle the EMS and activated alarm incidents.

1. How do you submit a request for a new type of service?
2. What first-year and continuing year budget considerations must be included?
3. How do you describe the value of this new service?

Access Navigate for more practice activities.

Introduction

Successful administrative communication plays a vital role in an efficient fire department. A fire officer must be able to process several types of information to supervise and support the fire company members effectively. An almost overwhelming volume of information is available today, particularly regarding newly identified hazards. A fire officer should regularly access publications and social media to stay aware of trends and issues within public safety. In addition, the officer should share significant information with the company members.

Clearly, effective communication skills are essential for a fire officer. These skills are required to provide direction to the crew members, review new policies and procedures, and simply exchange information in a wide range of situations. Effectively transmitting radio reports requires a unique skill set. Communication skills are equally important when working with citizens, conducting tours, releasing public information, and preparing reports.

One communication skill of particular importance is that of reporting expenditures and budget. Budget preparation is both a technical and a political process. The technical part relates to the calculation of funds required to achieve different objectives, whereas the political part is related to elected officials making the decisions about which programs should be funded among numerous alternatives.

The fire department represents one of the larger sections of a municipal budget, along with schools, police, and public works. The elected officials determine tax rates to generate the needed revenue. They then allocate the available funds among the different organizations and programs. Federal, state, and local regulations also influence the budget process.

Finally, the fire officer must possess the ability to communicate and recommend departmental changes. Identifying the problem, exploring alternate and select solutions, and executing implementation are skills the fire officer must practice in order to guarantee the success of his or her station.

Written Communications

Written communication is needed to document both routine and extraordinary fire department activities. The resulting documents establish institutional history and serve as the foundation of any activity that the fire department wishes to accomplish. As a senior chief explained, "If it is not written down, it did not happen."

Informal Communications

Informal communications include internal memos, e-mails, instant messages, and messages transmitted via mobile data terminals. This type of communication is used primarily to record or transmit information that may not be needed for reference in the future. Even so, all communications from a company officer are retrievable within an information management system. Consequently, each message and report needs to be appropriate, accurate, and ethical.

Some informal reports are retained for a period of time and may become a part of a formal report or investigation. For example, a written memo could document an informal conversation between a fire officer and a fire fighter regarding a performance issue.

The memo might note that the employee was warned that corrective action is required by a particular date. If the performance is corrected, the warning memo may be discarded after 12 months. If the performance remains uncorrected, the memo becomes part of the formal documentation of progressive discipline.

Formal Communications

A **formal communication** is an official fire department document printed on business stationery with the fire department letterhead. If it is a letter or report intended for someone outside the fire department, it is usually signed by the fire chief or a designated staff officer to establish that the document is an official communication. Subordinates often prepare these documents and submit them for an administrative review to check the grammar and clarity. A fire chief or the staff officer who is designated to sign the document performs a final review before it is transmitted. When sending electronic messages with attachments, it is a good practice to double-check exactly what you are sending by opening the attachment(s) to make sure it is what you intended (Sovick, 2013).

The fire department maintains a permanent copy of all formal reports and official correspondence from the fire chief and senior staff officers. Formal reports are usually archived.

Standard Operating Procedures

Standard operating procedures (SOPs) are written organizational directives that establish or prescribe specific operational or administrative methods to be followed routinely for the performance of designated operations or actions. They are intended to provide a standard and consistent response to emergency incidents as well as personnel supervisory actions and administrative tasks. SOPs are a prime reference source for promotional exams and departmental training.

SOPs are formal, permanent documents that are published in a standard format, signed by the fire chief, and widely distributed. They remain in effect permanently or until they are rescinded or amended. Many fire departments conduct a periodic review of all SOPs so they can revise, update, or eliminate any that are outdated or no longer applicable. Any changes in the SOPs must also be approved by the fire chief.

Standard Operating Guidelines

Standard operating guidelines (SOGs) are written organizational directives that identify a desired goal and describe the general path to accomplish the goal, including critical tasks or cautions. Like SOPs, SOGs are formal, permanent documents that are published in a standard format and remain in effect until they are rescinded or amended. SOGs are often used in multijurisdictional activities, such as an automatic aid response plan, and provide flexibility for each responding agency to achieve a goal through various methods and resources.

The U.S. Fire Administration's *Developing Effective Standard Operating Procedures for Fire and EMS Departments* (USFA, 1999) looked at the difference between SOPs and SOGs. Its review of legal proceedings indicates that terminology is less important than content and implementation of SOPs/SOGs. The review revealed that the courts tend to assess liability based on factors such as the following:

- Systems in place to develop and maintain SOPs/SOGs
- Compatibility with regulatory requirements and national standards
- Consideration of unique departmental needs
- Adequacy of training and demonstration of competence
- Procedures used to monitor performance and ensure compliance

General Orders

General orders are formal documents that address a specific subject, policy, condition, or situation. They are usually signed by the fire chief and can remain in effect for various periods of time, ranging from a few days to permanently. These documents may also be called *executive orders*, *departmental directives*, *master memos*, or other terms. Many departments use general orders to announce promotions and personnel transfers. Copies of the general orders should be available for reference at all fire stations.

Announcements

The formal organization may use announcements, information bulletins, newsletters, websites, or other methods to share additional information with fire department members. The identification of these communications is different for each department. These types of announcements are used to distribute short-term and nonessential information that is of interest.

Legal Correspondence

Fire departments are often required to produce copies of documents or reports for legal purposes. Sometimes the fire department is directly involved in the legal action, whereas in other cases the action involves other parties but relates to a situation in which the fire department responded or had some involvement. It is not unusual for a fire department to receive a legal order demanding all of the documentation that is on file pertaining to a particular incident. Collecting and transmitting this information is a task that can require extensive time and effort. In any situation in which reports or documentation is requested, the fire department should consult with legal counsel.

The fire officer who prepared a report may be called on—sometimes years later—to sign an affidavit, appear at a deposition, or appear in court to testify that the information provided in the report is complete and accurate. The fire officer might also have to respond to an **interrogatory**, which is a series of written questions asked by someone from an opposing legal party (CLSHC, 2019) The fire department must provide written answers to the interrogatories, under oath, and produce any associated documentation. If the initial incident report or other documents provide an accurate, factual, objective, complete, and clear presentation of the facts, the response to the interrogatory is often a simple affirmation of the written records. The task can become much more complex and embarrassing if the original documents are vague, incomplete, false, or missing.

FIRE OFFICER TIP

Write a Letter for the Fire Chief

A popular assessment center exercise requires the candidate to prepare a letter for the fire chief's signature. This represents a common assignment for a supervising fire officer in many fire departments. In the scenario, the candidate is provided with all of the necessary information and must demonstrate the ability to prepare a properly formatted and worded letter. The letter is typically a response to a complaint or inquiry by a citizen or community interest group.

This exercise has two primary goals:
1. Assess the candidate's writing skills.
2. Assess the candidate's decision-making skills.

Preparing for this part of the assessment center requires the candidate to learn the proper format for an official letter or formal report. Generally, the assessment center exercise requires a business letter format. One method of preparing for this exercise is to contact the fire chief's administrative assistant and ask how this type of letter is typically prepared for the fire chief.

Writing a Report

Reports should be accurate and present the necessary information in an understandable format. The report could be intended to brief the reader, or it might provide a systematic analysis of an issue. If the fire chief wants only to be briefed on a new piece of equipment and the fire officer delivers a fully researched presentation, both parties would be frustrated. An executive summary would help in this case.

The fire officer must consider the intended audience. Consider a report that proposes replacing four engine and three ladder companies with three quints. If the report is internal, the technical information could include fire department jargon. In contrast, if the report will be submitted to the city council, with copies going to the news media, many of the terms and concepts would have to be explained in detail.

When preparing the report, the appropriate format must be selected. Options range from an internal memo format with one or two pages of text to a lengthy formal report that includes photographs, charts, diagrams, and other supporting information (Nelson and Stenberg, 2018).

Some reports require a recommendation. In such a case, the fire officer may need to analyze and interpret data that are maintained in a database or a records management system. The fire department's records management system typically contains NFIRS run reports, training records, occupancy inspection records, preincident plans, hydrant testing records, staffing data, and company activity records.

A computer-aided dispatch system uses a combination of databases to verify addresses and determine the units that should be dispatched; it also maintains information that can be retrieved relating to individual addresses. All of the transactions performed by a computer-aided dispatch system are recorded in another database. This information can be retrieved to examine the history of a particular incident or to identify all of the incidents that have occurred at a particular location.

When developing some reports, the fire officer might be looking at variances, such as differences between the projected budget and actual expenses in different accounts in a given year. When a variance is encountered, the analysis should consider both the cause and the effect. If the budget included an allocation of $10,000 for training and the expenditure records indicate that only $500 was spent, there would be a variance of $9500 that could be reallocated for some other purpose. This variance could also indicate that little training was actually conducted.

Recommendation Report

A **recommendation report** is a document that advocates a particular action or decision. Some reports incorporate both a chronological section and a recommendation section. For example, the fire fighter line-of-duty death investigations produced by the National Institute for Occupational Safety and Health (NIOSH) include a chronological report of the event, followed by a series of recommendations aimed at preventing the occurrence of a similar situation.

The goal of a decision document is to provide enough information and make an effective, persuasive argument so that the intended individual or body accepts your recommendation. This type of report includes recommendations for employee recognition or formal discipline, for a new or improved procedure, or for adoption of a new device. A decision document usually includes the following elements:

- Statement of problem or issue: Include one or two sentences.
- Background: Provide a brief description of how this became a problem or an issue.
- Restrictions: Outline of the restrictions affecting the decision. Factors such as a federal law, state regulations, local ordinances, budget or staff restrictions, and union contracts are all restrictions that could affect a decision.
- Options: Where appropriate, provide more than one option and the rationale behind each option. In most cases, one option is to do nothing.
- Recommendation: An explanation of why the recommended option is the best decision. The recommendation should be based on considerations that would make sense to the decision maker. If the recommendation is going to a political body and involves a budget decision, it should be expressed in terms of lower cost, higher level of service, or reduced liability. The impact of the decision should be quantified as accurately as possible, with an explanation of how much money it will save, which new levels of service will be provided, or how much the liability will be reduced.
- Next action: A clear statement of the action that should be taken to implement the recommendation. Some recommendations may require changes in departmental policy or budget. Others could involve an application for grant funds or a request for a change in state or federal legislation.

Presenting a Report

A fire officer should be prepared to make an oral presentation of a decision document or to speak to a community group on a fire department issue. Using the written report as a guide, a verbal presentation would consist of four parts:

1. Getting their attention: Have an opening that entices the audience to pay attention to your message.
2. Interest statement: Immediately and briefly explain why listeners should be interested in this topic.
3. Details: Organize the facts in a logical and systematic way that informs the listeners and supports the recommended decision.
4. Action: At the close of your presentation, ask the audience to take some specific action. The most effective closing statements are actions that relate directly to the interest statement at the beginning of the presentation.

FIRE OFFICER TIP

Getting in Front of a Controversial Issue

Dave Statter, a retired TV reporter who runs STATter911 Communications, has warned about trends in news stories that end up becoming issues for fire departments across the United States. Most of these stories have dealt with budget concerns such as claims of excessive overtime, sick leave, and shift swapping. Clear patterns are evident in how those stories evolved in multiple jurisdictions across the country. Often these reports have hurt the image of fire fighters even when there has been no wrongdoing. In the presentation "Effective Communications in a Digital Age" (Statter, 2019), Statter provides this advice about facing down controversial issues:

- Get your house in order now—before a political leader makes it an issue or a reporter starts asking questions.
- Take corrective action on any abuses you uncover.
- If you believe any problems you discovered are likely to become news, consider breaking the news yourself.
- Be able to defend your policies publicly.
- Change the policy if you are unable to defend it publicly.

Navigating the Budgetary Process: A Case Study

Using the alternative response unit (ARU) request described in "You Are the Fire Officer," let's navigate through the budgetary process. In the case study, the fire department is requesting to establish a new service, using a different type of vehicle to respond to EMS and activated alarm calls.

Developing a Budget Proposal

Because this equipment is a new vehicle providing a new service, the first step in the budget proposal process is to describe what the new unit will do and what the impact will be if the unit is not funded. The description should consider the audience that will review, authorize, and approve the application. Provide enough information in an understandable format to ensure understanding of the goals that the program will meet.

Overview of the Alternative Response Unit Program

The following is the narrative that is part of the budget submission. It describes what the ARU program will do, the impact of the ARU program on current operations, and the consequences if the new program is not funded.

The ARU is a two-person unit that responds to 911 calls for medical assistance, activated smoke or carbon monoxide alarms, and fire sprinkler and fire alarm pull station alerts. The ARU operates every hour of every day out of Fire Station 2 and responds to these calls in place of Engine 2.

The ARU has three roles as follows:

- As an emergency medical responder, appropriately treat the patient until the ambulance arrives. Calls for paramedic assistance if the situation is warranted.
- As a responder to an activated smoke or carbon monoxide alarm, assess the situation, identify the location of the alarm, and perform an investigation using proper personal protective clothing as the situation requires. Call for fire suppression assistance if there is evidence of an active fire or evolving emergency service situation.
- As a responder to a fire sprinkler alarm or pull station, assess the situation, identify the location of the alarm, and perform an investigation using proper protective clothing as the situation requires. Call for fire suppression assistance if there is evidence of an active fire or evolving emergency service situation.

If the ARU program is not funded, the department will continue dispatch Engine 2 to these events. This will continue the accelerated aging of the pumper where it will be required to replaced 5 years before the budgeted replacement apparatus. Engine 2 will miss 40 percent of the structure fires within its response district because it is handling an EMS first responder or activated alarm situation and cannot immediately clear from that incident.

ARU Annual Personnel and Operating Expenditures

This part of the budget submission shows the annual expenses to operate the ARU. Both personnel and operating expenses will be incurred in this example.

Personnel Expenses

The ARU will have a captain, three lieutenants, and four technicians. In this example, the personnel cost is described in two areas. The first area is the direct salary for each position. Most civil service pay classification systems use a pay level that includes steps tied to seniority. Budget protocol requires that new positions reflect the average seniority step that the incumbents occupy. That means that the captain is at seniority step 7; the lieutenant is at seniority step 5; and the technicians are at seniority step 3.

The second area is the calculation of the fringe benefits for each employee position. It is calculated as a percentage of the employee's salary. For the ARU example, fire fighter fringe benefits account for 28.7 percent of the salary. That means that every $100 of salary actually costs the city $128.70 (Menifield, 2017).

Operating Expenses

The cost of operating the ARU unit entails three general areas: vehicle cost, office operations, and personnel training. Jurisdiction fleet maintenance has determined the cost to operate and replace a small, all-wheel-drive emergency services hybrid SUV is $0.84 per mile. This amount covers the fuel and scheduled maintenance of the vehicle. It also includes a per-mile charge that is designed to cover the replacement of the vehicle when it reaches the end of its scheduled service life. Vehicle service life is determined in three ways: years of service, miles accumulated, and specific vehicle cost. It is estimated that the unit will handle 8687 responses and accumulate 17,000 miles per year. The city replaces small emergency service vehicles at 80,000 miles or after 10 years. The ARU will need replacement after 5 years.

Office operations include rent, telecommunication charges, and equipment maintenance. In this example, the ARU is operating out of a fire station, so rent is zero. The office has two hard-wired lines to provide service for the phones, Internet access, and fax machine. The municipality's information technology (IT) office has an annual fee per hard-wired line that covers telephones and Internet broadband access. IT also handles computer maintenance and repairs. As with the vehicle, the annual maintenance and operating charge covers the anticipated annual costs of maintenance, software upgrades, and repairs. It also includes an amount that will allow replacement of the computers on a 5-year cycle. The ARU has two desktop computer workstations, one vehicle-based desktop workstation, and four laptop computers.

Operating expenses also include the expenses of mandated annual fire fighter training. The cost reflects the per-person cost of training, for both the tuition and the position replacement overtime to cover the ARU person who is receiving the training. This expense covers training required by federal or state regulations, such as annual bloodborne pathogen and hazardous materials refresher training.

ARU Capital Budget

Many of the required items would be purchased via federal, state, or regional purchasing contracts. The budget summary shows the entire first-year capital expenditures to start the program, including the response vehicle, the administrative office, the monitoring equipment, and the impact of eight new fire department positions **FIGURE 12-1**.

ARU Vehicle

During the planning process with the fire department shop, the team determines the appropriate type of vehicle. The state offers a range of vehicles for purchase. The following options are considered for the ARU:

- Compact four-door sedan, $15,470
- Midsized hybrid four-door sedan, $17,786
- Small all-wheel-drive hybrid SUV, emergency services, $24,765
- Large four-wheel-drive SUV, emergency services, $33,948

The fire apparatus shop recommends that the ARU vehicle should be the same as other staff and command vehicles, a small all-wheel-drive hybrid SUV. In addition to the vehicle, equipment is needed to convert a small SUV into a fire department response vehicle, including emergency lights/siren, mobile fire radio and computer, two portable fire radios, two self-contained breathing apparatus (SCBA) units, fire extinguisher, and semi-automatic cardiac defibrillator. Depending on the local budget process, these items may need to be identified in a line-item format consistent with budget documentation protocol. More than half of the ARU vehicle cost is devoted to the specialized equipment that goes on or into the vehicle.

ARU Administrative Office

The workstations will go into an available space at Fire Station 2. The capital equipment represents all of the durable items needed to establish this office: desks, chairs, computers, and printer. The budget summary document has no cost of rent or real estate.

ARU Uniforms and Protective Clothing

The expense of fire fighter protective clothing and the initial uniform issue are included in the initial-year capital expenditures. Because the ARU program creates the new positions, the expense shows up here.

ARU Cost Recovery and Reduction

Any new or expanded service needs to show its impact on the annual budget. This project has an annual operating and personnel cost of $491,512. Points to consider when demonstrating impact include the following:

- Does making that expenditure recover costs or reduce the budget in any way?
- Does the new service provide an additional service or value to the community? (Watson and Hassett, 2015)

The motivation of the ARU was to avoid a premature replacement of the 11-year-old engine assigned to Station 2 by reducing its call volume by 20 responses per day, or 50 percent of the current workload. The maintenance and operating cost for a pumper is $3.75 per mile and the vehicle replacement is calculated at $2800 per month. The "back-of-the-envelope" calculation shows that the pumper monthly maintenance costs will be reduced by $5,313 per month or $63,756 per year. If the pumper had to be replaced as a 15-year-old instead of a 20-year-old rig, the department would incur $170,800 in unplanned capital expenses, as the replacement fund for the pumper was calculated for a 20-year life span. This puts a small dent in the annual operating and personnel cost of $491,512 to staff and operate ARU 2. Another approach is to determine the improved response time for Engine 2 and the reduced number of times a farther-away fire company needed to respond to an incident in Engine 2's district.

Fire Department - Operations
New program: **Alternative Response Unit**

Year 1: Capital Expenditures		Unit	Total	Source or comments
ARU Vehicle				
1	Small AWD SUV Hybrid, emergency services	$ 24,765	$ 24,765	State vehicle contract
1	Mobile fire radio and mobile computer terminal	$ 15,525	$ 15,525	Communications
1	Front and rear camera driving monitor	$ 349	$ 349	Apparatus
1	Fire emergency lighting package	$ 2,765	$ 2,765	Apparatus
1	Safety striping and graphics	$ 711	$ 711	Apparatus
2	Thermal imaging camera	$ 4,759	$ 9,518	Logistics
2	Forcible entry tools	$ 639	$ 1,278	Logistics
2	Self contained breathing apparatus	$ 3,511	$ 7,022	Logistics
12	Individual SBCA facepieces	$ 411	$ 4,932	EMS Division
2	Ambulance "Aide" bags, equipped	$ 366	$ 732	EMS Division
2	EMT Airway/Bag-Valve Mask kits	$ 333	$ 666	EMS Division
1	Portable oxygen delivery system, "E" bottle	$ 750	$ 750	EMS Division
1	Semi-automatic defibrillator	$ 911	$ 911	EMS Division
1	Fire company tool box, complete	$ 1,376	$ 1,376	Logistics
1	Safety step ladder	$ 226	$ 226	Logistics
1	Fire sprinkler control kit	$ 871	$ 871	Logistics
1	Water fire extinguisher	$ 163	$ 163	Logistics
2	Portable two-way radios	$ 7,746	$ 15,492	Communications
	Total APU vehicle capital cost:		**$ 88,052**	
ARU Office at Fire Station 2				
2	Cubicle work stations	$ 877	$ 1,754	Facilities catalog
2	Computer work stations - model A	$ 3,200	$ 6,400	IT catalog
2	Smartphones	$ 533	$ 1,066	IT catalog
1	Workstation scanner/printer/fax - model A	$ 1,457	$ 1,457	IT catalog
1	File/storage cabinet	$ 220	$ 220	Facilities catalog
	Total APU office capital cost:		**$ 10,897**	
New uniform positions - maintain a 2 person team 24 hours every day				
8	Protective clothing ensembles	$ 3,476	$ 27,808	Warehouse
4	Initial fire fighter uniform issue	$ 865	$ 3,460	Warehouse
4	Initial fire fighter uniform issue	$ 709	$ 2,836	Warehouse
	Initial uniform/PPE cost:		**$ 34,104**	
	Year 1 expense:		**$ 133,053**	
Annual Expenses				
Personnel				
1	Captain at pay step 28-7	$ 66,077	$ 66,077	Salary
3	Lieutenant at pay step 26-5	$ 46,785	$ 140,355	Salary
4	Technicians at pay step 22-3	$ 39,219	$ 156,876	Salary
	Fringe benefits		$ 104,269	28.7% of salary
	Total annual personnel expenditures:		**$ 467,577**	
Operational				
1	APU vehicle maintenance/operating/replacement	$ 14,280	$ 14,280	$.84 per mile @ 17K a year
2	Telecommunications fee per line	$ 375	$ 750	Intranet/phone service
2	Office supplies and paper per work cubicle	$ 365	$ 730	Logistics
8	Mandated fire department training	$ 275	$ 2,200	Academy, per position
5	Computer maintenance and replacement	$ 570	$ 2,850	Information Technology (IT)
2	Portable radio maintenance and replacement	$ 1,000	$ 2,000	Communications
1	Mobile fire radio / mobile computer maintenance/replace	$ 1,125	$ 1,125	Communications
	Total annual personnel expenditures:		**$ 23,935**	
	Annual operating expenses:		**$ 491,512**	
	Year One total expenditure:		**$ 624,565**	

FIGURE 12-1 Summary budget submission sheet for the ARU program.
Courtesy of Mike Ward.

The Purchasing Process

Most agencies have established a standardized method for making purchases. The fire officer must understand these policies and procedures of the organization—failure to do so can have significant consequences for the organization. Because most fire departments are either political entities or nonprofit organizations, they are accountable to the public for the effective use of funds. Audits are used to monitor expenditures and identify potential or actual budget problems. Purchasing violations are sometimes found during the audit process.

Petty Cash

Policies and procedures for purchasing typically vary based on how much an item costs. For example, the fire department may allow any items costing less than $100 to be purchased directly in a noncompetitive manner. Often this is done through a petty cash system. In the petty cash system, a member of the fire department acts as the custodian of an amount of cash that is provided by the organization. As members of the fire department purchase small-dollar-value items, the petty cash custodian reimburses them for the expense in exchange for the sales receipt.

When the petty cash custodian is low on cash, the accumulated receipts are turned over to the finance department in exchange for a like amount of cash. In small organizations, the receipts are turned over to the governing body, such as the fire board, which approves the purchases and then writes a check to the petty cash custodian. This check is then cashed and the funds are placed back in the petty cash box. The petty cash custodian should always have the authorized amount on hand in the form of either cash or receipts. The petty cash account is regularly audited to ensure that it is being used properly.

Some personnel tend to think of the petty cash account as an endless fund, which is not true. On each receipt, the appropriate notation is made to identify which account, or part of the budget, is covering the expenditure. Petty cash allows for small purchases to be made without having to obtain purchase orders. Each time petty cash is spent, it lowers the account balances.

Purchase Orders

A purchase order is a method of ensuring that a budget account contains sufficient funds to cover a purchase. As with petty cash expenditures, most organizations place a limit on the maximum amount of a purchase order; for example, an organization may allow purchase orders to be used for purchases up to $2000.

The fire officer's role is to acquire the item at the most reasonable cost to the organization. To facilitate this outcome, most organizations allow phone bids for these items. For example, if a fire officer needs to purchase a pair of tires, he or she first determines which size is needed and then calls at least three sources to get prices on the tires. The officer takes the lowest price and completes a purchase order (commonly called a "PO"). The purchase order typically requires an authorizing signature by an official who has control over the budget area from which the funds are paid. In this example, it might be the chief. The purchase order is then entered into the purchasing system, which debits the maintenance account.

The fire officer then takes the purchase order to the location that has the tires and the sale is transacted. A copy of the purchase order is given to the vendor, who attaches the sales receipt to it and sends it to the fire department's finance department for reimbursement. In effect, the vendor allows the fire department to charge the merchandise until the purchase order is paid. For its part, by issuing the purchase order, the fire department indicates to the vendor that there are funds available for the purchase and the purchase has been approved. Not all vendors accept purchase orders, however (Menifield, 2017).

Requisitions

For purchases that exceed a predetermined amount, such as $2000, a requisition is required, rather than a purchase order. This occurs for large purchases, such as a new hydraulic extrication tool. A requisition has even more stringent requirements than a purchase order.

The requisition differs from a purchase order in that the exact amount of the purchase is not known. Instead, the department requests that a specific amount of funds be encumbered, or set aside from the budget account, that will more than cover the cost of the purchase. In the case of the rescue tool, the department might set aside $15,000. A bidding process will be followed. Once it is complete, the requisition will allow for payment of the item.

The Bidding Process

Once funds have been encumbered, the fire department must go about the process of making the purchase. This is done in one of two ways. For smaller items, the fire department may develop specifications for bids. The second method, which is often used for very large or complex purchases, is a request for proposal (RFP).

To purchase an item based on specifications (commonly known as "specs"), the fire department writes up exactly what it desires in the product. For example, it might specify that a portable power unit for a fire truck must not exceed a size of 18 × 14 × 21 in. (46 × 36 × 53 cm). With a spec, every requirement must be met for the vendor to be considered. In this example, if a vendor wanted to bid but its power unit was 19 × 13 × 20 in. (48 × 33 × 51 cm) in size, its bid would be rejected. This type of restriction may be important if the compartment in which the power unit will be placed will accommodate only the specified size. Some fire departments use this method to ensure that only one vendor or product can meet the specifications; however, this is an illegal and unethical practice. Specs should be used to identify which criteria must be met for the purchase to be most effective and efficient for the fire department.

Once developed, a bid sheet listing the specifications is sent to vendors for pricing. Most organizations have a bidding list of vendors that wish to do business with them. The vendors place their names and addresses on the list so each time a bid comes up, a copy is sent to them for consideration. Local administrative rules may also require that a notice of bid be placed in the local paper. Other organizations place notices of bids in trade magazines and papers. The Internet also is becoming a method of getting information out to the public about upcoming bid processes.

Often, the use of specifications eliminates virtually every potential vendor or greatly increases the price because only more expensive items meet the specifications. Sometimes, a bid specification limits the bidders. For example, the purchase of a mobile computer system for use in the fire apparatus would likely use the RFP method rather than the bid specification method because of the complexity of the technology.

An RFP provides a formal process to solicit goods or services in a procedure that is determined by the authority having jurisdiction. In an RFP, the fire department gives the general information about what is desired and allows the vendor to determine how it will meet the need. For example, the mobile computer RFP might require that the system provide coverage to 99 percent of the city. One vendor might opt to place 10 antenna towers to deliver the required coverage, whereas another might opt for only 7 antenna towers but place them at a higher elevation. Both ideas would meet the need of the fire department. The fire department also determines how the proposals will be evaluated. For example, the price might be worth 50 percent of the overall consideration, whereas delivery time might be worth 5 percent. Once the RFP is developed, it is sent out to potential vendors for bidding.

Whether the process uses an RFP or bid specifications, vendors are given a specific amount of time to reply. Their replies are placed in a sealed envelope indicating the bid number on the outside of the envelope. At a time that was outlined in the bid specifications or RFP, all bids or proposals are opened in public view. With a bid process, the contract is then awarded to the lowest bidder and the amount is attached to the requisition for payment when delivery is made.

When an RFP is used, each proposal is evaluated to determine whether it best meets the needs of the department. The proposals are typically assessed based on performance as well as price. Once each performance requirement has been addressed and the price is considered, the RFP is awarded to the vendor with the highest score. The price of the winning proposal is added to the requisition for payment when the item is delivered (NASPO, 2019).

In recent years, there has been a trend toward developing intergovernmental or interagency cooperative arrangements to streamline purchases and leverage greater discounts. The federal government, for example, issues many purchasing contracts for a wide range of equipment. Most vendors honor these rates for other governmental agencies, which in turn streamlines the process because the bidding process has already occurred. Local organizations can also band together to purchase common items. For example, a group of fire departments may jointly develop an RFP to purchase SCBA. This practice ensures a lower bid because the group is purchasing items in greater quantity, and it allows all agencies to use like equipment (Kaiser, 2011)

Revenue Sources

The revenue stream depends on the type of organization that operates the fire department and the formal relationship between the organization and the local community. There is a wide diversity both in organizations and in revenue stream sources. Each type of organization has a different process for obtaining revenue and authorizing expenditures.

Most municipal fire departments operate as components of local governments, such as towns, cities, and counties. A fire district is a separate local government unit that is specifically organized to collect taxes that support fire protection. In some areas, fire departments are operated as regional authorities or an equivalent structure.

Grants

Municipal and volunteer fire departments can also obtain funds by applying for grants. One of the larger sources of grant funds is the Assistance to Firefighters Grant (AFG) program, which distributes federal funds to local jurisdictions.

Although the AFG program is the largest fire service–oriented program in the United States, hundreds of smaller grant programs are offered by the federal government, state governments, private institutions, nonprofit organizations, and for-profit companies (USFA, 2012). Grants are competitive and require the local fire department to meet specific eligibility and documentation requirements. Many grant programs also require the local jurisdiction to provide a percentage of the total funding in the form of matching funds.

Some state agencies offer grants to local responders from specially designated funds, such as a fee on motor vehicle permits, a surcharge on traffic fines, or a tax on fire insurance premiums.

The U.S. Fire Administration provides a four-step method to develop a competitive grant proposal:

1. Conduct a community and fire department needs assessment.
2. Compare community needs to the priorities of the grant program.
3. Decide what to apply for.
4. Complete the application.

The process of following the four steps in developing a grant proposal requires compliance with the grant instructions. To qualify, applicants must follow a specific set of procedures and provide a detailed level of financial and operational information. These requirements may seem excessive and intrusive, but the grant administrators have to ensure that the money is spent wisely and within legislative guidelines.

The AFG program applications are evaluated by a peer-based committee that ranks them on the basis of preestablished criteria. To be approved, the application must make a request that matches the goal of the grant program, show that the proposal is cost efficient, and demonstrate strong community and financial need for the requested goods or services (Cotter, 2017).

When seeking grant funding, the first step is to determine whether a grant program exists that covers your problem area. Once you have identified the program, determine whether your organization is eligible under the rules of the grant program and which process is used to apply for the grant.

A grant application must describe how the department's needs fit the priorities of the grant program. The grant applicant needs to analyze the community, conduct a risk assessment, evaluate the existing capabilities of the fire department, and identify the department's needs.

There is strong competition for the limited amount of grant funds available, so the applicant must provide a sound justification for why the organization should receive the funding. Use the information gathered in the needs assessment in a narrative to describe the problem in a concise but detailed manner.

FIRE OFFICER TIP

Justifying New Gas Monitors

A fire department is preparing a grant request to purchase four-channel gas meters to replace flammable vapor detectors assigned to ladder and rescue companies. The old detectors can measure the concentration of a flammable vapor only in relation to the lower explosive limit, whereas the newer detectors also measure the percentage of oxygen and carbon monoxide in the atmosphere. As part of the needs assessment to justify a grant for these new monitors, the fire department would describe the increase in service calls to check on carbon monoxide alarm activations.

In addition to the needs assessment, the grant application must demonstrate the department's financial need and explain why it needs outside assistance to acquire these resources. The application should show that efforts have been made to obtain the funds from other sources. In describing the current financial situation, include factors such as an eroding tax base, expanding community growth, local/state legislation that restricts taxes, and any significant local economic problems, such as the loss of a major employer.

The application also needs to show that this solution provides a benefit at the lowest possible amount of funding. Factors include collaborating with other organizations to share expensive equipment.

Nontraditional Revenue Sources

Fire departments have solicited donations from a targeted business or industry for a specific purpose. A resort hotel–motel association provided funding to purchase an additional aerial platform truck that serves the high-rise hotel district. Equipment for a hazardous materials team might be donated by a company in the community that produces hazardous materials. One fire department offered to sell advertising on its vehicles.

Reimbursement for Response

Local or state regulations may allow a fire department to recover the extraordinary expenses incurred in responding to a hazardous materials incident, particularly decontamination, spill clean-up, and recovery costs. The department can invoice the responsible party for extraordinary charges, such as overtime and replacement of expendable supplies (e.g., absorbents and neutralizing agents). In addition, the carrier may be responsible for the expense of repair or replacement of durable equipment, such as contaminated monitoring equipment, chemical protective suits, fire fighter protective clothing, and fire suppression tools.

Cost recovery also applies to some special services supplied by fire departments. For example, if a movie production company is shooting a film that involves pyrotechnics, the fire department might provide an engine company to stand by for several hours. The cost of overtime to staff the unit and an hourly rate for the use of the apparatus could be charged to the production company.

Expenditures

The annual budget for the fire department describes how the available funds are spent during the year. In the case of a municipality, the fire department budget is usually a part of the overall budget for that governmental entity. The format of the budget should comply with recommendations of the **Governmental Accounting Standards Board (GASB)**, which is the standards-setting agency for governmental accounting (Morgan, Robinson, Strachota, and Hough, 2015).

The local government budget typically includes a system of accounts that classifies all expenditures within certain categories and complies with the generally accepted accounting principles as outlined by the GASB. In most cases, the accounting system uses a database system to support a line-item budget. A **line-item budget** is a format in which expenditures are identified in a categorized line-by-line format. The accounting system may have a complex numbering system, such as 02-12301-876543, that includes a category for every type of expenditure. In this example, the line item describes the following:

- **02:** The first two numbers identify the fund. In the example, the 02 indicates that this money is coming from the general fund.
- **12301:** The second part (the five-digit number) identifies the department and the division or subdivision where the money is being spent. Called the "object code" in some systems, this part of the example says that the money is allocated to the Fire Department (12) and to the Suppression Division (301).
- **876543:** The third part (the six-digit number) provides a detailed description of what the money is being used for. Called the "subobject code" in some systems, this one says that the money is to be used for replacement fire hose.

The accounting system allows budget analysts and financial managers to keep track of expenditures throughout a municipal government. For example, law enforcement, public works, schools, and fire departments all purchase work uniforms. The database system would allow a budget analyst to determine how much the municipality spent on all uniforms and to break down expenditures by department and type of uniform. This could be valuable information when one is considering the savings that could be obtained by purchasing all uniforms from one vendor.

Fire department expenditures are generally divided into three general areas: personnel costs, operating costs, and capital expenditures. The accounting system allows budget analysts and managers to keep track of how much money is spent on fire fighter salaries (personnel costs), fuel for vehicles (operating costs), and construction of new fire stations (capital expenditures).

Personnel Expenditures

More than 90 percent of a career fire department's budget is likely to be allocated to salaries and benefits. This includes fringe benefits, such as pension fund contributions, worker's compensation, and life insurance. Civil service regulations, local administrative decisions, and labor contracts often determine the cost of these benefits. For example, the worker's compensation rate is determined by the history of claims and payments. Fire fighters have a high worker's compensation rate when compared to other municipal employees.

Fringe benefits also reflect the estimated cost of providing sick and annual leave benefits to the employee. Many municipalities have a sliding scale, in which the amount of leave earned per pay period increases as the employee gains seniority. The base cost of this fringe benefit equals the amount of the employee's accrued leave. In many cases, the actual cost to the fire department also includes overtime to pay for another employee to fill the vacancy (Menifield, 2017).

Operating Expenditures

The operating budget covers the basic expenditures that support the day-to-day delivery of municipal services.

Uniforms, protective clothing, telephone charges, electricity for the fire stations, flashlight batteries, fire apparatus maintenance, and toilet paper are all examples of purchases that would be classified as operating expenses. These funds are allocated in categories that allow for some flexibility throughout the year. For example, a fire department might be authorized to spend $250,000 on protective clothing during the year. The actual numbers of coats, pants, helmets, boots, and gloves that would be purchased would depend on the needs and the cost per item at the time of purchase. When a budget needs to be trimmed, operating expenditures are usually the first area to be considered as a target for cost-cutting efforts. For example, employee training, travel, and consulting expenses are often eliminated when revenue falls below expectations.

The cost of operating a vehicle is calculated on a per-mile basis that includes fuel, scheduled maintenance, and anticipated repairs. Maintenance costs for fire department vehicles are high because they are loaded with additional equipment, such as emergency lighting and siren systems, two-way radios, and computers. A large fire apparatus, which is expensive to repair and accumulates low annual mileage, may have an operating cost of several dollars per mile. If the actual fleet management costs exceed the budget projections because of high fuel costs or unanticipated repairs, the additional amount is sometimes charged back to the fire department at the end of the fiscal year. That adjustment is included in the third-quarter adjustments.

As part of their operating expenditures, many municipalities include a per-mile vehicle replacement fee applied to automobiles, police cars, buses, ambulances, and light-duty vehicles. The precise amount is calculated by estimating the anticipated replacement cost and the projected number of years that the vehicle will be used. This accounting technique is also used by fire departments that have dozens of pumpers and aerial apparatus and a fire apparatus replacement schedule. In most cases, the replacement of a pumper, aerial apparatus, or specialized suppression rig will be a capital item.

Replacement uniforms and protective clothing are another expense included in the operating budget **FIGURE 12-2**. For example, the fire department might budget a lump sum of $150 per fire fighter to cover the annual cost of cleaning and repairs to protective clothing and up to $400 per fire fighter for uniforms.

Mandated training, such as hazardous materials and cardiopulmonary resuscitation recertification classes, is also factored into the fire department's operating costs. That figure includes the cost of the training per fire fighter per class. The cost to pay another fire fighter overtime to cover the position during the mandated training could also be included.

FIGURE 12-2 Replacement protective clothing is an operating expense calculated in the budget.
© Jones & Bartlett Learning. Photographed by Glen E. Ellman.

FIGURE 12-3 Two examples of equipment purchased as capital items. **A.** Computer. **B.** Tool.
© Jones & Bartlett Learning. Photographed by Glen E. Ellman.

Capital Expenditures

Capital expenditures comprise purchases of durable items that cost more than a predetermined amount and will last for more than one budget year. Local jurisdictions differ on the cost and service period used to define capital expenditures. Items such as computers, hydraulic rescue tools, washing machines, and SCBA units are examples of equipment purchased as capital items **FIGURE 12-3**. In

> **FIRE OFFICER TIP**
>
> **Remember Annual Expenses**
>
> When discussing or responding to a budget proposal, remember to focus on the annual expenses in addition to the initial purchase cost. The annual costs fall into two areas: continuity and personnel.
>
> Consider the purchase of semi-automatic cardiac defibrillators for three fire companies. What happens when one of the devices breaks down? Continuity considers the need for a spare unit to be used when one of the new devices is broken or unavailable, as well as the cost of maintenance and repairs. Additional costs would include the initial training of fire fighters to use the devices, plus annual recertification classes.
>
> Personnel costs are the largest expense for career fire departments. An additional ladder truck might cost $800,000 to purchase, but the annual cost of additional personnel to provide staffing for that vehicle could easily exceed $1 million. This cost could be difficult to justify for a company that will make fewer than 100 responses per year.

general, the municipality establishes a unique inventory record that documents the purchase price, source, assignment, and disposal of each capital item.

Sedans, pick-up trucks, and sport-utility vehicles (SUVs) are usually part of the fire department's capital budget. Specialized vehicles, such as heavy fire apparatus, are also capital equipment, but they are often included in a special section of the capital budget because of their cost and complexity. In larger departments, replacement vehicles are purchased from a special set-aside fund, and only additional vehicles are included in the supplemental capital budget.

Capital improvement projects encompass the construction, renovation, or expansion of municipal buildings or infrastructure. These expensive projects are often funded through long-term loans or bonds issued by the municipality.

Bond Referendums and Capital Projects

Large capital improvement projects, such as new fire station construction or major renovations, are funded through bond programs. A bond is a certificate of debt issued by a government or corporation; the bond guarantees payment of the original investment plus interest by a specified future date. The same type of funding mechanism is also used to build roads, water distribution systems, sewer systems, parks, libraries, and similar public facilities. Voters must approve this type of expenditure through a local referendum. If the voters approve, the municipality is authorized to borrow money from investors by issuing bonds up to a set amount. The authorization to sell the bonds is usually valid over a period of 5 to 10 years.

The bonds are repaid over a period of 10 to 30 years and return a fixed interest rate to investors. Funds obtained by issuing bonds are classified as a special revenue source and require compliance with a different set of accounting controls and reporting procedures. Annual status reports are required, and specific restrictions are placed on which items can be purchased with the bond revenue. If a city is issuing $100 million in bonds to be repaid over 20 years, a fraction of a percentage difference in the interest rate can have a significant impact to the taxpayers.

Bond funding allows taxpayers to actually pay for the facility while it is being used, just as a mortgage allows a family to live in a home while they are paying off their loan. The repayment period should be at least equal to the anticipated life span of the facility. Bonds are repaid by collecting a special property tax or, if the facility will produce revenue, such as an airport or convention center, the bonds may be repaid from future income.

When bonds are used to build a new fire station, the authorized amount often includes the land, site improvements, building construction, permits, architectural and engineering fees, and furniture and equipment that go into the building, as well as the apparatus and equipment that will be assigned to the new station.

Occasionally, a bond program is used to fund a large fire department apparatus purchase. For example, the St. Louis Fire Department used a bond to cover the cost of an entire fleet of new apparatus in 1986. That large purchase allowed for the replacement of all existing pumpers and ladders with 30 new quints to establish the "Total Quint Concept." In fiscal year 2000, the city authorized another bond program so that it could purchase 30 replacement quints.

Recommending Change

The fire officer has the best opportunity to dramatically improve the safety and effectiveness of fire operations by implementing changes in policies and procedures. Reginald Freeman and David Matthew point out that fire service cultures are ever changing, driven by internal and external forces. They assert that culture change is a process, and there are six elements for success as follows (Freeman and Matthew, 2018):

- Transparency
- Teaching
- Inclusion
- Incentivizing

- Practicing ethics
- Procedural justice

It is the fire fighters, company officers, and chiefs that will make or break a change effort.

Identify the Problem

Filling the gap between what is desired and the current state is where the problem exists. This gap is identified after a postincident briefing, a near-miss incident, or a direct observation by the fire officer. Description of the gap should include description of external and internal aspects of the gap.

Explore Alternatives and Select Solution

The fire officer brainstorms alternative ways to close the gap from the current state to the desired state. Engage others in this process. The after-incident review process covers six areas: communication skills, teamwork, task allocation, critical decision making, situational awareness, and debriefing. Covering these areas during the brainstorming session assists in completely considering the dimensions of the gap.

Implement and Evaluate

The team will develop a proposal to address the gap. The captain will identify the organizational issues in implementing a new policy or procedure using the process put in place by the fire department.

Some departments will conduct a pilot program, where selected fire companies will utilize the new procedure, or the department will temporarily implement a new policy. During this pilot period, the captain will closely monitor the results and see if the new policy or procedure has closed the gap.

You Are the Fire Officer Conclusion

The ARU program is described as a new resource that will handle EMS first responder and activated fire alarm systems in place of Engine 2. The captain described the need using two criteria: accelerated deterioration of Engine 2 in responding to 34 EMS and fire alarm events every day and requiring 40 percent of the structure fire events in Engine 2's district to be handled by another fire company because Engine 2 is committed to an EMS or activated fire alarm incident. The ARU is proposed as a pilot program of alternative response to non-fire incidents.

The first-year costs include everything needed to get the program started: vehicle, offices, fire fighter equipment, and tools. The continuing year considerations include salaries with fringe benefits, fire fighter annual training, maintenance on all equipment, replacement fund for all equipment, and day-to-day costs of fuel and office supplies.

The value of the new service will be the improvement of response times of Engine 2 to fire incidents and the reduction on the workload on other fire companies. There is a projected improvement general response time to all incidents in Engine 2's district, the reduction of fire pumper maintenance, and the extension of the service life of the pumper.

After-Action REVIEW

IN SUMMARY

- Written communication is needed to document both routine and extraordinary fire department activities, establishing institutional history and serve as the foundation of any activity that the fire department wishes to accomplish.
- Informal communications include internal memos, e-mails, instant messages, and messages transmitted via mobile data terminals. This type of communication is used primarily to record or transmit information that may not be needed for reference in the future.
- Formal communication is an official fire department document printed on business stationery with the fire department letterhead.
- Standard operating procedures are written organizational directives that establish or prescribe specific operational or administrative methods to be followed routinely for the performance of designated operations or actions.

- Standard operating guidelines are written organizational directives that identify a desired goal and describe the general path to accomplish the goal, including critical tasks or cautions. Like SOPs, SOGs are formal, permanent documents that are published in a standard format and remain in effect until they are rescinded or amended.
- General orders are formal documents that address a specific subject, policy, condition, or situation.
- An interrogatory is a series of written questions asked by someone from an opposing legal party where the fire department provides written answers under oath and produces any associated documentation.
- Reports should match the needs of the intended audience, be accurate, and present the necessary information in an understandable format.
- A recommendation report is a document that advocates a particular action or decision, sometimes utilizing a chronological section that provides the necessary background so the reader can make a decision.
- A budget proposal for a new service will need to describe what the new service will do and what the impact will be if the unit is not funded.
- A budget proposal must include all of the first-year expenditures as well as the annual personnel and operating expenses.
- Annual operating expenses include vehicle cost, office operations, and personnel training.
- A purchase order is a method of ensuring that a budget account contains sufficient funds to cover a purchase.
- More expensive items require a requisition process that may require a bid or a request for proposal.
- In a request for proposal (RFP), the jurisdiction writes up exactly what it desires in the product in detailed specifications. Vendors respond to the RFP.
- Each RFP response is evaluated to determine whether it best meets the needs of the department.
- Municipal and volunteer fire departments can obtain funds by applying for grants from federal, state, or private resources.
- Grant applications often require a needs assessment and description of the fire department's financial situation.
- Budget format should conform to the Governmental Accounting Standards Board (GASB).
- A line-item budget identifies expenditures in a categorized line-by-line format.
- Personnel fringe benefits includes pension fund contributions, worker's compensation, and life insurance. These are calculated as a component of the personnel expenditures.
- Operating expenditures cover the basic expenditures that support the day-to-day delivery of municipal services.
- Capital expenditures comprise purchases of durable items that cost more than a predetermined amount and will last for more than one budget year.
- A bond is a certificate of debt issued by a government or corporation; the bond guarantees payment of the original investment plus interest by a specified future date. Large capital improvement projects are funded through bond programs.
- Recommending change starts with identifying the problem: the gap between what is desired and what the current state is.
- Recommending change includes exploring alternatives to close the gap.
- Once change is implemented, the team needs to monitor and evaluate the results.

KEY TERMS

Formal communication An official fire department letter or report that is presented on stationery with the fire department letterhead and generally is signed by a chief officer or headquarters staff member.

General orders Short-term directions, procedures, or orders signed by the fire chief and lasting for a period of days to 1 year or more.

Governmental Accounting Standards Board (GASB) An organization whose mission is to establish and improve the standards of state and local governmental accounting and financial reporting, thereby resulting in useful information for users of financial reports, and to guide and educate the public, including issuers, auditors, and users of those financial reports.

Informal communications Internal memos, e-mails, instant messages, and computer-aided dispatch/mobile data terminal messages. Informal reports have a short life and may not be archived as permanent records.

Interrogatory A series of formal written questions sent to the opposing side of a legal argument. The opposition must provide written answers under oath.

Line-item budget A budget format in which expenditures are identified in a categorized line-by-line format

Recommendation report A decision document prepared by a fire officer for the senior staff. Its goal is to support a decision or an action.

Standard operating guidelines (SOGs) Written organizational directives that identify a desired goal and describe the general path to accomplish the goal, including critical tasks or cautions.

Standard operating procedures (SOPs) Written organizational directives that establish or prescribe specific operational or administrative methods to be followed routinely for the performance of designated operations or actions.

REFERENCES

Civil Law Self-Help Center (CLSHC). 2019. *Responding to the Other Side's Request for Information*. Las Vegas, NV: Regional Justice Center.

Cotter, Heather. 2017, March 28. "Five Grant Writing Tips for Fire, EMS Agencies: Quality Writing, Demonstrated Need and Staying Within the Financial Parameters Are Critical in Determining Whether or Not Your Agency Wins an Award." *Praetorian Digital*. Accessed July 24, 2019. https://www.firerescue1.com/fire-grants/articles/225621018-5-grant-writing-tips-for-fire-EMS-agencies/.

Freeman, Reginald, and David Matthew. 2018, June 12. "Transformational Change in the Fire Service." Center for Public Safety Excellence. Accessed July 25, 2019. https://cpse.org/2018/06/12/transformational-change-and-the-fire-service/.

Kaiser, Frederick M. 2011, May 31. *Interagency Collaborative Arrangements and Activities: Types, Rationales, Considerations*. [R41803] Washington, DC: Congressional Research Service.

Menifield, Charles E. 2017. *The Basics of Public Budgeting and Financial Management: A Handbook for Academics and Practitioners, Third edition*. Lanham, MD: Hamilton Books.

Morgan, Douglas, Kent S. Robinson, Dennis Strachota, and James A. Hough. 2015. *Budgeting for Local Governments and Communities*. New York, NY: Routledge.

National Association of State Procurement Officials (NASPO). 2019. *State and Local Government Procurement: A Practical Guide, Third edition*. Lexington, KY: National Association of State Procurement Officials

Nelson, Kimberly L., and Carl W. Stenberg. 2019. *Managing Local Government: An Essential Guide for Municipal and County Managers*. Thousand Oaks, CA: Sage Publications.

Statter, Dave. 2019. "Effective Communications in a Digital Age." *STATter911 Communications*. Presented May 2019 at the Virginia Fire Prevention Association conference in Virginia Beach.

U.S. Fire Administration (USFA). 1999. "Developing Effective Standard Operating Procedures for Fire and EMS Departments, FA-197." Emmitsburg, MD: Federal Emergency Management Agency.

U.S. Fire Administration (USFA). 2012, April. "Funding Alternatives for Emergency Medical and Fire Services, FA-331." Emmitsburg, MD: Federal Emergency Management Agency.

Watson, Douglas J., and Wendy L. Hassett, eds. 2015. *Local Government Management: Current Issues and Best Practices*. New York, NY: Routledge.

Fire Officer in Action

1. Which of these are written organizational directives that prescribe operational methods to be followed routinely?
   A. General orders
   B. Standard operating procedures
   C. Announcements
   D. Standard operating guides

2. Which of these is a document that advocates a particular action or decision?
   A. Recommendation report
   B. White paper
   C. Incident report
   D. Operations guide

3. Which of these is a document exactly describing the specifications of an item or a service desired by the fire department and publicly shared?
   A. Request for proposal
   B. Purchase order
   C. Capital budget request
   D. Requisition

4. When applying for a grant to fund a program, the fire department must:
   A. calculate the efficiency of local government financial activity.
   B. critique the value of the other applications submitted.
   C. document the local and state political support for this grant application.
   D. provide sound justification for why the organization should receive the funding.

Fire Captain Activity

This chapter describes the administrative and communication aspects of the fire department

NFPA Fire Officer II Job Performance Requirement 5.4.1

Develop a policy or procedure, given an assignment, so that the recommended policy or procedure identifies the problem and proposes a solution.

Application of 5.4.1

1. Locally maintained private bridges that go from a county road to a single-family home or farm building are too narrow for fire apparatus to cross over. There are 27 such bridges in the district, and some of the houses are over one quarter of a mile from the bridge. None of the bridges can support the weight of a pumper with 750 gal (2800 L) of tank water. Identify the issues and propose a solution.

2. The venting of a mid-rise apartment building creates a haze of condensation that generates frequent good intent but false fire calls in the winter. Identify the issues and propose a solution.

NFPA Fire Officer II Job Performance Requirement 5.4.3

Describe the process of purchasing, including soliciting and awarding bids, given established specifications, in order to ensure competitive bidding so that the needs of the organization are met within the applicable federal, state/provincial, and local laws and regulations.

Application of 5.4.3

1. Using information from a jurisdiction you are familiar with, describe the process that the jurisdiction uses to purchase 20 replacement SCBA units.

2. Using information from a jurisdiction you are familiar with, describe the process the jurisdiction uses to replace a fire pumper or ambulance.

NFPA Fire Officer II Job Performance Requirement 5.4.6

Develop a plan to accomplish change in the organization, given an agency's change of policy or procedures, so that effective change is implemented in a supportive manner.

Application of 5.4.6

1. Using the information in the chapter on the alternative response unit (ARU), develop a SOP for the ARU response to an activated sprinkler alarm in a commercial building.

2. Using the information in the chapter on the ARU, develop a SOP for the ARU response to mass-casualty EMS incident.

 *Access Navigate for flashcards to test your key term knowledge.*

CHAPTER 13

Fire Officer II

Fire Cause Determination

KNOWLEDGE OBJECTIVES

After studying this chapter, you will be able to:

- Identify the common causes of fire. (pp. 341–343)
- Explain when to request a fire investigator. (p. 335)
- Describe how to find the area of origin of a fire. (**NFPA 1021: 5.5.1**) (pp. 335–336)
- Discuss the legal considerations of fire cause determination). (p. 347)

SKILLS OBJECTIVES

After studying this chapter, you will be able to:

- Determine the area of origin of a fire. (pp. 335–336)
- Demonstrate how to secure the scene to prevent unauthorized persons from entering the incident scene. (pp. 335–336)

ADDITIONAL NFPA STANDARDS

- NFPA 921, *Guide for Fire and Explosion Investigations*
- NFPA 1033, *Standard for Professional Qualifications for Fire Investigator*

You Are the Fire Officer

A mechanic working in an illegal and unregistered garage accidently short-circuits the battery pack of an electric car he was working on. The cascades of sparks burned the car, the garage, and a nearby fence. The mechanic suffered second-degree burns to his face and arms. The owner of the car was also moderately burned.

1. Does this incident require on-scene attention of a fire investigator?
2. Who determines if a crime has occurred?
3. What role does the fire company officer play in fire cause determination?

 Access Navigate for more practice activities.

Introduction

The company officer needs to conduct a fire cause investigation in order to obtain an idea of how the fire started. This evaluation is undertaken once the fire is extinguished and before the property is turned back over to the owner. One of the best starting points for efforts to prevent future fires is understanding the causes of fires that have occurred in the past. In addition, a formal fire investigation is important to establish the responsibility for fires that could have been caused by criminal acts, negligence, or fire code violations. The results of an investigation may lead to legal actions, ranging from prosecution for arson or related crimes to civil litigation over deaths, injuries, or property damage.

The incident commander (IC) is responsible for completing the National Fire Incident Reporting System (NFIRS) documents or a local equivalent report. Sometimes, the IC delegates this responsibility to another fire officer, typically the officer in charge of the first-arriving company.

The first goal of the fire officer assigned this responsibility is to determine whether a formal fire investigation is needed. The authority having jurisdiction (AHJ) provides criteria for requesting an investigator. The fire officer must evaluate all of the circumstances as well as the probable cause of the fire. Any fire that results in a serious injury or fatality meets the criteria for a formal investigation. Any fires that appear to be arson or related to a criminal act also meet the criteria. Some agencies automatically dispatch a fire investigator to all working structure fires.

The legal responsibility for conducting fire investigations is defined by state or local legislation or regulations. The fire officer should determine which laws apply to fire investigations and which agency is responsible for conducting fire investigations in different circumstances. This information is available from the fire marshal's office or from an equivalent agency.

In situations where no formal investigation occurs and a fire investigator does not respond to the scene to determine the cause and origin, the IC is responsible for determining and reporting the fire cause.

The Fire Officer's Role in Fire Cause Determination

The primary goal of fire cause determination by the local fire company is to memorialize conditions found on arrival of the incident, identify and preserve potential physical evidence, document activity performed during fire suppression, and note any unusual situations or human encounters during the incident (IAAI, 2012).

A **fire investigation** is the process of determining the origin, cause, and development of a fire or explosion. The credentialed fire investigator conducts a systematic approach using technology, science, knowledge, and skill. Due to the required technical expertise and legal requirements, most jurisdictions use a fire investigator certified to meet the minimum job performance requirements of NFPA Standard 1033, *Standard for Professional Qualifications for Fire Investigator*. Fire and explosion investigations follow NFPA Standard 921, *Guide for Fire and Explosion Investigations*; the guide establishes guidelines and recommendations for the safe and systematic investigation or analysis of fire and explosion incidents.

Suspicious Fire Considerations

Once the fire is controlled, the fire officer needs to evaluate and consider delaying overhaul operations if the fire incident requires the service of a credentialed fire investigator or can be identified as a possible crime scene (National Fire Academy, 2014). The goal will be to perform a preliminary scene examination to identify, secure, and preserve physical evidence and avoid contaminating the crime scene. This chapter will provide information and recommended practices to recognize when you need to call for an investigator and when you need to halt overhaul operations to facilitate a fire cause determination.

Many jurisdictions have established a protocol on when a fire investigator is required and when the IC or fire company officer should complete a fire cause determination. A metro-sized fire department has provided this standard operating procedure (SOP), identifying the following incidents that will not require a fire investigator (Phoenix, 2010):

- Minor fires where the cause is determined to be accidental
- Scalding burns, electrical accidents, and minor accidental burn injuries
- Car fires originating in the engine area during vehicle operation, abandoned vehicles, or vehicle arsons with no identified subjects
- Minor grass, fence, or trash fires with no witnesses or suspects
- Traffic accident fatalities involving fire subsequent to collision (investigated by law enforcement)
- Minor fires caused by juveniles playing with fire, matches, etc. (referred to the Youth Firesetter program)

FIRE OFFICER TIP

Coordinated Overhaul and Salvage

Although it is important not to destroy evidence during fire suppression operations, the fire officer's first responsibility is to maintain a safe fire ground. That means promptly controlling and extinguishing the fire. After the searches are completed and the fire is declared under control, the fire officer must consider evidence preservation. Overhaul and salvage operations should be coordinated with the fire investigator, if the investigator is on the scene. For example, a fire investigator may wish to photograph or document the condition of a ceiling before it is removed during overhaul operations. If the investigator's response is delayed, the fire officer should attempt to identify and protect the area of origin, limiting overhaul in that area to the absolute minimum required to ensure that the fire does not rekindle. NFPA 921 makes the following recommendations:
- Use caution with straight- or solid-stream water patterns; they can move, damage, or destroy physical evidence.
- Restrict the use of water for washing down, and try to avoid possible areas of fire origin.
- Refrain from moving any knobs or switches.
- Use caution with power tools in the fire scene.
- Refuel this equipment away from the fire scene.
- Limit the number of fire fighters performing overhaul and salvage until the fire investigator is finished documenting the scene.

This department requires a fire investigator respond to the following:

- All first alarm or greater structural fires
- Fires with serious civilian burn injury or fatality
- Attempted arsons
- Bombings
- Explosions
- Car, field, or dumpster fires that have an identified lead or suspect in custody
- Fires whose cause cannot be determined by the on-scene IC.

Each jurisdiction will have a policy based on their resources and organization of fire investigations, with some communities using a state or regional fire investigator.

Fire Company Fire Cause Evaluation

The company officer should consider the needs of the fire investigator's use of the scientific method when evaluating a fire scene. The **scientific method** is the systematic pursuit of knowledge involving the recognition and definition of a problem; the collection of data through observation and experimentation; analysis of the data; the formulation, evaluation, and testing of hypotheses; and, where possible, the selection of a final hypothesis **FIGURE 13-1** (IAFC, IAAI, and NFPA, 2019). With few exceptions, the proper methodology for a fire or explosion investigation is to first determine and establish the origin(s), then investigate the cause: circumstances, conditions, or agencies that brought the ignition source, fuel, and oxidant together (NFPA 921). The fire officer should use the same methodology when evaluating the fire scene.

Finding the Area of Origin

The first step in fire cause determination is to identify the **area of origin**, the exact physical location where a heat source and a fuel first interact, resulting in a fire or explosion. It is usually determined by examining fire damage and fire pattern evidence at the fire scene. Flames, heat, and smoke leave distinct patterns that can often be traced back to identify the area or the specific location where the heat ignited the fuel and the burning first occurred. A fire investigator usually starts in the area where the least amount of damage occurred and follows the patterns back toward the area of greatest fire damage. Eyewitnesses can often provide valuable information that can assist in determining the area of origin. Sometimes, a witness

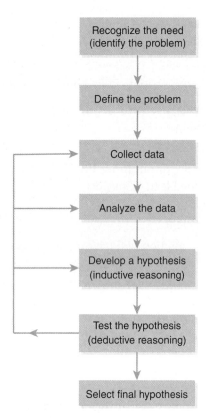

FIGURE 13-1 The scientific method.
Reproduced from NFPA 921-2017, Guide for Fire and Explosion Investigations, Copyright © 2017, National Fire Protection Association. This reprinted material is not the complete and official position of the NFPA on the referenced subject, which is represented only by the standard in its entirety.

FIGURE 13-2 Often, the point of a V-shaped pattern is near or at the point of a fire's origin.
Courtesy of Eddie D. Smith/Unified Investigations & Sciences, Inc.

actually saw what happened and can describe where it occurred. In other cases, witness accounts can identify the area where the fire was first observed or where indications such as visible smoke or a burning odor were noted. Witnesses may also be able to describe what was in the suspected area of origin before the fire and what normally occurred in that area.

Determining the area of origin requires the analysis of information from four sources. First, the physical marks, or fire patterns, left by the fire. Second, the observations reported by persons who witnessed the fire or were aware of conditions present at the time of the fire. Third, analysis of the physics and chemistry of fire initiation, development, and growth as an instrument to related known or hypothesized fire conditions capable of producing those conditions. Fourth, noting the location where electrical arcing has caused damage, as well as the electrical circuit involved. An electrical arc is a luminous discharge of electricity from one object to another, typically leaving a blackening of objects in the immediate area.

Fire Patterns

The area of origin can be identified by interpreting fire patterns. The fire pattern provides the fire officer with a history of the fire. Any flaming fire produces a plume of smoke, heat, and flame. As a fire burns up against a wall, it spreads up and out, creating a V- or U-shaped pattern. The origin of the fire is typically at the base of this V- or U-shaped pattern **FIGURE 13-2**. This type of pattern is also known as a movement pattern because it allows the fire officer to trace the fire and smoke patterns back to their origin.

The second type of pattern indicates the intensity of the fire. The intensity pattern indicates how much heat (energy) was transferred to the surrounding area and objects. The intensity is indicated by the response of various materials to the fire's rate of heat release and heat flux. It may produce a line of demarcation, which can indicate the area closest to the point where the greatest amount of heat was produced.

The analysis of char is closely related to the fire intensity pattern. **Char** is the carbonaceous material that has been burned or pyrolyzed and has a blackened appearance. For the fire officer, the depth of char can assist in determining the direction of fire spread; generally, the deeper the char, the longer the fire burned and therefore the closer the area of origin. Depth of char is only one indicator of the apparent duration and intensity of a fire **FIGURE 13-3**. It may be influenced by a variety of factors, including different types of wood and the fire's intensity, such that deep charring can occur in locations that are remote from the area of origin. For example, deep charring may occur at a location where a fire vents through a window or doorway because of the intensity of the fire as it combines with the fresh air in that area.

Evaluating the Cause of the Fire

Cause is defined as the circumstances, conditions, or agencies that brought about or resulted in the fire or explosion incident, damage to property resulting

FIGURE 13-3 Depth of char.
© Vladimir Zanadvorov/Shutterstock.

from the fire or explosion incident, or bodily injury or loss of life resulting from the fire or explosion incident. Cause determination can be approached as a three-step process:

1. Determine the source of ignition. This generally includes some device or piece of equipment that was involved in the ignition. A competent ignition source must be present to ignite the fuel.
2. Determine the fuel that was first ignited. Both the type of material and the form of the material should be identified.
3. Determine the circumstances or human actions that allowed the ignition source and the fuel to come together, resulting in a fire.

This three-step process describes only those factors that must be determined to conclusively establish the cause of a fire. It is not sufficient to identify and focus on a possible cause that fits the circumstances that were noted; the scientific method requires the fire officer or fire investigator to eliminate any alternative theories or explanations.

Before the investigator can classify a fire cause as incendiary or intentional, all accidental causes must be considered and eliminated. Potential causes should be ruled out only if definite evidence shows that they could not have caused the fire. For example, an electric heater can be ruled out if it was unplugged prior to the fire event.

The fire officer does not have to consider every possible fire cause in relation to every fire that occurs. The facts of the situation can be considered in relation to previous knowledge to identify a list of potential causes for the fire. The fire officer must then apply deductive reasoning, considering each possible cause one by one and sequentially eliminating each cause that is not supported by the evidence.

For example, when evaluating a house fire where an outside wall was ignited from the exterior in proximity to the electrical service, the fire officer must consider any potential source of ignition that could start a fire in that location. For instance, the fire officer might identify lightning as a potential cause of the fire. The officer would then determine whether lightning was occurring in the area at the time of the fire. If it can be established that there was no lightning, then lightning can be eliminated from the list of possible causes. The officer might also consider that the fire could have been caused by a short circuit. If careful examination of the electrical equipment indicates that no short circuit occurred, that possibility can be eliminated. The cause of a fire cannot be established until all potential causes have been identified and considered and only one cannot be eliminated. After all investigative possibilities have been exhausted, if two or more potential causes remain, the cause of the fire is considered undetermined.

Source and Form of Heat Ignition

The **source of ignition** is the energy source that caused a material to ignite. It must have been located at or near the area of origin. The fire officer may be able only to infer a probable ignition source; some sources of ignition remain at the area of origin in recognizable form, whereas others may be altered, destroyed, or removed.

If the equipment that provided the source of ignition was a cigarette lighter, for example, the form of the heat of ignition would be an open flame. The person who started the fire could have removed the source of ignition by placing it back into his or her pocket and leaving the scene. If the pilot light flame from a gas-fueled water heater ignited the fire, however, the source of ignition is likely to be found unaltered at the area of origin.

The source of ignition must have enough energy and must remain in contact with the fuel long enough to cause it to ignite. A competent ignition source has three components:

- Generation: The ignition source must produce sufficient heat energy to raise the fuel to its ignition temperature.
- Transmission: Sufficient heat energy must be transmitted from the source to the fuel to raise the fuel to its ignition temperature. Heat can be transferred through conduction, convection, or radiation.
- Heating: The heat transfer from the source to the fuel must continue long enough for the fuel to be heated to its ignition temperature

Material First Ignited

The **type of material** first ignited refers to the nature of the material itself. For example, the type of material might be cotton. The **form of material** tells how that material is used. For instance, cotton could be in the form of cotton plants in a field, baled cotton fibers, rolls of cotton thread, woven cloth, or clothing (National Fire Data Center, 2010).

The physical configuration of the fuel is an important characteristic in ignition. A 12 by 12-in. (30 by 30-cm) beam and a pile of shavings are both wood; however, the beam is much more difficult to ignite. This information is significant when the investigator is trying to determine which type of heat source could have ignited the material. A pile of wood shavings could be ignited by a dropped match, whereas the wooden beam would have to be exposed to a more powerful source of heat.

Ignition Factor and Cause

The third important factor in determining the cause of a fire is the sequence of events that brought together the source of ignition and the fuel. The cause of a fire could be a human act that was either accidental or deliberate. In the case of an accidental cause, negligence—that is, failure to exercise appropriate care to avoid an accident—could be a factor. The fire cause could also be related to a mechanical failure, a poor design or improper assembly of a device, a worn-out piece of equipment, or a natural force, such as lightning.

Failure analysis is a logical, systematic examination of an item, component, assembly, or structure and its place and function within a system, conducted to identify and analyze the probability, causes, and consequences of potential and real failures. If the cause of a fire is determined to be related to some malfunction that occurred within a device or system, the fire investigator performs a failure analysis to identify what happened and why. This analysis could identify a design error, malfunctioning component, inadequate maintenance, operator error, or some other factor.

Fire Analysis

Fire analysis is the process of determining the origin, cause, development, responsibility, and, when required, a failure analysis of a fire or explosion. Fire analysis brings together all of the available information that can be obtained to examine what happened. In many cases, the fire investigator may need to construct a timeline of events to establish the sequence of the events that led up to the fire **FIGURE 13-4**.

Conducting Interviews

The fire investigator may have to interview fire victims, witnesses, fire fighters, and suspected perpetrators. In doing so, the investigator should make every effort to conduct separate interviews to provide the most accurate representation of what each person observed or experienced. When interviews are done in a group, it is human nature for individuals to change their stories to match what others have said.

Each individual may have relevant information that will help to identify the cause and origin of a fire. For example, a witness might indicate that she saw lightning strike the building a few seconds before seeing smoke and flames. In another case, a witness might report that a can of gasoline was spilled accidentally and a few minutes later the basement erupted in flames.

The first-arriving fire company might report that the front door was broken open when they arrived. They also have valuable information about the color of the smoke or flames or the area where the fire was burning when they arrived. The fire officer should memorialize these observations in the fire incident report narrative.

During the interview, the investigator determines whether the witness is corroborating known information or providing conflicting information. The fact that the witness tells a different version of the story does not necessarily mean that the person is lying; the person could have made a different observation or have an inaccurate memory. Alternatively, the original information might have been inaccurate, and the witness is providing good information. The collection of conflicting information means that further evaluation is required for the actual facts to

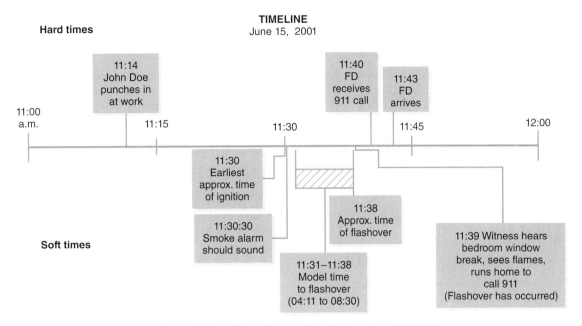

FIGURE 13-4 Example of a scaled timeline, from NFPA 921 section 22.2.5.5.
Reproduced from NFPA 921, Guide for Fire and Explosion Investigations, Copyright © 2008, National Fire Protection Association. This reprinted material is not the complete and official position of the NFPA on the referenced subject, which is represented only by the standard in its entirety.

be determined. A witness statement should be disregarded only if it can be established with certainty that the information is incorrect—for example, if a witness indicates that he saw a man running from the back of a house before the fire started, yet there are no footprints in the snow.

Open-ended questions allow witnesses to tell what they saw or know, whereas questions that seek a "yes" or "no" answer limit the exchange of information. For example, the fire officer should ask the first-arriving crew to describe what they saw when they arrived, not ask whether there were flames showing through the front windows. After a witness provides an overall description, more direct questions can be asked to clarify the facts.

Cause and origin interviews are typically conducted at the fire scene. As noted earlier, interviews are normally conducted with one individual at a time so that the comments made by each person do not influence the responses of the others. Some witnesses may be reluctant to make a full disclosure of information. A special type of interview, called an interrogation, is used when an individual who is attempting to conceal information is being questioned. Interrogations require special knowledge and skills that are beyond the scope and duties of this text. Only a trained fire or police investigator should conduct interrogations.

Interview information must be documented. The documentation could be in the form of the interviewer's notes or a handwritten statement from the witness. An alternative is to record the witness statement using an audio recorder or video camera.

Vehicle Fire Cause Determination

Approximately one in eight fires that a fire department responds to is a highway vehicle fire. No fire cause was determined in 23 percent of highway vehicle fires (National Fire Data Center, 2018). Mechanical failure or malfunction was identified as an ignition factor in 45 percent of incidents where a fire cause was determined **FIGURE 13-5**. Chapter 27 of NFPA 921 provides a standardized procedure to conduct a vehicle fire investigation.

The fire investigator traces fire development back to the area of origin, looking for the area where the most fire damage occurred and the areas of low burn. As in structural fires, there is often a distinctive burn pattern, such as a V-shape, that helps identify the area of origin; however, this pattern should be viewed with caution because of the characteristics of the materials that are present within most vehicles. Fires involving the fuel or tires tend to be particularly intense.

Once the area of origin is established, the specific cause must still be determined. If the cause is related to a mechanical malfunction, the exact cause may be difficult to establish without the assistance of an automotive expert. Additional factors should be considered to determine whether more extensive investigative techniques are required. Most fire departments do not conduct extensive evaluations on vehicle fires unless arson is suspected because of the time and expense involved.

The first objective in most vehicle fire investigations is to look for indications of arson. Because some expensive vehicles are burned when the owners are unable to keep up with the payments, insurance companies often conduct a more thorough investigation if they suspect that the fire was not accidental. The vehicle should be examined for its general condition. Is it a new vehicle burning at the end of a dead-end street in an unpopulated area? Have items of value, like a sound system or navigation system, been removed? Are there indications of accelerant use inside the vehicle? Fuel lines should be checked to see whether they have been loosened. The condition of the tires might indicate that new tires were removed and replaced with old tires before the fire.

Several potential sources of accidental ignition must be considered for a vehicle fire. Sources may include the electrical system, exhaust system, catalytic converter, and turbocharger. There are special considerations for fires involving electric-powered or alternative fuel vehicles. Smoking materials can cause passenger compartment fires.

Note the make, model, and year of the vehicle, as well as the vehicle identification number (VIN). This information allows for a review of previous fires that have occurred in vehicles of the same make, model, and year. Some makes and models have a history of fires, and the reporting of this information can be useful in identifying product defects.

A diagram of the scene should be drawn. Photographs should be taken before the vehicle is removed.

Interview the driver/owner to determine when the vehicle was last driven and how far, what the total vehicle mileage is, whether there have been operating abnormalities, when the last service took place, when the vehicle was last fueled, how the vehicle was equipped (e.g., sound system, custom wheels), and which personal items were in the vehicle.

FIGURE 13-5 Mechanical failures associated with internal combustion engines are frequent causes of vehicle fires.
© AlexKalashnikov/Shutterstock.

If the vehicle was being driven, determine the speed and any loads that were being pulled, any abnormal indications before the fire, the time and location at which the smoke and/or fire was observed, actions taken by the driver, and the amount of time that elapsed before the fire department arrived.

Wildland Fire Cause Determination

The combination of the increase of residents in the wildland–urban interface (Radeloff et al., 2018) and the exponential growth of the size and severity of wildfires in the late 2010s (NICC, 2018) has resulted in a significant increase in the amount and cost of wildland fire suppression efforts. Additional factors include changes in weather conditions, fuel build-up, and growth in the wildland–urban interface.

The characteristics of wildland fires are quite different from those of structural fires. Wildland fires are influenced by environmental conditions, including topography, fuel load, wind, and weather. Such fires tend to spread vertically through convection, from lower vegetation to taller vegetation, and horizontally through radiation. The rate of spread varies, depending on the type of material burning and its density, the wind speed and direction, the humidity and fuel moisture content, the slope of the terrain, and natural features, such as valleys.

When a fire burn pattern on the side of a hill is investigated, the area of origin is most likely on the lower part of the slope, but not necessarily at the lowest point. The fire burns up the hill very rapidly, but it also burns down the hill, albeit at a slower rate. The wind also dramatically affects fire progression, either on flat ground or on a slope **FIGURE 13-6**.

Evaluating the degree of burn on the fuels helps establish the direction of travel. Ash residue also can be evaluated to establish the direction in which it was blown, which indicates the wind direction. When a tree burns and falls, the remaining trunk is usually burned at an angle, creating a point. This point generally appears on the side of the stump opposite the direction of fire approach (IAFC, IAAI, and NFPA, 2019).

Once the area of origin has been determined, the fire investigator can look for clues to help determine the cause of the fire. The following are examples of evidence that might be found at the area of origin:

- Campfire remains
- Time-delay devices
- Cigarette remains
- Lighters

FIGURE 13-6 Fire burns faster up a hill than it does down a hill, but is also dramatically affected by wind.
© AbleStock.

- Multiple ignition points
- Splintered trees (indicating lightning strikes)
- Fulgurites (glassy, rootlike residue resulting from lightning strikes)
- Piles of fuel that are prone to spontaneous combustion
- Barrels used to burn trash
- Fallen electrical wires
- Trees on power lines
- Railroad tracks

Although none of these observations definitively confirms the exact cause of a fire, they are helpful in identifying potential causes. The National Wildfire Coordinating Group has developed professional qualification standards for wildland fire investigators.

Fire Cause Classifications

Once the facts are known, the circumstances of the fire are generally divided into four classifications according to NFPA 921:

1. **Accidental fire cause:** Fire for which the proven cause does not involve an intentional human act to ignite or spread fire into an area where the fire should not be.
2. **Natural fire cause:** Fire caused without direct human intervention or action, such as fire resulting from lightning, earthquake, and wind.
3. **Incendiary fire cause:** Fire that is intentionally ignited in an area or under circumstances where and when there should not be a fire.
4. **Undetermined fire cause:** Classification of fire when the cause cannot be proven to an acceptable level of certainty.

Accidental Fire Causes

Many accidental fires result from human activities that are not intended to start or spread a fire. Finding evidence of candle use near the area of origin, for example, may suggest that an unattended open flame could be an ignition factor.

The most frequent ignition cause in residential fires is unattended cooking (NFPA, 2019). When a fire originates on the stove, a pan is found on one of the burners, and the burner knob is on, the likely cause of the fire is accidental **FIGURE 13-7**. The second most common cause is smoking materials. Heating is the third leading cause of residential fires, with the two primary ignition causes being improper maintenance and combustibles located too close to the heating device. Other accidental fire causes include refueling of gasoline-powered equipment near an ignition source, placement of fireplace ashes in a combustible container, and sparks from welding. (NFPA, 2017)

When wood is continuously subjected to a moderate level of heat below its normal ignition temperature, but over a long period, it starts to break down into

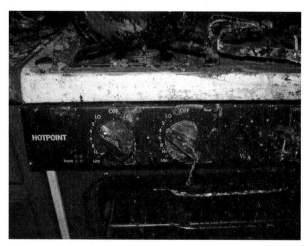

FIGURE 13-7 Note the positions of the knobs on this range.
Courtesy of the NJ State Fire Marshal's Office, Arson/K-9 Unit.

carbon. This chemical decomposition, called **pyrolysis**, a process in which material is decomposed, or broken down, into simpler molecular compounds by the effects of heat alone; pyrolysis often precedes combustion. Pyrolysis should be considered if the area of origin includes steam pipes, fluorescent light ballasts, flue pipes for a fireplace, or a wood-burning stove. The use of zero-clearance fireplace flues has caused an increase in this type of ignition.

The most common electrical fire scenario is misuse by the occupant, such as overloading electrical circuits, using lightweight extension cords for major appliances, or operating too many devices for the electrical service. Any of these errors can overload the wiring or the device, causing heat and eventually igniting a fire.

Large appliances, such as dishwashers and clothes dryers, may fail if their high-temperature controls or timing mechanisms fail. Excessive lint buildup that impedes air exchange and overheats the dryers is a frequent accidental scenario at multiple-family buildings with community laundry rooms. Dishwashers are susceptible to fire caused by plastic dishes and utensils falling onto the heating element—a consideration that should be taken into account when the dishwasher is the area of origin of a fire.

Electrical devices and appliances that start fires usually produce evidence of electrical damage on their power supply cord relatively close to the device or appliance. The wire insulation near the appliance often melts because of the heat produced by the burning appliance or other materials ignited by it. The melted insulation usually results in energized conductors contacting each other, generating a short circuit in the cord. An exception would be an appliance or device malfunction that results in the circuit breaker or fuse tripping so that the power cord is de-energized before the insulation melts.

Natural Fire Causes

A lightning strike in a forest can ignite a tree, and a strike on a building can have a tremendous destructive effect, sending several thousand volts through the electrical wiring system. This surge can destroy televisions, computers, telephone systems, and anything else that is plugged into the electrical system. Lightning can also follow antenna wires, telephone wires, metal plumbing, or the steel structure of a building and ignite combustibles that come in contact with those conduits.

When lightning is suspected, the fire officer should look for a contact point, usually near the top of the structure, concentrating on roof peaks with metal edging, antennas, or large metal objects, such as air-handling units. If the building's electrical system is involved, the surge usually fuses or melts the main fuses or circuit breakers. The surge can travel through television cables, phone lines, power lines, or even natural gas or propane lines, so the point of entry for each utility should be examined for scorching. The joints of aluminum rain gutters may show the same signs (Ahrens, 2013).

> **FIRE OFFICER TIP**
>
> **Energized Aluminum Siding**
>
> Be cautious with aluminum-sided houses where the electrical service is mounted on the side of the house. The entire house exterior could be energized if a lightning strike has fused the circuit breaker box.

If there are indications that lightning could have caused a fire, the fire investigator should check with weather services to determine whether any lighting strikes were recorded in the area. Some fire investigators commonly include a notation, such as "It was a clear and calm night with no lightning observed," in the introduction to their reports to establish that this potential source of ignition was eliminated.

Earthquakes, tornadoes, floods, and hurricanes can all cause fires. As a result of these natural forces, power lines may contact trees or fall onto structures, or sheets of metal may be blown across power lines. Earthquakes may also cause gas mains to break and release flammable gases into structures, where they find an ignition source. Lava from a volcano can easily start a forest fire or incinerate a community.

Incendiary Fire Causes

Because an incendiary fire is an intentional occurrence, the direct cause is a person; however, the methods and the reasons for starting fires vary tremendously. Given the complexity in identifying a fire as having an incendiary origin, a later section of this chapter is devoted to explaining why people intentionally set fires and which evidence might indicate that a fire was set intentionally.

An incendiary fire is one that is intentionally started when the person knows it should not be started. An incendiary fire is not necessarily arson. **Arson** is the crime of maliciously and intentionally or recklessly starting a fire or causing an explosion. The legal definition of arson as a criminal act is uniquely defined within different jurisdictions. The fire investigator will classify the fire's cause and origin as incendiary, but ultimately the prosecuting attorney or grand jury will decide whether the finding should result in arson charges.

FIRE OFFICER TIP

NFPA Statistical Snapshot

1. An estimated 261,330 intentional fires of all types were set in 2010–2014.
2. These fires resulted in 440 deaths, 1,130 injuries, and $1 billion in property damage.
3. Three-fourths of intentional fires were set outside.
4. An estimated 49,690 structure fires were deliberately set or suspected of having been deliberately set.
5. The number of intentional fires and associated losses have shown a steady decline, from a high of 327,100 fires in 2002 to a low of 248,500 fires in 2014 (Campbell, 2017).

Undetermined Fire Causes

No matter how much training and experience a fire investigator has acquired, how many resources are available, or how much effort is expended, sometimes the cause of a fire cannot be determined. Perhaps the extensive damage caused by the fire or the firefighting activities makes it impossible to say with certainty what happened. This may occur when there are multiple reasons that the fire could have started that cannot be ruled out. In such a case, the evidence may not exist to make a definitive determination of the cause.

For example, if a gas can is found next to the water heater, it may be impossible to determine whether the vapors were accidentally ignited by the pilot light or intentionally ignited by an arsonist. Without additional evidence, either explanation may be feasible, so the cause cannot be determined with certainty.

The absence of any logical cause is another reason for classifying the cause of a fire as undetermined. Something caused the fire, to be sure, but the lack of evidence or knowledge precludes the fire investigator from determining the cause.

Indicators of Incendiary Fire Causes

When the fire officer evaluates a fire incident, one important role is to identify indicators of an incendiary fire. If found, it is important to both call for a fire investigator and to preserve the crime scene. Once the other causes have been eliminated, many conditions or factors may indicate that incident involved an intentional fire. These typically fall under five general categories:

- Disabled built-in fire protection
- Delayed notification or difficulty in getting to the fire
- Accelerants and trailers
- Multiple points of origin
- Tampered or altered equipment

Accelerants and Trailers

Accelerant is a fuel or oxidizer, often an ignitable liquid, intentionally used to initiate a fire or increase the rate of growth or spread of fire. Trained canines or survey instruments can detect ignitable liquid accelerants. In addition, ignitable-liquid–fueled fires often leave distinct burn and char patterns **FIGURE 13-8**.

FIRE OFFICER TIP

No Fuel-Powered Firefighting Tools Inside the Fire Incident

When the incident is suspicious, ensure that no petroleum-powered devices are operating inside the fire scene. The exhaust gases from the devices with contaminate the fire scene, obliterating the technical analysis of accelerant traces by machine or dog.

FIGURE 13-8 Ignitable-liquid–fueled fires leave distinct burn and char patterns.
Courtesy of Robert A. Corry/SceneInvestigator.com

Trailer is a solid or liquid fuel used to intentionally spread or accelerate the spread of a fire from one area to another. The following materials are often used as trailers:

- Paper towels
- Black gunpowder
- Film wrapped in paper rags
- Kerosene or other combustible liquids
- Gasoline or other flammable liquids
- Decorative streamers
- Cotton batting
- Paper
- Sheets or rolls of fabric softener
- Newspapers
- Combinations of these items

Trailers usually leave a distinct fire pattern that resembles the material's shape and often run from one room to the next. Liquids typically have irregular edges and areas that are burned where the liquids pooled because of low spots. Sometimes, a trailer of ignitable liquid is poured from one electrical outlet to another to simulate an electrical fire in an attempt to mislead the investigator.

Multiple Points of Origin

Arsonists frequently utilize multiple ignition points inside a building. This is done in case one fire burns out prematurely as well as to maximize the amount of fire growth before the fire department can respond. In other cases, the arsonist is attempting to cause confusion or to trap the occupants by blocking access to all of the exits. Igniting fires in multiple buildings at the same time presents the fire department with a complicated problem. The responding units may be sufficient to manage one fire, but two or three buildings on fire create a challenge.

Multiple points of origin do not necessarily mean that a fire was intentional. In some cases, material falling from the ceiling and burning at the floor level can create a secondary U or V pattern, resembling an additional area of origin. There may be nothing at that location that could have caused the fire to ignite.

Multiple points of origin may also occur when an electrical surge causes ignition at different locations in a building's electrical system. In one instance, an overpressurized natural gas system started fires simultaneously in several kitchens and water heater closets, where each pilot light became a torch.

Arson

Arson is the crime of maliciously and intentionally, or recklessly, starting a fire or causing an explosion. National statistics indicate that 6 percent of structure fires are intentional (National Fire Data Center, 2017). In 2017, law enforcement agencies reported 41,171 arsons. Participating agencies provided expanded offense data regarding 38,764 arsons. Of reported arsons, 45 percent involved structures (FBI, 2018).

Arson Motives

There are six basic motives for arson (USFA, 2016):

1. Profit
2. Crime concealment
3. Excitement
4. Spite/revenge
5. Extremism
6. Vandalism

Profit

Most often, the plan behind arson committed for profit is to collect insurance money. Indicators of insurance fraud include inability to meet payments, failure to complete business contracts, poor sales volume, or lack of supplies and inventory. Investigators should verify the insurance coverage amounts and look for recent changes in the policy. In some cases, the insured makes claims for burned inventory that was actually removed before the fire or that never existed.

Elaborate schemes have been conducted to manipulate supply and demand for various products by destroying inventory or manufacturing facilities to cause a shortage and drive up prices. Arson has also been used for extortion or to eliminate competition. Sometimes, a contractor or individual starts a fire and then offers to repair the damaged property. Many vacant buildings have been burned to save the cost of their demolition.

Crime Concealment

A business owner or employee may elect to burn the business to destroy records that show embezzlement of cash, supplies, or inventory. Burglars might set fire to a building to destroy evidence of their entry, eliminate fingerprints, or even conceal the fact that a theft occurred before the fire. Arson is also used to destroy evidence of other crimes, including murder, or to create a distraction while a crime is taking place in another location.

Voice of Experience

As a fire officer, it is important to remember that your initial fire scene assessment will set the stage for the rest of the investigation. In fact, it is probably the most important part of any investigation. On many occasions as a fire investigator, when I arrived at the scene to begin my investigations, what the fire officer had done prior to my arrival made my job either much easier or much harder. The fire officer, in the performance of his or her duties, will make or break the investigation.

One of the primary responsibilities of the fire officer is the protection of evidence and security of the scene. Once the scene has been compromised, the value of evidence is lost. On many investigations, what greeted me was a pile of smoldering furniture on the front lawn. When I asked the fire officer what happened, I was told that they were doing overhaul. Excessive overhaul operations can cause the investigator to be unable to properly investigate the fire for origin and cause. As an investigator, we used to refer to the suppression crews as "evidence eradication teams."

When suppression and overhaul may destroy evidence, the fire officer must decide whether it is possible to preserve the evidence and, if so, how it can be done. Once the evidence has been removed or altered, fire investigators lose continuity in the chain of custody. Once the chain of custody is broken, we may have to say the cause of the fire is undetermined.

The favorite question that I always asked the fire officer once I arrived on the scene was, "So what happened?" I got many varied answers, some of which I cannot mention here. The fire officer should have a sound knowledge of how to determine the origin and cause of the fire. Establishing the room of origin and the possible points of origin are essential. As an investigator I really didn't care whether the fire officer was able to distinguish between a V- or U-pattern. As long has the fire officer could say that there is a pattern in this room and that the room was preserved was all I needed.

As a fire officer you are setting the investigation in motion. How you preserve evidence and reduce the destruction of evidence during overhaul operations will go a long way in determining the origin and cause of a fire. To be able to identify patterns and areas of origin are critical first steps in any investigation. Often a fire officer, prior to conducting overhaul operations, can stop for 2 minutes and take photos inside the room of origin. That officer is preserving evidence, and when those photos reach the fire investigator, he or she might not have to say the cause is undetermined.

What you do during suppression and overhaul operations will in many cases make or break the investigation.

Doug Goodings
Manager of Academic Standards and Evaluations
Office of the Fire Marshal/Ontario Fire College
Gravenhurst, Ontario, Canada

Excitement

Sometimes, a fire is started for the excitement of the arsonist, who could be seeking thrills, attention, or recognition. The arsonist could be planning to make a dramatic rescue to gain praise or to obtain recognition for discovering or extinguishing the fire. The arsonist may also simply enjoy the excitement and spectacle that are generated by a fire.

Spite/Revenge

Spite and revenge fires involve intense emotions. The arsonist may intentionally set a fire in the exit stairway of an occupied building in the middle of the night. Frequently, this action is triggered by hatred, jealousy, or other uncontrolled emotions following a lover's quarrel, divorce, or bar fight. In many cases, the individual has consumed alcohol or drugs before starting the fire.

Extremism

Extremism on behalf of a variety of causes has been the rationale for setting fires for many centuries. Abortion clinics, religious institutions, businesses that were ecologically damaging, and businesses undergoing labor disputes have all been arson targets. An extremist may want to cause a monetary loss to the person or business or to bring attention to a cause. Radical activists know that a large loss of life focuses attention on their cause. Incendiary devices are frequently used in these crimes.

Vandalism

The motive for vandalism is simply to cause damage for its own sake. Vandalism is most often directed toward schools, abandoned structures, vegetation, and trash containers. In most cases, the fire-setter commits this crime within walking distance of his or her home (South, 2017).

Documentation and Reports

All fires must be properly documented and reported according to the fire department's standard procedures. Most fire departments use the NFIRS reporting system or some variation of it (NFDC, 2010). The basic report includes the incident number, alarm time and date, location of the incident, property ownership, building construction and occupancy type, weather conditions, responding units and personnel, and numerous other factors that are required for full documentation of an incident.

The data collected through NFIRS provide important information that the fire department can use to identify risk factors and trends and to plan the most efficient utilization of resources to prevent fires and respond to emergencies. This information also flows into the state and national database systems to provide a better understanding of the overall fire problem. The fire service and elected officials may also use these data to determine where resources should be directed and when laws should be changed.

Preliminary Investigation Documentation

In addition to the basic incident report, the fire officer often writes up a narrative report if the cause of the fire is incendiary or if unusual circumstances are involved. Such a narrative report is particularly valuable if the fire officer is later called to testify in court about the incident. The fire officer could use the following format to document activities and observations for the fire investigator:

A. Receipt of the alarm
  1. Who reported the fire?
  2. What time was the alarm received?
  3. Who discovered the fire?

B. Response to the incident
  1. Did the fire company encounter any suspicious activity while responding to the fire?
  2. How much time elapsed from dispatch to arrival?
  3. What were the observed weather conditions?

C. Accessibility at the scene
  1. What was the general condition of the fire?
    a. What were the extent and the intensity of the fire?
    b. What was the location of the fire or fires?
    c. Did anyone meet the fire fighters on arrival?
    d. Were any familiar spectators at the scene?
  2. Circumstances on arrival
    a. Was anything unusual, considering the fire load?
    b. Where were smoke and fire coming from?
    c. What were the rapidity and the spread of the fire?
  3. Gaining access to the building
    a. How did the fire fighters get into the structure?
    b. If entry was forced, how did this occur and who forced entry?

D. Fire suppression
   1. What were your immediate observations on entry?
      a. Where was the fire centered?
      b. Were there any unusual flames, smoke, or odors?
   2. What were the conditions of fire extinguishment?
      a. Did the fire flash when it was hit by water?
      b. Was the fire difficult to extinguish?
   3. Were there obstructions to fire suppression?
      a. Were fire protection systems tampered with?
   4. Was the alarm system functioning properly?
E. Civilian contacts
   1. Did witnesses make statements to fire fighters?
   2. Was the owner at the scene?
   3. What were the names of persons who were allowed into the fire scene?
F. Scene integrity
   1. Were any physical evidence or artifacts removed from the scene?
   2. Were any photos or videos taken during fire suppression?
   3. Did any fire department members make holes in walls or ceilings?

The report must be clear, complete, and factual. The fire officer must make no assumptions or speculations. The best report is a narrative that accurately and completely describes what the fire officer observed and what the fire company did (NFA, 2014).

Legal Proceedings

A company officer may occasionally be called on to testify in court proceedings as a witness. Although a witness can provide the court with testimony based only on his or her personal knowledge and observations, an expert witness has scientific, technical, or other specialized knowledge that can be relied upon to interpret the facts. The role of an expert witness is to assist the judge and jurors to understand the evidence or to determine the true facts in an issue **FIGURE 13-9**. An expert witness is allowed to give an opinion based on the facts or the data of the case.

The first step in serving as a witness is to fully prepare for giving testimony. This includes reviewing all reports, photographs, and diagrams of the incident. Also, review any previous depositions or testimony that you have given regarding the incident. The

FIGURE 13-9 An expert witness can assist the judge and jurors in understanding the evidence or to determine the facts.
© Corbis/age fotostock.

prosecutor will review your testimony and qualifications. When testifying, do the following:

- Dress appropriately.
- Follow the prosecutor's directions.
- Sit up with both feet on the floor.
- Avoid gesturing.
- Keep answers short and to the point.
- Use language a jury can understand.
- Be courteous and patient.
- Be honest.
- Do not hesitate or avoid answering questions.
- Speak clearly and loudly.
- If you do not remember, do not guess.

Remember that the defense counsel's job is to try to discredit you and your statements. It is essential to remain calm and cool, even when the lawyer asks questions or makes statements that are designed to make you look bad. Your role is to present factual information that you can support. Every fire officer should be familiar with NFPA 921.

After the Fire Officials Are Gone

Many fire investigations continue long after the fire department has cleared the event. For example, the insurance company may engage in a continuing investigation. This evaluation is not intended to undermine or question the fire investigator's conclusions, but rather to look at the fire from the insurance company's point of view.

The roles of the fire investigator and the insurance company investigator differ in many cases. The fire investigator could be interested in determining the fire cause and origin to help prevent future fires or to help prosecute criminal actions. The insurance investigator might also be looking at the factors that contributed to the loss, such as the absence or inadequacy of fixed fire protection systems and adherence to or disregard of the applicable codes.

The insurance company has a financial interest in any fire loss. Consequently, it is often interested in determining which other parties could have liability for the loss, even if there is no criminal responsibility. Although a criminal prosecution requires proof beyond a reasonable doubt, a civil action involving the insurance company is decided by a preponderance of the evidence. Many insurance cases are not settled for several years after the fire is extinguished.

You Are the Fire Officer Conclusion

A fire with moderate injuries would usually require the on-scene attention of a fire investigator. In this incident, the fire occurred in a commercial business that lacked a fire prevention permit and was not registered by the AHJ.

The AHJ's attorney general will make a determination of whether a crime occurred after reviewing the fire investigator's report and related documentation. The role of the company officer is to complete a fire cause evaluation and determine, based on existing SOPs, if a fire investigator is required.

After-Action REVIEW

IN SUMMARY

- One of the best starting points for efforts to prevent future fires is understanding of the causes of fires that have occurred in the past.
- The leading cause of home structure fires is cooking.
- The fire officer must evaluate all of the circumstances as well as the probable cause of the fire.
- Be alert for conditions or situations that might delay the fire department's ability to get to the fire. Malfunctioning keys and key cards, vandalized doors, and materials blocking access are conditions to note.
- Document unusual conditions noticed at the fire scene. A professional arsonist wants to try to make a fire appear accidental.
- A fire officer who conducts a preliminary fire cause investigation and suspects that a crime has occurred should immediately request the response of a fire investigator and secure the scene to protect any evidence.
- The first step in fire cause determination is to identify the area of origin—the exact physical location where a heat source and a fuel came in contact with each other.
- U- and V-shaped fire patterns can lead the fire officer to the area of origin for a fire.
- Cause refers to the particular set of circumstances and factors that were necessary for the fire to have occurred.
- The source of ignition must have been located at or near the area of origin. The fire officer may be able only to infer a probable ignition source.
- The physical configuration of the fuel is an important characteristic in ignition.
- Failure analysis is a logical, systematic examination of an item, component, assembly, or structure and its place and function within a system, conducted to identify and analyze the probability, causes, and consequences of potential and real failures.
- Fire analysis is the scientific process of examining a fire occurrence to determine all of the relevant facts, including the origin, cause, and subsequent development of the fire, as well as to identify the responsibility for whatever occurred.

- As part of the fire investigation, the fire officer may have to interview fire victims, witnesses, fire fighters, and suspected perpetrators.
- Fire departments respond to more vehicle fires than structure fires.
- Wildland fires are influenced by environmental conditions, including topography, fuel load, wind, and weather.
- The four fire cause classifications are accidental, natural, incendiary, and undetermined.
- Many accidental fires result from human activities that are not intended to start or spread a fire.
- If lightning might have caused a fire, the fire officer should check with weather services to determine whether any lighting strikes were recorded in the area.
- An incendiary fire is a fire that is intentionally started when the person knows it should not be started, but it is not necessarily arson.
- No matter how much training and experience a fire investigator has acquired, how many resources are available, or how much effort is expended, sometimes the cause of a fire cannot be determined.
- When considering whether a fire might have been set intentionally, look for disabled built-in fire protection, delayed notification/difficulty in getting to the fire, accelerants and trailers, multiple points of origin, and tampered/altered equipment.
- Arson is the crime of maliciously and intentionally, or recklessly, starting a fire or causing an explosion.
- There are six basic motives for arson: profit, crime concealment, excitement, spite/revenge, extremism, and vandalism.
- All fires must be properly documented and reported according to the fire department's standard procedures. Most fire departments use the NFIRS reporting system or a variation of it for this purpose.
- In addition to the basic incident report, the fire officer often writes up a special narrative report if the cause of the fire is incendiary or if unusual circumstances are involved.
- Whereas a witness can provide the court with testimony based only on his or her personal knowledge and observations, an expert witness has scientific, technical, or other specialized knowledge that can be relied on to interpret the facts.
- Many fire investigations continue long after the fire department has cleared the event.

KEY TERMS

Accelerant A fuel or oxidizer, often an ignitable liquid, intentionally used to initiate a fire or increase the rate of growth or spread of fire. (NFPA 921)

Area of origin A structure, part of a structure, or general geographic location within a fire scene, in which the "point of origin" of a fire or explosion is reasonably believed to be located. (NFPA 921, revised 2017)

Arson The crime of maliciously and intentionally or recklessly starting a fire or causing an explosion. (NFPA 921)

Cause The circumstances, conditions, or agencies that brought about or resulted in the fire or explosion incident, damage to property resulting from the fire or explosion incident, or bodily injury or loss of life resulting from the fire or explosion incident. (NFPA 921)

Char Carbonaceous material that has been burned or pyrolyzed and has a blackened appearance. (NFPA 921)

Failure analysis A logical, systematic examination of an item, component, assembly, or structure and its place and function within a system, conducted to identify and analyze the probability, causes, and consequences of potential and real failures. (NFPA 921)

Fire analysis The process of determining the origin, cause, development, responsibility, and, when required, a failure analysis of a fire or explosion. (NFPA 921)

Fire investigation The process of determining the origin, cause, and development of a fire or explosion. (NFPA 921)

Form of material How the material was used. (NFIRS, 2010)

Pyrolysis A process in which material is decomposed, or broken down, into simpler molecular compounds by the effects of heat alone; pyrolysis often precedes combustion. (NFPA 921)

Scientific method The systematic pursuit of knowledge involving the recognition and definition of a problem; the collection of data through observation and experimentation; analysis of the data; the formulation, evaluation and testing of hypotheses; and, where possible, the selection of a final hypothesis. (NFPA 921)

Source of ignition Energy source that causes a material to ignite. (NFIRS, 2010)

Trailer Solid or liquid fuel used to intentionally spread or accelerate the spread of a fire from one area to another. (NFPA 921)

Type of material The nature of the material itself. (NFIRS, 2010)

REFERENCES

Ahrens, Marty. 2013, June. *Lightning Fires and Lightning Strikes.* Quincy, MA: National Fire Protection Association.

Ahrens, Marty. 2017, January. *Trends and Patterns of U.S. Fire Loss.* Quincy, MA: National Fire Protection Association.

Campbell, Richard. 2017, July. *Intentional Fires.* Quincy, MA: National Fire Protection Association.

Federal Bureau of Investigation (FBI). 2018. "Crime in the United States, 2017: Arson." Accessed July 26, 2019. https://ucr.fbi.gov/crime-in-the-u.s/2017/crime-in-the-u.s.-2017/topic-pages/arson.

International Association of Arson Investigators (IAAI). 2012. "The First Responder's Role in Fire Investigation." [22:50 video]. *Stonehouse Media.* Accessed July 26, 2019. https://vimeo.com/45526647.

IAFC, IAAI, and NFPA. 2019. *Fire Investigator: Principles and Practice to NFPA 921 and 1033, Fifth edition.* Burlington, MA: Jones and Bartlett Learning.

National Fire Academy (NFA). 2014. *Initial Fire Investigation for First Responders: Instructor's Guide.* Emmitsburg, MD: Federal Emergency Management Agency.

National Fire Data Center. 2010. *National Fire Incident Reporting System: Complete Reference Guide.* Emmitsburg, MD: Federal Emergency Management Agency.

National Fire Data Center. 2017. *Fire in the United States 2006–2015, 19th edition.* Emmitsburg, MD: Federal Emergency Management Agency

National Fire Data Center. 2018, July. "Highway Vehicle Fires (2014–2016)." *Topical Fire Report Series* (19): 2. Emmitsburg, MD: Federal Emergency Management Agency.

National Fire Protection Association (NFPA). 2019. "Cooking." Accessed July 27, 2019. https://www.nfpa.org/Public-Education/Fire-causes-and-risks/Top-fire-causes/Cooking.

National Interagency Coordination Center (NICC). 2018. *Wildland Fire Summary and Statistics Annual Report 2018.* Boise, ID: National Interagency Fire Center.

Phoenix Fire Department. 2010. "Fire Cause Investigation, M.P. 202.13." Accessed July 26, 2019. https://www.phoenix.gov/firesite/Documents/074735.pdf.

Radeloff, Volker C., David P. Helmers, H. Anu Kramer, Miranda H. Mockrin, Patricia M. Alexandre, Avi Bar-Massada, Van Butsic, et al. 2018. "Rapid Growth of the US Wildland-Urban Interface Causes Wildfire Risk." *Proceedings of the National Academy of Sciences (USA)* 115 (13): 3314–3319.

South, Rhena. 2017. "Understanding Arson through Community Resilience." Master's thesis, University of Central Florida. Accessed August 5, 2019. http://stars.library.ucf.edu/etd/5558.

U.S. Fire Administration (USFA). 2016, April 12. "Arson Motives." Accessed July 27, 2019. https://www.usfa.fema.gov/prevention/outreach/wildfire_arson/arson_motives.html.

Fire Officer in Action

Your fire company was the first arriving unit to an apartment fire where someone ignited a pile of clothes in the public hallway of an apartment building. Consider the role of the fire company officer at an intentional fire incident.

1. Once the fire is under control, the fire company officer needs to take what step?
    A. Initiate an arson investigation.
    B. Photograph physical evidence.
    C. Complete a preliminary scene examination.
    D. Start decontamination of room of origin.

2. The company should consider the fire investigator's use of the _____ method when evaluating a fire scene.
    A. scientific
    B. Kennedy
    C. grid analysis
    D. fire flow

3. Which of these is the first step in fire cause determination?
    A. Identify the prevailing winds.
    B. Rule out arson.
    C. Establish a search perimeter.
    D. Identify the area of origin.

4. The base of a V- or U-shaped pattern identifies:
    A. an area where burning material fell from ceiling or wall.
    B. an air gap between the floor and the wall.
    C. the origin of fire.
    D. the presence of an accelerant.

Fire Captain Activity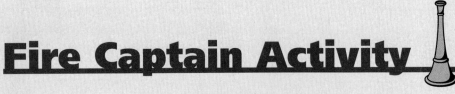

NFPA Fire Officer II Job Performance Requirement 5.5.1

Determine the area of origin and preliminary cause of a fire, given a fire scene, photographs, diagrams, pertinent data, and/or sketches, to determine if arson is suspected so that law enforcement action is taken.

Application of 5.5.1

1. Using a jurisdiction you are familiar with, describe the incidents that require the response of a fire investigator and the responsibilities the fire officer has on the events when a fire investigator does not respond or is unavailable.

2. Is a fire company officer without credentials as a fire investigator considered an expert witness on fire cause determination in a jurisdiction you are familiar with? Identify the training, experience, and credentials the district attorney will require for a fire officer to testify or be interrogated as a fire cause determination expert witness.

3. It is 3:15 AM on Saturday morning and a suspicious fire is declared under control. The state fire investigator cannot get to the incident scene until 10 AM on Monday morning. Describe how you will secure the scene and control rekindles.

4. Your fire company arrives at an explosion in an apartment building. A resident is seriously injured by the blast and needs to go to the burn center. Describe how you will document the incident scene for the fire investigator.

 Access Navigate for flashcards to test your key term knowledge.

CHAPTER 14

Fire Officer II

Managing Major Incidents

KNOWLEDGE OBJECTIVES

After studying this chapter, you will be able to:

- Describe the National Response Framework (NRF). (p. 354)
- Compare and contrast the Incident Command System (ICS), National Incident Management System (NIMS), and the National Response Framework (NRF). (**NFPA 1021: 5.6.1**) (p. 354)
- Explain the functions of ICS division, group, and unit supervisors. (**NFPA 1021: 5.6.1**) (pp. 360–361)
- Identify additional duties a company-level officer might perform on a large incident scene. (**NFPA 1021: 5.6.1**) (pp. 359–360)
- Explain company management duties at an expanded incident. (**NFPA 1021: 5.6.1**) (p. 362)
- Compare and contrast managing fires in multifamily dwellings to managing a single-family dwelling fire. (**NFPA 1021: 5.6.1**) (pp. 363–364)
- Describe the tactical level of incident management. (**NFPA 1021: 5.6.1**) (pp. 360–362)
- Develop and conduct a postincident analysis (**NFPA 1021: 5.6.2**) (pp. 365–367)

SKILLS OBJECTIVES

After studying this chapter, you will be able to:

- Manage a multiunit incident using the National Incident Management System. (**NFPA 1021: 5.6.1**) (pp. 354–362)
- Serve in an ICS unit supervisor capacity at an incident. (**NFPA 1021: 5.6.1**) (p. 361)

ADDITIONAL NFPA STANDARDS

- **NFPA 1600**, *Standard on Continuity, Emergency, and Crisis Management*
- **NFPA 1700**, *Guide for Structural Firefighting*
- **NFPA 1710**, *Standard for the Organization and Deployment of Fire Suppression Operations, Emergency Medical Operations, and Special Operations to the Public by Career Fire Departments*
- **NFPA 1720**, *Standard for the Organization and Deployment of Fire Suppression Operations, Emergency Medical Operations, and Special Operations to the Public by Volunteer Fire Departments*

You Are the Fire Officer

Pulling out of the station for a reported building on fire, the chauffer notes a black column of smoke against the blue night sky. Dispatch reports that they are getting many calls reporting this fire.

1. What is the role of the first-arriving fire officer after the incident management team arrives?
2. Who calls for federal help?
3. Does the command practice change once the president declares a disaster?

Access Navigate for more practice activities.

Introduction

A fire officer is expected to perform the duties of a first-arriving officer at any incident, including assuming initial command of the incident, establishing the basic management structure, and following standard operating procedures. A fire officer must also be fully competent at working within the incident command system (ICS) and functioning as a unit, group, or division leader. This chapter looks at the resources and issues that are involved when commanding a major incident.

National Response Framework

The federal government established the **National Response Framework (NRF)** in March of 2008. NRF is a comprehensive, national, all-hazards approach to domestic incident response that describes specific authorities and best practices for managing incidents (FEMA, 2016). NRF rules the collection and allocation of federal and federal contractor resources to support the **National Incident Management System (NIMS)** (FEMA, 2017), which provides a consistent template for managing incidents. The Homeland Security Act charged the Department of Homeland Security administrator with building a comprehensive national incident management system with federal, state, and local government personnel, agencies, and authorities to respond to attacks and disasters; consolidate existing federal emergency response plans into a single, coordinated national response plan; and administer and ensure the implementation of NRF, including coordinating and ensuring the readiness of each emergency support function under NRF.

Federal, state, and some private-sector and nongovernmental entities organize their resources and capabilities under 15 emergency support functions (ESFs). ESFs align categories of resources and provide

TABLE 14-1 ESF Annexes Applicable to Fire Officers

Annex	Primary Agencies
ESF 4: Firefighting	Department of Agriculture/Forest Service
ESF 5: Emergency Management	Department of Homeland Security (DHS)/Federal Emergency Management Agency (FEMA)
ESF 6: Mass Care, Emergency Assistance, Housing and Human Services	DHS/FEMA
ESF 8: Public Health and Medical Services	Department of Health and Human Services
ESF 9: Search and Rescue	DHS/FEMA DHS/U.S. Coast Guard (USCG) Department of the Interior (DOI)/National Park Service (NPS) Department of Defense (DOD)
ESF 10: Oil and Hazardous Materials Response	Environmental Protection Agency, DHS/USCG
ESF 13: Public Safety and Security	Department of Justice (DOJ)

© Jones & Bartlett Learning.

strategic objectives for their use. The ESF annexes that are of interest to a fire officer are listed in **TABLE 14-1**.

The Stafford Act

How the federal government deploys resources after an incident occurs remains an evolving process that has shaped how fire departments respond to catastrophic

Voice of Experience

Perhaps the most common ICS position for the company officer is that of incident commander (IC). On every company response, the company officer is the IC. During larger, more complex responses, the company officer may find himself or herself functioning in a subordinate ICS position assigned by the IC. Sometimes this happens during one event.

Engine 5 was finishing another routine shift when the company received an alarm for an unknown fire in an industrial district. The fire fighters responded as the first-in engine company and discovered that the fire was a grass/vegetation fire on some abandoned industrial property. While conducting the initial size-up, all indications identified the scene as a routine grass fire and suggested that no additional help would be needed. Additional units en route were told to disregard the call, and confinement and extinguishment actions were implemented by the crew of Engine 5.

The company officer was the IC and responsible for all decisions related to selecting strategy, implementing tactical options, supervising various work tasks, and ensuring the safety of everyone involved. Everything was routine. All command and control responsibilities were being performed by the company officer, and his crew was implementing his tactical plan of attack to extinguish the fire. Everything was going well.

Then things changed. Smoke from the burning grass/vegetation became much darker and thicker than one would normally expect from this type of fire. As part of his continuing scene size-up process, the IC spoke with a security representative from an adjacent property and learned that the property involved was a registered Superfund site. The fire scene was really a hazardous materials scene.

Offensive fire suppression operations were immediately abandoned. A defensive strategy was adopted, and additional units were dispatched to the scene to aid in safely confining and controlling the fire. When the battalion chief arrived on the scene, he was briefed, and command was transferred. The chief officer became the new IC, and the company officer assumed duties as a group supervisor. The scene continued to evolve as more engine companies became involved in operations.

One operational period transitioned into another. Additional command staff and city emergency management officials arrived on scene, as did representatives from various state and federal agencies, and a few mutual aid partners who were requested to respond. All of these additional parties were integrated into the incident management structure. The fire was extinguished, but the focus had shifted to assessing the hazardous materials threat. As the scene grew, many company officers found themselves reassigned to fulfill various ICS positions such as incident safety officer, operations section chief, hazardous materials branch director, and decontamination group supervisor. Eventually, a comprehensive scene hazard and risk analysis was completed, and it was determined that the scene had been stabilized. The incident transitioned from an emergency mode to a recovery operation, where clean-up and mitigation efforts became the focus. Units that were no longer needed were demobilized, and the incident command structure began to collapse and get smaller and smaller.

Continues

Voice of Experience

Eventually, all the chiefs had cleared the scene and the additional units that were needed for emergency operations went back into service. In the end, the same engine company that initially responded and established command was left as the last unit on scene and terminated the incident by returning to service. After 48 hours of continuous emergency operations, Engine 5 could honestly claim they were "first in, last out."

Robert L. Havens, CFPS
Captain/Paramedic
Port Arthur Fire Department
Port Arthur, Texas

incidents or events that overwhelm local and state resources. The 1974 Robert T. Stafford Disaster Relief and Emergency Assistance Act, known as the "Stafford Act," was amended in 2007 to provide federal government disaster and emergency assistance to state and local governments, tribal nations, eligible private nonprofit organizations, and individuals affected by a declared natural disaster or emergency. Adoption and implementation of NIMS is required for eligibility for Stafford Act Funding. NIMS has five components: preparedness, communications and information management, resource management, command and management, and ongoing management and maintenance (**FIGURE 14-1**). The scope of the Stafford Act covers all hazards, including terrorist events and natural disasters. The activity of the Act is to supplement state and local efforts to save lives and to protect property and public health and safety, or to lessen or avert the threat of catastrophe (Lindsay, 2019).

Incident Action Plan

The **incident action plan (IAP)** is a basic component of the ICS; all incidents require an action plan. The IAP outlines the strategic objectives and states how emergency operations will be conducted. At most incidents, the IAP is relatively simple and can be expressed by the IC in a few words or phrases. A written IAP is required for large or complex incidents that have an extended duration. The IAP for a large-scale incident can be a lengthy document that is regularly updated and used for daily briefings of the command staff.

Staffing the Incident Management System

The incident management system can readily expand to handle larger and more complex incidents. Managing and supervising fire officers may be assigned to section, branch, division, and group assignments within the incident management system.

Command Staff

After command has been transferred from the initial IC, the company-level officer may be assigned to the **command staff**. Individuals on the command staff perform functions that report directly to the IC. The safety officer, liaison officer, and information officer are always part of the command staff; these duties cannot be delegated to other sections of the incident organization (FEMA, 2017).

Aides, assistants, and advisors may be assigned to work directly for the IC. An aide is a fire fighter (sometimes a fire officer) who serves as a direct assistant to a command officer. In many fire departments, operational-level command officers have regularly assigned aides who drive the command vehicle and perform administrative support duties, as well as functioning as aides at incident scenes. In other departments, aides are assigned only as needed at incident scenes. The IC may assign more than one aide to perform support functions at a major incident. Aides can also be assigned to other officers in the command structure.

Components of NIMS			Levels of Training		
			Awareness	Advanced	Practicum
Preparedness			IS 800 IS 705		
Communications and Info Management		IS-700	IS 704		
Resource Management			IS 703 IS 706 IS 707		
Command & Management	ICS		ICS 100 ICS 200	ICS 300 ICS 400	Position-specific courses
	MACS		ICS 701		
	Public Info		ICS 702		
Ongoing Management & Maintenance					

FIGURE 14-1 The Stafford Act process to deploy federal resources.
Courtesy of the National Response Plan (NRP)/United States Army Combined Arms Center.

Safety Officer

The **safety officer** is responsible for ensuring that safety issues are managed effectively at the incident scene. The safety officer acts as the eyes and ears of the IC by identifying and evaluating hazardous conditions, watching out for unsafe practices, and ensuring that safety procedures are followed. Normally, the safety officer is appointed early during an incident. As the incident becomes more complex, additional qualified personnel can be assigned as assistant safety officers to subdivide the responsibilities.

The safety officer is an advisor to the IC but has the authority to stop or suspend operations when unsafe situations occur. This authority is clearly stated in national standards, including NFPA 1500, *Standard on Fire Department Occupational Safety, Health, and Wellness Program*; NFPA 1521, *Standard for Fire Department Safety Officer Professional Qualifications*; and NFPA 1561, *Standard on Emergency Services Incident Management System and Command Safety*. Several state and federal regulations require the assignment of a safety officer at hazardous materials incidents and certain technical rescue incidents. Making such an assignment is a good practice even if it is not explicitly required by a regulatory agency. The safety officer should be a qualified individual who is knowledgeable in fire behavior, building construction and collapse potential, firefighting strategy and tactics, hazardous materials, technical rescue practices, and departmental safety rules and regulations. A safety officer should also have considerable experience in incident response and specialized training in occupational safety and health. Many fire departments have full-time safety officers who perform administrative functions relating to health and safety when they are not responding to emergency incidents (Dodson, 2016).

Liaison Officer

The **liaison officer** is the IC's point of contact for representatives from outside agencies and is responsible for exchanging information with representatives from those agencies. During an active incident, the IC may not have time to meet directly with everyone who comes to the command post. The liaison officer position takes the IC's place by obtaining and providing information or directing people to the proper location or authority. The liaison area should be adjacent to, but not inside, the command post.

The IC can also assign a liaison officer to represent the fire department directly with another agency. At a complex incident involving extensive interaction between the police and fire departments, where unified command has not been established, the fire department could assign a liaison officer to work with the police, or the police commander could assign a liaison officer to the fire department command post (Walsh et al., 2012).

Public Information Officer

The **public information officer** is responsible for gathering and releasing incident information to the news media and other appropriate agencies. At a major incident, the public wants to know what is being done. The public information officer serves as the contact person for media requests so that the IC can concentrate on managing the incident. A media briefing location should be established that is separate from the command post (Fleming, 2013).

General Staff Functions

When an incident is too large or too complex for one person to manage effectively, the IC may appoint officers to oversee major components of the operation **FIGURE 14-2**. Four standard components are defined in the ICS model. Everything that occurs at an emergency incident can be divided among these four major functional components:

1. Operations
2. Planning
3. Logistics
4. Finance/administration

The IC decides which (if any) of these four components need to be activated, when to activate them, and who should be placed in each position. Recall that the blocks on the ICS organization chart refer to functional areas or job descriptions, not to positions that must always be staffed. The positions are assigned only when they are needed and, if a position is not assigned, the IC is responsible for managing that function.

The persons in charge of the four major sections are known as the **ICS general staff**. The four people on the ICS general staff, when they are assigned, may conduct their operations either from the main command post or from a different location. At a large incident, the four functional organizations may operate from different locations, but the general staff are always in direct contact with the IC (FEMA, 2017).

Operations

The **operations section** is responsible for the management of all actions that are directly related to controlling the incident. The operations section fights the fire, rescues any trapped individuals, treats the

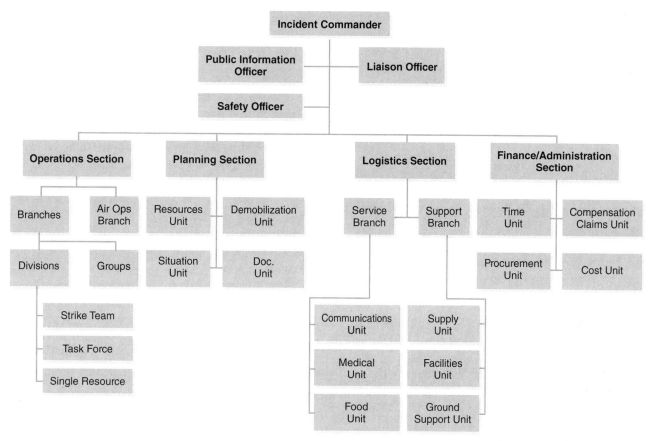

FIGURE 14-2 Expanded ICS organizational chart.
Modified from: FEMA. (April 2012) "ICS 400 – Advanced ICS" Student manual.

patients, and does whatever else is necessary to deal with the emergency situation. This part of the organization produces the most visible results. For most structure fires, the IC directly supervises the functions of the operations section. A separate **operations section chief** is used at complex incidents so that the IC can focus on the overall situation while the operations section chief deals with the strategy and tactics that are required to get the job done.

Planning

The **planning section** is responsible for the collection, evaluation, dissemination, and use of information relevant to the incident. To do so, it works with status boards and preincident plans, as well as building construction drawings, maps, aerial photographs, diagrams, and reference materials.

The planning section is also responsible for developing and updating the IAP. In many ways, this component resembles the scheduling department at a company. It plans what needs to be done by whom and identifies which resources are needed. To perform this role, the planning section must be in close and regular contact with all of the other sections.

The IC activates the planning section when information needs to be obtained, managed, and analyzed. The **planning section chief** reports directly to the IC. Individuals assigned to planning functions examine the current situation, review available information, predict the probable course of events, and prepare recommendations for strategies and tactics. The planning section also keeps track of resources at large-scale incidents and provides the IC with regular situation and resource status reports.

The planning functions may be delegated to subunits. These include a resources unit, a situation unit, a documentation unit, a demobilization unit, and technical specialists.

Logistics

The **logistics section** is responsible for providing supplies, services, facilities, and materials during the incident. The **logistics section chief** reports directly to the IC. Among the responsibilities of this section are keeping apparatus fueled, providing food and refreshments for fire fighters, obtaining the foam concentrate needed to fight a flammable liquids fire, and arranging for a bulldozer to remove a large pile of debris.

In essence, the logistics section is similar to the purchasing and support services departments of a large company—that is, it ensures that adequate resources are always available and functional.

Logistical functions are routinely performed by personnel assigned to a support services branch. These groups work in the background to ensure that the members of the operations section have whatever they need to get the job done. Resource-intensive or long-duration situations may require assignment of a logistics section chief because service and support requirements are so complex or extensive that they need their own management component.

The logistics section may use subunits to provide the necessary support for large incidents. These units may include a supply unit, rehabilitation unit, facilities unit, ground support unit, communications unit, food unit, and medical unit.

Finance/Administration

The **finance/administration section** is the fourth major ICS component under the IC. This section is responsible for the administrative, accounting, and financial aspects of an incident, as well as any legal issues that may arise. This function is not activated at most incidents, because cost and accounting issues are usually addressed before or after the incident. Nevertheless, a finance/administration section may be needed at large-scale and long-term incidents that require immediate fiscal management, particularly when outside resources must be procured quickly.

A finance/administration section is usually established during a natural disaster, when state or federal expense reimbursements are expected, or during a hazardous materials incident in which reimbursement may come from the shipper, carrier, chemical manufacturer, or insurance company. The finance section is equivalent to the finance department in a company; it accounts for all activities of the company, pays the bills, ensures that there is enough money to keep the company running, and keeps track of the costs.

The finance/administration section may incorporate subunits to handle its duties efficiently. These include a time unit, procurement unit, compensation and claims unit, and cost unit.

Tactical-Level Incident Management

Divisions, groups, and units are where the strategy specified in the IAP is converted into action. Supervising and managing fire officers will be commanding these incident management assignments.

Divisions, Groups, and Units

Divisions, groups, and units are tactical-level management elements that are used to assemble companies and resources for a common purpose **FIGURE 14-3**. The flexibility of the ICS enables organizational units to be created as needed, depending on the size and the scope of the incident. In the early stages of an incident, individual companies are often assigned to work in different areas or perform different tasks. As the incident grows and more companies are assigned to areas or functions, the IC can establish tactical-level units to place multiple resources under one supervisor. This organization keeps the IC's span of control within desired limits and provides direct coordination of common efforts.

A **division** represents a geographic operation, such as one floor or one side of a building. A **group** represents a functional operation, such as ventilation. **Unit** is a generic term that can be applied to either a geographic or functional component. An officer is assigned to supervise each division, group, or unit as it is created.

These organizational elements are particularly useful when several resources are working near each other, such as on the same floor inside a building, or are doing similar tasks, such as providing ventilation. The assigned supervisor can directly observe and coordinate the actions of several crews (Brunacini, 2002).

Divisions

A division is composed of the resources responsible for operations within a defined geographic area. This area could be a floor inside a building, the rear of the fire building, or a geographic area at a brush

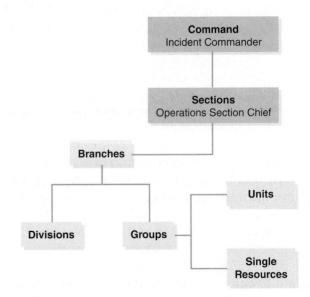

FIGURE 14-3 Divisions, groups, and units.
© Jones and Bartlett Learning.

fire. Divisions are most often used during routine fire department emergency operations. When all of the units in one area are assigned to one supervisor, the IC has to communicate with just that one individual. The assigned supervisor, in turn, coordinates the activities of all resources that are working within that area, in accordance with the IC's strategic plan.

The division structure provides effective coordination of the tactics being used by different companies working in the same area. For example, the **division supervisor** might coordinate the actions of a crew that is advancing a hose line into the fire area, a crew that is conducting search-and-rescue operations in the same area, and a crew that is performing horizontal ventilation.

Groups

An alternative way of organizing resources is by function rather than by location. A group is composed of resources assigned to a specific function, such as ventilation, search and rescue, or water supply. The officer assigned to supervise the ventilation group, for example, would use the radio designation "ventilation group" to refer to this function and is required to coordinate activities with other division, group, or unit supervisors.

Groups are responsible for performing an assignment, wherever it may be required, and often work within more than one division. Sometimes groups are established with both functional and geographic designations, such as a west wing search-and-rescue group and an east wing search-and-rescue group (Brunacini, 2002).

Units

A unit is an organizational element with functional responsibility for a specific incident activity, such as planning or logistics, or a specific geographic assignment. The officer assigned to supervise the air resupply unit, for example, would use the radio designation "air resupply" to refer to it and is required to coordinate activities with other division, group, or unit supervisors.

The unit is the smallest organizational element within the incident management system. Units are more frequently used in large or complex incidents where many specialized groups must work together.

Division/Group/Unit Supervisor Responsibilities

In many cases, a company-level officer is assigned by the IC to function as a division/group/unit supervisor. When an incident is rapidly escalating, the company-level officers on the first alarm are often assigned to geographic or functional areas and designated as the division/group/unit supervisor. For example, the company officer in charge of the first-arriving ladder company could be assigned as the "ventilation group" supervisor. Additional companies assigned to ventilation would then become part of this group. When additional command officers arrive, the ventilation group supervisor might be relieved by a chief officer and revert to functioning as a company-level officer (Brunacini, 2004).

As a division/group/unit supervisor, the company-level officer directly supervises and monitors the operations of assigned resources and provides updates to command. Duties of a division/group/unit supervisor include the following:

- Using an appropriate radio designation, such as roof division, division A, or rescue group
- Completing the objectives assigned by command
- Accounting for all assigned companies and personnel
- Ensuring that operations are conducted safely
- Monitoring work progress
- Redirecting activities as necessary
- Coordinating actions with related activities and adjacent division/group/unit supervisors
- Monitoring the welfare of assigned personnel
- Requesting additional resources as needed
- Providing the IC with essential and frequent progress reports
- Reallocating or releasing assigned resources

Division, group, and unit supervisors all function at the same level within the ICS, regardless of the rank of the individual who is assigned. Divisions do not report to groups, and groups do not report to divisions; instead, the supervisors are required to coordinate their actions and activities with each other. A division supervisor must be aware of everything that is happening within the assigned geographic area, so a group supervisor must coordinate with the division supervisor when the group enters the division's geographic area, particularly if the group's assignment will affect the division's personnel, operations, or safety. Effective communication among divisions and groups is critical during emergency operations.

Branches

A **branch** is a supervisory level established in either the operations or logistics function to provide an appropriate span of control. At a major incident, several different activities may occur in separate geographic areas or several distinct but related functions may be performed. The span of control might still be a problem, even after the IC has established divisions/groups/units. In these situations, the IC can establish branches to place a **branch director** in charge of a number of divisions/groups/units. For example, after a tornado, if damage occurred in two separate areas across town, the IC could establish two branches, one for each area affected. If numerous injuries occurred, the IC could establish a medical branch to coordinate all aspects of caring for the victims, including triage, treatment, and transportation within each branch.

Fire Officer Greater Alarm Responsibilities

The IC calls for additional resources or greater alarms when more resources are needed at an incident. In many cases, the IC calls for a greater alarm to have stand-by resources immediately available, in case they are needed or in anticipation of a future need. A company-level officer may be called upon to perform many duties when responding as an additional resource:

- Reinforcing a fire attack strategy by adding resources.
- Relieving an exhausted crew from the first alarm and performing the fire-ground task that they were performing.
- Performing support activities, such as salvage, overhaul, moving equipment to a forward staging area, going door-to-door to evacuate a neighborhood, assisting occupants in recovering personal items, or providing runners to support a face-to-face communications system.
- Maintaining a ready reserve in a staging area.
- Performing additional related duties. This catchall phrase means that the officer and company may be performing tasks that are not traditional fire company tasks, but are time or mission critical. For instance, they could be assigned to an evacuation center to provide assurance and assessment of the occupants who fled from an apartment fire.

Additional resources may be dispatched as strike teams or task forces, as opposed to individual companies. In such a case, the companies are directed to combine forces to respond and operate as a group. The group members could be assigned to meet at a designated location and travel to the scene together or to assemble at the staging area (Walsh et al., 2012).

Staging

Staging refers to a standard procedure to manage uncommitted resources at the scene of an incident. Instead of driving up to the incident scene and committing apparatus and crews without direction from the IC, staged resources stand by uncommitted and wait for instructions at a predetermined location, usually some distance away from the incident. These resources are immediately available for use where and when the IC needs them. This organization allows the IC to determine the most appropriate assignment for each unit.

Many fire departments have policies that automatically commit one or two first-arriving companies to the incident and direct all others to level I staging. In level I staging, only the predesignated units respond directly to the scene, and later-arriving units remain uncommitted and wait for instructions. For example, the second-arriving aerial ladder would stop about a block short of the fire building, where it would have the flexibility to go to whichever side it was needed. Standard operating procedures guide when units stage, where they stage, and how they communicate with the IC to indicate they are available for an assignment.

Level II staging, which is generally used for greater alarm incidents, directs responding companies to a designated stand-by location away from the immediate incident scene. A designated officer supervises the level II staging area with the "staging" radio identification and assigns units from the staging area as they are requested by the IC. The IC might direct the staging officer to maintain a minimum level of resources in the staging area (Brunacini, 2002).

Task-Level Incident Management

Individual companies, sometimes referred to as single resources, operate at the task level. A company is considered a single resource because it is the basic work unit assigned to perform most emergency scene tasks. Certain types of tasks are routinely assigned to engine and ladder companies. The normal role of a company-level officer is to supervise a group of fire fighters who are operating as a company at the task level.

At a simple incident, the company-level officers report directly to the IC. When divisions/groups/units are established, company-level officers report to their assigned division/group/unit supervisor and continue to supervise the operations of their respective companies in performing assigned tasks. Sometimes, the assigned supervisor is one of the company-level officers.

Company-level officers advise the division/group/unit supervisor of work progress, preferably through face-to-face contact. All company-level officer requests for additional resources or assistance must go through the division/group/unit supervisor.

Task Forces and Strike Teams

Task forces and strike teams are groups of single resources that have been assigned to work together for a specific purpose or for a period of time. Grouping resources reduces the span of control by placing several units under a single supervisor.

A **task force** includes two to five single resources that are assembled to accomplish a specific task. For example, a task force could be composed of two engines and one truck company, or two engines and two brush units, or one rescue company and four ambulances. A task force operates under the supervision of a **task force leader**. In many cases, the designated task force leader is one of the company-level officers. All communications for the separate units within the task force are directed to the task force leader.

A **strike team** consists of five units of the same type with an assigned leader **FIGURE 14-4**. A strike team could be five engines (engine strike team), five trucks (truck strike team), or five ambulances (EMS strike team). This team operates under the supervision of a **strike team leader**. In most cases, the strike team leader is a command officer (FEMA, 2017).

Strike teams are commonly used for wildland fires, where dozens of companies may respond. During wildland fire season, many departments establish strike teams of five engine companies that are dispatched and work together on major fires. The assigned companies are assigned to rendezvous at a designated location and then respond to the scene together. Each engine has an officer and a crew of fire fighters, and one officer—typically a battalion chief—is designated as the strike team leader. All communications for the strike team are directed to the strike team leader.

FIGURE 14-4 A strike team is five units of the same type under one leader.
© Keith D. Cullom/www.fire-image.com

EMS strike teams, consisting of five ambulances and a supervisor, are often organized to respond to multicasualty incidents or disasters. Rather than requesting 15 ambulances and establishing an organizational structure to supervise 15 single resources, the IC can request three EMS strike teams and coordinate their operations through the three strike team leaders—a far more efficient arrangement (Florida Ambulance Association, 2018).

Greater Alarm Infrastructure

Small fire departments have limited infrastructure support. When they are working at a major incident, a considerable delay might occur before a rehabilitation unit is established with rest and recovery supplies. Fire officers should prepare for this possibility by making sure that their apparatus carries enough water and food to support the fire company for a reasonable period of time. In wildland areas, many companies plan for a 2-day excursion, carrying 5 gallons of water and six energy bars per fire fighter. In the summer, 10 gallons of drinking water per fire fighter per day could be required (Ryan, 2014).

Selected Structure Major Incidents Considerations

There are command considerations based on occupancy. Single-family dwellings, mid-rise dwellings, and high-rise structures all require special considerations due to differences in structure.

Single-Family Dwellings

The most frequent location of civilian fire deaths and fire-related injuries is in the home (Evarts, 2018). These structures may range from a single-story frame house of less than 1000 ft^2 (305 m^2) to a mansion with more than 15,000 ft^2 (4572 m^2) on a private estate in a non-hydrant area. They are not subject to fire prevention inspections. Balloon frame construction, which was common until the 1930s, incorporated void spaces in the exterior walls, interconnected with the void spaces in the floors and ceilings, where a fire could spread rapidly from the basement to the attic. Single-family dwellings constructed since 1980 often have an open geometry design with lightweight structural components supporting floors and roofs. Rapid fire spread will occur in both modern and balloon frame buildings (Kerber, 2012).

A variety of hazards can be found in single-family dwellings, including faulty electrical wiring, structural changes, and hoarder housing (Knox, 2017). Each of these conditions increases the probability that a fire will occur at that home. Many homes contain storage areas for a retail business in the basement, attic, garage, or spare bedroom. Other home-based businesses, such as vehicle repair businesses or home-cleaning services, store chemical products in the home.

A trend in some older communities is to convert buildings into smaller or additional apartments. Often the renovations are not code-compliant (Pickett, 2010). Another trend is several families living together in one crowded house. In such a case, a fire company could respond to a fire in a single-family dwelling and encounter a dozen trapped residents (National Fire Data Center, 1997).

Mid-Rise Multifamily Dwellings

Newer three- to seven-story wood-frame multifamily dwellings are often Type V (wood-frame) construction, with lightweight wood truss components in the floors and roof. They are often of "podium" construction, which includes a first floor of Type I (fire resistive) construction for retail occupancies or parking (Hoesly, 2019). This type of construction has become very popular, with 187,000 new housing units consisting of buildings with 50 or more units completed in 2017. This is the most since the U.S. Census Bureau started keeping track in 1972 (Fox, 2019). Creative building designs often create voids as well as architectural features like cathedral ceilings that limit the ability of fire companies to open up ceilings to check for fire extension. Newer multifamily dwellings have extensive insulation.

Often, the first-arriving fire companies responding to an activated fire alarm in a garden-style apartment report nothing showing on arrival. When they force the door of the apartment that has the activated smoke alarm, they encounter thick smoke boiling out of the apartment.

High peak heat release furniture and tighter insulation mean that this type of building retains much more heat than older buildings. Internal exposures will suffer more heat damage, and there is a higher chance of fire spread. Lightweight wood trusses require minimal fire damage to collapse—an event that often occurs suddenly when the first-arriving company starts its operation. Multifamily dwellings with NFPA-compliant 13R automatic sprinkler systems provide greater protection to the occupants but may increase the degree of difficulty for the fire fighters. Buildings with a 13R system will be taller with longer hallways. Apparatus access may be restricted due to creative landscaping. Even with these features, a fire within the 13R-protected area will be small with less damage. The preincident plan should identify the best apparatus access positions and the hose lays required to get to every space in the building. Make sure that the fire apparatus can actually get into gated communities, and make provisions to allow prompt access to the property during every hour of every day (Norman, 2019).

FIRE OFFICER TIP

Massive Fire Losses of Mid-Rise Apartments Under Construction

Eight of the 22 largest-loss structure fires in 2017 were in mid-rise apartment buildings under construction. Many were within weeks of completion (Badger, 2018). The built-in fire protection elements—sprinklers and standpipes—had not yet been finished. The incidents required significant fire suppression resources to control the blaze; all of the buildings were declared total losses.

High-Rise Considerations

A working fire within a high-rise structure requires more fire fighters and an expanded ICS (Griffiths, 2017). The IC is responsible for the overall management of the incident. Because of the complexity of a high-rise fire, the IC often expands the ICS to include planning, operations, and logistics sections. The IC also usually appoints staff to fill the positions of safety officer, liaison officer, and public information officer on the command staff.

High-rise fires require a tremendous number of personnel to be deployed in a carefully concerted effort to achieve the strategic objectives. The National Institute of Standards and Technology (NIST) *Report on High-Rise Fireground Field Experiments: Technical Note 1797* (Butler, 2013) documents the dramatic effect crew size has on the ability to accomplish tasks. In responding to a medium-growth-rate fire on the 10th floor, the three-person crews confronted an environment where the fire had released 60 percent more heat energy than the fire encountered by the six-person crews doing the same work. Put simply, the smaller crew encountered a larger fire. It took the three-person crews 1 hour to complete their fire response, while six-person crews required 40 minutes to accomplish the same response tasks.

Managing a high-rise fire requires dividing the incident into manageable units (i.e., branches), each with a supervisor. The last operations section position that a fire officer might be called on to fill at a high-rise incident is the staging area manager. The staging area manager is responsible for managing all activities within the staging area, including layout, check-in functions, and traffic control within the area. All fire officers must understand their relationship within the ICS. The large-scale operation required to respond to a high-rise fire necessitates a large ICS structure to maintain a manageable span of control and a margin of safety and accountability (Norman, 2019).

The fire officer may also be called on to fill a position within the logistics section. Like the operations section, the logistics section is divided into branches. Typically, it includes a support branch and a **service branch**. The service branch is headed by the service branch director and is responsible for communications and fire fighter rehabilitation. The **support branch** is headed by the support branch director and is responsible for ensuring that adequate supplies, personnel, and equipment are available. A high-rise incident requires large amounts of personnel and support equipment, so a support branch director must be established as an early part of the command staff (FEMA, 2017).

The support branch director is often located at a base. A **base** is an area where the primary logistics functions are coordinated and administered. The incident command post may be co-located with the base, and there is only one base per incident. Spare self-contained breathing apparatus (SCBA) cylinders are the first priority for the support branch (Walsh et al., 2012).

In a high-rise fire, it is important to gain control of the lobby quickly. The lobby control officer controls the entry and exit of both civilians and fire fighters in the lobby. In addition, the lobby control officer oversees the use of the elevators; operates the local building communication system; and assists in the control of the heating, ventilating, and air-conditioning systems. The lobby control officer reports to the logistics section chief or, if that position is not established, to the IC.

A particularly labor-intensive task is undertaken by the **stairwell support group**. These fire fighters move equipment and water supply hose lines up and down the stairwells of the high-rise building. The stairwell support unit leader reports to the support branch director or the logistics section chief. One practice is to position fire fighters on every third floor. Equipment moves up the high rise in an assembly-line fashion, as the third-floor fire fighter moves equipment up to the sixth-floor fire fighter (Griffiths, 2017).

Residential high-rise buildings present extra challenges to fire officers; three wind-driven fires have killed members of the first-arriving fire company since 1983. Recommended practices include the following:

- Comply with your organization's standard operating procedures.
- Consider bringing the big attack line first.
- Beware of weather conditions.
- Assemble an adequate crew.

Postincident Analysis

Situations that involve multicompany operations should be reviewed from the perspective of how well the overall team performed, in addition to individual fire officers conducting company-level reviews. The format of the multicompany review depends on the nature and the magnitude of the incident, whether it was a one-room house fire that was addressed by three or four companies or a five-alarm blaze that involved dozens of companies and activated every component of the fire department. It is relatively easy to gather the companies that were involved in a one-alarm fire to conduct a basic review of the incident soon after the event. The critique of a large-scale incident, by comparison, can be a major event involving extensive planning, scheduling, and preparations.

Some organizations conduct critiques only for multialarm or unusually large-scale incidents or for situations where something obviously went wrong. This creates the impression that nothing can be learned from ordinary-sized incidents, unless someone made a mistake. Whether it is conducted on a large or small scale, the primary purpose of a review process should always be to serve as an educational and training tool, not to place blame for improper or deficient actions. Approach every postincident review in a nonthreatening manner and provide an open format for discussion as an aid to future training.

The review process should be conducted routinely and in a systematic manner. The same basic factors should be examined and discussed. Many fire departments have a standard critique format and procedure.

Preparing Information for an Incident Review

A multicompany incident review is conducted by the IC. The preparatory work, however, is often delegated to one of the company officers. An administrative fire officer might assign the officer in charge of the first-due company to prepare the information and make the necessary arrangements for the review of a one-alarm incident. The battalion chief could be responsible for preparing the information for a critique of a multialarm situation that will be presented by a higher-ranking chief officer; however, the work of gathering and assembling the information might still be delegated to a company officer or staff officer.

FIRE OFFICER TIP

Addressing Poor Performance

Reviews should always focus on lessons learned and positive experiences; however, in some situations, errors must be addressed. The serious discussion between the fire officer and his or her immediate supervisor should occur in private, not in front of a room full of observers. At the company level, poor performance by one particular crew member should be discussed one-on-one in the office, not at the kitchen table. Areas where the whole crew needs to make improvements should be discussed with the entire crew.

Before the actual review occurs, information about the situation leading up to the incident and whatever occurred before the arrival of the first fire department unit should be obtained. This background sets the stage for a discussion of the actual event from an operational perspective and often reveals important factors that were unknown before the incident occurred. The presentation of this information allows individuals to understand what happened and what they observed at the time but did not have the opportunity to interpret.

In the case of a fire involving a building, this process should begin with a review of the building, including its size and arrangement, construction type, date of construction, known modifications or renovations, and any known history that could be significant. All built-in fire protection features, including firewalls, automatic sprinklers, standpipes, and alarm and detection systems, should be identified, with as much detail as possible being provided. Any structural factors that could affect fire spread, such as unsealed openings around pipes, open shafts, and concealed spaces, should be identified as well.

Other questions to be addressed deal with the nature of the occupancy. Was the building occupied, unoccupied, vacant, abandoned, or undergoing renovation? If it was a business, what did it do, make, or store on the premises? What was the fire load? Did the fire grow and spread more quickly because of the nature of the contents?

Some questions may not have answers at the time of the postincident review but should still be considered in planning for this critique. Where, when, and how did the fire start? What was burning initially and how did the fire spread? Did the fire involve the contents or the structure, or both? Who discovered the fire, what did they observe, and when was the fire department called?

If the desired information about the building and occupancy is not readily available, the officer who is gathering this information can refer to several sources of information, such as the inspection file at the fire prevention bureau or the city department that issues building permits and keeps copies of the plans on file. Valuable information can often be obtained from preincident plans. Indeed, it is an excellent quality-control technique to compare the preincident plan information with the situation that was actually encountered. The postincident review is also a good opportunity to determine whether the information on file was useful to the IC.

The fire investigator can usually provide information about the fire's cause and origin, and the communications center can provide information about the source of the alarm and a complete listing of the resources that responded. These data should include all apparatus used, as well as the staffing level on each vehicle, if this is a variable. A recording of the incident radio communications should be obtained, as should photographs or videos of the incident.

If water supply was a significant factor in the incident, additional information about the water system should be obtained and evaluated. The locations of hydrants and available flow data should be obtained. The main sizes and whether the area is within a distribution grid or at the end of a dead-end main could be important.

All of this information should set the stage for the analysis of the events that transpired from the time the first unit arrived at the scene. The analysis of the situation before arrival often reveals factors or circumstances that were unknown to the companies involved in the incident. The officer who is preparing for the incident review should be prepared to present all of this information in a manner that is factual and informative.

Conducting a Critique

The actual critique should be scheduled as soon as convenient after the event, while allowing adequate time to gather and prepare the necessary information. It is best to capture the information while it is still fresh in the minds of the participants. With a small-scale incident, it may be possible to have all of the involved companies in attendance at the review so that all of the crew members can be involved in the discussion and learning experience. The attendance at a critique of a larger-scale incident could be limited to company officers and command officers; however, it is a good idea to invite the entire crew of any company that played a significant role or was involved in some unusual occurrence.

Start the critique with an overview presentation of the background and basic information about the incident, including the timeline and the units that were dispatched. The first-arriving officer should then be asked to describe the situation as it was presented on arrival and the actions that were taken. This description should include the first officer's role as the initial IC as well as the tactical operations performed by that company.

Each successive company should then take a turn explaining what they saw and what they did. An effective visual method is to draw a plot layout of the area on a dry-erase board and then have the various company-level officers locate where they positioned their apparatus and where they operated and identify which actions they took. This input should gradually produce a complete diagram of the incident. Photographs, videos, and radio recordings can be inserted at pertinent points in the discussion. If a large number of companies were involved in the incident, the presentation could be made by division/group/unit officers instead of individual company officers.

During this process, the moderator should keep the analysis directed at key factors, including what the initial strategy was and how it changed (if such changes occurred), how the command structure was developed, how resources were allocated, and which special or unusual problems were encountered. The discussion should also focus on standard operating procedures, including whether they were followed and how well they worked in relation to the actual situation. The officer acting as moderator should help the participants differentiate between problems that occurred because procedures were not followed and areas where procedures should be changed or updated **TABLE 14-2**. The best way to evaluate the effectiveness of procedures is to determine whether following them actually produced the anticipated results. If a deviation from standard procedures occurred, the reasons and the impact should be discussed.

TABLE 14-2 Incident Review Questions

- Did the preincident plan provide accurate and useful information?
- Are there factors that could have or should have been addressed by fire prevention before the incident?
- Were the appropriate units dispatched based on procedures and the information that was received?
- Were the units dispatched in a timely manner?
- Was the appropriate information obtained and transmitted to the responding units?
- What was the situation on arrival?
- What was the initial strategy as determined by the initial incident commander?
- How did the strategy change during the incident?
- How was the Incident Command structure developed?
- Were the resources provided adequate for the situation?
- How were the resources allocated and assigned?
- Were standard operating procedures followed?
- Do any standard operating procedures need to be changed?
- Which unusual circumstances were encountered, and how were they addressed?
- Is additional training needed?
- Did all support systems function effectively?

© Jones & Bartlett Learning.

Once every company-level officer has had a chance to make a presentation, the officer directing the critique should provide his or her perspective on and assessment of the operation, including both positive and negative factors. This officer should make a point of evaluating the outcome in terms of where and when the fire was stopped and how much damage occurred. If the outcome was positive and everything met expectations, praise should be widely distributed. Where there is room for improvement, that fact should be noted in terms of valuable lessons learned.

Have the attendees identify the positive and negative points and list the actions that should be taken to improve future performance. These points should be listed on the board for everyone to see as the final product of the critique.

Documentation and Follow-Up

The last step in conducting an incident review is to write a summary of the incident for departmental records. The results of this review should be documented in a standard format; many fire departments use a form to document the essential findings. Use of a form allows the information to be stored in a consistent manner and ensures that all relevant information is collected. The form should also list recommendations for changes in procedures, if necessary.

The completed package should then be forwarded to the appropriate section or sections within the department for review and follow-up. The training division should receive a copy of every incident review to determine training deficiencies and needs. Recommendations for policy changes should also be forwarded to the appropriate personnel for consideration.

The completed report should include all of the basic incident report information included in the National Fire Incident Reporting System (NFIRS) as well as the operational analysis information.

You Are the Fire Officer Conclusion

The first-arriving officer at a major incident is expected to take initial command of the incident, complete a size-up, announce strategy, and call for additional resources. The fire officer will need to start the organizational structure of a larger incident management team as well as identifying special hazards or needs. Once the command hand-off is made with the chief who is taking over as the IC, the fire officer will be assigned to a division, group, or unit based on the needs and resources.

The governor activates the state emergency operations center and, if needed, requests a presidential declaration. The federal resources will start after the declaration is made. Under NIMS, there is no change in command practice after the event is declared a major disaster.

After-Action REVIEW

IN SUMMARY

- The National Response Framework (NRF) is a comprehensive, national, all-hazards approach to domestic incident response that describes specific authorities and best practices for managing incidents.
- The National Incident Management System (NIMS) provides a consistent template for managing incidents.
- Federal, state, and some private-sector and nongovernmental entities organize their resources and capabilities under 15 emergency support functions (ESFs).
- The "Stafford Act" is the administrative regulation that describes federal government disaster and emergency assistance to state and local governments, tribal nations, eligible private nonprofit organizations, and individuals affected by a declared natural disaster or emergency.
- The incident action plan (IAP) outlines the strategic objectives and states how emergency operations will be conducted.
- Command staff under NIMS requires a minimum of an incident commander and safety officer.
- The complete command staff positions under NIMS include incident commander, safety officer, liaison officer, and public information officer.

- The four major components of emergency scene operations under NIMS are operations, planning, logistics, and finance/administration.
- The individuals in charge of the four components of emergency scene operations are known as ICS general staff.
- Tactical-level is where strategy specified in the IAP is converted into action.
- Tactical-level management elements are dispersed into divisions, groups, and units.
- Division represents a geographic operation, such as one floor or one side of a building.
- Group represents a functional operation, such as ventilation.
- Unit is a generic term can be applied to either a geographic or functional component.
- Company-level officers can be assigned by the incident commander to function as a division/group/unit supervisor.
- A branch is a supervisory level to provide an appropriate span of control.
- Staging is a standard procedure to manage uncommitted resources at the scene of an incident.
- Individual companies operate at the task level.
- A task force includes two or five single resources assembled to accomplish a specific task.
- A strike team consists of five units of the same type with an assigned leader.
- The most frequent location of civilian deaths and fire-related injuries is in the home.
- Wood-frame, multifamily, 3- to 7-story dwellings of Type V construction have been the largest type of housing units recently constructed.
- High-rise fires require a tremendous number of personnel and resources to make an effective fire attack, requiring dividing the incident into manageable branches.
- A high-rise stairwell support group practice is to place a fire fighter on every three floors to move hose and equipment up in an assembly-line fashion.
- The primary purpose of a postincident analysis is to serve as an educational and training tool.

KEY TERMS

Base The location at which the primary logistics functions are coordinated and administered.

Branch Supervisory level established in either the operations or logistics function to provide an appropriate span of control.

Branch director A person in a supervisory-level position in either the operations or logistics function to provide a span of control.

Command staff In an incident management organization, positions consisting of the incident commander, public information officer, safety officer, liaison officer, and other positions as required.

Division A supervisory level established to divide an incident into geographic areas of operations.

Division supervisor A person in a supervisory-level position responsible for a specific geographic area of operations at an incident.

Finance/administration section The section responsible for all costs and financial actions of the incident or planned event, including the time unit, procurement unit, compensation/claims unit, and the cost unit.

Group A supervisory level established to divide the incident into functional areas of operation.

ICS general staff The group of incident managers composed of the operations section chief, planning section chief, logistics section chief, and finance/administration section chief.

Incident action plan (IAP) A verbal plan, written plan, or combination of both that is updated throughout the incident and reflects the overall incident strategy, tactics, risk management, and member safety requirements approved by the incident commander.

Liaison officer A member of the command staff, the point of contact for assisting or coordinating agencies.

Logistics section The section responsible for providing facilities, services, and materials for the incident or planned event, including the communications unit, medical unit, and food unit within the service branch and the supply unit, facilities unit, and ground support unit within the support branch.

Logistics section chief A supervisory position that is responsible for providing supplies, services, facilities, and materials during the incident. The person in this position reports directly to the incident commander.

National Incident Management System (NIMS) A comprehensive guide to prevent, protect against, mitigate, respond to, and recover from incidents.

National Response Framework (NRF) A comprehensive, national, all-hazards approach to provide federal resources in support of a major multijurisdictional incident.

Operations section The section responsible for all tactical operations at the incident or planned event, including up to 5 branches; 25 divisions/groups; and 125 single resources, task forces, or strike teams.

Operations section chief A supervisory position that is responsible for all tactical operations at the incident or planned event. This position deals with the strategy and tactics that are required to get the job done and reports directly to the incident commander.

Planning section The section responsible for the collection, evaluation, dissemination, and use of information related to the incident situation, resource status, and incident forecast.

Planning section chief A supervisory position that is responsible for the collection, evaluation, dissemination, and use of information relevant to the incident. This position reports directly to the incident commander.

Public information officer A member of the command staff responsible for interfacing with the public and media or with other agencies with incident-related information requirements.

Safety officer A member of the command staff responsible for monitoring and assessing safety hazards or unsafe situations and for developing measures for ensuring personnel safety.

Service branch A branch within the logistics section responsible for service activities at an incident or planned event, including the communications unit, medical unit, and food unit.

Staging A specific function in which resources are assembled in an area at or near the incident scene to await instructions or assignments.

Stairwell support group A group of fire fighters who move equipment and water supply hose lines up and down the stairwells at a high-rise fire incident. The stairwell support unit leader reports to the support branch director or the logistics section chief.

Strike team Specified combinations of the same kind and type of resources, with common communications and a leader.

Strike team leader A supervisory position that is in charge of a group of similar resources.

Support branch A branch within the logistics section responsible for providing the personnel, equipment, and supplies to support incident operations, including the supply unit, facilities unit, and ground support unit.

Task force A group of resources with common communications and a leader that can be pre-established and sent to an incident or planned event or formed at an incident or planned event.

Task force leader A supervisory position that is in charge of a group of dissimilar resources

Unit An engine company, truck company, or other functional or administrative group.

REFERENCES

Badger, Stephen. 2018, November 1. "Large-loss Fire and Explosions in 2017." *NFPA Journal*. Accessed July 29, 2019. https://www.nfpa.org/News-and-Research/Publications-and-media/NFPA-Journal/2018/November-December-2018/Features/Large-Loss-Fires-2017.

Brunacini, A. V. 2002. *Fire Command: The Essentials of IMS*. Phoenix, AZ: Heritage Publishers.

Brunacini, A. V. 2004. *Command Safety: The IC's role in Protecting Firefighters*. Stillwater, OK: Fire Protection Publications.

Butler, Kathryn M., ed. 2013, April. *Report on High-Rise Fireground Field Experiments. NIST Technical Note 1797*. Gaithersburg, MD: National Institute of Standards and Technology.

Dodson, David W. 2016. *Fire Department Incident Safety Officer, Third edition*. Burlington, MA: Jones and Bartlett Learning.

Evarts, Ben. 2018, October. *Fire Loss in the United States during 2017*. Quincy, MA: National Fire Protection Association.

Federal Emergency Management Agency (FEMA). 2016. *National Response Framework, Third edition*. Washington, DC: Department of Homeland Security.

Federal Emergency Management Agency (FEMA). 2017. *National Incident Management System, Third edition*. Washington, DC: Department of Homeland Security.

Fleming, Robert S. 2013. *Emergency Incident Media Coverage*. Tulsa OK: Pennwell Corporation.

Florida Ambulance Association. 2018, July 2. *Florida Ambulance Deployment Standard Operating Procedure*. McLean, VA: American Ambulance Association.

Fox, Justin. 2019. "Why America's New Apartment Buildings All Look the Same." *Bloomberg Businessweek*. Accessed July 29, 2019. https://www.bloomberg.com/news/features/2019-02-13/why-america-s-new-apartment-buildings-all-look-the-same.

Griffiths, James S. 2017. *Fire Department of New York – an Operational Reference, 11th edition*. Little Silver, NJ: James S. Griffiths.

Hoesly, Nathan. 2019. "Cost-Effective Mid-Rise Podium Construction." *Niskian.com*. Accessed July 29, 2019. https://www.nishkian.com/cost-effective-mid-rise-podium-construction/.

Kerber, S. 2012. "Analysis of Changing Residential Fire Dynamics and Its Implications on Firefighter Operational Timeframes." *Fire Technology* 48 (4): 865–891.

Knox, Allison G. S. 2017, August. "Hoarding: A Public Safety Issue That Needs to Be Addressed." *Emergency and Disaster Management Digest*. Accessed July 29, 2019. https://edmdigest.com/opinion/hoarding-public-safety/.

Lindsay, Bruce R. 2019, January 16. *Stafford Act Assistance and Acts of Terrorism, Report 44801*. Washington, DC: Congressional Research Service.

National Fire Data Center. 1997. *Socioeconomic Factors and the Incidence of Fire*. Emmitsburg, MD: United States Fire Administration.

National Transportation Safety Board (NTSB). 2018, October 11. "Preliminary Report Pipeline: Over-pressure of a Columbia Gas of Massachusetts Low-pressure Natural Gas Distribution System." Accessed July 29, 2019. https://www.ntsb.gov/investigations/AccidentReports/Pages/PLD18MR003-preliminary-report.aspx.

Norman, John. 2019. *Fire Officer's Handbook of Tactics, Fifth ed*. Tulsa, OK: Fire Engineering Books and Videos.

Pickett, William. 2010. "Scooped Out Living Space: New Challenge for Operations." *FDNY PRO*. Accessed July 28, 2018. http://www.fdnypro.org/scooped-out-living-space-new-challenge-for-operations/.

Ryan, Tim. 2014, June 1. "Packing a Red Bag" *FireRescue-Magazine.com*. Accessed July 28, 2019. https://www.firerescuemagazine.com/articles/print/volume-9/issue-6/wildland-urban-interface/packing-a-red-bag.html.

Walsh, Donald W., Hank T. Christen, Christian E. Callsen, Geoffrey T. Miller, Paul M. Maniscalco, Graydon C. Lord, and Neal J. Dolan. 2012. *National Incident Management System: Principles and Practice, Second edition*. Burlington, MA: Jones and Bartlett Learning.

Fire Officer in Action

You are the first-arriving fire officer at what is going to be a major incident. A huge warehouse that stores hazardous materials has suffered an explosion with a partial building collapse. An intense fire with concentrated and odd-colored flames is issuing from the collapsed area.

1. Which of these is the document that establishes the rules for the deployment of federal resources to an incident?
   A. National Incident Management System
   B. Incident Command System
   C. National Response Framework
   D. National Mutual Aid Plan

2. Every major incident requires a(n):
   A. incident action plan.
   B. signed joint interim command agreement.
   C. geospatial representation of the incident-related hazards.
   D. payroll master.

3. If a flood requires evacuation of an apartment community, command will assign a _____ officer to assist the apartment management team.
   A. community risk reduction
   B. finance
   C. liaison
   D. logistics

4. The team of fire fighters assigned to the seventh floor of an office building that is on fire will be identified as a:
   A. task force.
   B. group.
   C. division.
   D. strike team.

Fire Captain Activity

NFPA Fire Officer II Job Performance Requirement 5.6.1
Produce operational plans, given an emergency incident requiring multi-unit operations; the current editions of NFPA 1600, NFPA 1700, NFPA 1710, and NFPA 1720; and AHJ-approved safety procedures, so that required resources and their assignments are obtained and plans are carried out in compliance with NFPA 1600, NFPA 1710, and NFPA 1720 standards as well as the NFPA 1700 guidelines along with the AHJ-approved safety procedures to mitigation the incident.

Application of 5.6.1

1. Using a preincident plan of a hospital or nursing home in a jurisdiction you are familiar with, develop a multi-alarm incident action plan (IAP) to handle a liquid-oxygen-fed fire on the third floor that is burning all of the contents of two rooms and has spread into the common hallway. You have three residents and two healthcare workers with serious burns.

2. It is 4 AM, and you are the first-arriving fire company at a reported apartment fire. You observe heavy fire in the lobby of the three-story garden-style apartment building, cutting off occupant access to the interior stairway. You see people at the second and third floor balconies on side A. You also see black smoke and a red glow coming from the B-C corner of the building. There are 12 apartments in the fire building. Document your IAP. Establish command and implement your IAP.

3. In 2018, an overpressure of a natural gas distribution system damaged 131 structures and destroyed 5 homes in Merrimack Valley, Massachusetts (NTSB, 2018). Review the information about this incident and develop an IAP for a community you are familiar with if a similar situation were to develop.

4. It is 11:40 AM on a Thursday morning, and your fire company is responding as an emergency medical first responder to the high school cafeteria for a sick student. On arrival you observe four students in distress, there is vomit on the floor, and the assistant principal reports that some type of gas was released during the first lunch period. There are dozens of sick students and staff. Establish command and develop your IAP.

 Access Navigate for flashcards to test your key term knowledge.

Continuous Case Study

PART 2: Captain (Fire Officer II)

Two days after the incident, the lieutenant meets with the captain and provides a written report of the argument and the initial investigation. The report notes that it appears that many fire fighters are members of the private social media page, and he is concerned about the vile, demeaning, and threatening comments.

The captain performs follow-up interviews:

The male fire fighter who was involved in the personal relationship admits to posting the pictures and adding derogatory remarks. He said that he was mad and "blowing off steam" within a private social media site. The male fire fighter was evasive and unresponsive when asked if the pictures were taken when he was on duty and whether or not the private social media site was accessed using a fire department computer or digital connection.

The female fire fighter who was involved in the personal relationship complains that she is continuing to receive vile and demeaning comments on the private social media site, including some comments that threaten her with sexual assault and on-the-job sabotage. She provides screen shots of the message threads. She says she is worried about her safety and will be notifying the police.

Questions for discussion or assignment:

1. Should the captain suspend or change the assignment of the male fire fighter who was in the argument at the training session and who posted pictures and derogatory comments on a private social media page?
2. What obligations does the captain have when a part-time employee expresses concern about on-duty safety?
3. Does the fire department have any jurisdiction over what is posted in a private (password access-controlled) social media site?
4. When should the deputy chief be notified? Should the deputy chief be notified when the captain has completed the investigation, or sooner, when a high-liability situation appears to be evolving?

Appendix A
Preparing for Promotion

Introduction

The purpose of this appendix is to provide a general description of the civil service promotional examination process that is used by most fire departments. Many variations of the testing procedures and promotional processes exist, so you need to understand the specific process that is used in your organization. Many organizations change elements or assessment tools from examination to examination.

The Origin of Promotional Examinations

Prior to the Civil War, most government jobs were awarded according to the patronage or **spoils system**—that is, those in power could appoint people to public office based on a personal relationship or political affiliation, rather than on merit. The best jobs went to political supporters and often required a payment to the individual who had the power to make appointments.

When the State of New York created the Metropolitan Fire Department in 1865, Tammany Hall was the seat of political power. William March "Boss" Tweed was the chairman of the Democratic Party and the Grand Sachem (leader) of the Society of St. Tammany, which had been founded in 1789 as a club for patriotic and fraternal purposes. Between 1865 and 1871, Boss Tweed and Tammany Hall exploited the spoils system in New York City, swindling an estimated $75 million to $200 million from the city. In response to this widespread corruption, the public demanded political reform (Ackerman, 2011).

Congress enacted the Pendleton Civil Service Reform Act in 1883 in response to the extensive corruption at Tammany Hall in New York City and the many other publicized abuses of the patronage system (Glass, 2018). The Pendleton Act established the civil service system within the federal government and provided a model for the civil service systems that were developed by many states and cities in subsequent years. The spoils system was gradually replaced by merit selection and promotion for most government employees. The process of developing promotional examinations for fire officer positions, based on testing for specific knowledge and skills, was derived directly from this act.

When the International Association of Fire Fighters (IAFF) was organized in 1918, one of their principle objectives was to ensure that appointments and promotions were based on individual merit, not political affiliation (IAFF, 2018).

Sizing Up The Promotional Process

The fire officer promotional process remains an inexact science. Many variations of examination and testing procedures are in use, with differences in emphasis and on the weights assigned to particular **dimensions** (attributes or qualities that can be described and measured during a promotional examination). The rank (sergeant, lieutenant, or captain) and the classified job description determine the technical, theoretical, or behavioral emphasis of the examination (Thiel, 2012).

In most fire departments, completion of a promotional examination process creates an eligibility list that lasts from 2 to 6 years. Depending on local practice, the list may be either rank ordered or banded. On a rank-ordered list, the highest-scoring candidate is ranked number 1, the second-highest-scoring candidate is ranked number 2, and so forth. In most cases, individuals are promoted in the order in which they are placed on the eligibility list.

Other departments use a banded list, in which the candidates are placed into bands, or groups, of promotional candidates. The bands are usually identified as

"Highly Qualified," "Qualified," and "Not Qualified." In this case, all of the candidates on the Highly Qualified list are considered equally qualified for promotional consideration, followed by all of the Qualified candidates (Murtaugh, 1993).

Postexamination Promotional Considerations

Regardless of the eligibility-ranking scheme, the jurisdiction will make promotions to meet departmental and community needs (Krieger and Masten, 2012). A basic requirement is that the candidate must be medically qualified to assume the new job. Most promotional job announcements require successful candidates to have an appropriate medical or physical performance rating **FIGURE A-1**. A candidate who is medically restricted or cannot complete the required physical ability assessment may not be considered for promotion until these issues are resolved.

In addition, the candidate must not be the subject of active formal discipline. Departments often identify a required time period that must pass between a formal disciplinary action and consideration for an officer promotion. This period, which varies from a month to a full year, will be described in the jurisdiction's personnel regulations, the department's administrative procedures, or the current labor contract. For example, a candidate who receives a 2-day suspension without pay for repeated tardiness in reporting for duty may not be considered for promotion for 365 days, even after ranking number 1 on a lieutenant eligibility list (Robinson, 2011).

The increasing complexity of the fire department mission adds variables to the decision of who should be promoted. Some specialty supervisory positions require additional certifications or work experience.

FIGURE A-1 Most promotional job announcements require that the candidate have an appropriate physical performance rating.
© Jones & Bartlett Learning. Courtesy of MIEMSS.

For instance, to be promoted into the position of arson squad captain, the candidate also needs to have certification as an NFPA 1033 Fire Investigator and 2 years of experience as a fire investigator. Other assignments may include preferred qualifications. For example, the department may prefer promoting bilingual candidates in fire companies that serve diverse communities.

Each jurisdiction has a promotional process that evolved as a result of community needs, consent decrees, lawsuit settlements, arbitration decisions, grievance settlements, labor contracts, and memoranda of understanding. It is important to learn which rules, goals, or practices affect your department's promotional practice. Notably, results of the once-a-decade census document a community's diversity and may influence public safety promotion practices.

When Fire Officers Are Voted In

The American fire service started with neighbors helping neighbors using bucket brigades and long hooks. During that era, the community of fire fighters voted for their company foreman.

Today, some states do not require certification training for volunteer fire officers, instead depending on the authority having jurisdiction to establish experience and training requirements for fire company and command officers. Completing a fire officer training program will benefit the newly elected fire officer, providing information and knowledge of procedures that can assist the volunteer fire department in serving its community (LoSasso, 2012).

Preparing a Promotional Examination

The preparation of a promotional examination usually involves a combined effort between the fire department and the municipality's human resources section. Some jurisdictions may contract with a testing organization or a consultant to organize, develop, or deliver a promotional exam. Public safety promotional examinations are the most complex and detailed examinations conducted by a municipality.

The fire department usually establishes a test preparation committee consisting of three or more officers, chaired by a chief officer. These individuals are the subject-matter experts and are responsible for validating the technical content of the promotional examination. If the promotional examination is developed within the agency, the test committee members write the questions and establish its content.

Charting the Required Knowledge, Skills, and Abilities

The municipality's personnel or human resources department uses two documents to define the **knowledge, skills, and abilities (KSAs)** that are required for every classified position within the municipality: a narrative job description and a technical class specification. Fire officers need to be familiar with both of these documents to prepare a promotional examination (Mahoney, 1983).

A narrative **job description** summarizes the scope of the job and provides examples of the typical tasks a person holding that position would be expected to perform (San Francisco, 2013). The job description also lists the KSAs needed at the time of appointment and any special requirements that apply. Examples of special requirements could include possession of a valid driver's license and an emergency medical technician (EMT) certificate. Other requirements might include a Class "A" medical rating, passing a work performance test, or achievement of technical or professional certifications such as NFPA Fire Officer I. When the human resources department posts a promotional announcement, a full or partial narrative job description is usually included.

Human resources departments also prepare a technical **class specification** worksheet to quantify the KSA components of every classified municipal job. The classification system is a core component of the civil service system and is used to determine the compensation level for a position. For example, the classification details would show why the base salary for a battalion chief is set 35 percent higher than that for a lieutenant, based on the KSAs (Filer and Filer, 1977).

The worksheet also provides a map for the promotional examination to identify the factors that need to be evaluated. A promotional examination should focus on the unique, high-importance tasks that distinguish one position from another.

Periodically, a classification specialist from the human resources department surveys the individuals within a classification to rank the frequency and importance of a wide range of job tasks. This survey is performed to validate and update the job description and the KSA technical worksheet.

Promotional Examination Components

There is no perfect promotional exam. Through trial and error, research, grievances, and court decisions, the following components are frequently utilized components of a promotional exam (Carr, 2017).

The decision regarding which components to use is influenced by time requirements, expense, staff expertise, and past experience. These components can be locally developed or provided by a vendor, consultant, or assessment specialist.

Multiple-Choice Written Examination

Multiple-choice written examinations are widely used in the promotional process because they can be structured to focus on specific subjects and factual information. There is no element of style or creativity in answering a multiple-choice question; the candidate simply has to select the appropriate answer. The scoring is equally straightforward because an answer is either right or wrong.

The multiple-choice written examination concentrates on facts that can be found within the materials on a reading list. The reading list generally includes textbooks, reference books, standard operating procedures manuals, department rules and regulations, and other locally developed reference materials.

Using the KSA technical worksheet, the test committee determines the number of questions to include from each knowledge area.

The lieutenant examination usually includes many technical questions covering engine, truck, and rescue company operations. These questions are directed toward a candidate's ability to demonstrate the basic knowledge that would be important for a lieutenant. In general, the first-level supervisory examination has the most diverse reading list, and as much as 70 percent of this test focuses on the technical aspects of a supervising fire officer. Typical technical questions cover building construction, incident management, hydraulics, emergency medical care, and firefighting tactics. Many examinations also test the candidate's knowledge in specialized company tasks, such as technical/heavy rescue, truck company operations, and rapid intervention teams.

Supervisory questions tend to focus on the immediate requirements affecting fire company preparedness, such as a subordinate unfit for duty or injured on the job. For example, a lieutenant examination might include a question about how to handle a fire fighter reporting late for duty. The hierarchical nature of the fire department requires that the first-level supervisor identify the problem, stop the behavior, and report to or consult with a senior officer or chief before taking any significant supervisory action.

A captain examination usually includes fewer technical questions and more administration questions because the captain position involves more management responsibilities. The promotion process would assume that the candidate has already met the qualifications for the lower-level position. The second-level supervisor might also be expected to develop a fire fighter's work improvement plan, prepare a budget proposal, evaluate fire company performance, or develop a department-wide training plan. For example, a captain candidate could be asked questions that relate to evaluating a fire fighter's reporting-time performance over a year. The supervisor should be able to identify patterns and determine the underlying causes for tardy reporting.

There are three options for constructing a multiple-choice written examination. Sometimes the local exam committee develops the test and the committee members write the questions. Alternatively, private companies may develop generic examinations that are used by many fire departments. The department has the third option of hiring a consultant to write a more specific examination that is directed toward local priorities.

Larger fire departments are more likely to develop an examination internally because they have the necessary resources. For example, the fire chief in a large city can probably assign a command officer and five to seven company officers to write the questions for a supervising fire officer examination.

The committee developing the examination first determines how many questions are needed to assess the candidate's knowledge within each particular area. For example, the committee might decide that nine building construction questions should be included in an examination. Two of the committee members are assigned to work together and develop 15 multiple-choice questions on building construction. The whole committee would then meet to select the nine best questions for the examination.

The selection process includes evaluating each question for job-content/criterion-referenced validity and reliability. **Job-content/criterion-referenced validity** means that the committee has certified that the knowledge being measured by the question is required on the job and referenced to known standards. The job requirement is identified through the KSA technical worksheet developed from the job description and class specification. The criterion comes from the appropriate NFPA standard and is linked to a published reference document. The best test questions selected by the committee are the ones with superior job-content/criterion-referenced validity.

Reliability is the characteristic wherein a test measures what it is intended to measure on a consistent basis. Test-and-item analysis can be performed to determine whether the more able candidates were less likely to select an incorrect answer. With a poor question, high scorers tend to answer the item incorrectly more frequently than low scorers (OPR, 2000).

A smaller department would be more likely to go to an external source for the examination questions, simply because of the time and effort that would be required to write and validate them internally. Writing good examination questions is difficult work. Each question has to be fully researched and carefully worded. A question can be challenged and invalidated if the wording is not clear, if there is no correct answer, or if more than one of the answers provided could be correct.

In many fire departments, multiple-choice written examinations are the only assessment tools used for promotional examinations. The simplicity of a multiple-choice examination can also be viewed as a weakness, however. A candidate who knows all of the facts can do very well on this type of examination, even if the individual is unable to apply the information in a real-life situation. The description "book smart, street dumb" has often been used in complaints about a process that relies solely on written multiple-choice questions to promote officers.

Assessment Centers

Some public safety agencies began to use **assessment centers** in promotional examinations in the late 1970s. An assessment center comprises a series of simulation exercises that are used to evaluate a candidate's competence in performing the actual tasks associated with a job. The assessment center process was developed in the officer corps of the German army in the 1920s. The goal of the program was to select future officers based on their predictive performance in a 2- to 3-day assessment procedure (Tielsch and Whisenand, 1978). The same type of process can be developed for fire officers.

In-Basket Exercises

The most common assessment center activity is the **in-basket exercise**. It asks the candidate to deal with a stack of correspondence and related items that have accumulated in a fire officer's in-basket. This timed exercise measures the candidate's ability to organize, prioritize, delegate, and follow up on administrative tasks. Those individuals who do well on in-basket exercises tend to demonstrate successful managerial performance in real life.

A typical in-basket contains the following items:

- Instructions for the exercise
- A calendar

- An organizational chart or list of personnel
- Ten to 30 exercise items

An in-basket assessment generally begins by positioning the candidate as a newly promoted officer reporting to the fire station for the first time. The former officer suddenly retired or left the fire department, leaving a stack of issues that require attention in the in-basket. The candidate has a specified amount of time to go through the officer's in-basket and decide what to do with each item.

The situation is contrived so that the candidate is unable to call or contact anyone else directly for advice or assistance. The candidate might be provided with copies of the department's standard operating procedures and reference manuals. Hidden within the in-basket are surprises and time-critical issues, such as an important report that is overdue or an activity that has to be performed immediately. Frequently, the candidate encounters a scheduling conflict, such as an order from the operations chief to have the engine company report to a multiple-company drill at the same time the apparatus is scheduled for an annual pump test at the shop. Many in-basket assessments also include a writing exercise that could require the candidate to prepare a letter for the chief's signature or complete an incident report using correct formatting, spelling, and grammar.

From the test administrator's viewpoint, in-basket exercises are generally easy to organize and grade. This type of test can be difficult to develop for the lieutenant level, however, because many of the typical job tasks carried out by this fire officer in a fire station do not readily lend themselves to an in-basket assessment.

The following is a suggested method for handling promotional in-basket exercises:

1. **Review.** Look at every item and determine which items are important (critical) and urgent (involving time constraints), which are important but not urgent, and which are unimportant.
2. **Prioritize.** Handle the important and urgent items first, followed by the important and nonurgent items. When these are finished, you will have handled all items or will have only unimportant items left.
3. **Identify resources/options/alternatives.** Determine who can handle the items, which options you may have in how they can be handled, and which alternative courses of action are available.
4. **Follow up.** Provide some form of follow-up on all delegated items.
5. **Make notifications.** Ensure that appropriate persons are notified of necessary and critical information (Terpak, 2008).

Candidates can prepare for an in-basket exercise by practicing on sample exercises obtained from study material publishers. Both lieutenant and captain in-basket exercises are available; these include answers with explanations and descriptions of the dimensions assessed.

Emergency Incident Simulations

Emergency incident simulations are often included in a promotional examination to test the candidate's ability to perform in the role of officer at a fire or some other type of situation. Emergency incident management usually occupies a small amount of an officer's actual time, yet the required skill sets can be the most critical for evaluation in a promotional examination. An officer must be able to lead, supervise, and perform a set of essential skills at an emergency incident scene. Emergency incident simulations will take one of four formats (Kastros, 2018).

In the first format, the candidate is provided with information concerning an emergency situation, usually in a written format that includes pictures or preincident plan information. The candidate has to explain which actions would be taken in the situation described and which factors would be considered in the making of those decisions. This kind of exercise is known as a **"data dump" question** because it provides an opportunity for the candidate to demonstrate a depth of knowledge about a particular subject. For example, the evaluators might find out how much expertise the candidate can demonstrate regarding basement fires in mercantile properties.

In the second format, the candidate is provided with a set of basic information concerning an emergency incident to begin the exercise. As the simulation progresses, the candidate is provided with additional updates of the situation, such as the following: "Immediately after you arrive on Engine 1, the store manager reports an acrid, unknown odor coming from a leaking package in the rear of the store. It has made a dozen people sick. Which actions will you take? What will you report to dispatch?" The candidate has to react to the unfolding situation, which can change based on the answers to the previous questions.

In the third format, the candidate participates in an interactive emergency scene simulation in a classroom or incident simulation trainer. The candidate is presented with a dynamic emergency incident through a multimedia format that typically includes full-color pictures depicting the incident. The graphics often include simulated smoke and flames superimposed on photos of an actual building in the community. Using a hand-held radio, the candidate is expected to command the event using the local jurisdiction's standard operating procedures and incident command tools (accountability boards, command post clipboards, incident management system worksheets) that are appropriate to what a person at that rank would use on the street. This type of exercise is complex and can be expensive because the test-site administrator might need to arrange for role players and technical support staff to run the simulation event for every candidate.

The fourth simulation format attempts to make the conditions as realistic as possible by actually having the candidate don protective clothing, climb into the officer's seat on an apparatus, and respond to a realistic scenario with a crew of fire fighters. This type of exercise is usually performed at a large urban or regional training academy, where there are buildings and props that allow for a consistent simulation for all of the candidates. The real-life format allows a candidate to function fully in the role of an officer; however, it requires extensive preparations and a large supporting cast.

Interpersonal Interaction

An interpersonal interaction exercise is designed to test a candidate's ability to perform effectively as a supervisor. In a typical interpersonal interaction scenario, the candidate has to deal with a role player who has a problem, complaint, or dilemma. The candidate is usually provided with background information to prepare for the situation, followed by 10 to 20 minutes of face-to-face interaction with the role player.

© Jones & Bartlett Learning. Photographed by Kimberly Potvin.

Many of these assessments require the candidate to deal with a poorly performing or troubled employee. The background information could include the employee's last evaluation report, a sick leave use printout, or recent disciplinary action. In many cases, the role player has important information related to the performance issue that is revealed only if the candidate asks the right questions during the interview phase. After the interview, the candidate may have to write up an employee improvement plan or a discipline letter.

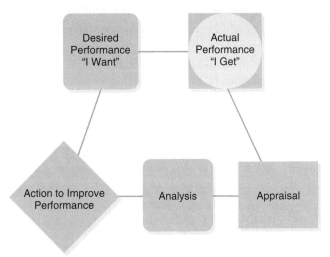

© Jones & Bartlett Learning

In an interpersonal interaction assessment, the key points are to meet the following criteria:

1. Maintain control of the interview.
2. Tell the employee the exact behavior you expect (e.g., arrive on time, wear your seat belt).
3. Give the employee a deadline to demonstrate a consistent behavioral change.
4. Specifically arrange for follow-up meetings. ("We will meet once a week to review your progress in meeting this goal.")
5. Attempt to get the employee to buy into or take personal responsibility for the improvement plan.
6. Be an empathetic listener, but remain focused on the reason for the interview.
7. Clearly explain the consequences if the employee's behavior does not change or improve.
8. Try to finish the session on a positive note ("You are a valuable member of this company and I am counting on you . . .").

A successful approach for an interpersonal interaction exercise is for the candidate to demonstrate the qualities of an "extreme supervisor." When this approach is used, any issue is handled in strict accordance with the rules and regulations of the organization.

An alternative scenario for this type of assessment is to have the role player confront the candidate as an angry or frustrated citizen. This exercise would assess how well the candidate can handle a customer service problem or a citizen complaint. This type of exercise is usually based on some situation that has actually occurred.

Writing or Speaking Exercise

A fire officer candidate should expect to deliver a short oral presentation or write a memo or report as part of the assessment center process. The oral presentation assesses oral communication, planning and organizing skills, and persuasiveness. Sometimes the candidate is required to write a report or a letter for a superior officer's signature. The written report assesses the candidate's written communication, problem analysis, and decision-making skills.

The oral presentation could be incorporated into one of the other exercises. For example, a candidate might be required to make a brief oral presentation to explain the size-up considerations of the emergency incident simulation.

Many assessment centers include an oral interview session in which the candidate is asked questions by an interview board. One of the most popular questions at this type of interview is to ask the candidate why he or she should be selected for promotion. The response provides an opportunity for the candidate to demonstrate skills or knowledge in a particular area or even to correct an error made in a previous exercise. In some cases, a candidate's ability to recognize an error and correct a response could be viewed as equal to having the right answer to every question (Kastros, 2018).

Technical Skills Demonstration

Fire officers are expected to be working supervisors and to be skilled in both task- and tactical-level activities. For some fire officer examinations, the candidates might also be required to demonstrate specific technical skills. This type of exercise is often included in promotional tests for officers who would command highly specialized teams, such as urban search and rescue, hazardous materials, fire investigations, and paramedic teams.

Mastering the Content and the Process of a Promotional Exam

To prepare for an examination, the candidate must master both the content of the examination (technical and reference information) and the process (answering multiple-choice questions or role playing as an officer). Most of the content comes from the materials on the reference reading list; however, you should also be prepared for questions that relate to current issues and recent local history. For example, if your department recently had a problem with on-duty fire fighters taking emergency incident scene videos and posting them on a website, do not be surprised to find that issue somewhere in the examination. You should stay current by reading fire service publications and periodicals to keep up with trends and issues.

Mastering the content for an examination requires a personal study plan. Many tests require fire officer candidates to absorb a tremendous quantity of factual information, which can require a major investment of time and effort. It is not unusual for candidates to spend more than a year preparing for an examination, setting aside a period of time every day for reading and answering practice questions.

Fire fighters who participate in structured programs and study groups often perform better than fire fighters who study on their own. Many fire departments have officially organized programs to help candidates prepare for exams.

Building a Personal Study Journal

One valuable tool in studying for a promotional examination is a **personal study journal**. You can use this journal to set up your personal study schedule, keep track of your progress, and make notes about confusing, interesting, or important facts you discover. The first section could include all of the information about the promotional examination—a copy of the official announcement and copies of all of the documents the candidate was required to submit. Any test-related documents from the department can also be placed in this section.

The second section would contain a calendar, usually in the one-month-per-page format. The calendar covers the period from the start of the preparation process through the last promotional test activity. Important deadlines are highlighted, and the candidate can create a study plan, working backward from the examination date.

Voice of Experience

Following a tough workday, and several months into my career as a promoted fire captain, the phone rang. The voice on the other end of the line was all too familiar as I heard the greeting, "Were you prepared for this?" I had heard this exact question asked in past firehouse stories and would continue to hear it as fire fighters prepare for promotion. Yes, I was prepared, and having learned how to handle situations to help others grow was extremely satisfying.

Preparing for promotion was one of the biggest investments in my career. After preparing for several promotional test and assessment interviews, and after watching others succeed, I have observed a great deal of what can successfully assist an individual in his or her own career enhancement. Proper preparation will take some monetary expense, but the quality time spent studying and gaining the necessary experience, though it can be overwhelming, will also reward you with excellent leadership abilities. You will be prepared to handle the many situations that the dynamics of the fire service will place in front of you. You will be able to mitigate hazards, control company operations on scene, supervise and manage a firehouse responsibly, and understand the necessity of time and performance management. I have seen many prepared individuals walk out of the fire chief's office with a new bugle, yet fail to have that winning lottery ticket expression. Fast forward several months, however, and these individuals are steps beyond those who "got lucky." Sure, some people can be successful at a test without putting forth the effort, but I have seen too many of these same individuals sweat over the minor details that face an officer on a day-to-day basis.

When your fire department provides the information for a future promotional process, you need to be "in the zone." Obtain personal copies of all of the promotional information, and know the process. Fire fighters tend to listen to the rumor mill, and they often end up with false information. Know what your department expects of you and all of its potential officers. Preparing yourself for the promotional process means obtaining all of the knowledge, skills, and abilities that you can, and utilizing as many resources as possible.

Take fire and educational courses outside of what your fire department has to offer. These courses will provide you with valuable information and skills that will remain with you throughout your career. Some of these courses may count toward a promotional process, but do not hang your hat on just those classes. Remember that this is an investment. Diversify your portfolio. A good company officer should provide you with the opportunity to "ride in the front seat" as the acting officer. Making decisions on incidents, and even throughout a station workday, will allow you to make decisions in a safe environment and with assistance from your officer.

Much of an officer's rank will lend itself to time management. Learning how to do the paperwork and company reporting will give you a chance to practice for a possible assessment center in-basket exercise and also provide you with the knowledge needed for your written exam.

Obtaining textbooks and finding the time to study is what tends to be most overwhelming for fire fighters. Other than a few hundred dollars for textbooks and a few bucks for highlighters and note cards, it is the study time that will be your biggest investment. Explaining to family members and friends what you are trying to accomplish will assist you in your endeavors. Let them know what you want and what you are willing to sacrifice—for them and for your future.

Voice of Experience

Being able to answer "yes" to the question of whether you were prepared for promotion will end up being one of your most rewarding moments. This will be when you realize that all of your time, money, and sacrifice were worth it. Utilize your resources, practice officer skills, and invest in your future. The fire service is always looking for those who come prepared to lead their men and women.

Chris Riley
Fire Captain
Portsmouth Fire, Rescue, and Emergency Services
Portsmouth, Virginia

The third section would contain the written reference materials. Subdivided by reference, the parts of this section would cover the items used to master the written material. Some candidates outline each chapter; others make up practice tests. You should make a note of any questions you miss in practice tests to be sure that you go back and obtain the correct answers. This information can be extremely valuable during your final review before the examination date.

The fourth section would cover information about the announced or anticipated components of the promotional examination. Some departments provide information to the candidates about topics that may be emphasized or areas of the references that will be excluded. For example, the department may clarify that candidates are expected to conform to the mayday procedure that was in effect before July 1. In such a case, a new procedure may be in effect on the day of the promotional test, but the exam will test the candidate on the earlier procedure. The fourth section of your personal study journal is also where you would store information about the exam structure and process.

Preparing for Role Playing

The most effective candidates in role-playing exercises are the ones who act naturally during the examination, while making supervisory decisions that are strictly based on policies and regulations. This behavior is sometimes described as performing in the "extreme supervisor" role. Many successful candidates model their performance in the examination, as well as afterward, on the behavior of successful and respected officers. It is a good idea to pay attention to the supervisors and officers who are known for being efficient and effective.

One of the best ways to prepare for role playing is to experience working in one of the busier or larger fire stations within your department. More supervisory issues come up in a fire station with 14 fire fighters and 3 officers on each shift than in a station with 3 fire fighters and 1 officer. Similarly, a fire company that runs 15 calls per day encounters more problem-solving opportunities than a company that runs 3 calls per day. Working under one of the widely respected officers is also a good way to develop supervisory skills. Candidates who have worked for a supervisor who demonstrates excellent leadership and serves as a positive role model have a significant advantage over candidates who have never had that experience.

After-Action REVIEW

IN SUMMARY

- Promotional examinations were a product of the 1883 Pendleton Civil Service Reform Act.
- In most fire departments, completion of a promotional examination process creates an eligibility list that lasts from 2 to 6 years. Depending on local practice, this list may be either rank ordered or banded.
- Each jurisdiction has a promotional process that evolved as a result of community needs, consent decrees, lawsuit settlements, arbitration decisions, grievance settlements, labor contracts, and memoranda of understanding.
- The preparation of a promotional examination usually represents the combined effort of the fire department and the municipality's human resources section.
- The municipality's personnel or human resources department uses two documents to define the KSAs that are required for every classified position within the municipality: a narrative job description and a technical class specification.
- The decision regarding which components to use in a promotional examination is influenced by time requirements, expense, staff expertise, and past experience.
- Multiple-choice examinations concentrate on facts from the reading list.
- Assessment centers provide a variety of role-playing exercises.
- During an in-basket assessment, you should remember to review; prioritize; identify resources, options, and alternatives; follow up; and notify.
- Emergency incident simulations take one of four formats: the "data dump" question, a mock emergency incident, an interactive emergency scene simulation, or a full-scale incident simulation.

- During an interpersonal assessment, remember to remain in control, exactly state the desired behavior, give the employee a deadline, arrange follow-up meetings, get the employee to buy into or take personal responsibility for the improvement plan, be empathetic but focused, explain consequences, and finish the session on a positive note.
- Fire officer candidates may be required to deliver a short presentation or write a memo or report as part of the assessment center process.
- Technical skills may be evaluated during promotional tests for highly specialized positions.
- The candidate needs to develop a personal study plan to master the content for a promotional examination. Study techniques include keeping a study journal and participating in role-playing activities.

KEY TERMS

Assessment centers A series of simulation exercises that are intended to identify a candidate's competency to perform the job that is offered in the promotional examination.

Class specification A technical worksheet that quantifies the knowledge, skills, and abilities (KSAs) by frequency and importance for every classified job within the local civil service agency.

"Data dump" question A promotional assessment question that asks the candidate to write or describe all of the factors or issues covering a technical issue, such as suppression of a basement fire in a commercial property.

Dimensions The tributes or qualities that can be described and measured during a promotional examination. On average, 5 to 15 dimensions are measured on a promotional examination. The six most commonly addressed are oral communication, written communication, problem analysis, judgment, organizational sensitivity, and planning/organizing.

In-basket exercise A promotional examination component in which the candidate deals with correspondence and related items that have accumulated in a fire officer's in-basket.

Job-content/criterion-referenced validity A type of validity obtained through the use of a technical committee of job incumbents who certify that the knowledge being measured is required on the job and referenced to known standards.

Job description A narrative summary of the scope of a job. It provides examples of the typical tasks.

Knowledge, skills, and abilities (KSAs) The traits required for every classified position within the municipality. KSAs are defined by a narrative job description and a technical class specification.

Personal study journal A personal notebook to aid in scheduling and tracking a candidate's promotional preparation progress.

Reliability The characteristic where a test measures what it is intended to measure on a consistent basis.

Spoils system Also known as the patronage system; the practice of making appointments to public office based on a personal relationship or affiliation rather than because of merit. The spoils system scandals of the New York City "Tweed Ring" and the Tammany Hall political machine (1865–1871) resulted in Congress passing the Pendleton Civil Service Reform Act of 1883.

REFERENCES AND ADDITIONAL RESOURCES

NFPA reprinted material is not the complete and official position of the NFPA on the referenced subject, which is represented only by the standard in its entirety.

Ackerman, Kenneth D. 2011. *Boss Tweed: The Corrupt Pol Who Conceived the Soul of Modern New York*. Falls Church, VA: Viral History Press.

Carr, Bryan. 2017. *Examination of the Promotional Process within the Fire Service*. U.S. Fire Administration Executive Fire Officer Program. Accessed July 2, 2019. https://www.usfa.fema.gov/downloads/pdf/mo_capstone_bryan_carr.pdf.

Filer, Robert J., and Randall K. Filer. 1977. *Assessment Centers: Development and Use*. Richmond, VA: Psychological Consultants, Inc.

Glass, Andrew. 2018, January 16. "Pendleton Act Inaugurates U.S. Civil Service System, Jan 16, 1883." *Politico.com*. Accessed July 1, 2019. https://www.politico.com/story/2018/01/16/pendleton-act-inaugurates-us-civil-service-system-jan-16-1883-340488.

International Association of Fire Fighters (IAFF). 2018. "IAFF History: Honor, Struggle and Action." Accessed June 30, 2019. http://services.prod.iaff.org/ContentFile/Get/21342.

Kastros, Anthony. 2018. *Mastering the Fire Service Assessment Center, Second edition*. Tulsa, OK: Pennwell/Fire Engineering

Krieger, Aisha, and Tirzah Masten. 2012. *Beyond the Consent Decree: Gender and Recruitment in the San Francisco Fire Department*. Emmitsburg, MD: U.S. Fire Administration, Executive Fire Officer Program.

LoSasso, C. 2012. *Mentoring Volunteer Officers Pre- and Post-promotion*. Emmitsburg, MD: U.S. Fire Administration, Executive Fire Officer Program.

Mahoney, Gene. 1983. *Firefighter Promotional Exams: Practice and Review*. New York: Arco Publications.

Murtaugh, M. 1993. *Fire Department Promotional Tests. A New Direction: New Testing Components, New Testing Formats*. Pearl River, NY: Fire Tech Promotional Courses.

Office of Policy and Research (OPR). 2000. *Testing and Assessment: An Employer's Guide to Good Practices*. Washington DC: U.S. Department of Labor.

Robinson, Graham J. 2011. *Lasting Leadership: Preparing for the Transition to Company Officer*. Emmitsburg, MD: U.S. Fire Administration, Executive Fire Officer Program.

San Francisco City and County. 2013. "Multiple Choice Test Preparation Manual." Accessed July 2, 2019. https://sfdhr.org/sites/default/files/FileCenter/Documents/18426-Multiple%20Choice%20Test%20Prep%20Manual.pdf.

Terpak, Michael A. 2008. *Assessment Center Strategy and Tactics*. Tulsa, OK: Pennwell/Fire Engineering

Thiel, Adam K. 2012. Professional Development. In *Managing Fire and Rescue Services*, edited by Adam K. Thiel and Charles R. Jennings, 247–72. Washington, DC: International City/County Management Association.

Tielsch, G. P., and P. M. Whisenand. 1978. *Fire Assessment Centers: The New Concept in Promotional Examinations*. Santa Cruz, CA: Davis Publishing.

Appendix B

NFPA 1021, *Standard for Fire Officer Professional Qualifications*, 2020 Edition Correlation Guide

NFPA1 1021, Job Performance Requirement	Corresponding Chapter(s)	Corresponding Page(s)
FIRE OFFICER I		
4.1 General. For qualification at Fire Officer Level I, the candidate shall meet the requirements of Fire Fighter II as defined in NFPA 1001, Fire Instructor I as defined in NFPA 1041, and the job performance requirements defined in Sections 4.2 through 4.7 of this standard.	1, 2, 5	See below
4.1.1 General Prerequisite Knowledge. The organizational structure of the department; geographical configuration and characteristics of response districts; departmental operating procedures for administration, emergency operations, incident management system, and safety; fundamentals of leadership; departmental budget process; information management and recordkeeping; the fire prevention and building safety codes and ordinances applicable to the jurisdiction; current trends, technologies, and socioeconomic and political factors that affect the fire service; cultural diversity; methods used by supervisors to obtain cooperation within a group of subordinates; the rights of management and members; agreements in force between the organization and members; generally accepted ethical practices, including a professional code of ethics; and policies and procedures regarding the operation of the department as they involve supervisors and members.	1, 2	pp. 4–6, 10–22, 26–31, 40–41
4.1.2 General Prerequisite Skills. The ability to effectively communicate in writing utilizing technology provided by the AHJ; write reports, letters, and memos; operate in an information management system; and effectively operate at all levels in the incident management system utilized by the AHJ.	1, 5	pp. 12, 147–149
4.2 Human Resource Management. This duty involves utilizing human resources to accomplish assignments in accordance with safety plans and in an efficient manner. This duty also involves evaluating member performance and supervising personnel during emergency and nonemergency work periods, according to the following job performance requirements.	2, 3, 5, 8	See below
4.2.1 Assign tasks or responsibilities to unit members, given an assignment at an emergency incident, so that the instructions are complete, clear, and concise; safety considerations are addressed; and the desired outcomes are conveyed.	2, 3	pp. 41–47, 74–80, 86–90

NFPA1 1021, Job Performance Requirement	Corresponding Chapter(s)	Corresponding Page(s)
(A) Requisite Knowledge. Verbal communications during emergency incidents, characteristics of leadership, techniques used to make assignments under stressful situations, and methods of confirming understanding.	2, 3	pp. 41–47, 74–80, 86–90
(B) Requisite Skills. The ability to condense instructions for frequently assigned unit tasks based on training and standard operating procedures.	2, 3	pp. 41–47, 74–80, 86–90
4.2.2 Assign tasks or responsibilities to unit members, given an assignment under nonemergency conditions at a station or other work location, so that the instructions are complete, clear, and concise; safety considerations are addressed; and the desired outcomes are conveyed.	2, 3	pp. 41–47, 74–80, 86–90
(A) Requisite Knowledge. Verbal communications under nonemergency situations, characteristics of leadership, techniques used to make assignments under routine situations, and methods of confirming understanding.	2, 3	pp. 41–47, 74–80, 86–90
(B) Requisite Skills. The ability to issue instructions for frequently assigned unit tasks based on department policy.	2, 3	pp. 41–47, 74–80, 86–90
4.2.3 Direct unit members during a training evolution, given a company training evolution and training policies and procedures, so that the evolution is performed in accordance with safety plans, efficiently, and as directed.	3, 8	pp. 74–78, 89–96, 98–99, 231–233
(A) Requisite Knowledge. Verbal communication techniques to facilitate learning.	3, 8	pp. 74–78, 89–96, 98–99, 231–233
(B) Requisite Skills. The ability to distribute issue-guided directions to unit members during training evolutions.	3, 8	pp. 74–78, 89–96, 98–99, 231–233
4.2.4 Recommend action for member-related problems, given a member with a situation requiring assistance and the member assistance policies and procedures, so that the situation is identified and the actions taken are within the established policies and procedures.	3, 5	pp. 74–78, 99–108, 141–145
(A) Requisite Knowledge. The signs and symptoms of member-related problems (such as behavioral health issues), causes of stress in emergency services personnel, adverse effects of stress on the performance of emergency service personnel, and awareness of AHJ member assistance policies and procedures.	3, 5	pp. 74–78, 99–108, 141–145
(B) Requisite Skills. The ability to recommend a course of action for a member in need of assistance.	3, 5	pp. 74–78, 99–108, 141–145
4.2.5 Apply human resource policies and procedures, given an administrative situation requiring action, so that policies and procedures are followed.	3, 5	pp. 74–83, 99–108, 138–144

NFPA1 1021, Job Performance Requirement	Corresponding Chapter(s)	Corresponding Page(s)
(A) Requisite Knowledge. Human resource policies, procedures, applicable laws, and legal concepts.	3, 5	pp. 74–83, 99–108, 138–144
(B) Requisite Skills. The ability to communicate and to relate interpersonally.	3, 5	pp. 74–83, 99–108, 138–144
4.2.6 Coordinate the completion of assigned tasks and projects by members, given a list of projects and tasks and the job requirements of subordinates, so that the assignments are prioritized, a plan for the completion of each assignment is developed, and members are assigned to specific tasks and both supervised during and held accountable for the completion of the assignments.	2, 3	pp. 41–47, 49, 51–52, 55–56, 59, 74–78, 80, 84–90
(A) Requisite Knowledge. Principles of supervision, leadership, and basic human resource management.	2, 3	pp. 41–47, 49, 51–52, 55–56, 59, 74–78, 80, 84–90
(B) Requisite Skills. The ability to plan and to set priorities.	2, 3	pp. 41–47, 49, 51–52, 55–56, 59, 74–78, 80, 84–90
4.3 Community and Government Relations. This duty involves dealing with inquiries of the community and communicating the role, image, and mission of the department to the public and delivering safety, injury prevention, and fire prevention education programs, according to the following job performance requirements.	4	See below
4.3.1 Implement a community risk reduction (CRR) plan at the unit level, given an AHJ CRR plan and policies and procedures, so that a community need is addressed.	4	pp. 116–119
(A) Requisite Knowledge. Community demographics and service organizations, verbal and nonverbal communication, and the role and mission of the department and its CRR plan.	4	pp. 116–119
(B) Requisite Skills. Familiarity with public relations and the ability to communicate.	4	pp. 116–119
4.3.2 Initiate action to a citizen's concern, given policies and procedures, so that the concern is answered or referred to the correct individual for action and all policies and procedures are complied with.	4	pp. 120–122
(A) Requisite Knowledge. Interpersonal relationships and verbal and nonverbal communication.	4	pp. 120–122
(B) Requisite Skills. Familiarity with public relations and the ability to communicate.	4	pp. 120–122

NFPA1 1021, Job Performance Requirement	Corresponding Chapter(s)	Corresponding Page(s)
4.3.3 Respond to a public inquiry, given policies and procedures, so that the inquiry is answered accurately, courteously, and in accordance with applicable policies and procedures.	4	pp. 121–124
(A) Requisite Knowledge. Written and oral communication techniques.	4	pp. 121–124
(B) Requisite Skills. The ability to relate interpersonally and to respond to public inquiries.	4	pp. 121–124
4.4 Administration. This duty involves general administrative functions and the implementation of departmental policies and procedures at the unit level, according to the following job performance requirements.	1, 2, 3, 5	See below
4.4.1 Recommend changes to existing departmental policies and/or implement a new departmental policy at the unit level, given a new departmental policy, so that the policy is communicated to and understood by unit members.	5	pp. 129–130
(A) Requisite Knowledge. Written and oral communication.	5	pp. 129–130
(B) Requisite Skills. The ability to relate interpersonally and to communicate change in a positive manner.	5	pp. 129–130
4.4.2 Execute routine unit-level administrative functions, given forms and record-management systems, so that the reports and logs are complete and files are maintained in accordance with policies and procedures.	1, 5	pp. 18–22, 145–151
(A) Requisite Knowledge. Administrative policies and procedures and records management.	1, 5	pp. 18–22, 145–151
(B) Requisite Skills. The ability to communicate.	5	pp. 145–151
4.4.3 Prepare a budget request, given a unit-level need, so that the request is in the proper format and is supported with data.	5	pp. 130–134
(A) Requisite Knowledge. Policies and procedures and the revenue sources and budget process.	5	pp. 130–134
(B) Requisite Skill. The ability to communicate.	5	pp. 130–134
4.4.4 Explain the purpose of each management component of the organization, given an organization chart, so that the explanation is current and accurate and clearly identifies the purpose and mission of the organization.	1, 2	pp. 13–14, 52
(A) Requisite Knowledge. Organizational structure of the department and functions of management.	1, 2	pp. 13–14, 52

NFPA1 1021, Job Performance Requirement	Corresponding Chapter(s)	Corresponding Page(s)
(B) Requisite Skills. The ability to communicate verbally in a clear and concise manner.	1, 2	pp. 13–14, 52
4.4.5 Explain the needs and benefits of collecting incident response data, given the goals and mission of the organization, so that incident response reports are timely and accurate.	3, 5	pp. 75, 147–150
(A) Requisite Knowledge. The agency's records management system.	3	p. 75
(B) Requisite Skills. The ability to communicate.	3	p. 75
4.5 Inspection and Investigation. This duty involves conducting inspections to identify hazards and address violations, conducting preincident plans, performing a fire investigation to determine area of origin and preliminary cause, securing the incident scene, and preserving evidence, according to the following job performance requirements.	6	See below
4.5.1 Describe the procedures of the AHJ for conducting fire inspections, given any of the following occupancies, so that all hazards, including hazardous materials, are identified, approved forms are completed, and approved action is initiated: (1) Assembly (2) Educational (3) Health care (4) Detention and correctional (5) Residential (6) Mercantile (7) Business (8) Industrial (9) Storage (10) Unusual structures (11) Mixed occupancies	6	pp. 159–166, 168–171, 177–185
(A) Requisite Knowledge. Inspection procedures; fire detection, alarm, and protection systems; identification of fire and life safety hazards; and marking and identification systems for hazardous materials.	6	pp. 159–166, 168–171, 177–185
(B) Requisite Skills. The ability to communicate and to apply the appropriate codes and standards.	6	pp. 159–166, 168–171, 177–185
4.5.2 Identify construction, alarm, detection, and suppression features that contribute to or prevent the spread of fire, heat, and smoke throughout the building or from one building to another, given an occupancy, and the policies and forms of the AHJ so that a preincident plan for any of the following occupancies is developed: (1) Assembly (2) Educational (3) Institutional (4) Residential (5) Business (6) Industrial (7) Manufacturing (8) Storage (9) Mercantile (10) Special properties (11) Mixed Occupancies	6	pp. 160–166, 168–185
(A) Requisite Knowledge. Fire behavior; building construction; inspection and incident reports; detection, alarm, and suppression systems; and applicable codes, ordinances, and standards.	6	pp. 160–166, 168–185
(B) Requisite Skills. The ability to use evaluative methods and to communicate.	6	pp. 160–166, 168–185

NFPA1 1021, Job Performance Requirement	Corresponding Chapter(s)	Corresponding Page(s)
4.5.3 Direct unit-level personnel to secure an incident scene, given rope or barrier tape, so that unauthorized persons can recognize the perimeters of the scene and are kept from restricted areas, and all evidence or potential evidence is protected from damage or destruction.	6	pp. 185–186
(A) Requisite Knowledge. Types of evidence, the importance of fire scene security, and evidence preservation.	6	pp. 185–186
(B) Requisite Skills. The ability to issue instruction for securing an incident scene.	6	pp. 185–186
4.6 Emergency Service Delivery. This duty involves supervising emergency operations and deploying assigned resources in accordance with the local emergency plan and according to the following job performance requirements.	2, 7, 8	See below
4.6.1 Develop an initial action plan, given size-up information for an incident and assigned emergency response resources, so that resources are deployed to control the emergency.	7	pp. 198–204, 206–207, 210–214
(A) Requisite Knowledge. Elements of a size-up, standard operating procedures for emergency operations, and fire behavior.	7	pp. 198–204, 206–207, 210–214
(B) Requisite Skills. The ability to analyze emergency scene conditions; to activate the local emergency plan, including localized evacuation procedures; to allocate resources; and to communicate orally.	7	pp. 198–204, 206–207, 210–214
4.6.2 Implement an action plan at an emergency operation, given assigned resources, type of incident, and a preliminary plan, so that resources are deployed to mitigate the situation.	7	pp. 198–203, 214–217
(A) Requisite Knowledge. Standard operating procedures, resources available for the mitigation of fire and other emergency incidents, an incident management system, scene safety, and a personnel accountability system.	7	pp. 198–203, 214–217
(B) Requisite Skills. The ability to implement an incident management system, to communicate orally, to manage scene safety, and to supervise and account for assigned personnel under emergency conditions.	7	pp. 198–203, 214–217
4.6.3 Develop and conduct a postincident analysis, given a single unit incident and postincident analysis policies, procedures, and forms, so that all required critical elements are identified and communicated, and the approved forms are completed and processed in accordance with policies and procedures.	2, 8	pp. 64–66, 241

NFPA1 1021, Job Performance Requirement	Corresponding Chapter(s)	Corresponding Page(s)
(A) Requisite Knowledge. Elements of a fire- or rescue-related postincident analysis; basic building construction; basic fire protection systems and features; basic water supply; basic fuel loading, basic fire growth and development; and departmental procedures relating to dispatch response tactics and operations, the source of any emergency operations controlling authority, including EMS protocols, if applicable, and customer service.	2, 8	pp. 64–66, 241
(B) Requisite Skills. The ability to write reports, to communicate, and to evaluate skills.	2	pp. 64–66
4.7 Health and Safety. This duty involves integrating health and safety plans, policies, procedures, and standards into daily activities as well as the emergency scene, including determining appropriate levels of personal protective equipment to ensure a work environment that is in accordance with health and safety plans for all assigned members, according to the following job performance requirements.	8	See below
4.7.1 Apply safety regulations at the unit level, given safety policies, procedures, and standards, so that required reports are completed, in-service training is conducted, and member responsibilities are conveyed.	8	pp. 226, 228–231, 233–234, 236–240
(A) Requisite Knowledge. The most common causes of personal injury and accident to members; safety policies and procedures; basic workplace safety; the components of an infectious disease control program; and the selection, care, and maintenance of personal protective equipment in accordance with Chapter 7 of NFPA 1851.	8	pp. 226, 228–231, 233–234, 236–240
(B) Requisite Skills. The ability to identify safety hazards and exposures, communicate, and complete documentation necessary for exposure reporting and tracking.	8	pp. 226, 228–231, 233–234, 236–240
4.7.2 Conduct an initial accident investigation, given an incident and investigation process, so that the incident is documented and reports are processed in accordance with policies and procedures of the AHJ.	8	pp. 239, 241
(A) Requisite Knowledge. Procedures for conducting an accident investigation and safety policies and procedures.	8	pp. 239, 241
(B) Requisite Skills. The ability to communicate and to conduct interviews.	8	pp. 239, 241
4.7.3 Explain the benefits of being physically and medically capable of performing assigned duties and effectively functioning during peak physical demand activities, given current fire service trends and agency policies, so that the need to participate in wellness and fitness programs is explained to members.	8	pp. 226, 230–231

NFPA1 1021, Job Performance Requirement	Corresponding Chapter(s)	Corresponding Page(s)
(A) Requisite Knowledge. National death and injury statistics, suicide prevention initiatives, fire service safety and wellness initiatives, and agency policies.	8	pp. 226, 230–231
(B) Requisite Skills. The ability to communicate.	8	pp. 226, 230–231
FIRE OFFICER II		
5.1 General. For qualification at Level II, the Fire Officer I shall meet the job performance requirements defined in Sections 5.2 through 5.7 of this standard.	9	See below
5.1.1 General Prerequisite Knowledge. The organization of local government; enabling and regulatory legislation and the law-making process at the local, state/provincial, and federal levels; and the functions of other bureaus, divisions, agencies, and organizations and their roles and responsibilities that relate to the fire service.	9	pp. 252–253, 256–265
5.1.2 General Prerequisite Skills. Intergovernmental and interagency cooperation.	9	pp. 258–259
5.2 Human Resource Management. This duty involves evaluating member performance, according to the following job performance requirements.	10	See below
5.2.1 Initiate actions to maximize member performance and/or to correct unacceptable performance, given human resource policies and procedures, so that member and/or unit performance improves or the issue is referred to the next level of supervision.	10	pp. 274–275, 277–280, 282–284
(A) Requisite Knowledge. Human resource policies and procedures, problem identification, organizational culture, group dynamics, leadership styles, and types of power.	10	pp. 274–275, 277–280, 282–284
(B) Requisite Skills. The ability to communicate to solve problems, to increase teamwork, and to counsel members.	10	pp. 274–275, 277–280, 282–284
5.2.2 Evaluate the job performance of assigned members, given personnel records and evaluation forms, so that each member's performance is evaluated accurately and reported according to human resource policies and procedures.	10	pp. 273–274, 277–283
(A) Requisite Knowledge. Human resource policies and procedures, job descriptions, objectives of a member evaluation program, and common errors in evaluating.	10	pp. 273–274, 277–283
(B) Requisite Skills. The ability to communicate and to plan and conduct evaluations.	10	pp. 273–274, 277–283

NFPA1 1021, Job Performance Requirement	Corresponding Chapter(s)	Corresponding Page(s)
5.2.3 Create a professional development plan for a member of the organization, given the requirements for promotion, so that the individual acquires the necessary knowledge, skills, and abilities to be eligible for the examination for the position.	10	pp. 287, 290–291
(A) Requisite Knowledge. Development of a professional development guide and job shadowing.	10	pp. 287, 290–291
(B) Requisite Skills. The ability to communicate.	10	pp. 287, 290–291
5.3 Community and Government Relations. This duty involves dealing with inquiries of allied organizations in the community and projecting the role, mission, and image of the department to other organizations with similar goals and missions for the purpose of establishing strategic partnerships and delivering safety, injury prevention, and fire prevention education programs, according to the following job performance requirements.	11	See below
5.3.1 Supervise multiunit implementation of a community risk reduction (CRR) program, given an AHJ CRR plan, policies, and procedures, so that community needs are addressed.	11	pp. 305–307
(A) Requisite Knowledge. Community demographics and service organizations, verbal and nonverbal communication, and the role and mission of the department and its CRR plan.	11	pp. 305–307
(B) Requisite Skills. Familiarity with public relations and the ability to supervise and communicate.	11	pp. 305–307
5.3.2 Explain the benefits to the organization of cooperating with allied organizations, given a specific problem or issue in the community, so that the purpose for establishing external agency relationships is clearly explained.	11	pp. 304
(A) Requisite Knowledge. Agency mission and goals and the types and functions of external agencies in the community.	11	pp. 304
(B) Requisite Skills. The ability to develop interpersonal relationships and to communicate.	11	pp. 304
5.4 Administration. This duty involves preparing a project or divisional budget, news releases, and policy changes, according to the following job performance requirements.	12	See below
5.4.1 Develop a policy or procedure, given an assignment, so that the recommended policy or procedure identifies the problem and proposes a solution.	12	pp. 317, 328–329
(A) Requisite Knowledge. Policies and procedures and problem identification.	12	pp. 317, 328–329

NFPA1 1021, Job Performance Requirement	Corresponding Chapter(s)	Corresponding Page(s)
(B) Requisite Skills. The ability to communicate in writing and to solve problems.	12	pp. 317, 328–329
5.4.2 Develop a project or divisional budget, given schedules and guidelines concerning its preparation, so that capital, operating, and personnel costs are determined and justified.	12	pp. 319–322, 326,
(A) Requisite Knowledge. The supplies and equipment necessary for ongoing or new projects; repairs to existing facilities; new equipment, apparatus maintenance, and personnel costs; and appropriate budgeting system.	12	pp. 319–322, 326,
(B) Requisite Skills. The ability to allocate finances, to relate interpersonally, and to communicate.	12	pp. 319–322, 326,
5.4.3 Describe the process of purchasing, including soliciting and awarding bids, given established specifications, in order to ensure competitive bidding so that the needs of the organization are met within the applicable federal, state/provincial, and local laws and regulations.	12	pp. 323–324
(A) Requisite Knowledge. Purchasing laws, policies, and procedures.	12	pp. 323–324
(B) Requisite Skills. The ability to use evaluative methods and to communicate.	12	pp. 323–324
5.4.4 Prepare a media release, given an event or topic, so that the information is accurate and formatted correctly.	11	pp. 305–307
(A) Requisite Knowledge. Policies and procedures and the format used for media releases by various media outlets, including the use of social media in accordance with AHJ policies and procedures.	11	pp. 305–307
(B) Requisite Skills. The ability to communicate.	11	pp. 305–307
5.4.5 Prepare a concise report for transmittal to a supervisor, given fire department record(s) and a specific request for details such as trends, variances, or other related topics, so that the information required for the AHJ is accurate and documented.	12	pp. 317–318
(A) Requisite Knowledge. The data processing system.	12	pp. 317–318
(B) Requisite Skills. The ability to communicate in writing and to interpret data.	12	pp. 317–318
5.4.6 Develop a plan to accomplish change in the organization, given an agency's change of policy or procedures, so that effective change is implemented in a supportive manner.	12	pp. 317, 328–329
(A) Requisite Knowledge. Planning and implementing change.	12	pp. 317, 328–329

NFPA1 1021, Job Performance Requirement	Corresponding Chapter(s)	Corresponding Page(s)
(B) Requisite Skills. The ability to clearly communicate.	12	pp. 317, 328–329
5.5 Inspection and Investigation. This duty involves conducting fire investigations to determine origin and preliminary cause, according to the following job performance requirements.	13	See below
5.5.1 Determine the area of origin and preliminary cause of a fire, given a fire scene, photographs, diagrams, pertinent data, and/or sketches, to determine if arson is suspected so that law enforcement action is taken.	13	pp. 335–336
(A) Requisite Knowledge. Indications of arson, common causes of fire, methods to preserve and protect the general area of origin, basic origin and cause determination, fire growth and development, and documentation of preliminary fire investigative procedures.	13	pp. 335–336
(B) Requisite Skills. The ability to investigate a fire scene and identify the general area of origin, implement procedures to preserve and protect potential sources of ignition within that general area of origin, and communicate.	13	pp. 335–336
5.6 Emergency Service Delivery. This duty involves supervising multiunit emergency operations, conducting preincident planning, and deploying assigned resources, according to the following job requirements.	14	See below
5.6.1 Produce operational plans, given an emergency incident requiring multiunit operations; the current editions of NFPA 1600, NFPA 1700, NFPA 1710, and NFPA 1720; and AHJ-approved safety procedures, so that required resources and their assignments are obtained and plans are carried out in compliance with NFPA 1600, NFPA 1700, NFPA 1710, and NFPA 1720, and approved safety procedures resulting in the mitigation of the incident.	14	pp. 354–364
(A) Requisite Knowledge. Standard operating procedures; national, state/provincial, and local information resources available for the mitigation of emergency incidents; an incident management system; and a personnel accountability system.	14	pp. 354–364
(B) Requisite Skills. The ability to implement an incident management system, to communicate, to supervise and account for assigned personnel under emergency conditions, and to serve in command staff and unit supervision positions within the Incident Management System.	14	pp. 354–364
5.6.2 Develop and conduct a postincident analysis, given multiunit incident and postincident analysis policies, procedures, and forms, so that all required critical elements are identified and communicated and the approved forms are completed and processed.	14	pp. 365–367

NFPA1 1021, Job Performance Requirement	Corresponding Chapter(s)	Corresponding Page(s)
(A) Requisite Knowledge. Elements of a fire- or rescue-related postincident analysis; basic building construction; basic fire protection systems and features; basic water supply; basic fuel loading; basic fire growth and development; and departmental procedures relating to dispatch response, strategy tactics and operations, sources of any emergency operations controlling authority, including EMS protocols, if applicable, and customer service.	14	pp. 365–367
(B) Requisite Skills. The ability to write reports, to communicate, and to evaluate skills.	14	pp. 365–367
5.6.3 Prepare a written report, given incident reporting data from the jurisdiction, so that the major causes for service demands are identified for various planning areas within the service area of the organization.	5	pp. 147–151
(A) Requisite Knowledge. Analyzing data.	5	pp. 147–151
(B) Requisite Skills. The ability to write clearly and to interpret response data correctly to identify the reasons for service demands.	5	pp. 147–151
5.7 Health and Safety. This duty involves reviewing injury, accident, and health exposure reports; identifying unsafe work environments or behaviors; and taking approved action to prevent reoccurrence, according to the following job requirements.	10	See below
5.7.1 Analyze a member's accident, injury, or health exposure history, given a case study, so that a report including action taken and recommendations made is prepared for a supervisor.	10	pp. 284–287
(A) Requisite Knowledge. The causes of unsafe acts; health exposures or conditions that result in accidents, injuries, occupational illnesses, or deaths; and requirements for reporting and receiving information related to health exposures.	10	pp. 284–287
(B) Requisite Skills. The ability to communicate and to interpret accidents, injuries, occupational illnesses, or death reports.	10	pp. 284–287

Glossary

100-year flood A frequency analysis performed by the U.S. Geological Survey to identify the 1 percent probability that a given event will be equaled or exceeded in any given year.

A

Accelerant A fuel or oxidizer, often an ignitable liquid, intentionally used to initiate a fire or increase the rate of growth or spread of fire. (NFPA 921)

Acceptable level of risk A community standard for the acceptable level of risk that citizens and government can tolerate and afford. This level of risk should be well-defined.

Accreditation Third-party attestation related to a conformity assessment body conveying formal demonstration of its competence to carry out specific conformity assessment tasks. These tasks include sampling and testing, inspection, certification, and registration. (ISO/IEC 17000, 2004)

Actionable item Employee behavior that requires an immediate corrective action by the supervisor; dozens of lawsuits have shown that failing to act in the face of such behavior will create a liability and a loss for the department.

Active failures Unsafe acts committed by people who are in direct contact with the situation or system that create voids within the defenses, barriers, and safeguards within the safety system that could lead to a loss.

Activity log An informal record maintained by the fire officer that lists fire fighter activities by date and includes a brief description; it is used to provide documentation for annual evaluations and special recognitions.

Acute stress disorder (ASD) An intense, unpleasant, and dysfunctional reaction beginning shortly after an overwhelming traumatic event and lasting less than a month.

Adaptive leadership How leaders encourage people to adapt when confronted with problems, challenges, and changes.

Adoption by reference Method of code adoption in which the specific edition of a model code is referred to within the adopting ordinance or regulation.

Adoption by transcription Method of code adoption in which the entire text of the code is published within the adopting ordinance or regulation.

Arbitration Resolution of a dispute by a mediator or a group rather than a court of law. Any civil matter may be settled in this way; some labor–management agreements include a binding arbitration clause.

Area of origin A structure, part of a structure, or general geographic location within a fire scene, in which the "point of origin" of a fire or explosion is reasonably believed to be located. (NFPA 921, revised 2017)

Arson The crime of maliciously and intentionally or recklessly starting a fire or causing an explosion. (NFPA 921)

Artifacts In the context of fire evidence, could include the remains of the material first ignited, the ignition source, or other items or components that are in some way related to the fire ignition, development, or spread.

Assessment centers A series of simulation exercises that are intended to identify a candidate's competency to perform the job that is offered in the promotional examination.

Assistant or division chiefs Midlevel chiefs who often have a functional area of responsibility, such as training, and who answer directly to the fire chief.

At-risk behavior Behavioral choice that increases risk when risk is not recognized, or is mistakenly believed to be justified.

Authority having jurisdiction (AHJ) An organization, office, or individual responsible for enforcing the requirements of a code or standard, or for approving equipment, materials, an installation, or a procedure. (NFPA 1951)

Autocratic A leadership style in which managers make decisions unilaterally, without input from subordinates.

Automatic sprinkler system A sprinkler system of pipes with water under pressure that allows water to be discharged immediately when a sprinkler head operates. (NFPA 853)

B

Base budget The level of funding required to maintain all services at the currently authorized levels, including adjustments for inflation, salary increases, and other predictable cost changes.

Base The location at which the primary logistics functions are coordinated and administered.

Battalion chiefs Usually the first level of fire chief. These chiefs are often in charge of running calls and supervising multiple stations or districts within a city. A battalion chief is usually the officer in charge of a single-alarm working fire.

Behavioral approach An emphasis on the personality characteristics of a leader.

Behavioral health Behaviors that affect physical or mental health.

Brainstorming A method of shared problem solving in which all members of a group spontaneously contribute ideas.

Branch director A person in a supervisory-level position in either the operations or logistics function to provide a span of control.

Branch Supervisory level established in either the operations or logistics function to provide an appropriate span of control.

Budget An itemized summary of estimated or intended expenditures for a given period, along with proposals for financing them.

C

Capital budget An allocation of money for the acquisition or maintenance of fixed assets such as land, buildings, and durable equipment.

Captain The second rank of promotion in the fire service, between the lieutenant and the battalion chief. Captains are responsible for managing a fire company and for coordinating the activities of that company among the other shifts.

Catastrophic theory of reform An approach in which fire prevention codes or firefighting procedures are changed in reaction to a fire disaster.

Cause The circumstance, condition, or agency that brought about or resulted in the fire or explosion incident, damage to property resulting from the fire or explosion incident, or bodily injury or loss of life resulting from the fire or explosion incident. (NFPA 921)

Central tendency An evaluation error that occurs when a fire fighter is rated in the middle of the range for all dimensions of work performance.

Chain of command A rank structure, spanning the fire fighter through the fire chief, for managing a fire department and fire-ground operations.

Char Carbonaceous material that has been burned or pyrolyzed and has a blackened appearance. (NFPA 921)

Chronological statement of events A detailed account of the fire company activities related to an incident or accident.

Class specification A technical worksheet that quantifies the knowledge, skills, and abilities (KSAs) by frequency and importance for every classified job within the local civil service agency.

Coaching A method of directing, instructing, and training a person or group of people with the aim to achieve some goal or develop specific skills.

Collective bargaining The process in which working people, through their unions, negotiate contracts with their employers to determine their terms of employment.

Command staff In an incident management organization, positions consisting of the incident commander, public information officer, safety officer, liaison officer, and other positions as required.

Community demographic profile Statistical description of economic and social issues of a community.

Community Emergency Response Team (CERT) A fire department training program to help citizens understand their responsibilities in preparing for disaster and increase their ability to safely help themselves, their families, and their neighbors in the immediate hours after a catastrophe.

Community risk assessment (CRA) A comprehensive evaluation that identifies, prioritizes, and defines the risks that pertain to the overall community. (NFPA 1300)

Community risk reduction (CRR) A process to identify and prioritize local risks, followed by the integrated and strategic investment of resources to reduce their occurrence and impact. (NFPA 1300)

Company journal A log book at the fire station that creates an up-to-date record of the emergency, routine activities, and special activities that occurred at the fire station. The company journal also records any fire fighter injuries, liability-creating events, and special visitors to the fire station.

Compensation and benefits Human resources system to identify and determine the pay, leave, and fringe benefits for each position in the organization.

Complaint Expression of grief, regret, pain, censure, or resentment; lamentation; accusation; or fault finding.

Conflict A state of opposition between two parties. This is often the manifestation of a complaint.

Construction type The combination of materials used in the construction of a building or structure, based on the varying degrees of fire resistance and combustibility. (NFPA 5000)

Contrast effect An evaluation error in which a fire fighter is rated on the basis of the performance of another fire fighter and not on the classified job standards.

Controlling Restraining, regulating, governing, counteracting, or overpowering.

Crew resource management (CRM) A program focused on improved situational awareness, sound critical decision making, effective communication, proper task allocation, and successful teamwork and leadership. (NFPA 1500)

Crisis communications Communications that seek to explain the specific event; identify likely consequences and outcomes; and provide specific harm-reducing information to affected communities in an honest, candid, prompt, accurate, and complete manner.

D

"Data dump" question A promotional assessment question that asks the candidate to write or describe all of the factors or issues covering a technical issue, such as suppression of a basement fire in a commercial property.

Decision making The process of identifying problems and opportunities and resolving them.

Defensive operation Actions that are intended to control a fire by limiting its spread to a defined area, avoiding the commitment of personnel and equipment to dangerous areas. (NFPA 1500)

Delegation An organizational process wherein the supervisor divides work among the subordinates and gives them the responsibility and authority to accomplish the respective tasks.

Democratic A leadership style in which managers reach decisions with the input of the employees but are responsible for making the final decision.

Demographics The characteristics of human populations and population segments, especially when used to identify consumer markets; generally includes age, race, sex, income, education, and family status.

Demonstrative evidence Tangible items that can be identified by witnesses, such as incendiary devices and fire scene debris.

Demotion A reduction in rank, with a corresponding reduction in pay.

Dimensions The tributes or qualities that can be described and measured during a promotional examination. On average, 5 to 15 dimensions are measured on a promotional examination. The six most commonly addressed are oral communication, written communication, problem analysis, judgment, organizational sensitivity, and planning/organizing.

Direct supervision A type of supervision in which the fire officer is required to observe the actions of a work crew directly; it is commonly employed during high-hazard activities.

Discipline The guidelines that a department sets for fire fighters to work within.

Diversity A characteristic of a fire workforce that reflects differences in terms of age, cultural background, race, religion, gender, and sexual orientation.

Division A supervisory level established to divide an incident into geographic areas of operations.

Division of labor Breaking down an incident or task into a series of smaller, more manageable tasks and assigning personnel to complete those tasks.

Division supervisor A person in a supervisory-level position responsible for a specific geographic area of operations at an incident.

Documentary evidence Evidence in written form, such as reports, records, photographs, sketches, and witness statements.

E

Education The process of imparting knowledge or skill through systematic instruction.

Employee assistance program (EAP) An employee benefit that covers all or part of the cost for employees to receive counseling, referrals, and advice in dealing with stressful issues in their lives. These problems may include substance abuse, bereavement, marital problems, weight issues, or general wellness issues.

Environmental noise A physical or sociological condition that interferes with the message in the communication process.

Ethical behavior Decisions and behavior demonstrated by a fire officer that are consistent with the department's core values, mission statement, and value statements.

Evidence Includes material objects as well as documentary or oral statements that are admissible as testimony in a court of law.

Expanded incident report narrative A report in which all company members submit a narrative on what they observed and which activities they performed during an incident.

Expenditure Money spent for goods or services.

F

Failure analysis A logical, systematic examination of an item, component, assembly, or structure and its place and function within a system, conducted to identify and analyze the probability, causes, and consequences of potential and real failures. (NFPA 921)

Finance/administration section The section responsible for all costs and financial actions of the incident or planned event, including the time unit, procurement unit, compensation/claims unit, and the cost unit.

Fire analysis The process of determining the origin, cause, development, responsibility, and, when required, a failure analysis of a fire or explosion. (NFPA 921)

Fire chief The top position in the fire department. The fire chief has ultimate responsibility for the fire department and usually answers directly to the mayor or other designated public official.

Fire code compliance inspection Periodic check of an occupancy by the authority having jurisdiction to identify if there are any fire code violations and to initiate a repair or restoration to return to fire code compliance.

Fire Ground Command (FGC) An incident management system developed in the 1970s for day-to-day fire department incidents (generally handled with fewer than 25 units or companies).

Fire investigation The process of determining the origin, cause, and development of a fire or explosion. (NFPA 921)

Fire Officer I The fire officer, at the supervisory level, who has met the job performance requirements specified in NFPA 1021 for Level I. (NFPA 1021)

Fire Officer II The fire officer, at the supervisory/managerial level, who has met the job performance requirements specified in NFPA 1021 for Level II. (NFPA 1021)

Fire prevention division or hazardous use permit A local government permit that is renewed annually after the fire prevention division performs a code compliance inspection. A permit is required if the process, storage, or occupancy activity creates a life-safety hazard. Restaurants with more than 50 seats, flammable liquid storage, and printing shops that use ammonia are examples of occupancies that may require a permit.

Fire scene reconstruction The process of recreating the physical scene during fire scene analysis investigation or through the removal of debris and the placement of contents or structural elements in their prefire positions. (NFPA 921)

FIRESCOPE Fire Resources of California Organized for Potential Emergencies; an organization of agencies established in the early 1970s to develop a standardized system for managing fire resources at large-scale incidents such as wildland fires.

Fire tax district A special service district created to finance the fire protection of a designated district.

Fiscal year A 12-month period during which an organization plans to use its funds. Local governments' fiscal years generally run from July 1 to June 30.

Floor plans Views of a building's interior. Rooms, hallways, cabinets, and the like are drawn in the correct relationship to each other.

Flow path The movement of heat and smoke from the higher pressure within the fire area towards the lower pressure areas accessible via doors, window openings, and roof structures. (NFPA 1410)

Followership The act or condition of following a leader; adherence. The characteristic that leaders can be effective only to the extent that followers are willing to accept their leadership.

Formal communication An official fire department letter or report that is presented on stationery with the fire department letterhead and generally is signed by a chief officer or headquarters staff member.

Formal written reprimand An official negative supervisory action at the lowest level of the progressive disciplinary process.

Form of material How the material was used. (NFIRS, 2010)

Frame of reference An evaluation error in which the fire fighter is evaluated on the basis of the fire officer's personal standards instead of the classified job description standards.

Fuel load The total quantity of combustible contents of a building, space, or fire area, including interior finish and trim, expressed in heat units or the equivalent weight in wood. (NFPA 921)

Fuel-limited fire A fire in which the heat release rate and fire growth are controlled by the characteristics of the fuel because there is adequate oxygen available for combustion. (NFPA 1410)

G

General orders Short-term directions, procedures, or orders signed by the fire chief and lasting for a period of days to 1 year or more.

Goal-setting The process of establishing specific goals over a designated period of time.

Good faith bargaining Legal requirement of both the union and the employer arising out of Section 8(d) of the National Labor Relations Act. Enforced by the National Labor Relations Board, the parties are required to meet regularly to bargain collectively for wages, hours, and other conditions of employment.

Governmental Accounting Standards Board (GASB) An organization whose mission is to establish and improve the standards of state and local governmental accounting and financial reporting, thereby resulting in useful information for users of financial reports, and to guide and educate the public, including issuers, auditors, and users of those financial reports.

Grievance A dispute, claim, or complaint that any employee or group of employees may have in relation to the interpretation, application, and/or alleged violation of some provision of the labor agreement or personnel regulations.

Grievance procedure A formal, structured process that is employed within an organization to resolve a grievance. In most cases, the grievance procedure is incorporated in the personnel rules or the labor agreement and specifies a series of steps that must be followed in a particular order.

Group A supervisory level established to divide the incident into functional areas of operation.

Group supervisor A person in a supervisory level position responsible for a functional area of operation. (NFPA 1561)

H

Halo and horn effect An evaluation error in which the fire officer takes one aspect of a fire fighter's job task and applies it to all aspects of work performance.

Hazard Capable of causing harm or posing an unreasonable risk to life, health, property, or environment. (NFPA 1072)

Health and safety officer The member of the fire department assigned and authorized by the fire chief as the manager of the safety and health program.

Health, safety, and security Human resources activities intended to provide and promote a safe work environment.

High-risk property Structure that has the potential for a catastrophic property or life loss in the event of a fire.

High-value property Structure that contains equipment, materials, or items that have a high replacement value.

Hostile work environment An environment that can result from the unwelcome conduct of supervisors, co-workers, customers, contractors, or anyone else with whom the victim interacts on the job, and the unwelcome conduct renders the workplace atmosphere intimidating, hostile, or offensive.

Hot wash Debriefing immediately after an incident to identify issues needing prompt attention as well as procedural or systemic problems needing later follow-up.

Human error An inadvertent action such as a slip, lapse, or mistake.

Human resources development All activities to train and educate employees.

Human resources management The process of managing employees in a company. The process may include hiring, firing, training, and motivating employees.

Human resources planning The process of having the right number of people in the right place at the right time who can accomplish a task efficiently and effectively.

I

ICS general staff The group of incident managers composed of the operations section chief, planning section chief, logistics section chief, and finance/administration section chief.

Immediately dangerous to life and health (IDLH) Any condition that would pose an immediate or delayed threat to life, cause irreversible adverse health effects, or interfere with an individual's ability to escape unaided from a hazardous environment. (NFPA 1670)

Impasse A situation in which the parties in a dispute have reached a deadlock in negotiations; also described as the demarcation line between bargaining and negotiation. A declaration of a deadlock in labor–management negotiations brings in a state or federal negotiator who will start a fact-finding process that will lead to a binding arbitration resolution.

In-basket exercise A promotional examination component in which the candidate deals with correspondence and related items that have accumulated in a fire officer's in-basket.

Incident action plan (IAP) A verbal plan, written plan, or combination of both that is updated throughout the incident and reflects the overall incident strategy, tactics, risk management, and member safety requirements approved by the incident commander. (NFPA 1026)

Incident commander (IC) The individual responsible for all incident activities, including the development of strategies and tactics and the ordering and release of resources. (NFPA 1026, 1500, 1072)

Incident command system (ICS) A system that defines the roles and responsibilities to be assumed by personnel and the operating procedures to be used in the management and direction of emergency operations; also referred to as an incident management system (IMS). (NFPA 1670)

Incident safety officer A member of the command staff responsible for monitoring and assessing safety hazards and unsafe situations and for developing measures for ensuring personnel safety. (NFPA 1500)

Incident scene rehabilitation The tactical-level management unit that provides for medical evaluation, treatment, monitoring, fluid and food replenishment, mental rest, and relief from climatic conditions of an incident.

Informal communications Internal memos, e-mails, instant messages, and computer-aided dispatch/mobile data terminal messages. Informal reports have a short life and may not be archived as permanent records.

Initial rapid intervention crew (IRIC) Two members of the initial attack crew who are assigned for rapid deployment to rescue lost or trapped members. (NFPA 1710)

Interrogatory A series of formal written questions sent to the opposing side of a legal argument. The opposition must provide written answers under oath.

Investigation A systematic inquiry or examination.

Involuntary transfer or detail A disciplinary action in which a fire fighter is transferred or assigned to a less desirable or different work location or assignment.

J

Job-content/criterion-referenced validity A type of validity obtained through the use of a technical committee of job incumbents who certify that the knowledge being measured is required on the job and referenced to known standards.

Job description A narrative summary of the scope of a job. It provides examples of the typical tasks.

Job instruction training A systematic four-step approach to training fire fighters in a basic job skill: (1) prepare the fire fighters to learn, (2) demonstrate how the job is done, (3) try them out by letting them do the job, and (4) gradually put them on their own.

Just culture A change in focus from errors and outcomes to systems design and management of behavioral choices of all employees.

K

Knowledge, skills, and abilities (KSAs) The traits required for every classified position within the municipality. KSAs are defined by a narrative job description and a technical class specification.

L

Laissez-faire A policy or attitude of letting things take their own course, without interfering.

Latent conditions Pre-existing conditions within the safety system defenses, barriers, and safeguards within the safety system that could lead to a loss through provoking errors or weakening system.

Leadership A complex process by which a person influences others to accomplish a mission, task, or objective and directs the organization in a way that makes it more cohesive and coherent.

Leading Guiding or directing in a course of action.

Legacy dwellings Single-family dwellings constructed before 1980.

Liaison officer A member of the command staff, the point of contact for assisting or coordinating agencies.

Lieutenant A company officer who is usually responsible for a single fire company on a single shift; the first in line among company officers.

Line-item budget A budget format in which expenditures are identified in a categorized line-by-line format

Logistics section The section responsible for providing facilities, services, and materials for the incident or planned event, including the communications unit, medical unit, and food unit within the service branch and the supply unit, facilities unit, and ground support unit within the support branch.

Logistics section chief A supervisory position that is responsible for providing supplies, services, facilities, and materials during the incident. The person in this position reports directly to the incident commander.

Loudermill hearing A predisciplinary conference that occurs before a suspension, demotion, or involuntary termination is issued. The term "Loudermill" refers to a U.S. Supreme Court decision.

M

Management The art of getting things done through and with people in formally organized groups. To forecast, plan, organize, command, coordinate, and control.

Managing fire officer The description from the IAFC *Officer Development Handbook* for the tasks and expectations for a Fire Officer II. In this role, the company officer is encouraged to acquire the appropriate levels of training, experience, self-development, and education to prepare for the Chief Fire Officer designation.

Masonry wall A wall that consists of brick, stone, concrete block, terra cotta, tile, adobe, precast, or cast-in-place concrete.

Mediation The intervention of a neutral third party in an industrial dispute.

Mentoring A developmental relationship between a more experienced person and a less experienced person (a protégé).

Mini/max codes Codes developed and adopted at the state level for either mandatory or optional enforcement by local governments; these codes cannot be amended by local governments.

Mistake An error or fault resulting from defective judgment, deficient knowledge, or carelessness; a misconception or misunderstanding.

Model codes Codes generally developed through the consensus process with the use of technical committees developed by a code-making organization.

Modern dwellings Single-family dwellings constructed since 1980; they are typically larger structures with an open house geometry, lightweight construction materials, and exponentially increased fuel load.

Moral injury The damage done to one's conscience or moral compass when that person perpetrates, witnesses, or fails to prevent acts that transgress one's own moral beliefs, values, or ethical codes of conduct.

N

National Fire Incident Reporting System (NFIRS) A nationwide database at the National Fire Data Center under the U.S. Fire Administration that collects fire-related data in an effort to provide information on the national fire problem.

National Fire Protection Association (NFPA) The association that delivers information and knowledge through more than 300 consensus codes and standards, research, training, education, outreach, and advocacy; and by partnering with others who share an interest in furthering its mission in helping save lives and reduce loss with information, knowledge, and passion.

National Incident Management System (NIMS) A system mandated by Homeland Security Presidential Directive-5 (HSPD-5) that provides a systematic, proactive approach guiding government agencies at all levels, the private sector, and nongovernmental organizations to work seamlessly to prepare for, prevent, respond to, recover from, and mitigate the effects of incidents, regardless of cause, size, location, or complexity, in order to reduce the loss of life or property and harm to the environment. (NFPA 1026)

National Response Framework (NRF) A comprehensive, national, all-hazards approach to provide federal resources in support of a major multijurisdictional incident.

Naturalistic decision making A framework of intuitive decision making when operating under limited time, uncertainty, high stakes, team and organizational constraints, unstable conditions, and varying amounts of experience. Recognition-primed decision making is a process within this decision-making framework.

O

Occupancy type The purpose for which a building or a portion thereof is used or intended to be used.

Offensive operation Actions generally performed in the interior of involved structures that involve a direct attack on a fire to directly control and extinguish the fire. (NFPA 1500)

Ongoing compliance inspection Inspection of an existing occupancy to observe the housekeeping and confirm that the built-in fire protection features, such as fire exit doors and sprinkler systems, are in good working order.

Operating budget An estimate of the income and expenditures of the jurisdiction.

Operations section chief A supervisory position that is responsible for all tactical operations at the incident or planned event. This position deals with the strategy and tactics that are required to get the job done and reports directly to the incident commander.

Operations section The section responsible for all tactical operations at the incident or planned event, including up to 5 branches; 25 divisions/groups; and 125 single resources, task forces, or strike teams.

Oral reprimand, warning, or admonishment The first level of corrective discipline. Considered informal, this discipline action remains with the fire officer and is not part of the fire fighter's official record.

Ordinance A law established by an authorized subdivision of a state, such as a city, county, or town.

Organizing Putting resources together into an orderly, functional, structured whole.

P

Performance management Oversight of process management to ensure that the ensemble of activities consistently meet organizational goals in an effective and efficient manner.

Personal bias An evaluation error that occurs when the evaluator's perspective skews the evaluation such that the classified job knowledge, skills, and abilities are not appropriately evaluated.

Personal power Power that reflects the effectiveness of the individual.

Personal study journal A personal notebook to aid in scheduling and tracking a candidate's promotional preparation progress.

Personnel accountability report (PAR) Periodic reports verifying the status of responders assigned to an incident or planned event. (NFPA 1026)

Planning Developing a scheme, program, or method that is worked out beforehand to accomplish an objective.

Planning section The section responsible for the collection, evaluation, dissemination, and use of information related to the incident situation, resource status, and incident forecast.

Planning section chief A supervisory position that is responsible for the collection, evaluation, dissemination, and use of information relevant to the incident. This position reports directly to the incident commander.

Plot plan A representation of the exterior of a structure, identifying doors, utilities' access, and any special considerations or hazards.

Policies Formal statements that provide guidelines for present and future actions. Policies often require personnel to make judgments.

Political action committee (PAC) An organization formed by corporations, unions, and other interest groups that solicits campaign contributions from private individuals and distributes these funds to political candidates.

Positional power Power that is defined by the role the individual has within the organization.

Posttraumatic stress disorder (PTSD) A condition characterized by symptoms of avoidance and nervous system arousal after experiencing or witnessing a traumatic event.

Power The capacity of one party to influence another party.

Preincident plan A document developed by gathering general and detailed data that is used by responding personnel in effectively managing emergencies for the protection of occupants, participants, responding personnel, property, and the environment. (NFPA 1620)

Pretermination hearing An initial check to determine if there are reasonable grounds to believe that charges against an employee are true and support the proposed termination.

Professional development A continuous process of training, education, knowledge, and skills enhancement.

Progressive negative discipline Process for dealing with job-related behavior that does not meet expected and communicated performance standards. The level of discipline increases from mild to more severe punishments if the problem is not corrected.

Psychological contract Refers to mutual unwritten expectations that exist between an employee and his or her employer regarding policies and practices in their organization.

Public information officer A member of the command staff responsible for interfacing with the public and media or with other agencies with incident-related information requirements.

Pyrolysis A process in which material is decomposed, or broken down, into simpler molecular compounds by the effects of heat alone; pyrolysis often precedes combustion. (NFPA 921)

R

Rapid intervention crew (RIC) A minimum of two fully equipped personnel on site, in a ready state, for immediate rescue of disoriented, injured, lost, or trapped rescue personnel. (NFPA 1500)

Reasonable person standard The actions that a similar person with the same background would take given the same situation.

Recency An evaluation error in which the fire fighter is evaluated only on incidents that occurred over the past few weeks rather than on the entire evaluation period.

Reckless behavior A behavioral choice to consciously disregard a substantial and unjustified risk.

Recognition-primed decision making (RPD) A model of how people make quick, effective decisions when faced with complex situations by generating a possible course of action based on intuition and comparing it to the constraints imposed by the situation, then selecting the first course of action that is not rejected.

Recommendation report A decision document prepared by a fire officer for the senior staff. Its goal is to support a decision or an action.

Regulations Official rules created by government agencies that detail how something should be done. (NFPA 475)

Rehabilitation The process of providing rest, rehydration, nourishment, and medical evaluation to members who are involved in extended or extreme incident scene operations.

Reliability The characteristic where a test measures what it is intended to measure on a consistent basis.

Rep (repetition) One complete motion of an exercise.

Resiliency The process of adapting well in the face of adversity, trauma, tragedy, threats, or significant sources of stress, such as family relationship problems, serious health problems, or workplace and financial stressors.

Restrictive duty Temporary work assignment during an administrative investigation that isolates the fire fighter from the public; this is usually is an administrative assignment away from the fire station.

Revenues The income of a government from all sources.

Right-to-work A worker cannot be compelled, as a condition of employment, to join or not to join or to pay dues to a labor union.

Risk assessment Identification of potential and likely risks within a particular community records management system (RMS).

Risk/benefit analysis A decision made by a responder based on a hazard identification and situation assessment that weighs the risks likely to be taken against the benefits to be gained for taking those risks. (NFPA 1006, 1670)

Risk communications The exchange of real-time information, advice, and opinions between experts and people facing threats to their health and economic or social well-being.

Risk management Identification and analysis of exposure to hazards, selection of appropriate techniques to control exposures, implementation of chosen techniques, and monitoring of results to ensure the health and safety of members.

Rules and regulations Directives developed by various government or government-authorized organizations to implement a law that has been passed by a government body.

S

Safety officer A member of the command staff responsible for monitoring and assessing safety hazards or unsafe situations and for developing measures for ensuring personnel safety.

Safety unit A member or members assigned to assist the incident safety officer; the tactical-level management unit that can be composed of the incident safety officer alone or with additional assistant safety officers assigned to assist in providing the level of safety supervision appropriate for the magnitude of the incident and the associated hazards.

Scientific method The systematic pursuit of knowledge involving the recognition and definition of a problem; the collection of data through observation and experimentation; analysis of the data; the formulation, evaluation and testing of hypotheses; and, where possible, the selection of a final hypothesis. (NFPA 921)

Servant leadership A leadership approach focusing on serving others, rather than accruing power or taking control.

Service branch A branch within the logistics section responsible for service activities at an incident or planned event, including the communications unit, medical unit, and food unit.

Set A group of consecutive repetitions.

Situational awareness The ongoing activity of assessing what is going on around you during the complex and dynamic environment of a fire incident. (NFPA 1410)

Situational leadership A leadership approach where a supervisor adjusts the style of leadership to fit the developmental level of subordinates. During emergency operations, it means adjusting the style of leadership based on the changing conditions of the incident.

Size-up The process of gathering and analyzing information to help fire officers make decisions regarding the deployment of resources and the implementation of tactics. (NFPA 1410)

Skills approach An emphasis on the skills or capabilities of a leader.

S.L.I.C.E.-R.S. An acronym intended to be used by the first arriving company officer to accomplish important strategic goals on the fire ground.

Source of ignition Energy source that causes a material to ignite. (NFIRS, 2010)

Span of control The maximum number of personnel or activities that can be effectively controlled by one individual (usually three to seven).

Special evaluation period Designated period of time when an employee is provided additional training to resolve a work performance/behavioral issue. The supervisor issues an evaluation at the end of the special evaluation period.

Spoils system Also known as the patronage system; the practice of making appointments to public office based on a personal relationship or affiliation rather than because of merit. The spoils system scandals of the New York City "Tweed Ring" and the Tammany Hall political machine (1865–1871) resulted in Congress passing the Pendleton Civil Service Reform Act of 1883.

Staffing The process of attracting, selecting, and maintaining an adequate supply of labor, as well as reducing the size of the labor force when required.

Staging A specific function in which resources are assembled in an area at or near the incident scene to await instructions or assignments.

Stairwell support group A group of fire fighters who move equipment and water supply hose lines up and down the stairwells at a high-rise fire incident. The stairwell support unit leader reports to the support branch director or the logistics section chief.

Standard operating guidelines (SOGs) Written organizational directives that identify a desired goal and describe the general path to accomplish the goal, including critical tasks or cautions.

Standard operating procedures (SOPs) Written organizational directives that establish or prescribe specific operational or administrative methods to be followed routinely for the performance of designated operations or actions.

Standard time–temperature curve A recording of fire temperature increase over time.

Standpipe system An arrangement of piping, valves, hose connections, and allied equipment installed in a building or structure, with the hose connections located in such a manner that water can be discharged in streams or spray patterns through attached hose and nozzles, for the purpose of extinguishing a fire, thereby protecting a building or structure and its contents in addition to protecting the occupants. This is accomplished by means of connections to water supply systems or by means of pumps, tanks, and other equipment necessary to provide an adequate supply of water to the hose connections. (NFPA 1620)

State police powers Authority of each state to govern matters related to the welfare and safety of the residents.

Strategic level Command level that entails the overall direction and goals of the incident.

Strike team Specified combinations of the same kind and type of resources, with common communications and a leader.

Strike team leader A supervisory position that is in charge of a group of similar resources.

Supervising fire officer The description from the IAFC *Officer Development Handbook* for the tasks and expectations for a Fire Officer I. In this role, the company officer is encouraged to acquire the appropriate levels of training, experience, self-development, and education to prepare for the Chief Fire Officer designation.

Supervisor's report A form that is required by most state worker's compensation agencies and that is completed by the immediate supervisor after an injury or property damage accident.

Supplemental budget Proposed increases in spending to provide additional services.

Support branch A branch within the logistics section responsible for providing the personnel, equipment, and supplies to support incident operations, including the supply unit, facilities unit, and ground support unit.

Suspension Negative disciplinary action that removes a fire fighter from the work location; he or she is generally not allowed to perform any fire department duties.

T

T-account Documentation system similar to an accounting balance sheet listing credits and debits, in which a single-sheet form is used to list the employee's assets on the left side and liabilities on the right side, so that the result resembles the letter "T."

Tactical level Command level in which objectives must be achieved to meet the strategic goals. The tactical-level supervisor or officer is responsible for completing assigned objectives.

Tactical worksheet A form that allows the incident commander to ensure all tactical issues are addressed and to diagram an incident with the location of resources on the diagram.

Task force A group of resources with common communications and a leader that can be pre-established and sent to an incident or planned event or formed at an incident or planned event.

Task force leader A supervisory position that is in charge of a group of dissimilar resources.

Task level Command level in which specific tasks are assigned to companies; these tasks are geared toward meeting tactical-level requirements.

Termination Situation in which the organization ends an individual's employment against his or her will.

Testimonial evidence Witnesses speaking under oath.

Trailer Solid or liquid fuel used to intentionally spread or accelerate the spread of a fire from one area to another. (NFPA 921)

Training The process of achieving proficiency through instruction and hands-on practice in the operation of equipment and systems that are expected to be used in the performance of assigned duties.

Trait approach An emphasis on the traits of an effective leader.

Transactional relationship A relationship where all parties are in it for themselves, and where individuals do things for each other with the expectation of reciprocation.

Transformational leadership A leadership approach that causes change in individuals and social systems. In its ideal form, it creates valuable and positive change in the followers with the end goal of developing followers into leaders.

Transitional attack An offensive fire attack initiated by an exterior, indirect handline operation into the fire compartment to initiate cooling while transitioning into interior direct fire attack in coordination with ventilation operations.

Two-in/two-out rule A guideline created in response to OSHA Respiratory Regulation (29 CFR 1910.134), which requires a two-person team to operate within an environment that is immediately dangerous to life and health (IDLH) and a minimum of a two-person team to be available outside the IDLH atmosphere to remain capable of rapid rescue of the interior team.

Type of material The nature of the material itself. (NFIRS, 2010)

U

Unconsciously competent The highest level of the Conscious Competence Learning Matrix developed by Dr. Thomas Gordon in the 1970s. At this level, the skill becomes so practiced that it enters the unconscious parts of the brain—it becomes second nature.

Unfair labor practices Employer or union practices forbidden by the National Labor Relations Board or state/local laws, subject to court appeal. It often involves the employer's efforts to avoid bargaining in good faith.

Unit An engine company, truck company, or other functional or administrative group.

Unity of command The concept by which each person within an organization reports to one, and only one, designated person. (NFPA 1026)

Use group A category in the building code classification system in which buildings and structures are grouped together by their use and by the characteristics of their occupants.

V

Ventilation-limited fire A fire in which the heat release rate and fire growth are regulated by the available oxygen within the space. (NFPA 1410)

W

Work improvement plan Written document that is part of a special evaluation period. The plan identifies performance deficiencies and lists the improvements in performance or changes in behavior required to obtain a "satisfactory" evaluation.

Y

Yellow dog contracts Pledges that employers required workers to sign indicating that they would not join a union as long as the company employed them. Such contracts were declared unenforceable by the Norris-LaGuardia Act of 1932.

Index

Note: Page numbers followed by *f* or *t* indicate material in figures, or tables, respectively.

A

academic accreditation, 287–288
accelerants, 343
acceptable level of risk, 303
accident investigation, 239
   and documentation, 239, 241
accidental fire causes, 341–342, 341*f*
accreditation
   in higher education, 288
   of skill certification, 288–290, 289*t*
ACE. *See* American Council on Education
actionable item, 29, 29*f*
activated fire protection system alarms, 256
active failures, 57–58
activity log, 273, 273*f*
acute stress disorder (ASD), 106
adaptive leadership, 47
administrative communications
   alternative response unit
      administrative office, 321
      capital budget, 321, 322*f*
      cost recovery and reduction, 321
      operating expenses, 320–321
      overview of, 320
      personnel expenses, 320
      uniforms and protective clothing, 321
      vehicle, 321
   bond referendums and capital projects, 328
   budgetary process, navigating, 319–320
      budget proposal, 320
   expenditures
      capital, 327–328
      operating, 326–327, 327*f*
      personnel, 326
   purchasing process
      bidding process, 323–324
      petty cash, 323
      purchase orders, 323
      requisitions, 323
   recommending change, 328–329
   revenue sources, 324
      grants, 325
      nontraditional revenue sources, 325–326
   writing reports, 318
      recommendation report, 319
   written communications
      formal communications, 317–318
      informal communications, 316–317

administrative controls, 225
administrative law, 253–254
adoption by reference, 167
adoption by transcription, 167
AFG. *See* Assistance to Firefighters Grant
AHJ. *See* authority having jurisdiction
air management, 231
air traffic controllers' strike 1981, 262
"all-hazards" approach, 256
allied organizations, cooperating with, 304
alternative disciplinary actions, 279
alternative response unit (ARU)
   administrative office, 321
   capital budget, 321, 322*f*
   cost recovery and reduction, 321
   operating expenses, 320–321
   overview of, 320
   personnel expenses, 320
   uniforms and protective clothing, 321
   vehicle, 321
aluminum-sided houses, 342
American Council on Education (ACE), 146
American Public University System (APUS), 288
American Society for Testing and Materials
      (ASTM), 203
announcements, 317
annual evaluation process, 282–283
annual operating expenses. *See* specific expenses
apartment buildings, 176
APUS. *See* American Public University System
arbitration, 260
area of origin, 335–336
arson, 342
   motives
      crime concealment, 344
      excitement, 346
      extremism, 346
      profit, 344
      spite/revenge, 346
      vandalism, 346
artifacts, 186
ARU. *See* alternative response unit
ASD. *See* acute stress disorder
assertive statement process, 59
assessment centers
   emergency incident simulations, 379–380
   in-basket exercises, 378–379
   interpersonal interaction, 380–381

technical skills demonstration, 381
writing/speaking exercise, 381
assigning tasks
emergency incidents, 88–89
nonemergency incidents, 89–90
safety considerations, 90
supervising, single company, 86–88, 86f
Assistance to Firefighters Grant (AFG), 325
assistant incident safety officers, 235
assistant/division chiefs, 11
ASTM. *See* American Society for Testing and Materials
at-risk behavior, 275
authentic leadership, 46
authority having jurisdiction (AHJ), 166, 334
authority-compliance management, 43
autocratic leadership, 43, 45
automatic sprinkler system, 168, 169f
awareness, 104

B

"back-of-the-envelope" calculation, 321
Baltimore Fire Department, 137
base, 365
base budget, 132
basic incident report, 346
battalion chief, 11, 363
behavioral approach, 42
behavioral health, 228
issues, 104–106
resources, 107
bidding process, 323–324
Blake and Mouton's managerial grid, 43f
authority-compliance, 43
country club, 43–44
impoverished management, 42
middle-of-the-road, 44
team management, 44–45
bond referendums and capital projects, 328
brainstorming, 84
branch director, 362
branches, 361–362
budget, 130, 319–320
cycle, 130, 131–132t
base and supplemental budgets, 132
budget request, 132–134
capital request form, 133f
project description and justification, 134
purchase information, 134
total cost of project, 134
proposal, 320
request, 132–134
building code *vs.* fire codes, 167
building stock evaluation, 301
built environment deterioration, 258
built-in fire protection system, 182
code violations, 159
fire alarm and detection systems, 171
special extinguishing systems
carbon dioxide, 170
dry/wet chemical, 171
foam systems, 171
Halon, 171
water-based, 168–170, 169f, 170f
business occupancy, 174, 175f, 183

C

CAAHEP. *See* Commission on Accreditation of Allied Health Education Programs
California Fire Protection District Law of 1987, 254
C-A-N process method, 79
cancer, reducing deaths from, 228–229
Candidate Physical Ability Test (CPAT), 227
capital budget, 130, 321, 322f
capital request form, 133f
captain, 5, 252
challenges for
catastrophic weather events, 258
deterioration of built environment, 258
diversity, 258
fire fighter emotional health, 256
increase in non-fire incidents, 256
supervision and motivation, 256
carbon dioxide system, 170
catastrophic theory of reform, 166
catastrophic weather events, 258
cause. *See also* fire cause
defined, 336–337
fire-ground injuries by, 216t
ignition factor and, 338
Center for Public Safety Excellence (CPSE), 146, 290, 302
fire officer credential, 290
central tendency, 282
CERT. *See* Community Emergency Response Team
chain of command, 11, 11f
char, 336, 337f
charismatic and visionary leadership, 46
CHEA. *See* Council for Higher Education Accreditation
chronological statement of events, 150
Civil War military deployment model, 256
class specification, 377
Cleveland Board of Education v. Loudermill, 278
closeness of supervision, 87
clothing, protective, 237
coaching, 90
CoAEMSP. *See* Committee on Accreditation for the Emergency Medical Services Professions
COL. *See* Company Officer Leadership
cold zone, 216
collective bargaining, legislative framework for
federal employees, 260–261
Landrum-Griffin Act of 1959, 260
Norris-LaGuardia Act of 1932, 259–260
state labor laws, 261–262, 261t
Taft-Hartley Labor Act of 1947, 260
Wagner-Connery Act of 1935, 260
COLS. *See* Company Officer Leadership Symposium
command mode, 201
command of initial emergency operations
developing incident action plan, 211–212
incident priorities, 212
location designators, 214
tactical priorities, 212–214
fire officer's role in incident management, 199, 199f
fire research and risk/benefit analysis
fire behavior graph, 203–204, 204f
flow path, 205, 205f
IAFC rules of engagement, 207–208
modern *vs.* legacy single family dwellings, 205–206, 206f
risk/benefit analysis, 206–207, 207f

command of initial emergency operations (*Continued*)
    incident command and National Incident Management System model
        ICS training, 197–198, 198f
        levels of command, 198–199
    incident command system, history of, 196–197
        developing one national system, 197
        FIRESCOPE and Fire Ground Command, 197
    initial incident command, 199
        command safety, 202–203
        establishing command, 200
        functions of command, 202
        initial operational modes, 200–201, 201f
        personnel accountability report, 203
        responsibilities of command, 200
    sizing up incident
        Lloyd Layman's five-step size-up process, 209–210
        national fire academy size-up process, 210–211
        on-scene observations, 208–209
        prearrival information, 208
    tactical safety considerations, 214–215
        fire fighter deaths by cause of injury, 215f, 215t
        fire-ground injuries by cause, 216t
        rapid intervention crews, 216–217
        scene safety, 215–216
    transfer of command, 217
command staff, 357
    assignments, 87–88
    liaison officer, 358
    public information officer, 358
    safety officer, 358
commander, 23–24, 23–24f
Commission on Accreditation of Allied Health Education Programs (CAAHEP), 287
Commission on Professional Credentialing (CPC), 290
Committee on Accreditation for the Emergency Medical Services Professions (CoAEMSP), 287
communication. *See also specific communications*
    cycle
        emergency incident, 78–80, 78f
        feedback, 77–78, 77f
        informing supervisor, 80–81, 81f
        medium, 75–77, 76f
        message, 75
        receiver, 77
        sender, 75
    skills, 8, 10f, 60
communications order model, 78
community, 116–117
community demographic profile, 300
Community Emergency Response Team (CERT), 120–121, 120f
    course schedule, 120–121
community fire
    risk, 224–225
    safety, fire officer's role in, 159, 159f
community risk assessment (CRA), 300
    gathering data
        building stock, 301
        community service organizations, 302
        critical infrastructure systems, 302
        demographics, 300
        economic factors, 302
        geography, 300–301
        hazards, 302
        past loss/event history, 302
        public safety response agencies, 301–302
community risk reduction (CRR), 300
    cooperating with allied organizations, 304
    gathering community risk assessment data
        building stock, 301
        community service organizations, 302
        critical infrastructure systems, 302
        demographics, 300
        economic factors, 302
        geography, 300–301
        hazards, 302
        past loss/event history, 302
        public safety response agencies, 301–302
    implementing and monitoring, 304
    media relations
        fire department public information officer, 305–306, 306f
        fire officer as spokesperson, 307–309, 308f
        press releases, 307, 308f
    plan
        developing, 303
        implementation, 119
        monitoring, evaluating and modifying, 119
        preparation, 119
        writing, 303–304
    public education, 119–120
    risks
        identification, 118
        likelihood and impact consequences, 302–303
        prioritization, 118
        strategies and tactics to mitigate, 118–119
    social media outreach, 309
        challenges, 309–310
community service organizations, 302
company journal, 147, 148f
company officer, definition of, 5
Company Officer Leadership (COL), 32
Company Officer Leadership Symposium (COLS), 291
company supervisor, Fire Officer I as
    as commander, 24–25, 23f
    credentialing and development, 31–32
    ethical behavior, 27–28
    fire department organization
        chain of command, 11, 11f
        function, 13
        geography, 13
        management principles, 12–13, 12f
        National Incident Management System, 12
        source of authority, 10–11
        staffing, 13
    fire service, 7–8
        communications, 8, 10f
        history of, 8, 9t
        paying for, 8, 10, 10f
    functions of management, 13–14
    integrity, 27, 27f
    policies, 14–16
    positive labor-management relations
        International Association of Fire Fighters, 17–18, 17f
        National Volunteer Fire Council, 18, 18f
    qualifications, 4–5
    rules and regulations, 14–16
    standard operating procedures, 14–16
    supervisor/employee relationship, 16–17

as supervisor, 18, 19f, 23, 23f
supervisor of, 26, 26f
tasks, 18–19
beginning of shift report, 19, 20f
decision making, 20–21, 21f
fire station workday, 21
notifications, 19–20
problem solving, 20–21, 21f
volunteer duty night, 21–22
trainer, 24–26, 24–25f
transition from fire fighter to, 22–23, 22f
workplace diversity, 28–31
company-level officer, 361
compensation and benefits, 283
complaint, 99
conflict management
conflict resolution model, 101–102, 101f
emotional confrontations, 102–104
follow-up, 102
investigation, 102, 103f
personnel conflicts and grievances, 100–101
take action, 102
conflict resolution model, 101–102, 101f
conflict, 99
conscious competence, 94–95
conscious incompetence, 94
consolidation, 137
construction type, 172. *See also specific types*
contract out services, 137
contrast effect, 282
controlling, meaning of, 14
cooking, accidental fire causes, 341
correcting unacceptable behavior
advance notice of substandard employee evaluation, 275
just culture approach, 274–275
negative discipline
alternative disciplinary actions, 279
formal written reprimand, 278
predisciplinary conference, 278–279
progression of, 277–278
suspension, 279–280
termination, 280
work improvement plan, 275–277, 276f
corrective discipline, 142–143
corrective maintenance, 146
cost recovery and reduction, 321
Council for Higher Education Accreditation (CHEA), 288
country club management, 43–44
CPAT. *See* Candidate Physical Ability Test
CPC. *See* Commission on Professional Credentialing
CPSE. *See* Center for Public Safety Excellence
CRA. *See* community risk assessment
crew integrity, maintaining, 230–231
crew resource management (CRM), 42
communication skills, 58–59
critical decision making, 62–63
origins of, 56–57
postincident analysis, 65–66
researching and validating, 57
situational awareness, 63–64
task allocation, 61–62, 62f
teamwork, 59–61
crisis communications, 305
critical decision making, 62–63

critical infrastructure systems, 302
CRM. *See* crew resource management
CRR. *See* community risk reduction
customer satisfaction, 122–123
customer service, 122

D

data analysis, 285–286
"data dump" question, 379
decision document, goal of, 319
decision making, 20–21, 21f, 210, 256
alternative solutions, 84
evaluation, 86
implementing solution, 84–86, 85t
problem, defining, 83–84, 84f
selecting solution, 84
defensive operation, 206, 207f
defer scheduled expenditures, 136–137
delegation, 56–57
deluge systems, 169
democratic leadership, 44, 44f
demographics, 116, 300
demonstrative evidence, 186
demotion, 279
Department of Health and Human Services (HHS), 147
departmental directives, 317
DETC. *See* Distance Education and Training Council
dimensions, 375
direct supervision, 284
directive and supportive behavior, 45
disaster declarations, 258
discipline, 13
Distance Education and Training Council (DETC), 288
diversity, 28, 258. *See also workplace diversity*
division of labor, 12
division supervisor, 361
divisions, 360–361, 360f
documentary evidence, 186
documentation, 368
accident investigation and, 239, 241
and reports, fire cause
legal proceedings, 347, 347f
preliminary investigation, 346–347
employee, 145
dormitories, 175–176
dry chemical system, 171
dry-pipe systems, 169
duty keys, 217

E

EAP. *See* employee assistance program
economic factors, 302
economic incentives, 303
education, 287
intervention, 303
training *vs.*, 287–288
educational occupancy, 175, 183
emergency incident
communications, 78f
communications order model, 78
initial situation report, 78–79
Project Mayday, 80
radio reports, 79
responding units, 79

emergency incident (*Continued*)
   injury prevention
      personal protective equipment, 236–237
      physical fitness, 236
   simulations, 379–380
emergency lighting, 181–182
emergency medical services (EMS), 256
emergency operation considerations, 165–166
emergency response interventions, 303
emergency scene leadership
   assigning tasks, methods of, 88
   critical situations, 88–89
emergency support functions (ESFs), 354
   annexes to fire officers, 354, 354t
emotional confrontation, 102–104
emotional health, 256
employee assistance program (EAP), 107–108
employee documentation, 145
empowerment, 142
EMS. *See* emergency medical services
engineering, 303
   controls, 225
environmental noise, 76–77, 76f
error management model, 58, 58f
ESFs. *See* emergency support functions
ethical behavior, 27–28
evaluation errors, 282
Everyone Goes Home program, 226–227
evidence, 186
   protecting, 186–187, 187f
executive orders, 317
exit interview, 180
exit signs, 181–182
expanded incident report narrative, 149
expenditures, 130
   capital, 327–328
   operating, 326–327, 327f
   personnel, 326
expert witness, 347, 347f
extinguishing systems. *See* specific systems
extremism, 346

F

factory industrial occupancies, 175, 175f, 183
failure analysis, 338
Family and Medical Leave Act (FMLA), 138
family dwelling units, 175
family problems, 105
fast-attack mode, operation, 201
Fatality Assessment and Control Evaluation program, 230–231
FBI. *See* Federal Bureau of Investigation
Federal Bureau of Investigation (FBI), 259
Federal Emergency Management Agency (FEMA), 197, 258
federal employees, collective bargaining for, 260–261
feedback on individual performance, 140, 141f
FEMA. *See* Federal Emergency Management Agency
FESHE. *See* Fire and Emergency Services Higher Education
FGC. *See* Fire Ground Command
finance/administration section, 360
financial problems, 105–106
fire
   patterns, 336, 336f
   prevention division, 182
   protection systems, 162

   pumps, 170, 170f
   research and risk/benefit analysis
      fire behavior graph, 203–204, 204f
      flow path, 205, 205f
      IAFC rules of engagement, 207–208
      modern *vs.* legacy single family dwellings, 205–206, 206f
      risk/benefit analysis, 206–207, 207f
   scene reconstruction, 187
fire alarm system, 171
fire analysis, 338, 339f
Fire and Emergency Services Higher Education (FESHE), 287
fire behavior graph, 203–204, 204f
fire burns, 340, 341f
fire cause
   area of origin, 335–336
   arson motives
      crime concealment, 344
      excitement, 346
      extremism, 346
      profit, 344
      spite/revenge, 346
      vandalism, 346
   classification
      accidental fire causes, 341–342, 341f
      incendiary fire causes, 342
      natural fire causes, 342
      undetermined fire causes, 343
   depth of char, 336, 337f
   documentation and reports
      legal proceedings, 347, 347f
      preliminary investigation, 346–347
   evaluating, 336–337
      conducting interviews, 338–339
      fire analysis, 338, 339f
      ignition factor and cause, 338
      material first ignited, 338
      source and form of heat ignition, 337–338
      vehicle fire cause determination, 339–340, 340
      wildland fire cause determination, 340–341, 341f
   fire company fire cause evaluation, 335
   fire officer's role in determination, 334–335
   fire patterns, 336, 336f
   incendiary fire, indicators of
      accelerants and trailers, 343–344, 343f
      multiple points of origin, 344
   scientific method, 335, 336f
fire chief, 10
fire code, 166–167
   building code *vs.*, 167
   local, 167
   model codes, 167–168
   retroactive code requirements, 168
   state, 167
fire code compliance inspections, 171–172
fire company
   inspections, 171–172
fire department
   authorization to function as
      administrative law, 253–254
      state police powers, 253
   budget cycle, 130, 131–132t
      base and supplemental budgets, 132
      budget request, 132–134
      capital request form, 133f

project description and justification, 134
purchase information, 134
total cost of project, 134
employee documentation, 145
human resources, 138–139
evaluation, 139–140, 140f, 141f
positive and corrective disciplinary actions, 141–143, 142f
organization
chain of command, 11, 11f
function, 13
geography, 13
management principles, 12–13
National Incident Management System, 12
source of authority, 10–11
staffing, 13
policies, 284
implementation, 128–130
recommendations and changes, 129
record-keeping systems
individual exposure history and medical records, 146–147
inspection and maintenance, 145–146
inventory, 145
training and credential records, 146
reducing size of, 137–138
reporting
using information technology, 150–152
types of, 147–152
written reports. See written reports
revenue sources, 134–135
local government sources, 135
lower revenue options, 136–138
volunteer fire departments, 135–136
fire detection system, 171
fire fighter
accountability, 202–203
annual goals, 273
deaths by cause of injury, 215, 215f, 215t
documenting performance
annual evaluation, conducting, 282–283
evaluation errors, 282
mid-year review, 280
injury and death trends
Everyone Goes Home, 226–227
national Firefighter Near Miss Reporting System, 227
reducing deaths from cancer, 228–229
reducing deaths from fire suppression operations, 230–231, 231f
reducing deaths from motor vehicle collisions, 229–230, 230f
reducing deaths from sudden cardiac arrest, 227–228, 228f
reducing deaths from suicide, 228
16 Firefighter Life Safety Initiatives, 226t
job description, 140f
keeping track of activity, 273–274, 273f
new, evaluation with, 139–140
organizing into labor unions, 262
practice working with hose lines and tools, 142, 143f
safety, habits to improve, 227
Fire Ground Command (FGC), 197
fire investigations, 334
legal responsibility for conducting, 334
fire officer
behavioral health resources, 107
first-arriving, 199
role in community fire safety, 159, 159f
role in fire cause determination, 334–335
role in incident management, 199, 199f
role in risk management
community fire risk, 224–225
five principal risk management steps, 225
as spokesperson, 307–309, 308f
training responsibilities
competence and confidence, 95–96, 96f
mentoring, 95
new/revised skill sets, 95
skill competence, 93–95
skill training, 90–93
Fire Officer I, 4
as company supervisor. See company supervisor, Fire Officer I as
roles and responsibilities of, 5–6
Fire Officer II, 5
additional roles and responsibilities for, 252–253
challenges for captain
catastrophic weather events, 258
deterioration of built environment, 258
diversity, 258
fire fighter emotional health, 256
increase in non-fire incidents, 256
supervision and motivation, 256
growth of IAFF as political influence, 264–265
labor actions in fire service, 262–263
negative impacts of strikes, 264, 264f
staffing, wages, and contracts, 263–264
striking for better working conditions, 263, 263f
striking to preserve wages and staffing, 263
Wisconsin challenge, 264
legislative framework for collective bargaining
collective bargaining for federal employees, 260–261
Landrum-Griffin Act of 1959, 260
Norris-LaGuardia Act of 1932, 259–260
state labor laws, 261–262, 261t
Taft-Hartley Labor Act of 1947, 260
Wagner-Connery Act of 1935, 260
requirements for, 252
roles and responsibilities of, 6–7
supervising multiple companies
assigning resources, 255–256, 255f
determining task assignments, 254–255
as transformational leader, 272–273
working with other organizations, 258–259
fire service, 7–8
communications, 8, 10f
history of, 8, 9t
labor actions in, 262–263
negative impacts of strikes, 264, 264f
staffing, wages, and contracts, 263–264
striking for better working conditions, 263, 263f
striking to preserve wages and staffing, 263
Wisconsin challenge, 264
paying for, 8, 10, 10f
Fire Service Occupational Cancer Alliance, 229
fire station safety
clothing, 237
housekeeping, 238
lifting techniques, 238
fire suppression operations, reducing deaths from
air management, 231
crew integrity, maintaining, 230–231
teams and tools, 231
thermal imaging devices, 231, 231f

fire tax district, 135
FireCARES. *See* Fire-Community Assessment/Response Evaluation System
Fire-Community Assessment/Response Evaluation System (FireCARES), 302
Firefighter Cancer Support Network, 228
fire-ground injuries, 216*t*
FIRESCOPE, 197
fiscal year, 130, 131–132*t*
floor plans, 161
flow path, 205, 205*f*
FMLA. *See* Family and Medical Leave Act
foam system, 171
followership, 47–48
form of material, 338
formal communications, 317–318
formal training program, 231, 232*f*
formal written reprimand, 278
frame of reference, 282
fringe benefits, 320, 326
fuel load, 165
    size-up, 209–210
fuel, physical configuration of, 338
fuel-limited fire, 203

G

GASB. *See* Governmental Accounting Standards Board
general orders, 317
geography, 300–301
Gilbert v. Homar, 278
goal setting, 273
good faith bargaining, 260
Governmental Accounting Standards Board (GASB), 326
grants, 325
grapevine, 81
Great Recession, 136
greater alarm infrastructure, 363
grid theory, 42
grievance, 81, 100–101
    procedure, 81–82, 83*f*
groups, 360, 360*f*, 361
    supervisor, 217

H

halo and horn effect, 282
Halon system, 171
harassment complaints, 30–31
Harvard Fire Executive Fellowship Program, 287
hazardous use permit, 182
hazards, 233, 302
    life-threatening, 181
    special, 162–165, 164*f*, 182
health and safety officer, 239
Health Insurance Portability and Accountability Act (HIPAA), 147
heat ignition, form of, 337–338
Helmreich's error management model, 58*f*
HFACS. *See* Human Factors Analysis and Classification System
HHS. *See* Department of Health and Human Services
high-rise fires, 364–365
high-risk property, 160
high-value property, 160
HIPAA. *See* Health Insurance Portability and Accountability Act
Homeland Security Act, 354
hostile work environment, 29
hostile workplace, 29–30
hot wash, 274
hot zone, 216
hotels, 175
housekeeping, 238
human error, 274–275
    active failures, 57–58
    error management model, 58
    latent conditions, 57–58
    reasons and ways, 57
Human Factors Analysis and Classification System (HFACS)
    level 1, 284–285
    level 2, 285
    level 3, 285, 285*t*
    level 4, 285
human resources, 138–139
    corrective discipline, 142–143
    development, 52–53
    evaluation
        with new fire fighter, 139–140, 140*f*
        providing feedback after incident or activity, 140, 141*f*
    four borders of
        federal and state laws, 283
        fire department policies, 284
        jurisdiction regulations, 283–284
        labor contract, 283
    management, 52–53
    planning, 52
    positive discipline, 141–142, 142*f*
    utilizing
        assignments, 53–56
        delegation, 56
        mission statement, 53

I

IAFC. *See* International Association of Fire Chiefs
IAFF. *See* International Association of Fire Fighters
IAP. *See* incident action plan
IC. *See* incident commander
ICS. *See* incident command system
IDLH. *See* immediately dangerous to life and health
IFSAC. *See* International Fire Service Accreditation Congress
IFSTA/FPP. *See* International Fire Service Training Association/Fire Protection Publications
ignitable-liquid–fueled fires, 343, 343*f*
ignition factor, 338
immediately dangerous to life and health (IDLH), 231
impasse, 264
impoverished management, 42
in-basket exercises, 378–379
incendiary fire causes, 341, 342
incident action plan (IAP), 211–212, 357
    incident priorities, 212
    location designators, 214
    tactical priorities
        RECEO-VS, 212, 213*t*
        S.L.I.C.E.-R.S, 212–214
incident command system (ICS), 5, 197, 233, 252
    FIRESCOPE and Fire Ground Command, 197
    general staff, 358
    history of, 196–197
    and National Incident Management System model
        ICS training, 197–198, 198*f*
        levels of command, 198–199
    training, 197–198
incident commander (IC), 198, 254, 255*f*, 334

incident management system, 233–234
  command staff, 357
    liaison officer, 358
    public information officer, 358
    safety officer, 358
  expanded organizational chart, 359f
  fire officer role in, 199, 199f
  general staff functions
    expanded organizational chart, 359f
    finance/administration section, 360
    logistics section, 359–360
    operations section, 358–359
    planning section, 359
incident reports, 149–150, 149f
incident review, 365–366, 367t
incident safety officer
  assistant incident safety officers, 235
  and incident management, 233–234
  incident scene rehabilitation, 235, 235f
  qualifications to operate as, 234–235
  typical tasks, 234–235
incident scene rehabilitation, 235, 235f
individual exposure history and medical record keeping, 146–147
infection control, 238
infectious disease exposure, 239, 240f
informal communications, 316–317
informal grievance, 82
informal work performance reviews, 274
informal written reprimand, 143
information technology, 150–152
infrequent reports, 150
initial emergency operations. *See* command of initial emergency operations
initial incident command, 199
  command safety, 202–203
  establishing command, 200
  functions of command, 202
  personnel accountability report, 203
  responsibilities of command, 200
initial operational modes, 200
  command mode, 201
  fast-attack mode, 201
  investigation mode, 201
  tactical worksheet, 201, 201f
initial radio reports, 202
initial rapid intervention crew (IRIC), 216
inspection
  conducting, 179–180
    exit interview, 180
    inspecting from outside in, bottom to top, 180
    meeting with representative, 180
  general requirements
    access and egress, 181
    built-in fire protection systems, 182
    exit signs and emergency lighting, 181–182
    portable fire extinguishers, 182
    special hazards, 182
  and maintenance record keeping, 145–146
  preparing for
    arranging visit, 179
    assembling tools and references, 179
    coordinating activity with fire prevention division, 177–179
    reviewing fire code, 177
    reviewing prior inspection reports, fire history, and preincident plans, 177
  writing inspection/correction report, 181
institutional occupancies, 175, 183
Insurance Service Office (ISO), 146
integrity, 27, 27f
International Association of Fire Chiefs (IAFC), 5
  Company Officer Leadership program, 32
  Company Officer Leadership Symposium, 291
  Safety, Health and Survival (SHS) Section, 207
International Association of Fire Fighters (IAFF), 17–18, 17f, 202, 226, 262–263, 375
  growth as political influence, 264–265
International Fire Service Accreditation Congress (IFSAC), 146, 289–290
International Fire Service Training Association/Fire Protection Publications (IFSTA/FPP), 287
interpersonal interaction, 380–381
interrogatory, 318
interview
  conducting, 338–339
  television, 308–309, 308f
inventory record keeping, 145
investigation, 102, 103f
  mode, operation, 201
investigator, requesting, 187–188
involuntary transfer or detail, 279
IRIC. *See* initial rapid intervention crew
ISO. *See* Insurance Service Office

J

job description, 377
job instruction training, 90
job-content/criterion-referenced validity, 378
jurisdiction regulations, 283–284
just culture concepts, 272
  for health and safety
    adding barriers, 286
    adding recovery, 286
    adding redundancy, 286–287
    controlling contributing factors, 286
    data analysis, 285–286
    designing better systems, 286f
    near-miss reports, 284–285, 285t
  to identify source of behavior
    at-risk behavior, 275
    human error, 274–275
    reckless behavior, 275

K

knowledge, skills, and abilities (KSAs), 377

L

labor actions in fire service, 262–263
  negative impacts of strikes, 264, 264f
  staffing, wages, and contracts, 263–264
  striking for better working conditions, 263, 263f
  striking to preserve wages and staffing, 263
  Wisconsin challenge, 264
labor contract, 283
labor–management relations, 262
  IAFF in, 17–18, 17f
  NVFC in, 18, 18f
LAFD. *See* Los Angeles Fire Department
laissez-faire leadership, 44
Landrum-Griffin Act of 1959, 260
latent conditions, 57–58

Layman, Lloyd, 209–210
leadership, 14. *See also* team leading
   accreditation of skill certification, 288–290, 289*t*
   challenges
      fire station as municipal work location *vs.* fire fighter home, 47–48
      in volunteer service, 48–49
   correcting unacceptable behavior. *See* correcting unacceptable behavior
   CPSE fire officer credential, 290
   documenting fire fighter performance
      annual evaluation, conducting, 282–283
      evaluation errors, 282
      mid-year review, 280
   IAFC company officer leadership symposium, 291
   job performance of assigned members
      annual fire fighter goals, 273
      informal work performance reviews, 274
      keeping track of fire fighter's activity, 273–274, 273*f*
   just culture concepts for health and safety
      data analysis, 285–286
      designing better systems, 286–287, 286*f*
      near-miss reports, 284–285, 285*t*
   and management, 40–41
      difference between, 41*t*
      theories. *See specific theories*
   National Professional Development Model, 290*f*
   NFA managing officer program, 290–291
   power as resource, 47
   process, 40
   professional development plan, 287, 290
   training *vs.* education, 287–288
   U.S. Marine Corps, 49, 52–53, 50–52*t*
leading, meaning of, 14
legacy dwellings, 205
legal considerations during investigations
   evidence, 186
      protecting, 186–187, 187*f*
   searches, 185
   securing scene, 185–186
legal correspondence, 317–318
legal proceedings, fire cause, 347, 347*f*
leniency/severity, 282
LEPC. *See* Local Emergency Planning Council
levels of command, 198–199
liaison officer, 358
lieutenant, 5
Life Safety Code, 167
lifting techniques, 238
line-item budget, 326
line-of-duty deaths, 231
local alarms, 169, 169*f*
local code, 167
Local Emergency Planning Council (LEPC), 160
Local Government 101, 142
local government sources, revenue, 135
location designators, incident action plan, 214
lodging/rooming houses, 175
logistics section, 359–360
logistics section chief, 359
Los Angeles Fire Department (LAFD), 288
Loudermill hearing, 278
lower revenue options
   defer scheduled expenditures, 136–137
   privatize or contract out services, 137
   reducing size of fire department, 137–138
   regional or consolidated services, 137
   workforce, reducing, 137
low-expansion foam system, 171

M

MABAS. *See* Mutual Aid Box Alarm System
major incidents
   fire officer greater alarm responsibilities, 362
   incident management system
      command staff, 357–358
      expanded organizational chart, 359*f*
      general staff functions, 358–360
   National Response Framework, 354
      ESF annexes to fire officers, 354
      incident action plan, 357
      Stafford Act, 354, 357*f*
   postincident analysis
      conducting a critique, 366–368, 367*t*
      documentation and follow-up, 368
      preparing information for an incident review, 365–366
   selected structure considerations
      high-rise considerations, 364–365
      mid-rise multifamily dwellings, 364
      single-family dwellings, 363–364
   tactical-level incident management
      branches, 361–362
      divisions, groups, and units, 360–361, 360*f*
   task-level incident management, 362–363
      greater alarm infrastructure, 363
      strike teams, 363, 363*f*
      task forces, 363
management, 40. *See also* crew resource management
   functions of, 12–13
   human resources, 52–53
   leadership and, 40–41
      difference between, 41*t*
      theories. *See specific theories*
managing fire officer, 5, 252
Managing Officer Program, 290–291
manufacturing facilities, 184
marital problems, 105
masonry wall, 173
mass-casualty events, 164–165
master memos, 317
mayday, 216
mechanical failures, causes of vehicle fires, 339, 340*f*
media relations
   fire department public information officer, 305–306
      measured responsiveness, 307
      proactive outreach, 307
      strong foundation building, 306, 306*f*
   fire officer as spokesperson, 307–309, 308*f*
   press releases, 307, 308*f*
mediation, 17
member-related problems
   behavioral and physical health issues, 104–106
   complaint, 99
   conflict, 99
   conflict management, 100–104
   mistake, 99
   resiliency to stress, 106–108
   types of problems, 99–100
mentoring, 95
mercantile occupancies, 175, 176*f*, 184

Metrofire, 137
Miami–Dade Fire Rescue Department, 137
Michigan v. Clifford, 185
Michigan v. Tyler, 185
middle-of-the-road management, 44
mid-rise multifamily dwellings, 364
mid-year review, 280
mini/max codes, 166, 167
mistake, 99
mixed-use property, 176, 185
model codes, 167–168
Model State Administrative Procedure Act (MSPA), 254
modern dwellings, 205
monthly activity and training report, 148–149
moral injury, 106
morning report to battalion chief, 147–148
motivation, 256
motor vehicle collisions, reducing deaths from, 229–230, 230*f*
movement pattern, fire, 336
MSPA. *See* Model State Administrative Procedure Act
multiple points of origin, 344
multiple-choice written examination, 377–378
municipal work location versus fire fighter home, 48
Murrah Federal Building bombing, 211
Mutual Aid Box Alarm System (MABAS), 137

N

National Board on Fire Service Professional Qualifications (NBFSPQ), 146, 289
National Fallen Firefighters Foundation (NFFF), 226, 284
National Fire Academy (NFA), 303
    Managing Officer Program, 290–291
    size-up process
        initial size-up, 211
        ongoing size-up, 211
        preincident information, 210–211
National Fire Academy Managing Fire Officer, 146
National Fire Incident Reporting System (NFIRS), 149, 256, 334, 368
National Fire Protection Association (NFPA), 4, 226
National Firefighter Near Miss Reporting System, 227
National Incident Management System (NIMS), 12, 197, 259, 354
National Incident Management System Consortium (NIMSC), 197
National Industrial Recovery Act (NIRA), 260
National Institute for Occupational Safety and Health (NIOSH), 226, 319
National Institute of Standards and Technology (NIST), 203, 364
    fire dynamics simulator model, 205*f*
National Professional Development Matrix (NPDM), 287
National Professional Development Model, 290*f*
National Response Framework (NRF), 198, 354
    ESF annexes to fire officers, 354
    incident action plan, 357
    Stafford Act, 354, 357*f*
National Volunteer Fire Council (NVFC), 18, 18*f*
natural fire causes, 341, 342
naturalistic decision making, 62
NBFSPQ. *See* National Board on Fire Service Professional Qualifications
near-miss reports, 284–285, 285*t*
negative feedback, 141
negligence, 338
NFA. *See* National Fire Academy
NFFF. *See* National Fallen Firefighters Foundation
NFIRS. *See* National Fire Incident Reporting System
NFPA. *See* National Fire Protection Association

NFPA 704 marking system, 176–177, 177*f*
NIMS. *See* National Incident Management System
NIMSC. *See* National Incident Management System Consortium
NIOSH. *See* National Institute for Occupational Safety and Health
NIRA. *See* National Industrial Recovery Act
NIST. *See* National Institute of Standards and Technology
noise
    environmental, 76–77, 76*f*
    physical, 76
    sociological, 76
non-fire incidents, increase in, 256
nontraditional revenue sources, 325–326
Norris-LaGuardia Act of 1932, 259–260
Northouse's classification
    behavioral approach, 42
    skills approach, 41, 42*f*
    trait approach, 41
"nothing showing" calls, 200
NPDM. *See* National Professional Development Matrix
NRF. *See* National Response Framework
NVFC. *See* National Volunteer Fire Council

O

object code, 326
occupancy, 174. *See also specific types*
    considerations, systematic approach, 161–162
    type, 187
    and use group, 174–176, 175–176*f*
        business occupancy, 174, 175*f*, 183
        educational occupancy, 175, 183
        factory industrial occupancies, 175, 175*f*, 183
        institutional occupancies, 175, 183
        manufacturing facilities, 184
        mercantile occupancies, 175, 176*f*, 184
        mixed-use property, 176, 185
        public assembly occupancy, 174, 182–183
        residential occupancies, 175–176, 184
        special properties, 176, 184
        storage occupancies, 176, 184–185
Occupational Safety and Health Administration (OSHA), 202, 225
offensive operation, 206
on-duty wallet, 217
100-year flood, 301
ongoing compliance inspection, 177
on-scene observations, 208–209
open-ended questions, 339
operating budget, 130
operating expenditures, 326–327
operating expenses, alternative response unit, 320–321
operating system, reports, 152
operations section, 358–359
operations section chief, 359
oral reprimand, warning, or admonishment, 143
Orange County Fire Authority, 137
ordinance, 167
organizing, meaning of, 14
Oriented-Vent-Enter-Isolate-Search (O-V.E.I.S.), 213
OSHA. *See* Occupational Safety and Health Administration
overhaul and salvage operations, 335

P

PAC. *See* political action committee
PAR. *See* personnel accountability report
past loss/event history, 302
PATCO. *See* Professional Air Traffic Controllers Organization

Pendleton Civil Service Reform Act, 375
performance evaluation, 140
   annual fire fighter goals, 273
   informal work performance reviews, 274
   keeping track of fire fighter's activity, 273–274, 273f
performance management, 284
personal bias, 282
personal competence, 60
personal power, 47
personal protective equipment (PPE), 229, 236–237
personal study journal, 381–384
personnel accountability report (PAR), 203
personnel conflicts, 100–101
personnel expenditures, 326
personnel protection control, 225
petty cash, 323
PHI. See protected health information
Phoenix Fire Department, 302
photovoltaic power system, 164f
physical elements, evaluation of, 161
physical fitness, 228, 228f, 236
physical health issues, 104–106
physical noise, 76
physical performance rating, 376, 376f
picketing, 264, 264f
PIO. See public information officer
plan of operation, 210
planning section, 359
planning section chief, 359
planning, meaning of, 14
plot plan, 161
PO. See purchase order
"podium" construction, 364
police department, 256
policies, 14
   implementation, 128–130
   recommendations and changes, 129
political action committee (PAC), 265
portable fire extinguishers, 182
portfolio management, 136
positional power, 47
positive discipline, 141–142, 142f
postal workers' strike 1970, 262
postexamination promotional considerations, 376, 376f
postincident analysis, 65t, 241
   conducting, 65–66, 366–368, 367t
   documentation and follow-up, 66, 368
   preparing information for incident review, 365–366
posttraumatic stress disorder (PTSD), 106–107
power, 47
PPE. See personal protective equipment
preaction sprinkler systems, 169
prearrival information, 208
predetermined disciplinary policy, 280
predisciplinary conference, 278–279
predisciplinary hearing, 278
preincident planning, 160–161, 160f
   built-in fire protection. See built-in fire protection
   classifying by building/occupancy
      construction type, 172–174, 172–174f
      NFPA 704 marking system, 176–177, 177f
      occupancy and use group, 174–176, 175–176f
   and code enforcement, 158–191
   fire code compliance inspections, 171–172
   fire codes, 166–167
      building code vs., 167
      local, 167
      model codes, 167–168
      retroactive code requirements, 168
      state, 167
   inspection. See inspection
   selected use group–specific concerns
      business occupancy, 183
      educational occupancy, 183
      factory industrial occupancies, 183
      institutional occupancies, 183
      manufacturing facilities, 184
      mercantile occupancies, 184
      mixed-use building, 185
      public assembly occupancy, 182–183
      residential buildings, 184
      special properties, 184
      storage facilities, 184–185
   systematic approach
      document findings, 166
      emergency operation considerations, 165–166
      occupancy considerations, 161–162
      physical elements and site considerations, 161
      special hazards, 162–165, 164f
      water supply and fire protection systems, 162, 163f
preliminary investigation, fire cause, 346–347
presentation software, 152
press releases, 307, 308f
pretermination hearing, 278
preventive maintenance, 146
privatization, 137
Pro Board, 289
proactive outreach, 307
probabilities, 209–210
probationary period, 139–140
problem solving, 20–21, 21f
Professional Air Traffic Controllers Organization (PATCO), 262
professional development, 287
   plan, 290
profit, 344
progressive negative discipline, 277
project description and justification, 134
Project Mayday, 80
promotional examination process
   assessment centers
      emergency incident simulations, 379–380
      in-basket exercises, 378–379
      interpersonal interaction, 380–381
      technical skills demonstration, 381
      writing/speaking exercise, 381
   mastering content and process of promotional exam
      personal study journal, 381–384
      preparing for role playing, 384
   multiple-choice written examination, 377–378
   origin of, 375
   preparing, 376
      knowledge, skills, and abilities, 377
   sizing-up, 375–376
      postexamination promotional considerations, 376, 376f
protected health information (PHI), 147
psychological contract, 102
PTSD. See posttraumatic stress disorder
public assembly occupancy, 174, 182–183
public education, 119–120
public information officer (PIO), 305–306, 358

measured responsiveness, 307
proactive outreach, 307
public inquiries
citizen concerns, 122, 123f
customer service vs. customer satisfaction, 122–124, 123f
responding to, 121–122
public safety response agencies, 301–302
purchase order (PO), 323
purchasing process
bidding process, 323–324
petty cash, 323
purchase orders, 323
requisitions, 323
pyrolysis, 342

R

ransomware attacks, 145
rapid intervention crew (RIC), 216–217
rapid intervention, 203
real estate, revenue for fire department operations, 136
reasonable person standard, 275
recency, 282
RECEO-VS, 212, 213t
reckless behavior, 275
recognition-primed decision making (RPD), 62
recommendation report, 319
record-keeping systems
individual exposure history and medical records, 146–147
inspection and maintenance, 145–146
inventory, 145
training and credential records, 146
regionalization, 137
regulations, 172
rehabilitation, 235, 235f
reimbursement for response, 326
relationship behaviors, 42
reliability, 378
rep (repetition), 95
replacement uniforms and protective clothing, 327, 327f
reporting
using information technology, 150–152
types of, 147–152
written reports, 318
incident reports, 149–150, 149f
infrequent reports, 150
monthly activity and training report, 148–149
morning report to battalion chief, 147–148
recommendation report, 319
request for proposal (RFP), 324
requisitions, 323
residential occupancies, 175–176, 184
resiliency, 106
resources assignment, 255–256, 255f
restrictive duty, 280
retroactive code requirements, 168
revenue sources, 134–135, 324
grants, 325
nontraditional, 325–326
revenues, 130
RFP. See request for proposal
RIC. See rapid intervention crew
right-to-work laws, 261–262, 261t
risk assessment, 119, 300
risk communications, 305
risk exposure, 225

risk likelihood, 302–303
risk management, 224, 285
accident investigation, 239
and documentation, 239, 241
control actions, 225
directing members during training evolutions, 231–232, 232f
reducing deaths during training, 232–233
evaluate risk exposure, 225
fire fighter injury and death trends
Everyone Goes Home, 226–227
National Firefighter Near Miss Reporting System, 227
reducing deaths from cancer, 228–229
reducing deaths from fire suppression operations, 230–231, 231f
reducing deaths from motor vehicle collisions, 229–230, 230f
reducing deaths from sudden cardiac arrest, 227–228, 228f
reducing deaths from suicide, 228
16 Firefighter Life Safety Initiatives, 226t
fire officer's role in
community fire risk, 224–225
five principal risk management steps, 225
identify risk exposure, 225
incident safety officer
assistant incident safety officers at large or complex incidents, 235
and incident management, 233–234
incident scene rehabilitation, 235, 235f
qualifications to operate as, 234–235
infection control, 238
infectious disease exposure, 239, 240f
postincident analysis, 241
rank and prioritize risks, 225
safe work environment
emergency incident injury prevention, 236–237
fire station safety, 237–238
safety policies and procedures, 236
risk/benefit analysis, 206–207, 207f
role playing exercises, 384
RPD. See recognition-primed decision making
rules and regulations, 14

S

safe work environment
emergency incident injury prevention
personal protective equipment, 236–237
physical fitness, 236
fire station safety
clothing, 237
housekeeping, 238
lifting techniques, 238
safety policies and procedures, 236
SAFER. See Staffing for Adequate Fire and Emergency Response
safety officer, 358
safety policies and procedures, 236
safety unit, 235
scene safety, 215–216
scientific method, 335, 336f
searches, 185
seat belt, 230, 230f
securing scene, 185–186
self-evaluation, 280
semi-automatic cardiac defibrillators, 328
servant leadership, 46–47
service branch, 365
Service Life Extension Program (SLEP), 136–137

set, 95
sexual harassment, 29–30
single resources, 362
single-family dwellings, 363–364
    modern *vs.* legacy, 205–206, 206*f*
site considerations, systematic approach, 161
situation assessment, 210
situational awareness, 63–64, 233
    maintaining, 63–64
situational leadership, 41, 45
16 Firefighter Life Safety Initiatives, 226, 226*t*
sizing-up, 208, 375–376
    Lloyd Layman's five-step size-up process
        decision making, 210
        facts, 209
        plan of operation, 210
        probabilities, 209–210
        situation assessment, 210
    National Fire Academy size-up process
        initial size-up, 211
        ongoing size-up, 211
        preincident information, 210–211
    on-scene observations, 208–209
    postexamination promotional considerations, 376, 376*f*
    prearrival information, 208
skill competence
    conscious competence, 94–95
    conscious incompetence, 94
    unconscious competence, 95
    unconscious incompetence, 94
skill training
    application, 92, 94*f*
    evaluation, 92–93
    preparation, 92, 93–94*f*
    presentation, 92, 94*f*
skills approach, 41, 42*f*
SLEP. *See* Service Life Extension Program
S.L.I.C.E.-RS, 212–214
SMACSS. *See* social media–assisted career suicide syndrome
S.M.A.R.T. format, goals, 273
smartphone, 217
social competence, 60
social media, 305
    outreach, 309
        challenges, 309–310
social media–assisted career suicide syndrome (SMACSS), 310
sociological environmental noise, 76
source of authority, 10–11
source of ignition, 337–338
span of control, 12
special evaluation period, 276
special properties, 176, 184
specs, 324
spite/revenge, 346
spoils system, 375
spreadsheets, reports, 152
Staffing for Adequate Fire and Emergency Response (SAFER), 138
staffing, 52
Stafford Act, 354, 357*f*
staging, 362
stairwell support group, 365
standard operating guidelines (SOGs), 16, 317
standard operating procedures (SOPs), 14–16, 254, 317
standard time–temperature curve, 203
standardized actions, 87

standpipe system, 170
state code, 167
state labor laws, 261–262, 261*t*
state police powers, 253
storage occupancies, 176, 184–185
strategic level, 198
strike
    defined, 262
    during national emergency, 260
    for better working conditions, 263, 263*f*
    negative impacts of, 264, 264*f*
    PATCO air traffic controllers, 262
    postal workers, 262
    staffing, wages, and contracts, 263–264
    to preserve wages and staffing, 263
strike team, 363, 363*f*
strike team leader, 363
subobject code, 326
substance abuse, 104–105
substandard employee evaluation, advance notice of, 275
success 102, 142
sudden cardiac arrest, reducing deaths from, 227–228, 228*f*
suicide, reducing deaths from, 228
supervising fire officer, 5
supervision, 256
supervisor
    and employee relationship, 16–17
    fire officer's role, 18, 19*f*, 26, 26*f*
    report, 150, 151*f*
    responsibilities, 361
    unpopular orders and directives, 23, 23*f*
supervisory tasks, team leading
    decision making, 82–86
    grievance procedure, 81–82, 83*f*
supplemental budget, 132
support branch director, 365
support branch, 365
suspension, 277, 279–280
Swiss cheese model, 58*f*, 286*f*
systematic approach
    document findings, 166
    emergency operation considerations, 165–166
    occupancy considerations, 161–162
    physical elements and site considerations, 161
    special hazards, 162–165, 164*f*
    water supply and fire protection systems, 162, 163*f*

T

T-account, 273, 273*f*
tactical level, 198
tactical priorities, incident action plan
    RECEO-VS, 212, 213*t*
    S.L.I.C.E.-RS, 212–214
tactical safety considerations, 214–215
    fire fighter deaths by cause of injury, 215*f*, 215*t*
    fire-ground injuries by cause, 216*t*
    rapid intervention crews, 216–217
    scene safety, 215–216
tactical worksheet, 201, 201*f*
tactical-level incident management
    branches, 361–362
    divisions, 360–361, 360*f*
    groups, 360*f*, 361
    supervisor responsibilities, 361
    units, 360*f*, 361

Taft-Hartley Labor Act of 1947, 260
task allocation, 61–62, 62f
task assignments, 254–255
task behaviors, 42
task force, 363
task force leader, 363
task level, 199
task-level incident management, 362–363
    greater alarm infrastructure, 363
    strike teams, 363, 363f
    task forces, 363
tax revenues, 135
team leading
    assigning tasks
        during emergency incidents, 88–89
        during nonemergency incidents, 89–90
        safety considerations, 90
        supervising single company, 86–88, 86f
    communication cycle
        emergency incident, 78–80, 78f
        feedback, 77–78, 77f
        medium, 75–77, 76f
        message, 75
        receiver, 77
        sender, 75
        supervisory responsibilities, 80–86
    fire officer training responsibilities
        competence and confidence, 95–96, 96f
        Four-Step Method of skill training, 90–93
        mentoring, 95
        new/revised skill sets, 95
        skill competence, 93–95
    member-related problems
        behavioral and physical health issues, 104–106
        complaint, 99
        conflict, 99
        conflict management, 100–104
        mistake, 99
        resiliency to stress, 106–108
        types of problems, 99–100
    new member training
        skills learning immediately, 96–97
        skills necessary for staying alive, 97–98
    specific training program
        assess needs, 98
        delivering, 98
        development, 98
        impact evaluation, 99
        lesson plans, 98
        objectives, 98
    supervisory tasks
        decision making, 82–86
        grievance procedure, 81–82, 83f
team management, 44–45
teamwork, 141
    handling conflict, 60
    leadership, 60, 60f
    mentoring, 60
    responsibility, 60–61, 61f
    self-assessment, 61
technical competence, 60
technical skills demonstration, 381
television interview, 308–309, 308f
termination, 277, 280
testimonial evidence, 186

thermal imaging devices, 231, 231f
TRADE. *See* Training, Resources and Data Exchange
trailer, 344
trainer
    knowing neighborhood, 25
    personal training library, 24–25, 25f
    problem-solving skills, 25–26
training program
    and credential record keeping, 146
    assess needs, 98
    competence and confidence, 95–96, 96f
    development, 98
    directing members during evolutions, 231–232, 232f
    vs. education, 287–288
    evaluation, 99
    lesson plans, 98
    mentoring, 95
    new fire fighter trainee, 96–98
    new/revised skill sets, 95
    objectives, 98
    reducing deaths during, 232–233
    skill competence, 93–95
    skill training, 90–93
    whole-skill, 232
Training, Resources and Data Exchange (TRADE), 287
trait approach, 41
transactional relationship, 272
transfer of command, 217
transformational leadership, 46, 272
    components of, 272–273
transitional attack, 207
two-in/two-out rule, 203
type I construction (fire resistive), 172–173, 172f
type II construction (noncombustible), 173, 173f
type III construction (ordinary), 173, 173f
type IV construction (heavy timber), 173, 174f
type of material, 338
type V (wood frame), 174, 174f

U

unconscious competence, 92, 95
unconscious incompetence, 94
undetermined fire causes, 341, 343
unfair labor practices, 260
uniforms and protective clothing, 321
units, 360, 360f, 361
unity of command, 12, 12f
unsafe supervision, 285, 285t
Urban Search And Rescue (USAR), 259
USAR. *See* Urban Search And Rescue
use group, 174. *See also under* occupancy
U.S. Fire Administration (USFA), 226, 287, 325
U.S. Marine Corps, 49, 52–53, 50–52t

V

vacant and abandoned buildings, 164
vacant building marking system, 164, 165f
vandalism, 346
VCOS. *See* Volunteer and Combination Officers Section
vehicle fire cause determination, 339–340, 340
vehicle, alternative response unit, 321
vehicle identification number (VIN), 340
ventilation-limited fire, 203
VIN. *See* vehicle identification number
visualization, size-up, 209

voluntary accreditation, 288
Volunteer and Combination Officers Section (VCOS), 229
volunteer fire departments, 135–136
volunteer fire service, 48–49

W

Wagner-Connery Act of 1935, 260
"walking the talk," 141–142, 142*f*
warm zone, 216
water supply, 162
water-based fire protection systems, 168–170, 169*f*, 170*f*
wet chemical system, 171
wildland fire cause determination, 340–341, 341*f*
work improvement plan, 275–277, 276*f*
workforce, reducing, 137
workplace diversity, 28
    fire officer's role in, 29
        actionable item, 29, 29*f*
        harassment complaint, 30–31
        hostile workplace, 29–30
        sexual harassment, 29–30
writing/speaking exercise, 381
written communications
    formal communications
        announcements, 317
        general orders, 317
        legal correspondence, 317–318
        standard operating guidelines, 317
        standard operating procedures, 317
    informal communications, 316–317
written reports, 318
    incident reports, 149–150, 149*f*
    infrequent reports, 150
    monthly activity and training report, 148–149
    morning report to battalion chief, 147–148
    recommendation report, 319
written reprimand, 278

Y

yellow dog contracts, 259
Yukl, Gary, 40